BIOPROCESS TECHNOLOGY

P T Kalaichelvan

Professor
CAS in Botany
University of Madras, Chennai

I Arul Pandi

Researcher
CAS in Botany
University of Madras, Chennai

MJP PUBLISHERS

Chennai 600 005

MJP PUBLISHERS

© Publishers, 2024   47, Nallathambi Street
All rights reserved   Triplicane
Printed and bound in India Chennai 600 005

Foreword

There has been a considerable thrust in recent years to utilize the potential of microbes in the production of compounds and molecules for human welfare. Microbes have been proved to be worthy of commercial exploitation in pharmaceutical, nutraceutical and fine chemical industries.

The process of understanding a microbe within the small culture vessel under defined laboratory conditions is totally different from the behaviour of the same organism in large volumes, under aseptic conditions. Basic research and the technology to utilize the findings are two inseparable aspects that need focus in the study of microbial technology.

The book on *Bioprocess Technology* by Prof. P T Kalaichelvan and associates is a very good effort to present in a consolidated form all relevant aspects pertaining to the title of the book. A wide range of topics are dealt with starting from the well known fermentation to the current topic of biodiesel. The authors have carefully chosen the topics that are of great relevance to human welfare.

This book could be one of those that is imperative for all the beginners of microbiology, biotechnology and biology in general and students of technology in particular.

I congratulate the authors for this sincere effort and wish that they should take up the other topics of current interest for publication in future.

Prof. N Anand, D.Sc.
Director
CAS in Botany
University of Madras
Chennai

PREFACE

The emerging field of Bioprocessing technology and engineering represents the integration of concepts and ideas from biological sciences, engineering disciplines, material sciences and clinical procedures. The bioprocess is a fast growing field in present decade which cannot be defined in single line because its application is versatile in most of the biology field.

The book is divided into eight units with a view of providing complete microbial bioprocess technology. The first unit, Fermentation Technology discusses the types of fermentation and fermenters, structure and anatomy of fermenters and process control in fermentation. The second unit, Food Bioprocessing discusses the food spoilage and its preservation, the production of fermented beverages, milk products and single cell proteins. The third unit, Pharmaceutical Bioprocessing deals with vaccines and antibiotic production. The fourth unit, Enzyme Bioprocessing deals with industrial enzymes, immobilization and enzyme kinetics. The fifth unit, Industrial Bioprocessing deals with organic acids, vitamins and microbial polysaccharides. The sixth unit, Agricultural Bioprocessing deals with biofertilizers and biopesticides. The seventh unit, Environmental Bioprocessing discusses bioremediation and biological waste treatment. The last unit, Energy Bioprocessing deals with biogas, biofuels, ethanol and biodiesel.

Every effort has been made to cover the detailed aspects of various applications of bioprocess. The book has been amply illustrated with figures and flowcharts to make the concepts clear. This would help the readers to follow the topics dealt with easily.

This book would serve as an essential handbook for the undergraduate and postgraduate students, research scholars and teachers of biological sciences and engineering students.

P T Kalaichelvan
I Arul Pandi

CONTENTS

UNIT I FERMENTATION TECHNOLOGY

Unit II Food Bioprocessing

Unit III   Pharmaceutical Bioprocessing

Unit IV   Enzyme Bioprocessing

Unit V   Industrial Bioprocessing

Unit VII Environmental Bioprocessing

UNIT VIII ENERGY BIOPROCESSING

FERMENTATION TECHNOLOGY

OVERVIEW OF BIOPROCESSING

1

The modern bioprocessing technology is an extension of ancient techniques of developing useful products from natural biological activities. Alcoholic beverages are good examples of bioprocessing. The combination of yeast cells and nutrients from grains forms a fermentation system in which the organisms consume the nutrients for their own growth and produce alcohol and carbon dioxide that helps to produce the beverage. The modern bioprocessing technology is more sophisticated, but its principle remains the same. Bioprocesses are widely used commercially in the production of enzymes, food products, pharmaceutical products, agricultural products, etc. With improved techniques and instrumentation, bioprocesses may have applications even in areas where chemical processes are used. Since bioprocesses use living organisms, they offer several advantages over chemical methods: (1) they usually require lower temperature, pressure and pH; (2) they use renewable raw materials; and (3) they can produce greater quantities with less energy consumption.

In most bioprocesses, enzymes are used to catalyse the biochemical reactions of the microorganisms or their cellular components. However, the biological catalyst does not change itself. In the fermentation tanks or fermenters, a series of reactions take place to change the initial raw materials into the desired end product. It may sound quite simple; however, there are two major challenges in this process.

First, the conditions under which the reactions take place must be rigidly maintained. The temperature, pressure, pH, oxygen level, and flow rate must be kept at specific levels. With the help of automated and computerized equipment, it is becoming easier to monitor reaction conditions and hence increased production and efficiency. Secondly, the reactions may produce many undesirable by-products. The presence of waste material often poses two problems: (a) how to recover the end product leaving residue as little as possible in the catalytic system (since enzymatic catalysts remain unchanged as they drive reactions, they can be used over and over again); and (b) how to isolate the desired product in pure form.

Bioprocessing done through laboratory procedures can produce only small amount of useful substances. As advances in bioprocess technology are made, especially separation and purification techniques, commercial firms can produce these substances in large amounts and, thus, make them available for use in medical research, food processing, agriculture, waste management and many other fields of science.

INDUSTRIAL BIOPROCESSING

In industrial bioprocessing, the techniques of modern molecular biology are applied to improve efficiency and reduce environmental impacts from industrial processes in textile, paper and pulp, and chemical manufacturing. For example, industrial bioprocessing units use biocatalysts, such as enzymes, to synthesize chemicals. A desired enzyme can be manufactured in commercial quantities from bioprocessing.

The traditional chemical synthesis involves large amounts of energy and often results in undesirable by-products, e.g. HCl. However, commodity chemicals and specialty chemicals can be produced using biocatalysts more economically and environmental friendly. An example would be the substitution of protease for other cleaning compounds in detergents. Detergent proteases, which can remove protein impurities, are essential in modern detergents. They can break down protein, starch and fatty acids. Protease production results in a biomass, which in turn yields a useful organic fertilizer. Bioprocessing is also used in the textile industry for the finishing of fabrics. It can also produce cotton that is warmer, stronger, has improved dye uptake and retention, enhanced absorbency, and wrinkle- and shrink-resistance.

Some crops, e.g. corn, can be used in the place of petroleum to produce chemicals. The sugar in the crop can be fermented to an acid, which can then be used to produce chemical feedstock for various products. It has been projected that 30% of the world's chemical and fuel needs could be supplied by such renewable resources in the first half of the next century. It has been demonstrated that biopulping requires much less electrical energy than wood pulping.

ENVIRONMENTAL BIOPROCESSING

Environmental bioprocessing is applied in waste treatment and pollution prevention. It can more efficiently clean-up wastes than by conventional methods and greatly reduce our dependence on land-based disposal. Every organism ingests nutrients to be alive and produces by-products as a result. Different organisms need different types of nutrients. Most bacteria thrive on the chemical components of waste products. Bioremediation is the broadest application of environmental bioprocessing. Environmental engineers introduce nutrients at a waste site to stimulate the activity of bacteria already present in the soil, or add new bacteria to the soil. The bacteria digest the waste, turn it into harmless by-products and finally die and return to their normal population levels in the environment.

Bioremediation is a field of increasing interest. Enzyme bioreactors are being developed, which will help pretreat industrial and food waste components and allow their removal through the sewage system rather than solid waste disposal mechanisms. Waste can also be converted to biofuel to run generators. Microbes can be induced to produce enzymes to convert plant and vegetable materials into building blocks for biodegradable plastics.

The by-products of the pollution-fighting microorganisms are useful in many cases. For example, methane can be derived from bacteria that degrade sulphur liquor, a waste product of paper manufacturing. Methane can be used as fuel or in other industrial processes.

HUMAN APPLICATIONS

Biotechnical methods are now used to produce several proteins for pharmaceutical and other specialized purposes. A harmless strain of *Escherichia coli*, given a copy of the gene of human insulin, can produce insulin. As these genetically modified (GM) bacterial cells mature, they produce human insulin, which can be purified and used to treat diabetes in humans. Microorganisms can also be modified to produce digestive enzymes. In future, these microorganisms could be colonized in the intestinal tract of patients with digestive enzyme deficiency. Products of modern bioprocessing include artificial blood vessels from collagen tubes coated with a layer of the anticoagulant heparin.

FERMENTATION TECHNOLOGY

Fermentation technology is the oldest of all biotechnological processes. The term was derived from Latin *fevere*, which means "to boil" that appears when yeast acts upon the fruit extracts or malted grain during the production of alcohol. Fermentation is a process of chemical change caused by organisms or their products, usually producing effervescence and heat.

In microbiology, the production of a thing by means of mass culture of microorganisms is considered fermentation. The biochemical metabolism in fermentation is an energy-generating aerobic process in which organic compounds act as electron donors. This microbiological concept is widely used in biotechnology. Industrial fermentation is different from laboratory cultivation of organisms in that the growth of organisms is optimized and maintained till the end of the process. Almost all products essential to us are produced from microbes. Hence, microbial fermentation technology is used in large-scale cultivation of microbes.

MICROBIAL GROWTH: NEED FOR ARTIFICIAL CULTURE

The growth of organisms involves complex energy-based processes. The rate of growth is dependent on several culture conditions, which provide the energy required for various chemical reactions. The production of a specific compound requires precise culture conditions at a particular growth rate. The rate of growth of microorganisms and hence the synthesis of various chemical compounds under artificial culture, requires organism-specific chemical compounds as the growth medium. The relative concentrations of the ingredients of the medium, the pH, temperature, purity of the cultured organism, etc. influence the microbial growth and hence the production of biomass (the total mass of cells or the organism being cultured). The nutrient sources for industrial fermentation are given in Table 1.1.

Table 1.1   Nutrient sources for industrial fermentation

Nutrient	Raw material
Carbon source	
Glucose	Corn sugar, starch, cellulose
Sucrose	Sugar cane, sugar beet molasses
Lactose	Milk whey
Fats	Vegetable oils
Hydrocarbons	Petroleum fractions

(Contd.)

Table 1.1   (Continued)

Nutrient	Raw material
Nitrogen source	
Protein	Soybean meal, corn steep liquor, distillers' soluble
Ammonia	Pure ammonia or ammonium salts
Nitrate	Nitrate salts
Nitrogen	Nitrogen gas

Phases of Microbial Growth

When an organism is introduced into a medium, i.e., the medium is inoculated with that organism, the growth of inoculum does not happen immediately because the inoculated medium is a new environment to that organism. This period of adaptation is called the lag phase. During the lag phase, the growth is very minimal. Following the lag phase, the rate of growth steadily increases for a certain period because the nutrients are abundantly available. This period is the log or exponential phase.

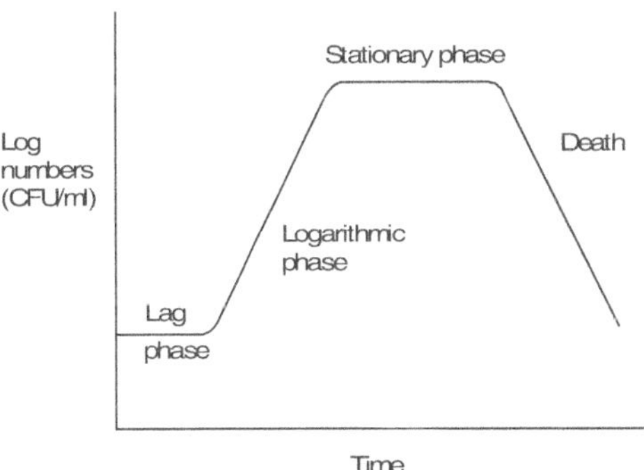

Figure 1.1   Growth curve of microbes

After the exponential phase, the rate of growth slows down due to the continuously falling concentrations of nutrients and continuously increasing cell numbers, and concentrations of metabolites and exhausted products. This phase, where the increase in the rate of growth is checked, is the stationary phase or steady state.

After the stationary phase, growth ceases and the culture enters a deceleration phase. The biomass remains constant unless certain accumulated chemicals in the culture lyse the cells (chemolysis). The chemical constitution remains unchanged unless other microorganisms contaminate the culture. Mutation of the organism in the culture can also be a source of contamination called internal contamination. Ultimately, the factors lead to the death of organism. The phases of microbial growth are shown in Figure 1.1.

Growth and the Monod Equation

There are many models of varying complexity that seek to relate specific growth rate to nutrition. The Monod equation is one such expression. It is particularly significant for this course because of its popularity, simplicity, and usefulness. While an in-depth analysis would explore the implications of the mathematics, the equation's basic precepts should be recognized easily.

Microbial populations increase until nutrients are exhausted. Analysis is aided by keeping focus on the limiting nutrient. The Monod equation relates limiting nutrient concentration to a population's growth rate.

The expression is:

$$\mu = \hat{\mu}\frac{S}{K_s + S}$$

where,

S = Concentration of limiting nutrient

μ = Specific growth rate coefficient

$\hat{\mu}$ = Maximum growth rate coefficient

K_s = Half saturation coefficient

Concepts of this expression:

The Monod equation is empirical but fits an actual data quite well.

We can solve for continuous culture at steady state where μ = D to get

$$S = \frac{D\,K_s}{\hat{\mu} - D}$$

where, D is the dilution rate defined as equal to the flow/volume. Basically, this implies that the growth rate is primarily a function of S. We have previously learned of different mechanisms that affect the operation and growth of a cell.

The maximum growth rate itself is independent of S and K_s.

Microbial Metabolites

Primary metabolites   During the log or exponential phase, organisms produce a variety of substances that are essential for their growth, namely nucleotides, nucleic acids, amino acids, proteins, carbohydrates and lipids or by-products of energy-yielding metabolism, namely ethanol, acetone and butanol. This phase is described as tropophase and the products are usually called primary metabolites (Table 1.2).

Table 1.2   Examples of commercially produced primary metabolites

Primary metabolite	Organism	Significance
Ethanol	*Saccharomyces cerevisiae, Kluyveromyces fragilis*	Alcoholic beverages
Citric acid	*Aspergillus niger*	Food
Acetone–butanol	*Clostridium acetobutyricum*	Solvents
Lysine, glutamic acid	*Corynebacterium glutamicum*	Nutritional additive, flavour enhancer
Riboflavin	*Ashbya gossipii, Eremothecium ashbyi*	Nutrient
Vitamin B_{12}	*Pseudomonas denitrificans, Propionibacterium shermanii*	Nutrient
Dextran	*Leuconostoc mesenteroides*	Industrial
Xanthan gum	*Xanthomonas campestris*	Industrial

Secondary metabolites   Organisms produce a number of products other than the primary metabolites. In the idiophase, the organisms yield products that have no role in the metabolism of culture organisms, and these products are called secondary metabolites. Secondary metabolites are mainly produced for survival during stationary phase. In reality, there is no proper distinction between primary and secondary metabolites. Many secondary metabolites are produced from intermediate and end products of secondary metabolism. Some like those of the enterobacteriaceae do not undergo secondary metabolism (Table 1.3).

Table 1.3   Examples of commercially produced secondary metabolites

Metabolite	Species	Significance
Penicillin	*Penicillium chrysogenum*	Antibiotic
Erythromycin	*Streptomyces erythreus*	Antibiotic
Streptomycin	*Streptomyces griseus*	Antibiotic
Cephalosporin	*Cephalosporium acrimonium*	Antibiotic
Griseofulvin	*Penicillium griseofulvin*	Antifungal agent
Cyclosporin A	*Tolypocladium inflatum*	Immunosuppressant
Gibberellin	*Gibberella fujikuroi*	Plant growth regulator

Secondary metabolism may be repressed in certain cases. For example, glucose represses the production of actinomycin, penicillin, neomycin and streptomycin; phosphate represses streptomycin and tetracycline production. Hence, the culture medium for secondary metabolite production should be carefully chosen.

REVIEW QUESTIONS

1.   Give the brief details about overview of bioprocessing.

2.   What are the uses of bioprocessing in industry?

3.   Give an account on environmental bioprocessing and human applications.

4.   Explain about microbial growth curve.

5.   Give the details about Monod equation.

6.   Explain about microbial metabolites.

TYPES OF FERMENTATION

Microbial fermentation can be broadly classified into two types:

1. Solid-state fermentation (SSF)

2. Submerged fermentation

This classification is mainly based on the nature of the substrate what we use. In SSF, solid or semi-solid medium is used to carry out fermentation, whereas in submerged fermentation, broth is used. SSF is preferable for cultivation of filamentous fungi because the fungi can cover the solid substrate with mycelium and utilize the nutrients completely. Submerged fermentation is mainly used to cultivate bacteria. However, some fungi can also be cultivated by submerged fermentation with certain conditions.

SOLID-STATE FERMENTATION

It is a fermentation technique in which a solid and semi-solid medium is used to carry out the fermentation. A solid porous matrix which can be biodegradable or not, but with a large surface area per unit volume, in the range of 10^3–10^6 m^2/cm^3, is used for microbial growth on the solid/gas interface. The matrix should absorb water several times its dry weight with a relatively high water activity on the solid/gas interface in order to allow high rates of biochemical processes. Air mixture of oxygen with other gases and aerosols should flow under a relatively low pressure and mix the fermenting mash. The solid/gas interface should be a good habitat for the fast development of specific cultures of moulds, yeasts or bacteria, either in pure or mixed cultures. The mechanical properties of the solid matrix should withstand compression or gentle stirring, as required for a given fermentation process. This requires small granular or fibrous particles, which do break or stick to each other.

The solid matrix should not be contaminated by inhibitors of microbial activities and should be able to absorb or contain available microbial foodstuff such as carbohydrates (cellulose, starch, sugars) nitrogen sources (ammonia, urea, peptides) and mineral salts.

Typical examples of SSF are traditional fermentations such as the following.

1. Japanese "koji" which uses steamed rice as solid substrate inoculated with solid strains of the mould *Aspergillus oryzae.*

2. Indonesian "tempeh" or Indian "ragi" which uses steamed and cracked legume seeds as solid substrate and a variety of non-toxic moulds as microbial seed.

3. French "blue cheese" which uses perforated fresh cheese as substrate and selected moulds such as *Penicillium roqueforti* as inoculum.

4. Composting of lignocellulosic fibres naturally contaminated by a large variety of organisms including cellulolytic bacteria, moulds and *Streptomyces sp*ecies.

In addition to traditional fermentations, new versions of SSF have been invented. For example, it is estimated that nearly one-third of industrial SSF and koji processes in Japan has been modernized for large-scale production of citric and itaconic acids. Furthermore, new applications of SSF have been suggested for the production of antibiotics, secondary metabolites or enriched foodstuff. Presently, SSF has been applied to large-scale industrial processes mainly in Japan. Traditional koji, manufactured in small wooden and bamboo trays, has changed gradually to more sophisticated processes: fixed-bed room fermentations, rotating drum processes and automated stainless steel chambers or trays with microprocessors, electronics sensors and servo mechanical stirring, loading and discharging.

Microorganisms for SSF

Bacteria, yeasts and fungi can grow on solid substrates, and find application in SSF processes. Filamentous fungi are the best adapted for SSF and dominate in research works. Some examples of SSF processes for each category of microorganisms are reported in Table 2.1. Bacteria are mainly involved in composting, anaerobic fermentation and some food processes (Doelle *et al.*, 1992). Yeasts can be used for ethanol, and food or feed production (Saucedo-Castañeda *et al.*, 1992a, 1992b).

But filamentous fungi are the most important group of microorganisms used in SSF process owing to their physiological, enzymological and biochemical properties.

The hyphal mode of fungal growth and their good tolerance to low water activity (A_w) and high osmotic pressure conditions make fungi efficient and competitive in natural microflora for bioconversion of solid substrates.

Koji and *Tempeh* The two most important applications of SSF with filamentous fungi are *Koji* and *Tempeh*. *Aspergillus oryzae* is grown on wheat bran and soybean for *Koji* production, which is the first step of soy sauce or citric acid fermentation. *Koji* is a concentrated hydrolytic enzyme medium required in the further steps of fermentation process. Tempeh is an Indonesian fermented food produced by the growth of *Rhizopus oligosporus* on soybeans. People consume the fermented product after cooking or toasting. The fungal fermentation allows better nutritive quality and degrades some anti-nutritional compounds contained in the crude soybean. The hyphal mode of growth gives a major advantage to filamentous fungi over unicellular microorganisms in the colonization of solid substrates and for the utilization of available nutrients. The basic mode of fungal growth is a combination of apical extension of hyphal tips and the generation of new hyphal tips through branching. An important feature is that, although extension occurs only at the tip at a linear and constant rate, the frequency of branching makes the kinetic growth pattern of biomass exponential, mainly in the first steps of the vegetative stage. That point is important for growth modelling and will be discussed further.

Table 2.1 Main groups of microorganisms involved in SSF processes

Microflora	SSF process
Bacteria	
Bacillus sp.	Composting, Natto, amylase
Serratia sp.	Composting
Streptococcus sp.	Composting
Lactobacillus sp.	Ensiling, food
Clostridium sp.	Ensiling, food
Yeast	
Endomycopsis burtonii	Tempeh, cassava, rice
Saccharomyces cerevisiae	Food, ethanol
Schwanniomyces castelli	Ethanol, amylase

(Contd.)

Table 2.1   (Continued)

Microflora	SSF process
Fungi	
Alternaria sp.	Composting
Aspergillus sp.	Composting, industrial, food
Fusarium sp.	Composting, gibberellins
Monilia sp.	Composting
Mucor sp.	Composting, food, enzyme
Rhizopus sp.	Composting, food, enzymes, organic acids
Phanerochaete chrysosporium	Composting, lignin degradation
Trichoderma sp.	Composting, biological control, bioinsecticide
Beauveria sp., *Metarhizium* sp.	Biological control, bioinsecticide
Aspergillus oryzae	Koji, food, citric acid
Rhizopus oligosporus	Tempeh, amylase, lipase
Aspergillus niger	Feed, proteins, amylase, citric acid
Pleurotus oestreatus, P. sajor kaju	Mushroom
Lentinus edodes	Mushroom
P. roquefortii, Penicillium notatum	Cheese, penicillin

The hyphal mode of growth gives the filamentous fungi the power to penetrate into the solid substrates. The cell wall structure attached to the tip and the branching of the mycelium ensure a firm and solid structure. The hydrolytic enzymes are excreted at the hyphal tip, without large dilution as in the case of LSF, what makes the action of hydrolytic enzymes very efficient and allows penetration into most solid substrates. Penetration increases the accessibility of all available nutrients within particles.

Fungi cannot transport macromolecular substrates, but the hyphal growth allows a close contact between hyphae and substrate surface. The fungal mycelium synthesizes and excretes high quantities of hydrolytic exoenzymes. The resulting contact catalysis is very efficient and the simple products are in close contact to enter the mycelium across the cell membrane promotes biosynthesis and fungal metabolic activities. This contact catalysis by enzymes can explain the logistic model of fungal growth commonly observed. This point will be discussed further.

Substrates

All solid substrates have a common feature which is their basic macromolecular structure. In general, substrates for SSF are composite and heterogeneous products from agriculture or by-products of agro industry. This basic macromolecular structure (e.g. cellulose, starch, pectin, lignocellulose, fibres, etc.) confers the properties of a solid to the substrate. The structural macromolecule may simply provide an inert matrix (sugar cane bagasse, inert fibres, resins) within which the carbon and energy source (sugars, lipids, organic acids) are adsorbed. But generally, the macromolecular matrix represents the substrate and also provides the carbon and energy source.

Preparation and pretreatment represent the necessary steps to convert the raw substrate into a form suitable for use, that include:

- size reduction by grinding, rasping or chopping.

- physical, chemical or enzymatic hydrolysis of polymers to increase substrate availability by the fungus.

- supplementation with nutrients (phosphorus, nitrogen, salts) and setting the pH and moisture content, through a mineral solution.

- cooking or vapour treatment for macromolecular structure pre-degradation and elimination of major contaminants. Pretreatments will be discussed under individual applications.

The most significant problem of SSF is the high heterogeneity, which makes difficult to focus one category of hydrolytic processes, and leads to poor trials of modelling. This heterogeneity is of different nature:

- non-uniform substrate structure (mixture of starch, lignocellulose, pectin).

- variability between batches of substrates, limiting the reproducibility.

- difficulty of mixing solid mass in fermentation, in order to avoid compaction, which causes non-uniform growth, gradients of temperature, pH and moisture, that makes representative samples almost impossible to obtain. Each macromolecular type of substrate presents different kind of heterogeneity.

Lignocellulose   Lignocellulose occurs within plant cell walls, which consist of cellulose microfibrils embedded in lignin, hemicellulose and pectin. Each category of plant material contains variable proportion of each chemical compound.

Two major problems can limit lignocellulose breakdown.

- ⮞ cellulose exists in four recognized crystal structures known as celluloses I, II, III and IV. Various chemical or thermal treatments can change the structure from crystalline to amorphous.

- ⮞ different enzymes are necessary in order to degrade cellulose, e.g. endo- and exocellulases plus cellobiase.

Pectins   They are polymers of galacturonic acid with different ratio of methylation and branching. Exo- and endopectinases and demethylases hydrolyse pectin into galacturonic acid and methanol. Hemicelluloses are divided into three major groups: xylans, mannans and galactans. Most of hemicelluloses are heteropolymers containing two to four different types of sugar residues.

Lignin   It represents between 26 to 29% of lignocellulose, and is strongly bounded to cellulose and hemicellulose, hiding them and protecting them from the hydrolase attack. Lignin peroxidase is the major enzyme involved in lignin degradation. *Phanerochaete chrysosporium* is the most recognized fungi for lignin degradation. So, lignocellulose hydrolysis is a very complex process. Effective cellulose hydrolysis requires the synergetic action of several cellulases, hemicellulases and lignin peroxidases. Despite this, lignocellulose is a very abundant and cheap natural renewable material, so a lot of work has been conducted on its microbial breakdown, specially with fungal species.

Starch   It is another very important and abundant natural solid substrate. Many microorganisms are capable to hydrolyse starch, but generally its efficient hydrolysis requires previous gelatinization. Some recent works concern the hydrolysis of the raw (crude or native) starch as it occurs naturally.

The chemical structure of starch is relatively simple compared to lignocellulose substrates. Essentially starch is composed of two related polymers in different proportions: amylose (16–30%) and amylopectin (65–85%). Amylose is a polymer of glucose linked by α-1,4 bonds, mainly in linear chains. Amylopectin is a large highly branched polymer of glucose including α-1,6 bonds at the branch points.

Within the plant, cell starch is stored in the form of granules located in amyloplasts, intracellular organelles surrounded by a lipoprotein membrane.

Starch granules are highly variable in size and shape depending on the plant material. Granules contain both amorphous and crystalline internal regions in respective proportions of about 30/70. During the process of gelatinization, starch granules swell when heated in the presence of water, which involves the breaking of hydrogen bonds, especially in the crystalline regions. Many microorganisms can hydrolyse starch, specially fungi which are then suitable for SSF application involving starch substrates. Glucoamylase, α-amylase, β-amylase, pullulanase and isoamylase are involved in the processes of starch degradation. Mainly α-amylase and glucoamylase are of importance for SSF.

α-amylase is an endo-amylase attacking α-1,4 bonds in random fashion which rapidly reduce molecular size of starch and consequently its viscosity producing liquefaction. Glucoamylase occurs almost exclusively in fungi including *Aspergillus* and *Rhizopus* groups. This exoamylase produces glucose units from amylose and amylopectin chains. Microorganisms generally prefer gelatinized starch. But large quantity of energy is required for gelatinization so it would be attractive to use organisms growing well on raw (ungelatinized) starch.

BIOMASS MEASUREMENT IN SSF

Biomass is a fundamental parameter in the characterization of microbial growth. Its measurement is essential for kinetic studies on SSF. Direct determination of biomass in SSF is very difficult due to problems of separation of the microbial biomass from the substrate. This is especially true for SSF processes involving fungi, because the fungal hyphae penetrate into and bind the mycelium tightly to the substrate. On the other hand, for the calculation of growth rates and yield it is the absolute amount of biomass, which is important.

Direct measurement of exact biomass in SSF is very difficult. For that we can consider the global stoichiometric equation of the microbial growth.

Respiratory Metabolism

Oxygen consumption and carbon dioxide release result from the respiration, the metabolic process by which aerobic microorganisms derive most of their energy for growth. As carbon compounds within the substrate are metabolized, they are converted into biomass and carbon dioxide. Production of carbon dioxide causes the weight of fermenting substrate to decrease during growth, and the amount of weight lost can be correlated to the amount of growth that has occurred.

The measurement of either carbon dioxide evolution or oxygen consumption is most powerful when coupled with the use of a correlation model. If both the monitoring and computational equipment is available then these correlation models provide a powerful means of biomass estimation since continuous on-line measurements can be made. Other advantages of monitoring effluent gas concentrations with paramagnetic and infrared analysers include the ability to monitor the respiratory quotient to ensure optimal substrate oxidation, the ability to incorporate automated feedback control over the aeration rate, and the non-destructive nature of the measurement procedure.

Production of Extracellular Enzymes or Primary Metabolites

Associated extracellular enzymes are other microbial products produced by metabolic activities. A good correlation between mycelial growth and organic acid production, is observed frequently which can be measured by the pH measurement or aposteriori correlated by HPLC analysis on extracts.

Protein content   The most readily measured biomass component is protein. The Folin method is more sensitive and allowed a greater dilution of the sample which avoided interference from the starch in the substrate.

Glucosamine   A useful method for the estimation of fungal biomass in SSF is the glucosamine method. This method takes advantage of the presence of chitin in the cell walls of many fungi. Chitin is a poly *N*-acetylglucosamine. Interference with this method may occur with growth on complex agricultural substrates containing glucosamine in glucoproteins.

ENVIRONMENTAL FACTORS FOR SSF

Environmental factors such as temperature, pH, water activity, oxygen levels and concentrations of nutrients and products significantly affect microbial growth and product formation. In submerged stirred cultures, environmental control is relatively simple because of the homogeneity of the suspension of microbial cells and of the solution of nutrients and products in the liquid phase.

Moisture Content and Water Activity (A_w)

SSF process can be defined as microbial growth on solid particles without presence of free water. The water present in SSF systems exists in a complexed form within the solid matrix or as a thin layer either absorbed to the surface of the particles or less tightly bound within the capillary regions of the solid. The optimum A_W for growth of a limited number of fungi used in SSF processes is at least 0.96 whereas the minimum growth A_W is generally greater than 0.9. This suggests that fungi used in SSF processes are not especially xerophilic. The optimum A_W values for sporulation by *Trichoderma viridae* and *Penicillium roqueforti* are lower than those for growth . Maintenance of the A_W at the growth optimum would allow fungal biomass to be produced without sporulation.

Temperature and Heat Transfer

Stoichiometric global equation of respiration is highly exothermic and heat generation by high levels of fungal activity within the solids lead to thermal gradients because of the limited heat transfer capacity of solid substrates. In aerobic processes, heat generation may be approximated from the rate or CO_2 evolution or O_2 consumption. Each mole of CO_2 produced during the oxidation of carbohydrates released 673 Kcal. Since the elevation risk of SSF temperature depends upon the evolution of CO_2, the CO_2 evolution is considered as a important parameter and monitored during SSF.

Heat removal is probably the most crucial factor in large-scale SSF processes, and conventional convection or conductive cooling devices are inadequate for dissipating metabolic heat due to the poor thermal conductivity of most solid substrates and result in non-acceptable temperature gradients. Only evaporative cooling devices provide sufficient heat elimination. Although the primary function of aeration during aerobic solid state cultivations was to supply oxygen for cell growth and to flush out the produced carbon dioxide, it also serves a critical function in heat and moisture transfer between the solid and the gas phase. The most efficient processes for temperature control consist of evaporating water, what needs in return to complete the loss to avoid desiccation.

pH Control and Risks of Contamination

The pH of a culture may change in response to microbial metabolic activities. The most obvious reason is the secretion of organic acids such as citric, acetic or lactic acids, which will cause the pH to decrease. On the other hand, the assimilation of organic acids which may be present in certain media will lead to an increase in pH, and urea hydrolysis result in alkalinization.

Oxygen Uptake

Aeration fulfils four main functions in solid state processes, namely (i) to maintain aerobic conditions, (ii) for carbon dioxide desorption, (iii) to regulate the substrate temperature and (iv) to regulate the moisture level. Solid state process allows free access of atmospheric oxygen to the substrate; aeration may be easier than in submerged cultivations because of the rapid rate of oxygen diffusion.

SOLID STATE BIOREACTORS

Traditionally, the SSF is carried out in metal trays, baskets, jars, bags, etc. (Figure 2.1). At present the sustainability, from different points of view (carbon balance, utilization of residues, recycling of wastes, low pollution of industrial processes, safe and self-maintaining of the ecosystems), is a very important aspect for Earth's health. A basic contribution to the sustainability is supplied by the complete utilization of plant materials, starch, vegetable oils, celluloses, hemicelluloses and lignins. Among these molecules the most difficult one to be utilized are the water-insoluble components of plant cell walls since their utilization needs their depolymerization to obtain water-soluble matter useful for the subsequent bioconversions.

The possibility of utilization of solid-state fermentation was checked with the first SSB, described in scientific literature, with which it is possible to treat about 100 g of solid-state matter contained in a central basket. After sterilization of the apparatus in a pressure cooker, the solid matter is inoculated with microorganisms growing in the broth contained in the upper funnel (Figure 2.2). Then the inoculum is pumped away and the apparatus is held in a thermostat at a chosen temperature. The possibility to change the inner atmosphere in a sterile manner allows to analyse the inner bioreactor atmosphere.

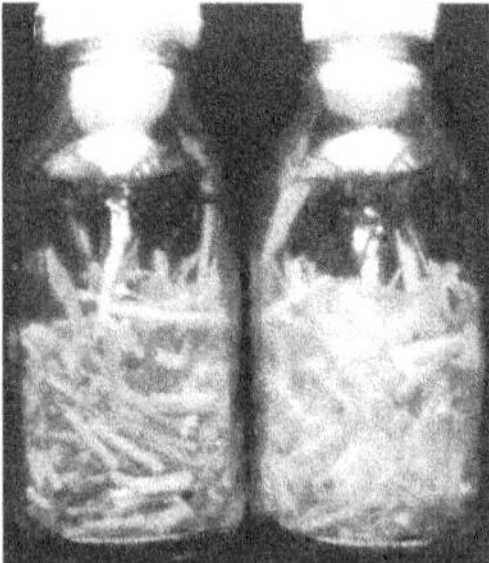

Figure 2.1 Traditional method

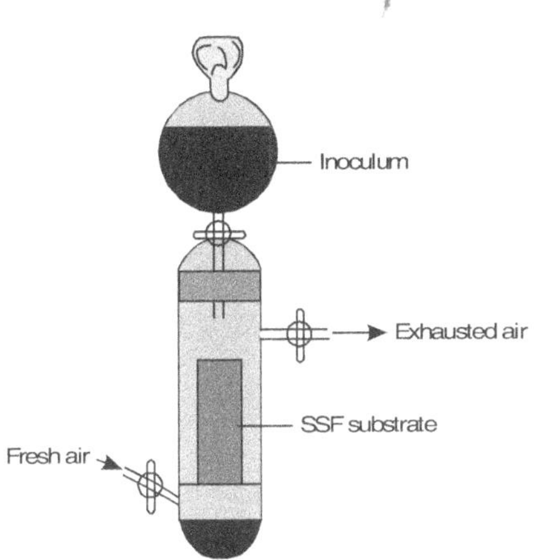

Figure 2.2   Principle designing of SSF bioreactor

SOLID-STATE FERMENTATION—PRODUCTION OF INDUSTRIAL ENZYMES

Solid-state fermentation (SSF) holds tremendous potential for the production of enzymes. It can be of special interest in those processes where the crude fermented product may be used directly as the enzyme source. In addition to the conventional applications in food and fermentation industries, microbial enzymes have attained significant role in biotransformation involving organic solvent media, mainly for bioactive compounds. Table 2.2 lists some of the possible applications of the enzymes produced in SSF systems. This system offers numerous advantages over submerged fermentation (SmF) system, including high volumetric productivity, relatively higher concentration of the products, less effluent generation, requirement for simple fermentation equipments, etc.

Table 2.2   Main applications of SSF processes in various economic sectors

Economic sector	Application	Examples
Agro-food industry	Traditional food fermentations	Koji, Tempeh, Rae, fermented cheeses
	Mushroom production and spawn	*Agaricus, Pleurotus,* Shn-take
	Bioconversion by-products	Sugar pulp, bagasse coffee pulp, silage osting, detoxication
	Food additives	Flavours, dye stuff, essential fat and organic acids

(Contd.)

Table 2.2   (Continued)

Economic sector	Application	Examples
Agriculture	Biocontrol, bioinsecticide	*Beauveria, Metarhizium, trichoderma*
	Plant growth hormones	Giberellins, *Rhizobium, Trichoderma*
	Mycorrhization, wild mushroom	Plant inactivation
Industrial fermentation	Enzymes production	Amylases, cellulases, proteases, pectinases, xylanases
	Antibiotic production	Penicillin, feed and probiotics
	Organic acid production	Ciric acid, fumaric acid, gallic acid, lactic acid
	Ethanol production	*Schwanniomyces* sp. starch malting and brewing
	Fungal metabolites	Hormones, alkaloids

Selection of Microorganisms

Microorganisms used for the production of enzymes in solid-state fermentation systems include bacteria, yeast and fungi. They produce different groups of enzymes. Selection of a particular strain, however, remains a tedious task, especially when commercially competent enzyme yields are to be achieved. For example, it has been reported that while a strain of *Aspergillus niger* produced 19 types of enzymes, α-amylase was being produced by several cultures. Thus, the selection of a suitable strain for the required purpose depends upon a number of factors, in particular upon the nature of the substrate and environmental conditions. Generally, hydrolytic enzymes, e.g. cellulases, xylanases, pectinases, etc. are produced by fungal cultures, since such enzymes are used in nature by fungi for their growth. *Trichoderma* spp. and *Aspergillus* spp. have most widely been used for these enzymes. Amylolytic enzymes too are commonly produced by filamentous fungi and the preferred strains belong to the species of *Aspergillus* and *Rhizopus*. Although commercial production of amylases is carried out using both fungal and bacterial cultures, bacterial α-amylase is generally preferred for starch liquefaction due to its high temperature stability. In order to achieve high productivity with less production cost, apparently, genetically modified strains would hold the key to enzyme production.

Selection of Substrate

Substrates used for the production of enzymes in SSF systems (Agro-industrial residues) are generally considered the best ones for the SSF processes, and use of SSF for the production of enzymes is no exception to that. A number of such substrates have been employed for the cultivation of microorganisms to produce host of enzymes. Some of the substrates that have been used included sugar cane bagasse, wheat bran, rice bran, maize bran, gram bran, wheat straw, rice straw, rice husk, soyhull, sago hampas, grapevine trimmings dust, saw dust, corncobs, coconut coir pith, banana waste, tea waste, cassava waste, palm oil mill waste, aspen pulp, sugar beet pulp, sweet sorghum pulp, apple pomace, peanut meal, rapeseed cake, coconut oil cake, mustard oil cake, cassava flour, wheat flour, corn flour, steamed rice, steam pre-treated willow, starch, etc. Wheat bran however holds the key, and has most commonly been used, in various processes.

The selection of a substrate for enzyme production in a SSF process depends upon several factors, mainly related with cost and availability of the substrate, and thus may involve screening of several agro-industrial residues. In a SSF process, the solid substrate not only supplies the nutrients to the microbial culture growing in it but also serves as an anchorage for the cells. The substrate that provides all the needed nutrients to the microorganisms growing in it should be considered as the ideal substrate. However, some of the nutrients may be available in sub-optimal concentrations, or even absent in the substrates. In such cases, it would become necessary to supplement them externally with these. It has also been a practice to pretreat (chemically or mechanically) some of the substrates before using in SSF processes (e.g. lignocellulose), thereby making them more easily accessible for microbial growth.

Factors Affecting the Substrate

Among the several factors that are important for microbial growth and enzyme production, using a particular substrate, particle size and moisture level/water activity are the most critical. Generally, smaller substrate particles provide larger surface area for microbial attack and, thus, are a desirable factor. However, too small a substrate particle may result in substrate acumination, which may interfere with microbial respiration/aeration, and therefore result in poor growth. In contrast, larger particles provide better respiration/aeration efficiency (due to increased inter-particle space), but provide limited surface for microbial attack. This necessitates a compromised particle size for a particular process.

SSF processes are distinct from submerged fermentation (SmF) culturing, since microbial growth and product formation occurs at or near the surface of the solid substrate particle having low moisture contents. Thus, it is crucial to provide an optimized water content, and control the water activity (A_w) of the fermenting substrate—for, the availability of water in lower or higher concentrations affects microbial activity adversely. Moreover, water has profound impact on the physico-chemical properties of the solids and this, in turn, affects the overall productivity process.

Design aspects of fermenter   Over the years, different types of fermenters (bioreactors) have been employed for various purposes in SSF systems. Laboratory studies are generally carried out in Erlenmeyer flasks, beakers, petri dishes, roux bottles, jars and glass tubes (as column fermenter). Large-scale fermentation has been carried out in tray-, drum- or deep-trough type fermenters. The development of a simple and practical fermenter with automation, is yet to be achieved for the SSF processes.

Factors affecting enzyme production in solid-state fermentation system   The major factors that affect microbial synthesis of enzymes in a SSF system include:

i.   selection of a suitable substrate and microorganism

ii.   pretreatment of the substrate

iii.   particle size (inter-particle space and surface area) of the substrate

iv.   water content and A_w of the substrate

v.   relative humidity

vi.   type and size of the inoculum

vii.   control of temperature of fermenting matter/removal of metabolic heat

viii.   period of cultivation

ix.   maintenance of uniformity in the environment of SSF system, and the gaseous atmosphere, i.e., oxygen consumption rate and carbon dioxide evolution rate.

SUBMERGED FERMENTATION

The fermentation process that is carried out using broth or liquid medium is known as submerged fermentation. Basically, three types of cultivation methods are carried out such as batch cultivation, fed-batch cultivation and continuous cultivation.

TYPES OF CULTURE SYSTEMS

Batch Processing Culture

The batch fermentation is a technique for large-scale production of microbes or microbial products in which, at a given time, the fermenter is stopped and the culture is worked up. Besides, the culture is allowed to cross all the phases of its growth and there is no addition of fresh medium or recovering of culture till the end of the process. At about the onset of the stationary phase, the culture is disbanded for the recovery of its biomass (cells, organism) or the compounds that accumulated in the medium (alcohol, amino acids), and a new batch is set up. The advantages of batch processing is that optimum levels of product can be recovered, it can be used for different reactions every day, and it can be properly sterilized. Little risk of infection or strain mutation and complete conversion of substrate is possible. The disadvantages are the wastage of unused nutrients, the peaked input of labour and the time lost between batches, much idle time for sterilization, growth of inoculum, cleaning after the fermentation and the safety problems when filling, emptying, and cleaning.

Continuous Culture

The culture medium may be designed such that growth is limited by the availability of one or two components of the medium. When the initial quantity of this component is exhausted, growth ceases and a steady state is reached, but growth is renewed by the addition of the fresh limiting component. A certain amount of the whole culture medium (aliquot) can also be added periodically, at the time of steady state sets in. The addition of nutrients will increase the volume of the medium in the fermentation vessel. It is so arranged that the increased volume will drain off as an overflow, which is collected and used for recovery of products. At each step of addition of the medium, the medium becomes dilute both in terms of the concentration of the biomass and the products. New growth, stimulated by the added medium, will increase the biomass and the products, till another log state sets in; and another aliquot of medium will reverse the process.

This is continuous culture or processing. Since the growth of the organism is controlled by the availability of growth limiting chemical component of the medium, this system is called a chemostat. The rate at which aliquots are added is the dilution rate that is in effect the factor that dictates the rate of growth.

The continuous processing offers the most control over the growth of cells.

The events in a continuous culture are:

i.   the growth rate of cells will be less than the dilution rate and they will be washed out of the vessel at a rate greater than they are being produced, resulting in a decrease of biomass concentration both within the vessel and in the overflow

ii.   the substrate concentration in the vessel will rise because fewer cells are left in the vessel to consume it

iii. the increased substrate concentration in the vessel will result in the cells growing at a rate greater than the dilution rate and biomass concentration will increase and

iv.   the log state will be re-established.

Hence, a chemostat is a nutrient-limited self-balancing culture system which may be maintained in a steady state over a wide range of sub-maximum specific growth rates.

Commercial adaptation of continuous processing is confined to biomass production, and to a limited extent to the production of potable and industrial alcohol.

The steady state of continuous processing is advantageous as the system is far easier to control. During batch processing, heat output, acid or alkali production, and oxygen consumption will range from very low rates at the start to very high rates during the late-exponential phase. In the continuous processing, the rates of consumption of nutrients and those of the output chemicals are maintainable at optimal levels. Besides, the labour demand is also more uniform.

Continuous processing may suffer from contamination, both from within and outside. The fermenter design, along with strict operational control, should actually take care of this problem. The production of growth associated products like ethanol is more efficient in continuous processing, particularly for industrial use.

Continuous culturing is highly selective and favours the propagation of the best-adapted organism in culture. A commercial organism is highly mutated such that it will produce very high amounts of the desired product. But physiologically such strains are inefficient and give way in culture to inferior producers—a kind of contamination from within. The draw backs are often disappointing like promised continuous production for months fail due to a infection, spontaneous mutation of microorganisms to non-producing strain may occur and inflexible that can rarely be used for other productions without substantial retrofitting.

Methods for Continuous Cultivation

The types of continuous culturing are:

- ⮞ **Chemostat**   Nutrient is fed at constant rate (the most popular method).

- ⮞ **Turbidostat**   Employs feedback control of pumping rate to maintain a fixed turbidity of the culture. The controller of a turbidostat slows feeding when the cell concentration is below the set point so that growth can restore the turbidity. If the set point is exceeded, the pump speeds up to dilute the cell concentration.

- ⮞ **Nustat/nutritstat/auxostat**   It is a method of feeding fresh nutrient based on the concentration of metabolic product and related parameter changes like pH, dissolved oxygen. When the product level is increased than the set point or as a result of production either increasing or decreasing pH and dissolved oxygen beyond their set point level, the fresh medium is added to maintaining the factors.

Chemostat   In this process the medium is pumped continuously, and the volume is constant because the excess medium overflows. Most laboratory research uses glass bioreactors so that observation is easy. Peristaltic (tubing squeezer) pumps are commonly used because simple connections with tubing can include some extra tubing to be placed in the pump. Magnetic stirring is used.

Some sources of concern are:

1. Foaming results in overflow with the volume of liquid not exactly constant.

2. Some very fragile cells are ruptured when caught between the magnetic stirring bar and the glass of the vessel. Suspending the stirring bar usually corrects this fault.

3. Changing pumping rate by turning the pump on and off over short time periods may not work because cells respond to sudden changes by altering their rates. Very short intervals of on/off gives good result.

4. Bacteria travel upstream quite easily. They will reach the reservoir of sterile medium quickly unless the liquid path is interrupted by an air break in which the medium falls in drops through air.

Turbidostat   A turbidostat is a continuous culturing method where the turbidity of the culture is held constant by manipulating the rate at which medium is fed. If the turbidity tends to increase, the feed rate is increased to dilute the turbidity back to its set point. When the turbidity tends to fall, the feed rate is lowered so that growth can restore the turbidity to its set point.

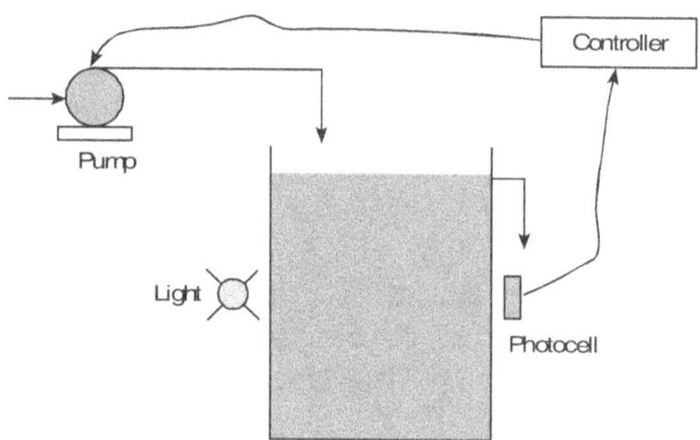

Figure 2.3   Principle of turbidostat control

The problem of growth or other materials fouling the optical surfaces of whatever method is used to measure turbidity has not been solved. While a turbidostat may operate well for a brief time, the control signal for turbidity soon becomes unreliable. Figure 2.3 shows a turbidostat.

Auxostats   In an auxostat, the microorganisms establish the feeding rate as it is adjusted to match their rate of metabolism. The turbidostat is partially auxostat. In turbidostat, fresh medium is added when the turbidity of the cell mass is raising over their set point. But the increasing turbidity is not only the growth of the organisms but also the presence of products and exhausted medium. So, when the turbidity is measured, the metabolite is also included and followed by addition of fresh medium. Here, since the metabolic products are partially considered for the tubidostat, it is also called as partial auxostat.

A device that uses the rate of feeding to control a state variable in continuous culture is termed an auxostat. The organisms establish their own dilution rate. While the well known chemostat is stable and simple for investigating continuous cultivation at low to moderate dilution rates, an auxostat tends to be much more stable at high dilution rates. Population selection pressures in an auxostat lead to cultures that grow rapidly. Practical applications include high-rate propagation, destruction of wastes with control at a concentration for maximum rate, open culturing because potential contaminating organisms cannot adapt before washing out, and operation of processes that benefit from careful balance of the ratios of nutrient concentrations.

This name was coined by Martin and Hempfling although earlier investigators had proposed the terms nutristat, nustat, and controlled-concentration-coupled-continuous-cultivation.

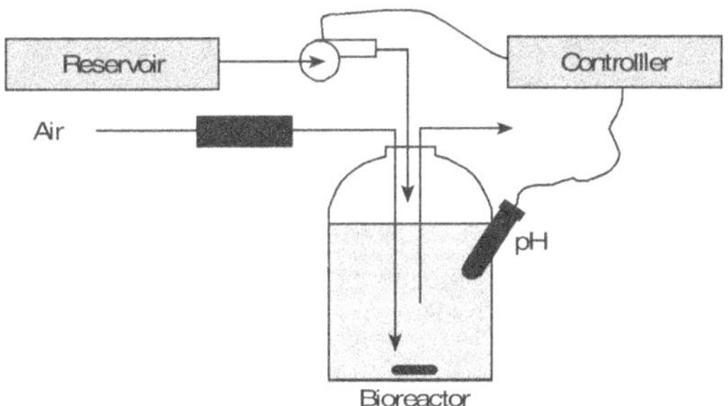

Figure 2.4   Principle model of auxostat

The most popular type of auxostat (Figure 2.4) is the pH-auxostat and dissolved oxygen-auxostat which is indirectly a nutrient controller. As the microorganisms consume a sugar (usually glucose), organic acid intermediates are produced that tend to lower the pH. This is countered by adding fresh medium that dilutes and buffers the acids. In other words, more acid equates to more medium. Because the volume of the vessel is fixed, the dilution rate increases and the residence time decreases when the feed rate rises.

pH-auxostat   The pH-auxostat couples the addition of fresh medium to control pH. As the pH drifts from a given setpoint, fresh medium is added to bring the pH back to the setpoint. The rate of medium addition is determined by the buffering capacity and the feed concentration of the limiting nutrient and not directly by the setpoint (pH) as in a traditional auxostat. The pH-auxostat is robust but controls nutrient concentration indirectly. The pH change is often an excellent indication of growth and meets the requirements as a growth-dependent parameter. However, the exact cause of pH change varies among organisms. It represents the summation of the production of different ionic species and ion release during substrate uptake. Therefore the pH can move either up or down as a function of growth. The most common situation is pH depression because of organic acid production and ammonium uptake. However for microorganisms growing on protein or amino acid-rich media, the pH will rise with growth because of the release of excess ammonia.

Dissolved oxygen-auxostat   Product and nutrient-based auxostats use direct measurements of chemical species that directly reflect metabolism. Their use is limited because of the lack of suitable sensors. Control based on the dissolved oxygen concentration (DO) was first proposed by Ohashi (1958). Operation is based on manipulating the liquid feed rate to maintain a certain DO level and not the manipulation of the air flow rate. Eventhough oxygen is a nutrient, it is normally not limiting in this type of operation and the absolute value is relatively unimportant. The DO-auxostat is similar in nature to a product-based auxostat because the dilution rate does not have a direct effect on DO. The flow of fresh air is not manipulated in simple operations but contributes along with the agitation rate to an overall mass transfer coefficient (k_1a), that is assumed to be constant. The governing equation for this type of auxostat is shown in the equation, that is based on an O_2 mass balance.

$$\frac{dC}{dt} = -RX + K_1a(C^* - C)$$

where,

C = concentration of oxygen,

C^* = concentration at saturation,

K_1a = overall oxygen transfer concentration,

R = rate of oxygen uptake by organisms and

X = concentration of organisms.

The term in the equation that has the coefficient times the concentration difference is generally termed the oxygen transfer rate (OTR) and the term with the minus sign is the oxygen uptake rate (OUR). By manipulating the feed rate of a limiting nutrient such as sugar, the OUR can be changed thereby effecting the DO. Oxygen consumption must be growth-related for effective control.

The first control of dissolved oxygen was reported with a yeast bioprocess using on/off control of the feed rate (Hospodka, 1966). The time dependency of the DO concentration and dilution rate were not reported. Yamada, *et al.*, (1979) used this technique for controlling the conversion of sorbitol to sorbose by *Acetobacter suboxydans*. They also implemented control based on the oxygen content of the bioreactor off-gas and obtained a more stable operation than with dissolved oxygen concentration. To improve the control, Konstantinov, *et al.*, (1990) used a more advanced technique where unmeasured, small, rapid disturbances were rejected through manipulation of the agitation rate, which adjusted the OTR. These disturbances were attributed to fluctuations in air flow rate and pressure. Feed rate was changed when the agitation rate moved outside a pre-determined window of operation to adjust for OUR changes.

Auxostat math   Feedback control of nutrient or product concentrations in bioprocesses has been steadily progressing for 40 years, but has suffered from the lack of reliable equipment. Continuous bioprocessing permits the observation of microorganisms at steady state. Although chemostats are in common use, the advantages of auxostats have been recognized for sometime. Auxostats can be operated in difficult or unstable conditions and can achieve steady state more quickly than can the open-loop chemostat. Operation near the maximum growth rate can be extremely unstable with a chemostat. The following equations are based on a bioprocess having Monod kinetics with mass balances for the liquid phase of the bioreactor.

$$z = DY\frac{S_0 - S}{D + MY}$$

$$\frac{dS}{dt} = D(S_F - S) - \frac{\mu X}{Y_{p/s}}$$

$$\frac{dP}{dt} = -DP + \mu XY_{p/x}$$

$$\mu = \hat{\mu}\frac{S}{K_s + S}$$

where,

μ = specific growth rate coefficient,

$\hat{\mu}$ = maximum growth rate coefficient,

S = concentration of limiting nutrient,

K_s = rate of oxygen uptake by organisms and

X = half-saturation coefficient.

$$DS_0 - DS = \frac{\mu z}{Y} + MX$$

where,

X = biomass concentration (g/l) ,

D = product concentration (g/l) and

X = dilution rate (L/hour).

Other important yield coefficients are:

$Y_{p/x}$ = product yield on biomass (g/g),

$Y_{p/s}$ = product yield on substrate (g/g) and

V_p = volumetric productivity (g/l).

Analysis of buffering capacity   Buffering capacity is defined as the equivalents of titrant required to change the medium pH to the reactor pH. The additional governing equation is the mass balance on the H^+ ion concentration in the bioreactor. By using fresh medium for pH control, the limiting nutrient level in the reactor can be manipulated through adjustments in buffering capacity and/or limiting nutrient concentration in the feed. Assuming steady state and that the difference between feed and bioreactor H^+ is very small (pH 7 means a very tiny amount of H^+), simplification gives the following:

$$X = \frac{BC}{h}$$

and

$$S = \frac{BC}{h Y_{x/c}}$$

An extensive study of buffering capacity by Rice and Hempfling (1985) showed that the specific growth rate was roughly ten times greater with low buffering capacity than with a well-buffered system.

The pH-auxostat is the most widely applied type of auxostat for efficient high-rate biomass production. Their improvement was separating the buffer from the nutrient medium to overcome inhibition of high buffer concentrations. The pH-auxostat is used to study aerobic production of ethanol by yeast operating near the maximum specific growth rate where specific ethanol productivity was highest. It also permitted an investigation of the coupling of energy and acid production, which was not possible in a chemostat.

Auxostat control based on specific gravity   When metabolism converts dissolved nutrients to carbon dioxide that escapes from the liquid, the specific gravity lessens. The change is small, and the method of measurement must be sensitive. The manometric method shown in the Figure 2.5 has one leg inclined at ten degrees from horizontal to magnify changes in the level. One side is connected to a glass tube positioned near the bottom of the bioreactor while the other side is connected to a glass tube submerged just below the surface of the medium. Each tube has its own valve to adjust its bubbling rate. The manometer is filled with dilute NaCl. An electrode is submerged in the vertical leg and another was adjustable in the inclined leg. Pumping of fresh medium counteracted the decrease in specific gravity due to metabolism; as the specific gravity decreased and the level in the inclined end rose, shorting the electrodes is sensed by a computer that turned the feed pump on.

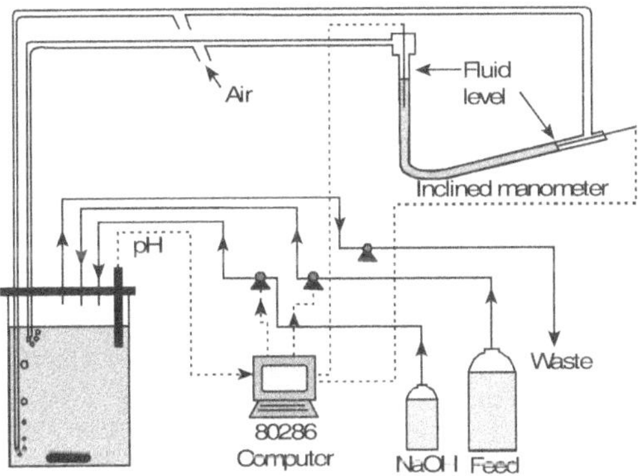

Figure 2.5   Manometric method

Runs for greatly extended periods of time, but dilution rate was erratic. A more sensitive measurement would be needed to improve control. This method would perform better in a deep bioreactor where the manometer pressure differences would be greater.

FED-BATCH CULTURE

In the fed-batch system, a fresh aliquot of the medium is continuously or periodically added, without the removal of the culture fluid. The fermenter is designed to accommodate the increasing volumes. The system is always at a quasi-steady state.

Advantages

- ➲ Production of high cell densities due to extension of working time (particularly important in the production of growth-associated products).

- ➲ Controlled conditions in the provision of substrates during the fermentation, particularly regarding the concentration of specific substrates, e.g. the carbon source.

- ➲ Control over the production of by-products or catabolite repression effects due to limited provision of substrates solely required for product formation.

- ➲ The mode of operation can overcome and control deviations in the organism's growth pattern as found in batch fermentation.

- ➲ Allows the replacement of water loss by evaporation.

- ➲ Alternative mode of operation for fermentations leading with toxic substrates (cells can only metabolize a certain quantity at a time) or low solubility compounds.

- ➲ Increase of antibiotic-marked plasmid stability by providing the correspondent antibiotic during the time span of the fermentation.

- ➲ No additional special piece of equipment is required as compared with the batch fermentation mode of operation.

Disadvantages

- ➲ It requires previous analysis of the microorganism, its requirements and the understanding of its physiology with the productivity.

- ➲ It requires a substantial amount of operator skill for the set-up, definition and development of the process.

- ➲ In a cyclic fed-batch culture, care should be taken in the design of the process to ensure that toxins do not accumulate to inhibitory levels and that nutrients other than those incorporated into the feed medium become limiting. Also, if many cycles are run, the accumulation of non-producing or low-producing variants may result.

- ➲ The quantities of the components to control must be above the detection limits of the available measuring equipment.

Fixed-volume Fed-batch

In this type of fed-batch, the limiting substrate is fed without diluting the culture.

The culture volume can also be maintained practically constant by feeding the growth limiting substrate in undiluted form, for example, as a very concentrated liquid or gas (e.g. oxygen).

Alternatively, the substrate can be added by dialysis or, in a photosynthetic culture, radiation can be the growth limiting factor without affecting the culture volume.

A certain type of extended fed-batch—the cyclic fed-batch culture for fixed volume systems—refers to a periodic withdrawal of a portion of the culture and use of the residual culture as the starting point for a further fed-batch process. Basically, once the fermentation reaches a certain stage, (for example, when aerobic conditions cannot be maintained anymore) the culture is removed and the biomass is diluted to the original volume with sterile water or medium containing the feed substrate. The dilution decreases the biomass concentration and results in an increase in the specific growth rate. Subsequently, as feeding continues, the growth rate will decline gradually as biomass increases and approaches the maximum sustainable level in the vessel once more, at which point the culture may be diluted again.

Variable-volume Fed-batch

As the name implies, a variable-volume fed-batch is one in which the volume changes with the fermentation time due to the substrate feed. The way this volume changes is dependent on the requirements, limitations and objectives of the operator.

The feed can be provided according to one of the following options:

i.   the same medium used in the batch mode is added;

ii. a solution of the limiting substrate at the same concentration as that in the initial medium is added; and

iii. a very concentrated solution of the limiting substrate is added at a rate less than (i) and (ii).

This type of fed-batch can still be further classified as repeated fed-batch process or cyclic fed-batch culture, and single fed-batch process.

The former means that once the fermentation reached a certain stage after which is not effective anymore, a quantity of culture is removed from the vessel and replaced by fresh nutrient medium. The decrease in volume results in an increase in the specific growth rate, followed by a gradual decrease as the quasi-steady state is established.

The latter type refers to a type of fed-batch in which supplementary growth medium is added during the fermentation, but no culture is removed until the end of the batch. This system presents a disadvantage over the fixed-volume fed-batch and the repeated fed-batch processes: much of the fermenter volume is not utilized until the end of the batch and consequently, the duration of the batch is limited by the fermenter volume.

Parameters Used to Control the Submerged Fermentation

All the parameters that are about to be described have an ultimate aim, to control a submerged fermentation, more specifically the growth rate and/or the flow through the central carbon metabolism and/or to reduce the overflow of carbon source to metabolic by-products. The control can be a one-step type in which only one of the parameters is used or it can be dual-level control in which two parameters are used, yielding a more "refined" control. For example, some processes require the control of different parameters at different stages of the fermentation. Such is the example of high density bacterial fermentation like the production of baker's yeast and penicillin. In these cases, as an example, two phases can be distinguished: (i) a phase in which the substrate needs to be controlled so as to avoid by-products formation and (ii) a second stage in which, due to the high cell density, oxygen transfer is limiting and so,this parameter is the one to be controlled above a critical value, under which the cellular metabolism changes. The control constraints switch from specific growth rate to oxygen concentration after some critical period of time in the fermentation process.

Calorimetry Calorimetry is an an excellent tool for monitoring and controlling microbial fermentations. Its main advantage is the generality of this parameter, since microbial growth is always accompanied by heat production, and the measurements are performed continuously on-line without introducing any disturbances to the culture. Moreover, the rate of heat production is stoichiometrically related to the rate of substrate consumption and product, including biomass formation. In many cases it can be replaced by exhaust gas analysis, although this approach cannot be considered in anaerobic processes which proceed without formation of gaseous products.

This technique has been proved successful to indirectly determine the substrate and product concentrations continuously during aerobic batch growth of *Saccharomyces cerevisiae* with glucose as the carbon and energy source. In the presence of this substrate, this yeast shows diauxic growth by initially consuming the glucose with concomitant production of ethanol and then, once glucose is depleted, using the produced ethanol as an energy source. Calorimetry can then be used to control the feed rate in such a way that ethanol formation is avoided.

Another interesting description of temperature-controlled reactors is maintaining the temperature by continuous addition of cooling medium. Normally, in continuous rectors when the cold fresh medium is added to the fermenter, the temperature is reduced gradually. Before starting the process, the equipment temperature and the reduction temperature due to the continuous addition of cold medium are calibrated. Based on the calibration the temperature is set and maintained.

Specific growth rate   For the production of a growth-associated product, the production of a certain product is related with the specific growth rate of the producing microorganism. Consequently, it is of interest to feed the fermenter in such a way that the specific growth rate remains constant. Such is the case of the production of hepatitis B surface antigen by *Saccharomyces cerevisiae*. The yield of the antigen is ten times more than that of the submerged cultivation for the same volume and total substrate added. Care should be given to the value of the chosen specific growth rate, because cells may not be "activated" easily, stress proteases can be produced that may degrade the product and also there might be a threshold value of specific growth rate above which there is production of by-products.

The substrates are important parameters which associate with growth of the organisms and production of products. When the concentration of substrate increased, especially carbon source, lead to reduce the production of by-products and increase the carbon dioxide production.

An example of adaptive feedback control of glucose concentration is found in the submerged production of thuringiensin, —an exotoxin that shows efficacious control against flies, beetles, bugs and mites—by *Bacillus thuringiensis*. Glucose concentration is estimated by using empirical correlation equations between the consumed glucose and the values of "oxygen uptake rate" (OUR) and "carbon dioxide production rate" (CPR). Then, the glucose concentration G(t) in the fermenter was expressed as

G(t) = [{initial mass amount} — {consumed mass} + {mass in feed}]/V(t)

V(t) is the volume of the fermenter for time t

By using this adaptive type of control, the production of thuringiensin was significantly improved, with readings that were ten times higher for submerged (according to the feed medium) as compared with other cultivations.

As a final comment the operator should be aware of the analytical equipment that is used so as to guarantee that the readings from the substrate concentrations fall within the limits of detection of the assay and/or analytical assay. At the same time, the operator should make sure that the concentrations are low enough to prevent by-product formation. Another aspect to consider are eventual interferences in the readings of this due to some other component that might be present in the growth medium.

By-product concentration   The production of by-products is undesirable because it reduces the efficacy of the carbon flux in a fermentation. The production of these components take place whenever the substrate is provided in quantities that exceed the oxidative capacity of the cells. This approach has been used in the fermentation of *Saccharomyces cerevisiae*, in which acid production rate is used to provide on-line estimates of the specific growth rate. Also, in modern submerged processes for yeast production, the feed is under strict control based on the measurement of traces of ethanol in the exhaust gas of the fermenter.

Inductive, enhancer or enrichment components   In certain fermentations it is of interest to continuously add either an inductive or fast consumed components and not only a limiting substrate. An example is the continuous addition of an antibiotic in recombinant microorganisms bearing an antibiotic marked plasmid. Another example is given by the production of glutathione by high-glutathione-accumulating *Saccharomyces cerevisiae*, the commonly used microorganisms for commercial production. Cysteine was found to be the only amino acid that enhanced glutathione formation. However, growth inhibition occurred and it was related to the concentration of cysteine. This problem was then resolved by an adequate addition of cysteine in exponential submerged culture without growth inhibition.

Submerged proves to be an appropriate mode of fermentation in microorganisms that are producing heterologous proteins and whose elevated protein expression results in product degradation by activation of proteases. A general insight on this subject was the study of a recombinant *E. coli* for production of chloramphenicol acetyltransferase. A gradual induction with IPTG and phenylalanine (rate limiting precursor) addition strategies were able to reduce the physiological burden imposed on the bacterium, thereby avoiding cellular stress responses and enhancing bioreactor productivity. In this case, IPTG and phenylalanine were the driving parameters that dominated the feed.

As a final note, the addition of precursors or inducers should take into account if the product of interest is growth associated or not. For example, the use of a tyrosine-deficient strain of *E. coli* in the production of phenylalanine requires a balance feed of tyrosine that, if not provided in low quantities is used as carbon source with subsequent production of excessive biomass synthesis at the expense of phenylalanine synthesis. This limitation on biomass production is possible because the phenylalanine production was not growth associated.

Respiratory quotient (RQ)

Gas analysers, especially mass spectrophotometers are relatively fast. Respiratory quotient, the ratio between the moles of carbon evolved per mole of oxygen consumed, has been a general method used to determine indirectly the lack of substrate in the growth medium. It is a fairly rapid method of measurement that is useful because the gas analysers can be related to crucial process variables. The method is not "universal"to all bioprocesses since some biosystems can produce by-products that affect the productivity of the process without affecting the RQ, such as the production of acetic acid by *E.coli.*

Based on the concept of respiratory quotient, there are the so called DO-stats, in which the feed is regulated in accordance with the dissolved oxygen. The analysis of the dissolved oxygen or carbon dioxide evolution rate can also be used to control or prevent the production of by-products. The respiratory quotient is often analysed to study the carbon flux, this is, the feed should be conditioned in such a way that it should prevent excess of carbon dioxide evolution caused by unnecessarily severe substrate limitation.

Parameters to Start and Finish the Feed, and Stop the Submerged (Fed-batch) Fermentation

The time at which the feeding should start and finish, as well as the criteria to stop a submerged fermentation is very much dependent on the specific cultivation kinetics and the operator's interest. For example, in substrate limited processes, the feed should start immediately after all substrate is consumed from the batch phase, otherwise the process may be difficult to control, for example, because of a lag phase due to previous starvation. The most common criteria to start the feed is the depletion of substrate, which can be measured by a multitude of techniques, from specific enzymatic assays, HPLC to indirect methods such as the exhaust gas analysis. Still related with the amount of substrate in the medium, the operator might not find necessary to reach the complete depletion but to be below a predetermined set-point (eventually related with historical data, growth models and known yields).

The submerged fermentation should be halted when the production slows down because of cell death, because the metabolic potential of the culture becomes inadequately low or because by-product excretion starts at significant levels. Some other criteria can be an increase in viscosity that implies an increased oxygen demand until the oxygen limitation is achieved, which is the case for penicillin production.

Limiting Nutrient and Growth of Submerged Fermentation

The concept of a limiting nutrient is essential to understand the biological processes. The nutrient in short supply relative to the others will be exhausted first and will thus limit cellular growth. The other ingredients may play various roles such as exhibiting toxicity or promoting cellular activities, but there will not be an acute shortage to restrict growth as in the case of the limiting nutrient becoming exhausted. The usual way of determining which nutrient is limiting is to increase the concentration of one nutrient. If there is no effect, some other nutrient is limiting.

Product Formation

When the cells or some constituent of cells that is proportional to cell mass is the product, the rate of formation of product directly relates to the rate of growth. Many other products are known to have zero or a small rate of formation until growth ceases. It is useful to think of this as the cell using resources to grow until some important nutrient such as the sugar falls to a low concentration. For several years, lactose was thought to be essential for the penicillin fermentation. It was found that lactose (glucose and galactose linked to form a disaccharide) was hydrolysed at a rate that supplied sugar for the synthesis of the antibiotic but did not provide concentrations that stimulated growth. Dramatic improvements in titre were realized by abandoning lactose and feeding glucose fast enough for optimum synthesis of penicillin while avoiding concentrations that gave growth instead. Actual situations, then, range from entirely growth-associated product formation, through cases where product formation is partially growth-associated, to cases where product formation is an alternative to growth.

Luedeking and Piret proposed the following simple equation:

$$\frac{dP}{dt} = \alpha \frac{dX}{dt} + \beta X$$

where,

P = concentration of limiting nutrient,

X = specific growth rate coefficient,

t = maximum growth rate coefficient and

α, β are coefficients

When tested with the microbial process for producing lactic acid, this equation fits actual data very well. This was somewhat fortuitous because the agreement is not so good with some other processes. Probably alpha and beta do not remain constant as the physiological age of the cells or the conditions of the process change. Nevertheless, the concept of distinguishing the relationship of product formation to growth rate is sound.

SYNCHRONOUS CULTURE

We tend to deal with properties of very large numbers of cells as if they were continuous functions. Actually, there is a broad spectrum of cell ages, and cell growth occurs as small increments that melt into smooth curves for anything that is measured. In other words, a single event for a cell is a tiny invisible blip for the many, many cells present. There are some natural situations in which cells become synchronized. Cells can all be stopped if darkness halts photosynthesis, and reillumination may start growth simultaneously. In the laboratory, restricted light or low temperature or some other factor can impair a step in division. The cells can be triggered to proceed together from that point with overall numbers that are stepwise with time. This is termed a synchronous culture when all the cells divide simultaneously. The steps are seldom distinct for more than a few generations unless the triggering event continues to be applied periodically. Synchronous culture is an important case where differential equations are not appropriate.

Research with synchronous cultures has been very enlightening in terms of understanding what goes on in cells. One cell is far too tiny to be analysed extensively, although there are techniques such as flow cytometry that provide some information as an individual cell passes through a laser beam. With synchronous culture, all the cells are the same age, and large samples analysed by any technique, complicated or simple, can be translated to information about a single cell just by dividing by the number of cells in the sample. Concentrations of biochemicals during the life cycle of a cell have been studied in this manner.

REVIEW QUESTIONS

1. Explain about different types of fermentation.

2. What is solid state fermentation and submerged fermentation?

3. Explain about substrate used in SSF.

4. Explain in detail the environmental factors for SSF.

5. Explain in detail about different types of submerged cultivations.

6. Define auxostat, turbidostat and chemostat.

7. What is continuous cultivation? How does is it differ from batch and fed-batch culture system.

8. Give an account on parameter control on submerged fermentation.

9. Define synchronous culture system.

STRUCTURE AND ANATOMY OF FERMENTER

INTRODUCTION

The fermenter is a device used to carry out submerged fermentation. Sometimes, fermenters are also called bioreactors. However, there is quite a difference between fermenters and bioreactors based on the nature of cultivation that is carried out using the device. If fermentation is carried out for the purpose of metabolites (e.g. production of antibiotics or enzymes), the device is used as fermenter. On the other hand, if fermentation is carried out for the purpose of cell biomass (e.g. cultivation of organisms for single cell proteins), the device is used as bioreactor. The shape, size and provisions present in the fermenter differ with the nature of the desired product. The stir tank batch fermenters are basic fermenters, which can be altered based on the nature of fermentation and facility needed. There are three modes of bioreactor operations: batch, continuous and fed-batch.

Fed-batch bioreactors are commonly used to produce biological products (Figure 3.1). Batch reactors are the second commonly used. Continuous reactors are rarely used for large scale production of biochemicals; however they are widely used in waste treatment processes. Most natural ecosystems, e.g. the rumen, gut, soil and rivers, are more closely represented by continuous culture system.

3

BIOREACTORS

Batch Cultures

In batch cultures, a bioreactor is filled with fresh medium and inoculated.

At the end of fermentation,

- ➲ the contents are removed for downstream processing
- ➲ the reactor is cleaned and sterilized
- ➲ it is refilled for next fermentation

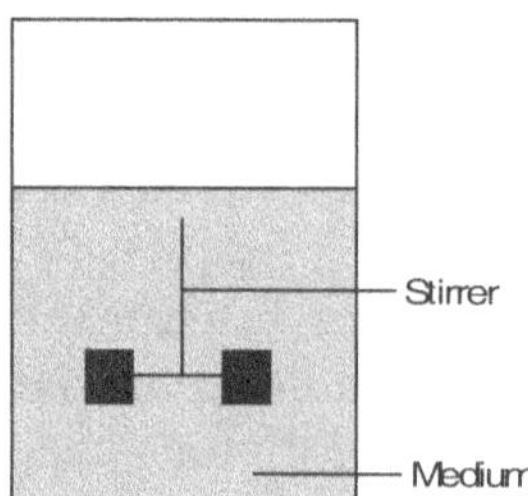

Figure 3.1   Batch reactor

Continuous Reactors

In continuous cultures, fresh media is continuously added into the bioreactor and, simultaneously, bioreactor fluid is continuously removed.

The cells continuously propagate on the fresh medium entering the reactor and, at same time, products, metabolic waste products and cells are removed in the effluent.

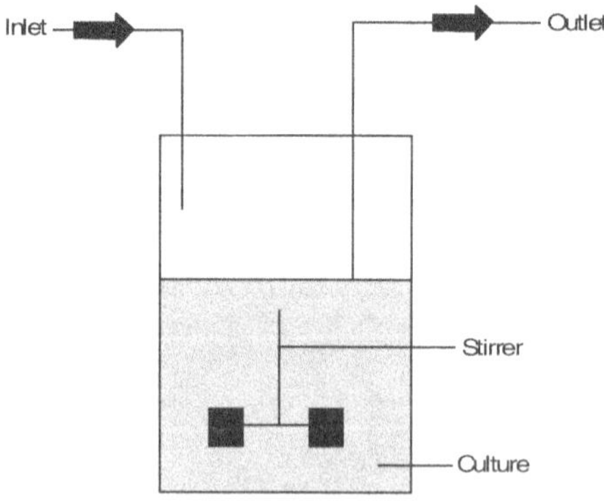

Figure 3.2   Continuous reactor

Continuous culture reactors need to be shut down less frequently than batch systems. In addition, the cells can be immobilized in the reactor to maximize their retention and thus increase productivity (Figure 3.2).

Fed-batch Bioreactors

In fed-batch cultures, the feed is continuously added until maximum liquid fermenter volume is reached. Then the fermenter may be allowed to continue or emptied partially or completely depending on the process (Figure 3.3).

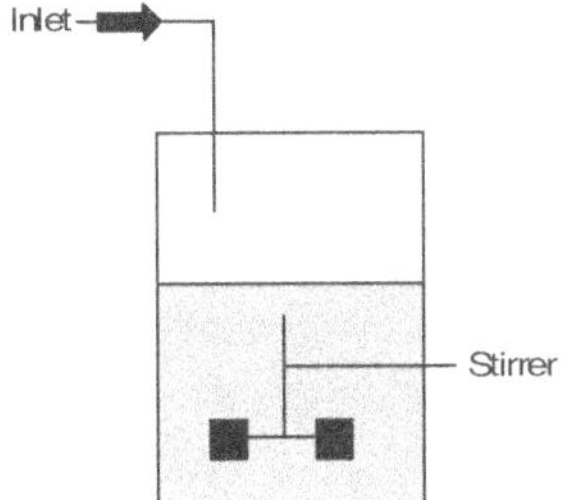

Figure 3.3   Fed-batch reactor

Comparison Between Fed-batch and Continuous Bioreactors

In both fermenters, it is possible to set and maintain specific growth rate and substrate concentration at optimal levels. The major difference between them is that the effluent is not continuously removed in fed-batch fermenter and thus washout does not occur. This provides fed-batch reactors a number of advantages over continuous reactors:

- Because cells are not removed during fermentation, fed-batch fermenters are well suited for the production of compounds at very slow or zero growth.

- The feed does not need to contain all nutrients to sustain growth. It could contain only a nitrogen source or a metabolic precursor.

- Contamination may not have a dramatic effect as the contaminant will not be able to completely take over the fermenter (unless contamination occurs during the early stages of fermentation).

- A fed-batch reactor can be operated in a number of ways. For example, the reactor must often be operated in the following sequence:

Batch → Fed-batch → Batch

The feed can also be manipulated to maximize product formation. One way to do this is that, during fermentation, the feed composition and feed flow rate are adjusted based on the growth stage of the organism.

STRUCTURE OF A FERMENTER

The fermenter (Figure 3.4) comprises several components, big and small. We shall discuss selective components that make fermentation feasible on a large scale, which include the following:

- ➲ Stirrers
- ➲ Spargers
- ➲ Baffles
- ➲ Compressors
- ➲ Heating and cooling systems
- ➲ Sensors
- ➲ Steaming unit

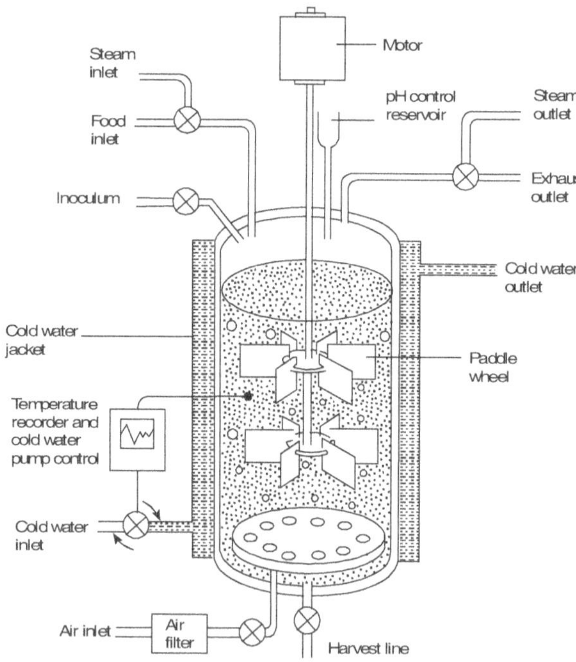

Figure 3.4   Basic structure of a fermenter

STIRRERS

The stirrer is used to agitate the medium constantly and allow even distribution of air throughout the medium. The stirrer comprises two parts, namely centre shaft and impellers. The impellers are fixed at the centre shaft and the shaft is connected to the electric motor, which rotates the impellers.

Impellers

Two types of impellers are used in industrial fermentation. The propeller agitator resembles marine propellers, the difference being that the vessel remains stationary while the fluid moves. Axial impellers are similar to the radial ones, except that the blades are pitched usually at a 45 degree angle, which causes flow to move downward, parallel to the shaft, and then upward along the wall of the tank. The impellers (Figure 3.5) are generally installed vertically in the tank so as to allow the fluid to circulate in one direction along the axis while flowing in the opposite direction along the walls. Propellers

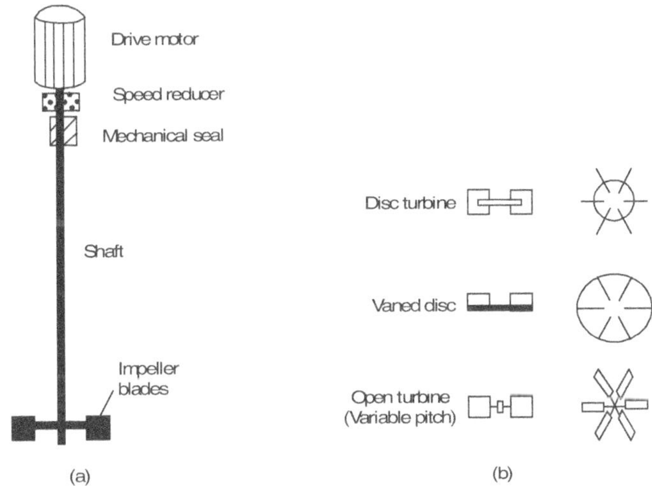

Figure 3.5   (a) Structure of impeller   (b) Some types of impellers

are generally employed in small-scale applications where their flexibility is desirable. They are also used to disperse gases or non-wetting solids into liquids due to the deep vortex they are able to create. They generally have diameters of not more than 1.5 meters and are characterized by high rotational speed. They use power in range up to 50 kW. This type of impeller can be applied to achieve bulk mixing and produce a homologous sample. Figure 3.6 illustrates the circulation pattern associated with marine type impeller.

Figure 3.6   Marine type impeller stirrer

The second type is turbine impeller. Turbine impellers are larger and mounted on shafts just like propellers, but they rotate at slower speeds. Turbines are more flexible and efficient than propellers in several ways. Radial-flow impellers are similar to centrifugal pumps, in that they discharge liquids at high velocities in the radial direction. It works like a jet mixer, causing entrainment of the surrounding fluid, while setting up two circulation systems, one above the impeller, the other below. The liquid flowing outward separates at the wall, with some flowing up to the surface and returning to the eye of the impeller along the shaft. The remainder flows down along the wall, across the bottom of the tank, and returns to the centre of the impeller.

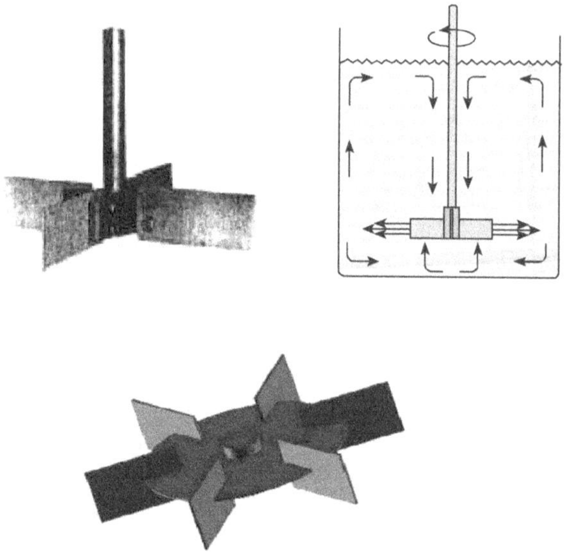

Figure 3.7   Disc type impellers

The diameters of practical turbine impellers are limited to approximately 5 metres, translating to multiple tanks or multiple agitators necessary for larger plant capacities. In fermentation plants, these impellers serve the purpose of creating mass transfer in order to break up oxygen bubbles. The high shear associated with the impeller necessitates 10–15 times power that of the propeller-type impeller. Figure 3.7 illustrates the circulation pattern associated with disc type impeller.

Seals

Shafts entering pressure vessels require seals. A simple seal (Figure 3.8) has a cavity in which there is a packing material. One type of packing is rope impregnated with graphite, but many materials that can be compressed can serve as packing material. An O-ring will work. Another type of seal has no packing. It is a labyrinth with plates attached to the shaft turning inside plates attached to the outer casing. Gas flow back and forth through the torturous space dissipates pressure. There is inherently some leakage of gas, but it can be negligible. Commercial fermentations are operated at slightly elevated pressure, so leaks will be outward. As you experiment with the spacing between fins, note that wide spacing would offer little resistance to the flow of gas. Close spacing might present construction problems and could have high resistance.

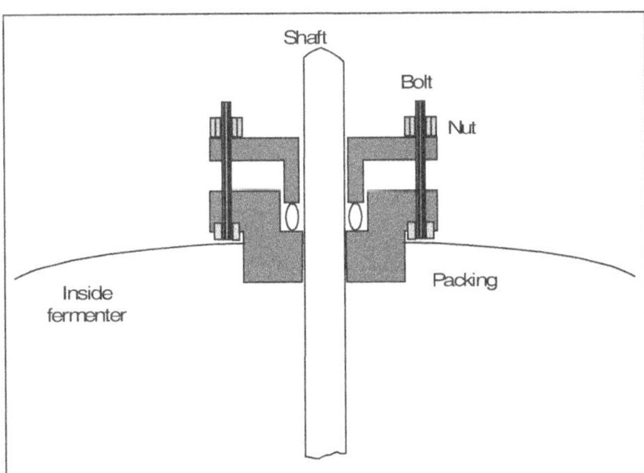

Figure 3.8   Basic diagram of seal

SPARGERS

Spargers are a type of equipment used to introduce smallest bubbles into a fermenter. The simplest sparger is the single-orifice sparger. As the size of fermenter increases, the gas requirements increase. To achieve this, the diameter of the orifice can be increased; however, the aim is to keep bubbles as small as possible. Therefore, for better aeration needs, the ring-sparger is incorporated. The ring-sparger has many holes to introduce the maximum expected air into the fermenter. Each hole is, therefore, considered to be a critical orifice. In modern systems, we can expect a diameter that will deliver air at about the speed of sound, which ensures smallest bubbles. The sparge line is located under the disc of the lower impeller so that the air stream is directed through the impellers. Spargers are of many types based on structure. The disc spargers are perforated discs; orifice spargers are perforated tubes; and nozzle spargers are tubes with a single pore. The combine spargers are a combination of agitator with sparger. Usually the perforated disc type spargers act as combine sparger. During agitation, sterile air is passed though the pores of the sparger which also aerate the medium. Thus aeration and agitation are carried out in a single device.

BAFFLES

In a fermenter, there is a need for agitation in both aerobic and anaerobic operations. Agitators, which are centrally mounted in the fermentation tanks, exhibit poor mixing characteristics if baffles are absent. Baffles reduce the time for reaching homogeneity. For example, when a dye is poured into the tank, it blends in lesser time if baffles are present. However, the drawback is that the agitator requires additional power to operate. Baffles enable the impellers to deliver power to the fluid by preventing it from swirling.

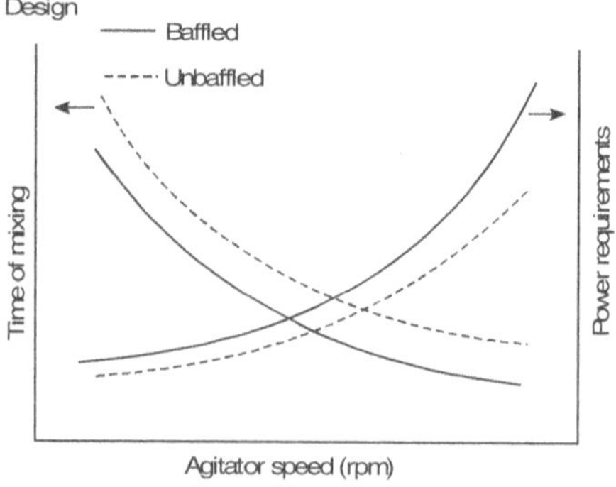

Figure 3.9   Agitation with and without baffles

The Figure 3.9 shows the relationship among the time taken to complete mixing the medium, impeller speed requirement and power consumption of baffled and unbaffled bioreactors. Comparatively, the baffled reactor shows that less agitator speed and power consumption is required to achieve the better agitation of the medium than unbaffled reactor.

COMPRESSORS

Compressors are necessary in industrial fermentation. They supply the fermentation tanks with the air necessary to run the process. Most tanks require a compressor capable of delivering 300 to 400 h.p. A fermentation plant may use one or several compressors. For plants that choose a single compressor, the power level rises to 3000 to 4000 h.p.

Single-compressor Plants

In single-compressor plants, the cost of operation is minimal. However, there is risk involved with operation of single compressor. If the compressor breaks down, the production gets disturbed until the compressor is fixed.

Multiple-compressor Plants

A compressor for each unit solves the problems associated with single-compressor. However, the cost of maintenance will be much higher than that of single-compressor plants.

COOLING JACKETS

Cooling jackets are present around the fermenter vessel. There are two types of cooling jackets 1. vessel type and 2. tube type. In vessel type, the jacket is like a vessel which is attached to the fermenter vessel and connected to the inlet and outlet (Figure 3.10). In tube type, (Figure 3.11) pipe is attached either in a spiral manner to the inner peripheral or outer peripheral region of the fermenter vessel. Based on temperature, either hot or cold water will be passed through the water jackets. In large fermenters, heating coils and cooling compressors will be present along with the pipes.

Fermenters for products such as antibiotics and vitamins are pressure vessels that are sterilized either before filling or with the contents present. Some heating comes from direct injection of steam, but all lines into or leaving the tank should also be steamed. Temperature control comes from heat transfer with water or steam. For small to moderate sized fermenters, a jacket is best because it leaves the inside of the vessel with fewer surfaces.

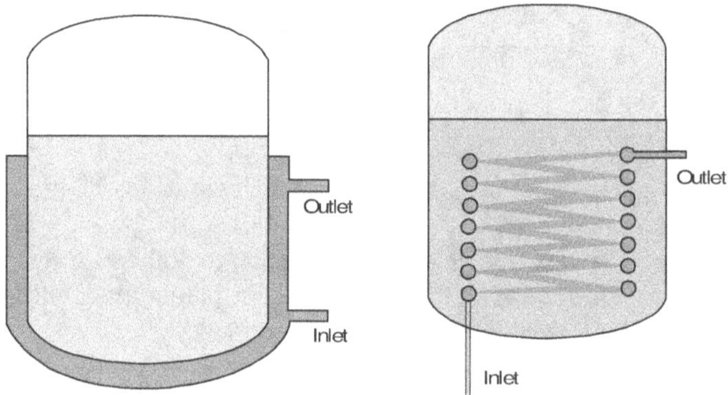

Figure 3.10 Schematic diagram of vessel type cooling jacket

Figure 3.11 Schematic diagram of (coil) type cooling jacket

Figure 3.12 Photograph of tube (coil) type

STEAMING UNIT

The steaming unit in the fermenter is a modified autoclave (Figure 3.13). Steam produced here is used to sterilize the unit before running the fermenters and after culture recovery. The steaming unit is automatically controlled when the maximum limitation of steam is produced and constant pressure of steam is passed through the heating jackets. Usually steam passing is controlled by the valve system, either controlled manually or automatically manner.

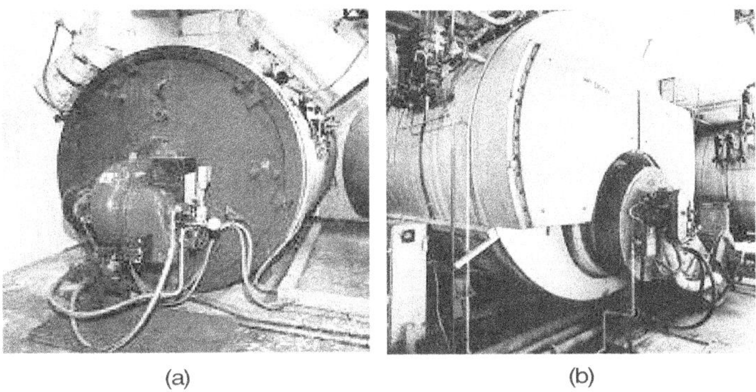

Figure 3.13   Steaming units in a fermenter

SENSORS

Sensors are devices that can detect and send the signals to the controlling units of a fermenter. These sensors are used to measure temperature, pH, dissolved oxygen, etc.

pH probe   It is mainly used to detect the pH of the fermentation medium. The pH electrode is always immersed inside the medium (Figure 3.14).

DO probe   It is an electrode used to detect the dissolved oxygen present in the fermentation medium. It should be kept immersed in the fermentation medium (Figure 3.15).

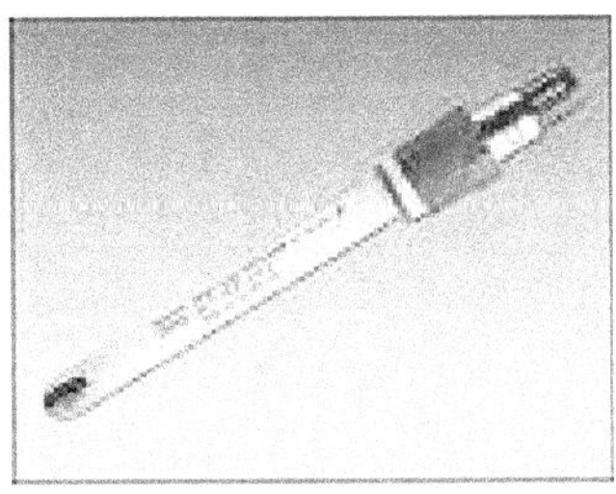

Figure 3.14   pH probe

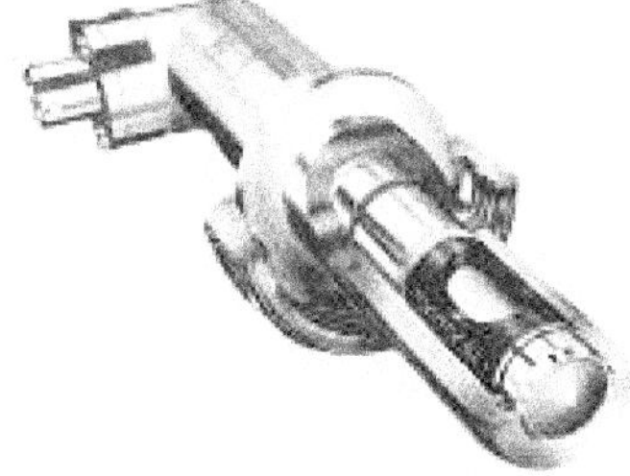

Figure 3.15   DO probe

Sensors Used in Fermentation

The sensors used for fermentation are quite different from the regular one, they can tolerate extreme heat and cold. The schematic diagram of a biosensor is shown in Figure 3.16.

- pH electrode

- redox electrode

- oxygen electrode

- ammonia electrode

- gas sensors (e.g. ethanol sensor)

- photocell (turbidity)

- viscosity gauges

- gravity gauges

- biosensors

In addition to the general properties, sensors used in fermentation control must meet some additional pre-requisites. They are:

- sterilizability

- fouling resistancy

- chemical compatibility with the microorganisms

- insensitivity to turbidity and coloured media

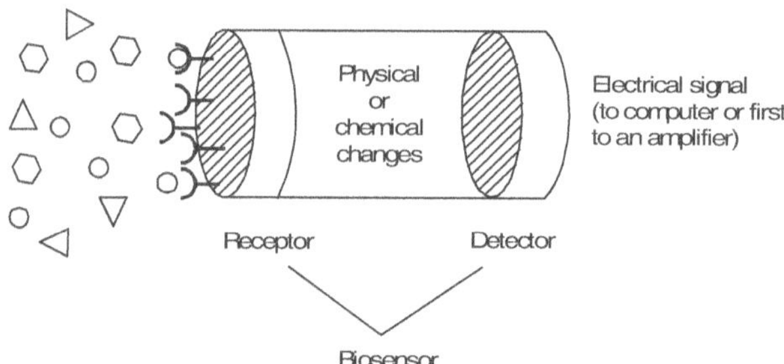

Figure 3.16   Schematic diagram of a biosensor

- **receptor**   The receptor is responsible for the selectivity of the sensor, e.g. enzymes, antibodies, lipid layers, immobilized cells.

- **detector**   It plays the role of a transducer, e.g. pH electrode, oxygen electrode, hydrogen peroxide sensor, thermistor, ion selective electrode, piezoelectric crystal. The fundamental advantage of biosensors over nearly all other sensor devices is their high selectivity and sensitivity. Depending on the actual design, different receptor–detector combinations are possible. For example an enzyme receptor can be combined with an oxygen electrode, a hydrogen peroxide sensor with a thermistor.

Only a few sensors satisfy all theses requirements, so that monitoring and control of industrial processes rely still on a relatively small number of measurements of variables.

Basically, biosensors can be divided into two main types:

1. Enzyme-based biosensors

2. Cell-based biosensors

Most biosensors rely on enzymes for the recognition of a certain substance. There are a few exceptions, for example, one which use optical or piezoelectric effects.

Enzyme-based Sensors

The enzyme-based sensors are working with the principle of enzyme–substrate reactions and the ions released. The specific membrane bound enzymes are used as detector. When the membrane bound enzyme set up is placed to the specific substrate, the enzymatic reaction takes place and the ions are released to membrane bound electrolytes which then can be measured by the power meter.

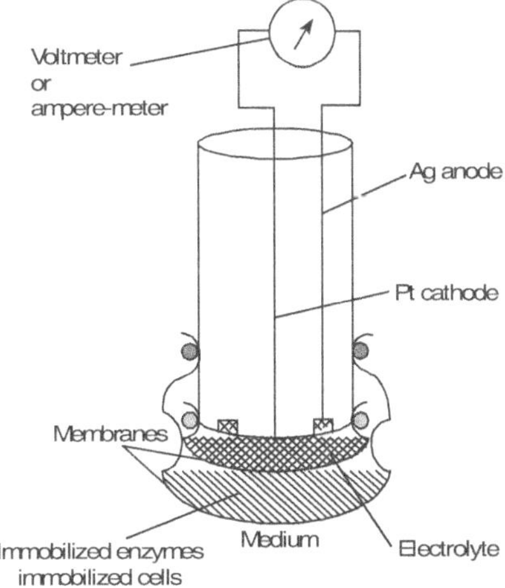

Figure 2.17   Basic enzyme-based biosensor

The rate of reaction and concentration of the substrate is calculated based on the electron flow rate. The enzyme sensors are specific and can be used for the specific substrate detection only.

The most widespread biosensor today is the glucose biosensor. It is very often used to measure the glucose concentration in blood. Its application for diabetic patients was also the driving force for the glucose sensor research. There are quite a few enzymes which could be used as glucose detectors. Many of them require cofactors, though. The following example shows the glucose enzyme reaction with glucose oxidase. This enzyme doesn't require cofactors and is in addition one of the most stable enzymes for glucose detection:

$$\text{Glucose} + O_2 \xrightarrow{\text{Glucose oxidase}} \text{Gluconic acid} + H_2O_2$$

$$H_2O_2 \xrightarrow{\text{Catalase}} \frac{1}{2}O_2 + H_2O$$

With the reaction shown above there are two possibilities to measure the glucose concentration: either by the oxygen consumption or the hydrogen peroxide production. As a third option it would be also possible to measure the temperature increase due to heat released by the two reactions. The presence of the second enzyme (catalase) although not really essential for the assay, enhances the first reaction. This happens due to the decomposition of the hydrogen peroxide.

Advantages

- more specific than cell-based sensors
- faster response due to shorter diffusion paths (no cell walls)

Disadvantages

- more expensive to produce due to the additional problem of isolating the enzyme
- enzymes are often unstable when isolated
- many enzymes need cofactors for the detection of substances

Cell-based Biosensors

In cell-based sensors, live immobilized organisms are used to detect the compounds. When immobilized cells are introduced into the substrates, the organisms metabolize the substrate and metabolites are released. Either released metabolites or metabolic activity are detected by other electrode. The presence of substrate in the medium is detected by detection of metabolites.

Just for an example, the bacterium *Acetobacter xylinium* is used to detect the ethanol in the medium. The *Acetobacter xylinium* along with oxygen electrode is introduced into the medium which contains ethanol. The *Acetobacter xylinium* aerobically metabolize the ethanol and convert into acetic acid which leads to reduction of oxygen in the medium. The oxygen reduction is measured by oxygen electrode and the oxygen reduction is considered to determine the presence of ethanol in the medium

$$CH_3CH_2OH + O_2 \xrightarrow{\text{\textit{Acetobacter xylinium}}} CH_3COOH + H_2O$$

$$\text{Ethanol} \hspace{6cm} \text{Acetic acid}$$

Advantages

- ➔ cheaper to isolate than individual cell components
- ➔ often less susceptible to changes in environmental conditions
- ➔ therefore often have an extended lifetime

Disadvantages

- ➔ longer response time (additional transport through cell wall)
- ➔ often not as specific as isolated enzyme sensors

Obstacles to Biosensor Utilization

Given the potential power and advantages of biosensors over other approaches, there are still only few examples of their use in bioprocesses. One of the main obstacles is sterilization. Most fermenters are sterilized with steam, which inactivates the biological component of biosensors. Biosensors have to be in direct contact with the medium. This leads to many problems: microbial fouling, poising through inhibition, degradation of the biocomponent. Most biosensors show a gradual drop-off in response with time of exposure to chemical environment. A further problem is the rather slow response time and the relatively limited time of stability of the signal. Both are severe drawbacks for a signal which is supposed to be used for control purposes.

REVIEW QUESTIONS

1. What is fermenter? Give the structure of a basic fermenter.

2. What is the difference between fermenters and bioreactors?

3. Explain in detail about components of fermenter.

4. Explain about different types of stirrers and spargers.

5. Give a detailed account on different types of bioreactors.

6. What are sensors? Add notes on different types of sensors.

TYPES OF FERMENTER

We know that fermenter is a device in which submerged fermentation is carried out. The size and configuration of the fermenter differ from one product to another. The commonly used fermenters are of six types.

1. Batch or stir tank reactors

2. Continuous reactors

3. Air-driven reactors

   - Bubble column reactors

   - Airlift reactors

4. Tower fermenters

5. Shake flask fermenter

BATCH OR STIR TANK REACTORS

Batch fermenter is the basic fermenter which is shown in Figure 4.1. It is available in various sizes, ranging from small laboratory type to pilot scale type with multi gallon capacity. The main parts of a fermenter are stirrer, sparger, baffle, sensors and cooling jackets (Figure 4.2). The stirrer is connected with a motor controlled by regulators. During aerobic cultivation, sterile air is passed through the spargers, which are present in the inner bottom region of the fermenter vessel. The sterile air is distributed by the stirrer all over the medium.

Figure 4.1 Batch fermenter

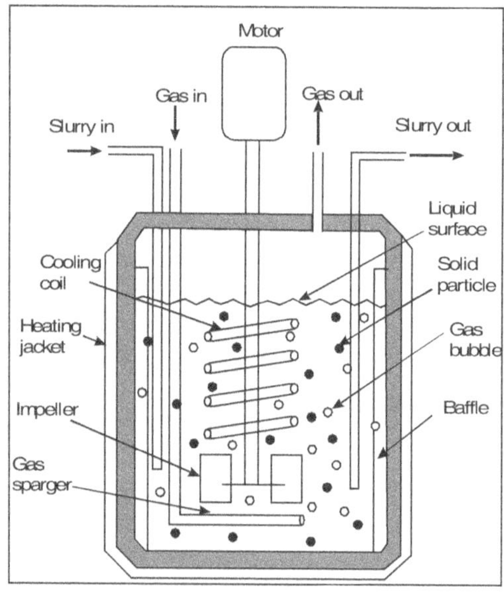

Figure 4.2 Principle model of a batch fermenter

In this kind of batch fermenter, the temperature is maintained through heating jackets. The heating jacket is present around the fermenter vessel, through which water or steam, as necessary, can be passed. Most batch reactors are sterilized by in-built sterilization methods, i.e., by passing steam. The cooling of the medium can be done by passing water through water jackets. Usually, the sensors available with batch fermenters are pH and temperature sensors. If necessary, other sensors like DO can be placed in the provision given in the reactor.

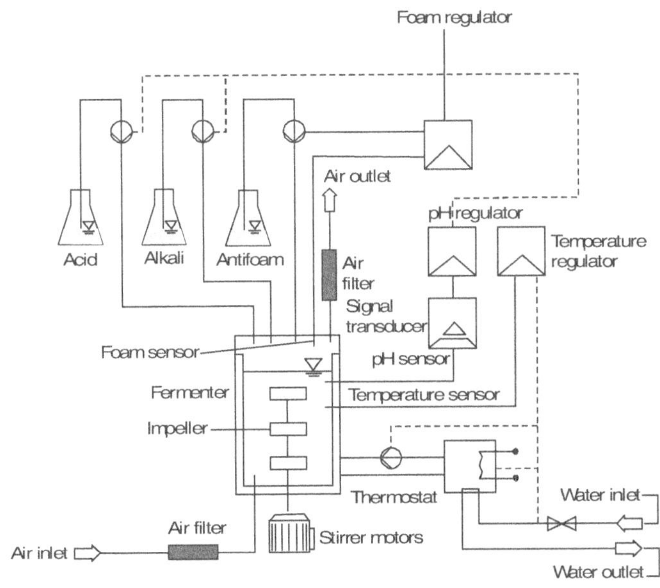

Figure 4.3 Fermenter with overall control

A fermenter with overall control in shown in Figure 4.3. Air compressor, boiling unit and peristaltic pump units are the main accessory units of batch reactor. The main function of the boiling unit is to generate steam for sterilization of the fermenter vessel, and the function of the compressor is to generate air. The air generated from the compressor is filtered through the millipore or HEPA filters and controlled by the blow meter. The air is then passed into the fermenter vessel through the sparger. During cultivation, sterile acid and alkali are prepared in separate vessels and are connected with peristaltic pumps. Based on the signal derived from pH electrode, acid or alkali will be poured automatically with the help of peristaltic pump. Antifoaming agents are also added in the same manner. The pressure of the fermenter vessel is measured by the pressure gauge attached to the fermenter vessel. The desired temperature is maintained by water jackets. The water pipe connected to the water jacket runs along the heating coils. Based on the signal from the temperature probe, either normal water will be passed through the water jackets or hot water will be passed through the pipe. The parameter optimization can be made either by automatic mode or manual mode.

CONTINUOUS REACTORS

There is no much structural variation in continuous reactors from batch reactors. All the provisions are same as in batch reactors. Continuous bioreactors have been designed to carry out submerged continuous fermentation. In continuous reactors, fresh medium is continuously added, and bioreactor fluid is continuously removed. As a result, cells continuously receive fresh medium and are simultaneously used for processing. The reactor can be operated for long time without having to be shut down. Continuous reactors can be more productive than batch reactors due to the fact that continuous reactors do not have to be shut down regularly and the growth rate of bacteria can be more easily controlled and optimized. In addition, cells can be immobilized in continuous reactors to prevent wastage, which further increases the productivity of these reactors.

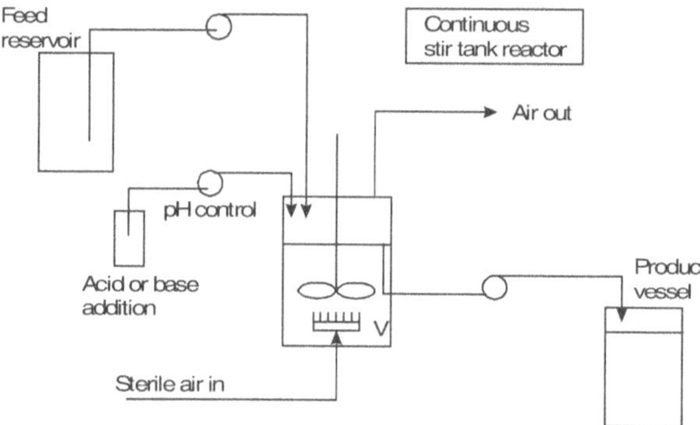

Figure 4.4   Structure of continuous flow stir tank reactor

The configuration of a typical well-mixed continuous reactor is shown in Figure 4.4. In this reactor, agitation is provided by an impeller or by the motion of the liquid phase created by the rising gas bubbles. In aerobic systems, oxygen is supplied to the cells via air sparging. The liquid phase is completely mixed, i.e., the liquid phase composition is uniform throughout the reactor. Similarly, the temperature is maintained constant by passing cool water in the vessel or in the jacket surrounding the vessel. In some cases, glycol is used in the jacket instead of cool water. Typically, the pH of the culture media is controlled by the addition of acids and bases.

CSTRs in Series

In contrast to batch cultures, the physiological state in continuous culture is no longer determined by the one preceding it, but by the dilution rate. In this scenario, the culture can be said to have no "history". In situations where the culture must pass through some physiological state in order to produce the desired product, a combination of reactors will provide optimal reactor configuration (defined as the smallest reactor volume for a specified degree of product conversion). In situations where the optimal rate of product formation may occur at temperature or pH different from that of growth, two or more stirred tank reactors operated in series permit optimized growth and product formation conditions in each reactor. When mixed substrates are used, the preferential substrate is consumed first. The use of two CSTRs in series enables the preferred substrate to be completely consumed in the first reactor and the second substrate in the second reactor, minimizing the required reactor volume. The configuration of a two-stage system is shown in Figure 4.5.

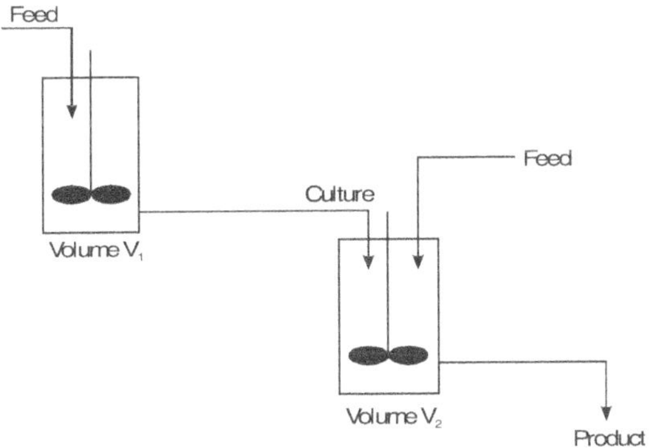

Figure 4.5   Two CSTRs in series

Plug-flow Reactor (PFR)

When the fluid moves through a large pipe or channel with sufficiently large Reynolds number, it approximates plug flow, which means there is no variation of axial velocity over the cross-section. It is assumed that the fluid movement through PFR can be described as plug flow, that is, there is no concentration or temperature gradients in the radial coordinate. Figure 4.6 shows the biomass and substrate concentration along a plug-flow bioreactor.

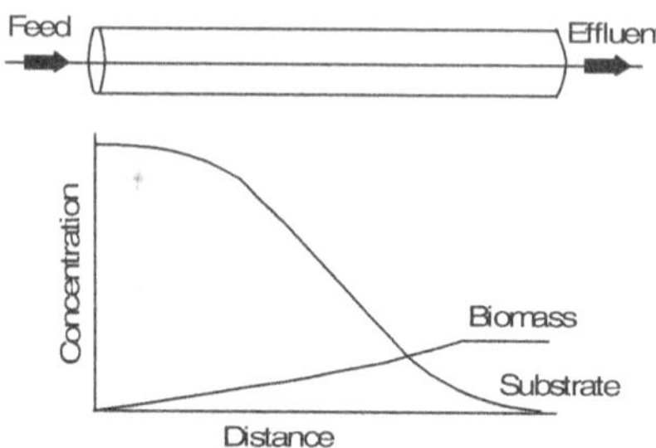

Figure 4.6   Biomass and substrate concentration along a plug-flow bioreactor

In contrast to CSTR, sterile feed to a PFR implies zero biomass concentration in the effluent. In CSTR, the medium flow will be continuous with constant flow which results in adequate mixing of fresh medium with culture already present in the fermenter. But in the plug-flow reactors, the fresh bulk amount of medium is added at once which will not completely mix with the culture previously present in the reactor.

One way to circumvent this problem is recycling, so that the incoming stream is inoculated before entering the vessel.

For a single reaction with ordinary kinetics, the PFR provides greater substrate conversion and higher product concentration than that of the CSTR. The contrary is true if the kinetics are autocatalytic (higher rates with decreasing substrate concentration). For microbial processes, the PFR typically maximizes the effluent product concentration. However, the requirement of continuous inoculation and practical difficulties with gas exchange in PFRs often result in the use of their analog (the batch reactor) when high final-product concentration is important. For exponential microbial growth, CSTR is more efficient than PFR.

AIR-DRIVEN COLUMN REACTORS

Generally, in stir tank fermenter agitation of the medium is done by stirrer and the aeration is done by sparging. The air sparged from the sparger is spread all over the medium. In the case of air-driven reactors, the sparging of air gives aeration and agitation of the medium without the presence of a stirrer. Sparging without mechanical agitation can also be used for aeration and agitation. There are two classes of air-driven bioreactors, namely bubble column fermenters and airlift fermenters.

BUBBLE COLUMN REACTORS

The bubble column reactors are air-driven reactors which do not have a mechanical stirrer. The passing of air bubbles from the sparger at the bottom of the reactor agitates the medium for adequate mixing.

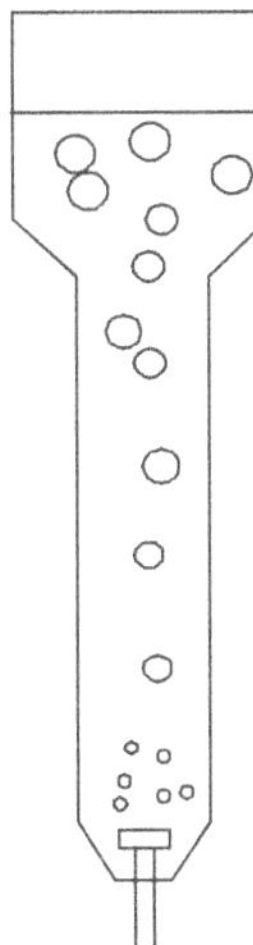

Figure 4.7 Bubble column reactor

Bubble-driven bioreactors are commonly used in the culture of shear sensitive organisms such as moulds and plant cells (Figure 4.7). An airlift fermenter differs from bubble-column bioreactor by the presence of a draft tube that provides better mass and heat transfer efficiencies and more uniform shear conditions. However, airlift fermenters are expensive than bubble column reactors. Several designs of airlift fermenters are available; the most commonly used design is the one with a central draft tube.

AIRLIFT BIOREACTOR

Airlift fermenters are specially designed for the cultivation of aerobic culture (Figure 4.8). When compared with a stir tank fermenter, an airlift fermenter lacks an agitating device. When the high pressure of filtered air is passed through the sparger present at the bottom of the reactor, the medium is raised enough along with airstream which agitate the medium enough. Besides, it provides adequate aeration also. An airlift reactor is divided into three parts: the air-riser (ascending column), downcomer (descending column) and disengagement zone.

The air-riser is the region into which bubbles are sparged. It may be inside or outside of the draft tube. The latter design is preferred in large-scale fermenters as it provides better heat transfer efficiencies.

The bubbles rising in the air-riser cause the liquid to flow in vertical direction. To counteract these upward forces through the ascending column, when the liquid is driven up due to pressure or air, the raising pressure is reduced gradually. When the liquid reaches the disengagement zone, it is free from pressure and flow in downward direction in the downcomer. This leads to liquid circulation and improved mixing efficiencies as compared to bubble columns. The enhanced liquid circulation causes bubbles to move in uniform direction at relatively uniform velocity. This bubble flow pattern reduces bubble coalescence and thus results in higher $k_L a$ values as compared to bubble column reactors.

The functions of the disengagement zone are to:

➲   add volume to the reactor

➲   reduce foaming and minimize recirculation of bubbles through the downcomer

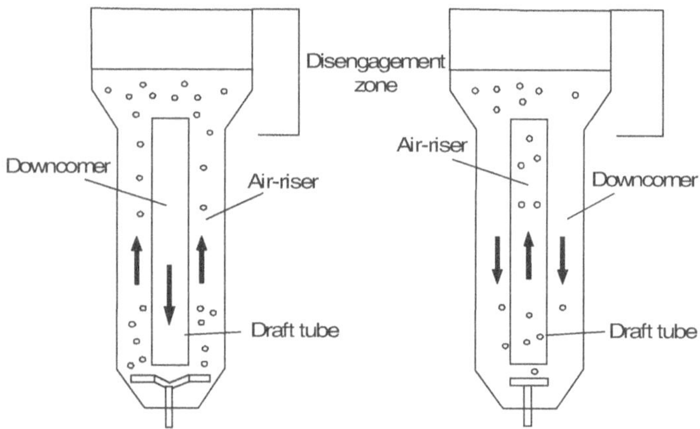

Figure 4.8   Structure of airlift bioreactor

Airlift fermenters are available with internal or external draft tubes (Figure 4.9a and 4.9b). In internal draft tube reactors, two ascending columns and one descending column are present. The liquid rises through two ascending columns and flows downward through single descending column. On the other hand, in external draft tube reactors, only single ascending and descending column is present, so the flow of fluid is in cyclic manner.

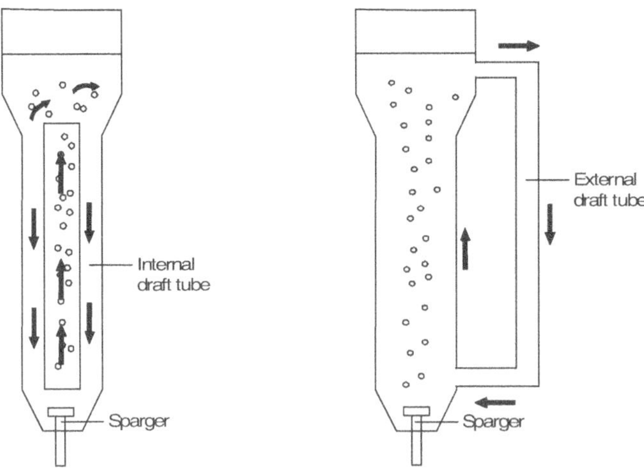

Figure 4.9a Airlift with internal   **Figure 4.9b** Airlift with external draft tube

Fluidized Bed Reactors

Fluidized bed bioreactors maintain high biomass concentrations as well as good mass transfer rates in continuous cultures. In a fluidized bed bioreactor (Figure 4.10), mixing is assisted by a pump.

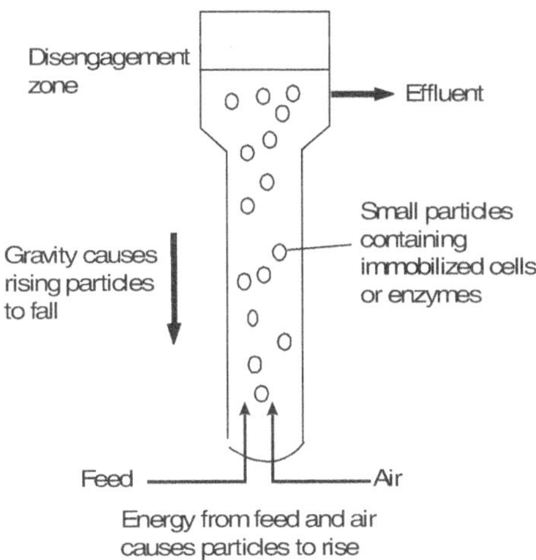

Figure 4.10 Fluidized bed reactor

In addition, the cells or enzymes are immobilized on the surface of light particles. The pump located at the base of the tank causes the immobilized catalysts to move with the fluid. The pump pushes the fluid and particles in vertical direction. The downward movement of the particles due to gravity balances the upward force of the pump. This results in good circulation. In aerobic microbial systems, sparging is used to improve oxygen transfer rates. A draft tube may be used to improve circulation and oxygen transfer. Both aerobic and anaerobic fluidized bed bioreactors have been developed for use in waste treatment. Fluidized beds can also be used with microcarrier beads which are previously attached to animal cell culture. Fluidized bed microcarrier cultures can be operated both in batch and continuous modes. In the former, the fermentation fluid is recycled in a pump-around loop.

TOWER FERMENTER

The tower fermenter (Figure 4.11) works with the same principle as fluidized bed reactor. It is mainly used in brewery industries. The inoculum and the fresh medium are pumped from the bottom, but the cells are not immobilized.

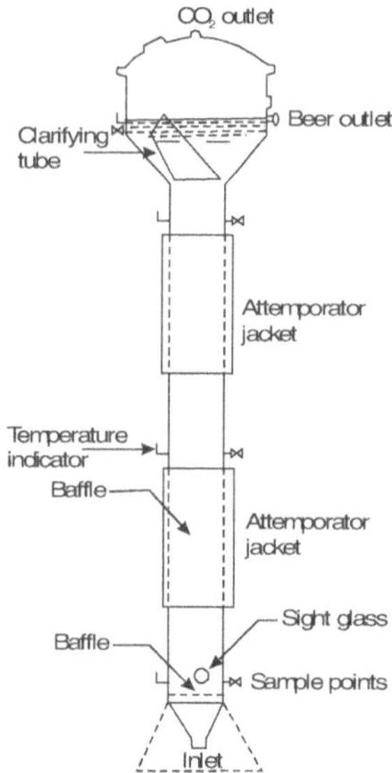

Figure 4.11   Tower fermenter

In the production of beer, yeast cells are inoculated inside the tower fermenter and wort (malt) is passed from the bottom with the help of a pump. At the top of the tower, a separator is available to exhaust the gas formed during the reaction. In addition, a clarifying grid helps to settle the yeast and collect beer. In this way, the yeast biomass is maintained in the tower fermenter. A significant feature of tower fermenter is the progressive and continuous fall in the specific gravity of the nutrient medium between the bottom and top of the tower. There is an initial, rapid fall at the bottom of the tower. It is followed by slower fall over the middle and top of the tower. This gradual fall in specific gravity is due to the fermentation of sugars.

SHAKE FLASK FERMENTER

Shake flask fermentation is carried out in conical flasks. Most of the parameter optimization studies like pH, temperature, nutrient requirement and inhibitory studies are preferably carried out in shake flask fermenters (Figure 4.12).

Figure 4.12   Shake flask fermenter

For example, for optimization of pH for an organism, the organism needs to be cultivated at different temperatures. If stir tank fermenters are used, it will be necessary to carry out the fermentation multiple times for all the pH values. It is a time-consuming and expensive process. But in a shake flask, at a stretch several flasks can be prepared and cultivation at different pH can be achieved very quickly. Unlike other fermenters, the pH maintenance is done manually. Aeration and temperature can be achieved by placing the flasks in mechanical shaking incubators.

REVIEW QUESTIONS

1.   Give brief account on different types of fermenters.

2.   Explain CSTRs in detail.

3.   How do airlift fermenters differ from bubble column reactors?

4.   Give the structural details and the applications of fluidized bed reactor and tower fermenter.

5.   Give an account on shake flask fermentation.

ISOLATION AND SCREENING OF INDUSTRIALLY IMPORTANT MICROBES

5

ISOLATION

Isolation is the first step in bioprocessing techniques, which involves detecting microorganisms of interest from the whole microbial population. The isolation of organisms may be done with conventional microbiological methods (e.g. agar dilutions) or direct sampling methods. However, the selection of sample is based on the desirable end product. When an organism is decided for use in production, it can be isolated from the related environment of the product. For example, the organism useful for the production of amylase enzymes can be isolated from the spoiled starch substrates because substrate spoilage occurs with the microbial action, that produces specific enzymes. These organisms should have the capability to produce those specific enzymes. Likewise, the microbes present in soil may have the ability to degrade hydrocarbons in oil wastes.

SCREENING

Screening is the next important step in bioprocessing, where the isolates are investigated for producing the desired product. Screening is basically of two types, namely primary screening and secondary screening. Primary screening is checking the quality of microbes. Most cases of primary screening are done on agar plates. Secondary screening is checking of isolates for their quantitative production in liquid media.

Primary Screening

Primary screening for enzyme is done by culturing the isolates in agar plates incorporated with specific substrate. The production of a specific enzyme can be detected in two ways: by observing the substrate utilization zone around the colonies of isolates or by adding reagents to the end product of the reaction, which shows the utilization zone.

The production of amylase and protease is the best example to illustrate. The screening for amylase production is carried out in starch agar plates (Figure 5.1). If the organism has produced amylase, the starch present around the colonies will have been utilized to form the utilization zone around the colony. But, since the starch agar plate is transparent, the zone will not appear visibly. To make it visible, iodine solution is layered over the cultured starch agar plates. Iodine reacts with starch to give deep blue colour all over the plates. No colour will appear around the colony since starch has been utilized by the organism.

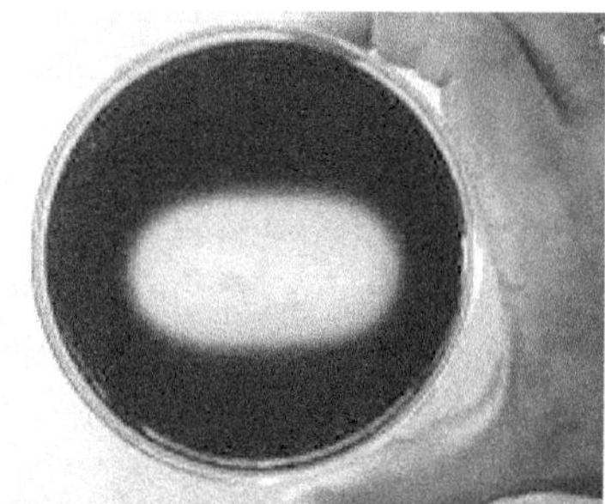

Figure 5.1   Screening for amylase production in starch agar plate

The screening for the production of protease is carried out in casein agar plates (Figure 5.2) and nutrient gelatin plates. The casein agar plate is opaque, and so the utilization zone will be visible. But, the nutrient gelatin plate is transparent. Hence, to make the zone visible, mercuric chloride solution is poured over cultured nutrient gelatin plates. Mercuric chloride reacts with protein (gelatin) and forms white precipitation. The precipitation may not form around the colonies if the organism has utilized the protein to produce protease enzyme.

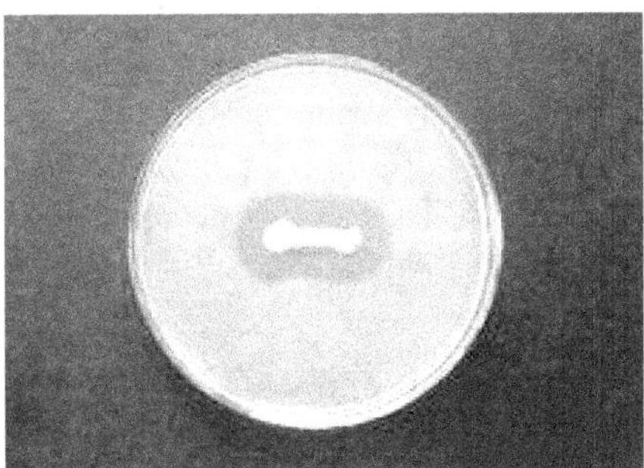

Figure 5.2   Screening for protease production in casein agar plate

In some cases, the utilization zone can be observed by adding dyes like congo red. For example, the plates used for screening chitinase and cellulase enzymes are transparent, and utilization zone may not be visible. In such a condition, the cultured agar plates are stained with congo red solution, and subsequently destaining can make the utilization zone visible.

Screening for antibiotic and toxin production differs from enzyme screening. The isolate is cross-streaked with test organisms to see the antagonistic activity of the isolate. If the isolate shows antagonistic activity, it will retard the growth of test organisms. The screening for degradation is quite different from other screening methods. The main difference is that the organisms are not isolated and screened separately. The isolation of organism is carried out in minimal agar plates impregnated with specific substrates. The colonies that grow in the medium will have the ability to degrade or utilize the substrates.

Screening of microorganisms that produce organic acids from various carbon substrates is often done in a medium incorporated with pH indicator dyes like neutral red and bromothymol blue. The production of acids will turn the colour of the dye. An alternative method for screening of organic acid producing strains is culturing the strain in calcium carbonate incorporated agar media. The acid produced will dissolve calcium carbonate and form a clear zone.

Secondary Screening

Secondary screening deals with production in liquid medium and quantification of products. The product can be evaluated with agar plates and other appropriate methods.

Secondary screening of enzymes is done through well-method in agar plates. The agar plates are incorporated with specific substrates and wells are made in them. The cells are separated from the cultured broth, and the broth is filled into the wells. After necessary incubation, the utilization zone around the well is observed. To improve visibility of the zone, appropriate techniques are used as described in primary screening (Figures 5.3 and 5.4).

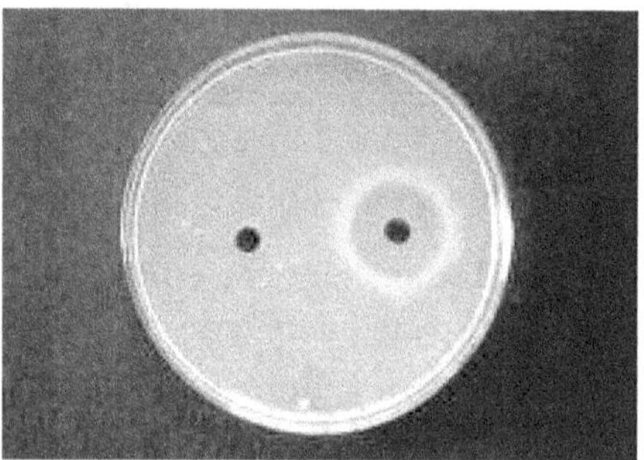

Figure 5.3   Protease activity in casein agar plate

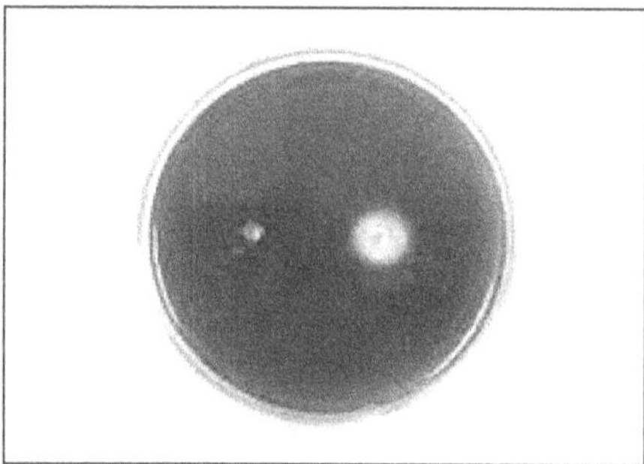

Figure 5.4   Amylase activity in starch agar plate

The activity of antibiotics that are produced using broth culture can be quantified through well-method. Mueller–Hinton agar is prepared and a well is made at the centre of the plate. The plate is swabbed with test organisms and the well is filled up with the culture filtrate of isolate. After necessary incubation, the inhibition zone of the culture filtrate is observed. Another method of antibiotic screening is disc diffusion method. In disc diffusion method (Figure 5.5), the paper disc is impregnated with culture filtrate and placed over the agar plate after swabbing with test organisms. The formation of inhibition zone can be measured and compared with standards to quantitate the inhibitory effects.

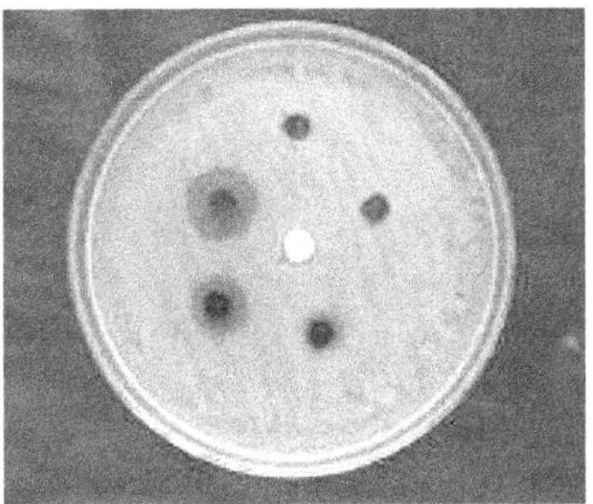

Figure 5.5   Antibiotic activity by disc diffusion method

Vitamins and amino acids produced from the new isolate in broth culture can be screened using auxotroph. The agar plates are prepared and supplemented with vitamin producing culture filtrate of new isolate. The specific vitamin lacking auxotroph is swabbed over the plate and necessary incubation is done. Observation is made on the growth of the auxotroph. Auxotrophs will not grow in the medium without necessary supplements. The growth in the agar plates indicates the presence of specific vitamins and amino acids in the culture filtrate.

REVIEW QUESTIONS

1.   What are primary and secondary screening of industrially important microbes?

2.   Give a brief account on isolation and screening of amylase and protease producing microbes.

3.   How will you isolate and screen the antibiotic producing microbes ?

4.   What is an agar plate assay ? Give its applications.

MEDIA FOR INDUSTRIAL FERMENTATION

The choice of potential medium is important for successful industrial production. Generally, the growth medium will have nutrients essential for the growth of the organisms, but the production medium will have all the nutrients for the growth of the organisms as well as the supplementation of compounds for biosynthesis of desired product. Two kinds of media are generally used for industrial fermentation, namely inoculum media and production media. The inoculum media are used to pre-enrich the culture and provide the production stage for the organisms. Before inoculating the organisms in production medium, the cultures are enriched in inoculum medium and subsequently inoculated into the production medium.

Table 6.1 Production medium composition for *E.coli*

Chemical	(g/l)	Trace metals formulation	(g/l)
KH_2PO_4	3.5	$FeCl_3$	1.6
K_2HPO_4	5.0	$CoCl_2.6H_2O$	0.2
$(NH_4)_2HPO_4$	3.5	$CuCl_2$	0.1
$MgSO_4 .7H_2O$	0.5	$ZnCl_2 .4H_2O$	0.2
Glucose	5.0	$NaMoO_4$	0.2
Yeast extract	5.0	H_3BO_4	0.05
Trace metals	1.0		
Antifoam	0.5		

In production medium, the main sources of nutrients are carbon and nitrogen. Table 6.1 shows the production medium composition for *E.coli*. In addition, the media contain inorganic salts, water vitamins, and other growth factors, precursor of fermentation product, buffering agent and anti-foaming agents. No single medium could bring successful production in fermentation. Using a variety of production medium with different compositions and standardization processes can bring about successful production in fermentation. The composition of ideal production medium is as follows.

1. Chemical Composition

The production medium should contain suitable carbon and nitrogen sources for the best growth of the isolate. Glucose or sucrose is a common carbon source for most organisms. Peptone is an organic nitrogen source used widely. In large-scale fermentations, inorganic nitrogen sources such as ammonium salts, urea and gaseous ammonia, are used as nitrogen source.

Yeast or meat extract is an excellent substrate for many organisms. They contain essential amino acids, peptides, water-soluble vitamins and carbohydrates. Typical composition of yeast extract (expressed in dry matter basis) is as follows:

- total nitrogen content is 8–12% among the total protein level from 50–75%
- nitrogen content 3 to 5.2%
- total carbohydrate content 4 to 13%
- no or very little lipid content

Soy meal is another choice of nutrient source for industrial fermentation. It is the residue of soy bean after extraction of oil. It contains carbohydrates like sucrose, arabinose, raffinose and acid polysaccharides.

Other than these nutrients, trace amounts of inorganic salts like calcium and magnesium are also needed in the production medium. Yet another important additive is a precursor or inducer of the product. In enzyme production, specific substrates are used as precursors or inducers. However, in lipase production, other than a substrate, glycerol is used as inducer.

2. Buffering Capacity of the Medium

The maintenance of pH during cultivation is an important factor for good production. Even though pH is optimized in the fermenter, the substances should not alter the pH and they should have some sort of buffering activity. In order to control the pH of the medium, salts like calcium carbonate are added to maintain the buffering activity of the medium. Protein, amino acids and peptides that are present in the medium usually have some sort of buffering activity. However, additional buffering activity can be brought out by phosphates like sodium phosphate, mono- and di-hydrogen phosphates.

3. Antifoam

Foam is a critical problem in the fermentation process, which causes spill out and contamination. It forms when the substrate is vortexed. Anti-foaming agents can be added as ingredients in the medium, or sterile anti-foaming agents can be added during foam formation. Vegetable oils and silicon products are widely used as anti-foaming agents in bioprocessing. Care should be taken that the anti-foaming agent should not interfere with the production and growth of organisms in any way.

4. Consistency

Consistency of the medium is important for aerobic cultivation. In aerobic cultivation, sterile air is passed through the sparger and distributed all over the media by the impeller. The consistency of the medium should be such that it ensures easy diffusion of air through the medium. High viscous and low diffusion rate cannot bring better production and growth of the organisms. The typical composition of baker's yeast production is given in Table 6.2.

Table 6.2   Composition of medium for baker's yeast

Compound	Concentration
$MgCl_2 \cdot 6H_2O$	0.52 g/l
$(NH_4)_2SO_4$	12.0 g/l
H_3PO_4 (85%)	1.6 ml/l
KCl	0.12 g/l
$CaCl_2 \cdot 2H_2O$	0.2 g/l
$NaCl$	0.06 g/l
$MnSO_4 \cdot H_2O$	0.024 g/l
$CaSO_4 \cdot 5H_2O$	0.0005 g/l
H_3BO_3	0.0005 g/l
$Na_2MoO_4 \cdot 2H_2O$	0.002 g/l
$NiCl$	0.0025 mg/l

(Contd.)

Table 6.2   (Continued)

Compound	Concentration
$ZnSO_4 \cdot 7H_2O$	0.012 g/l
$CoSO_4 \cdot 7H_2O$	0.0023 mg/l
KI	0.0001 g/l
$FeSO_4(NH_4)_2SO_4 \cdot 6H_2O$	0.035 g/l
myo-inositol	0.125 g/l
Pyridoxine-HCl (Vitamin B_6)	0.00625 g/l
Ca-n-Pantothenate	0.00625 g/l
Thiamine-HCl (Vitamin B_1)	0.005 g/l
Nicotinic acid	0.005 g/l
D-Biotin (vitamin H)	0.000125 g/l
Carbon source (e.g. glucose)	0–50 g/l
EDTA	0.1 g/l
Mineral stock solution	1 ml

The composition of mineral stock solution is given in Table 6.3.

Table 6.3   Mineral stock solution (100 X)

Compound	Weight/volume/ concentration
H_3PO_4 (85%)	160 ml
KCl	12.00 g
$CaCl_2 \cdot 2H_2O$	20.00 g
NaCl	6.00 g
$MnSO_4 \cdot H_2O$	2.40 g
$CaSO_4 \cdot 5H_2O$	0.05 g
H_3BO_3	0.05 g
$Na_2MoO_4 \cdot 2H_2O$	0.20 g
NiCl	0.25 mg
$ZnSO_4 \cdot 7H_2O$	1.20 g
$CoSO_4 \cdot 7H_2O$	0.23 mg
KI	0.01 g
Add water to make up to 1 litre	

RAW MATERIALS IN INDUSTRIAL FERMENTATION

Often, the construction of production medium using pure ingredients and economically may not be possible. So, plant and animal derived raw materials, which are cheaply available and rich in nutrients, can be used in industrial fermentation.

Molasses

Molasses is a main by-product from the processing of sugar cane or sugar beet. It has high carbon source and other nutrients, and hence it can be used as a substrate in industrial fermentation. On dry matter basis, molasses contain an average of 65.3 per cent total sugars, 0.90 per cent N, 0.07 per cent P, 1.15 per cent Ca, 0.61 per cent Mg, 0.10 per cent Na, 5.19 per cent K, 2.98 per cent Cl, 0.73 per cent S, 10.7 ppm Cu, 11.6 ppm Zn, 247 ppm Fe, 82 ppm Mn and 2.7 ppm Co. It has 13.6 per cent ash and 76.4 per cent dry matter.

Cane molasses   Molasses made from green sugar cane is treated with sulphur dioxide fumes. Sugar cane is harvested and stripped of its leaves. Juice is extracted from canes, usually by crushing or mashing. The juice is boiled to promote the crystallization of sugar. At the end of first boiling, sugar crystals are removed and the resulting substance is called first molasses, which has the highest sugar content. A second boiling and subsequent sugar extraction yields second molasses, which has a slight bitter taste.

The third boiling of sugar syrup gives blackstrap molasses. Although much of sucrose has been crystallized, blackstrap molasses still contains sugar by calories. But unlike refined sugars, it contains significant amounts of vitamins and minerals. Blackstrap molasses is a rich source of calcium, magnesium and iron. One tablespoon provides 20 per cent of the daily value of each of these nutrients. Blackstrap is often sold as health supplement as well as being used in the manufacture of cattle feed and for other industrial uses.

Sugar beet molasses   The syrup left after final crystallization is called molasses; intermediate syrups are referred to as high green and low green; these are further crystallized to maximize extraction. Beet molasses is different from cane molasses. It is about 50% sugar (sucrose) by dry weight and contains significant amounts of glucose and fructose. The non-sugar content includes salts such as calcium, potassium, oxalate and chloride. They are originally found in plant material or chemicals used in the processing. As it is unpalatable, it is mainly used as an additive to animal feed or as a fermentation feedstock.

It is possible to extract sugar from beet molasses through a process known as molasses desugarization. A technique called industrial scale chromatography is used to separate sucrose from non-sugar components. At present, this technique is practised in the US and some parts of Europe.

Cellulose

Cellulose is the most abundant, naturally occurring organic substance. It is the principal component of cell walls in higher plants, which provides the structural feature. Cotton is almost pure cellulose at 98%; flax is 80%. Wood is made up of 40–50% cellulose and the other 50% is made up of other polysaccharides like pectin, lignin, tannin, etc.

Cellulose is insoluble, but drastic chemical disintegration reveals that cellulose is a glucan in which the residues are linked beta(1-4). Light scattering methods reveal that the cellulose chains range from 5,000 to 10,000 glucose residues long, and there is no branching. The molecular structure of cellulose is clearly shown in Figure 6.1.

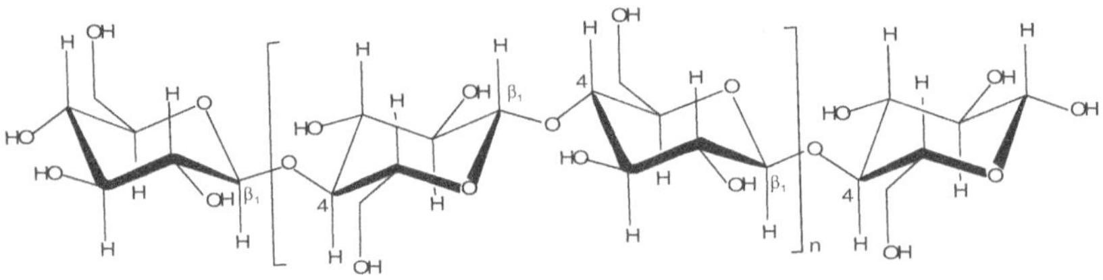

Figure 6.1   Molecular structure of cellulose

Generally, cellulosic materials are pretreated before used as substrate. Naturally occurring bagasse and straw are pretreated with alkali. The raw material is crushed and fed into a counter-current extractor in which delignification is effected through counter-current contact of the raw material with sodium hydroxide, followed by washing with warm water. High alkali concentration in the reaction mixture, however, causes simultaneous dissolution of pentosan, leading to reduction in sugar recovery. The optimum biomass to sodium hydroxide ratio for favourable saccharification is 1.0:0.12.

Lignocellulosic biomass contains cellulose, hemicellulose, lignin and ash combined in a complex structure (Table 6.4). Pretreatment increases the crystallinity of cellulose, while removing lignin and other inhibitors, thereby enabling its enzymatic degradation. In addition, pretreatment may increase the surface area of cellulose, thereby enhancing its reactivity with the enzyme and thus its transformation. The flowchart of pretreatment of cellulosic biomass is given in Figure 6.2.

Table 6.4　Composition of lignocellulosic biomass of various substrates

Cellulosic biomass	Cellulose	Hemicellulose	Lignin	Ash
Bagasse	41	24	18	2
Rice straw	35	35	6	8
Wood	40–55	20–35	25–30	0.2–2.0

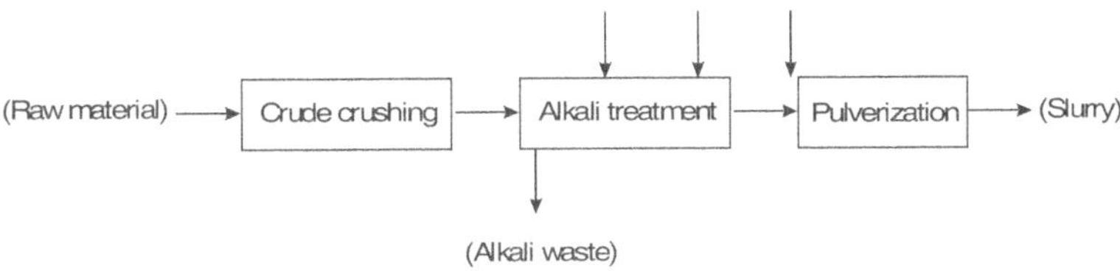

Figure 6.2　Flowchart for pretreatment of cellulosic biomass

Corn Steep Liquor

Corn steep liquor is the by-product of corn processing. It has pH between 3.7 and 4.1, a specific gravity of 1.25, 30% nitrogen, 1.45 to 1.65% amino nitrogen, 0.15 to 0.30% volatile nitrogen, 11% reducing sugar, 5.1% lactic acid, 0.1% volatile acid, 5% riboflavin, 81.9% niacin, 23.8% pantothenic acid, 19.1% pyridoxine and 0.125% biotin. The main disadvantage of corn steep liquor is its variable composition. On the other hand, corn steep liquor is an inexpensive alternative to expensive materials, such as yeast and peptone. Table 6.5 lists some of the applications of corn steep liquor.

Table 6.5　Some uses of corn steep liquor in microbial production

Yeast food
Bakers' yeast
Organic acids from cellulosic material
Beer
Bread dough
Brewing adjunct
Sorbose by *Acetobacter suboxydans*
Ketogluconic acids by *Acetobacter suboxydans* and an unnamed organism
Gluconic acid by *Aspergillus niger*

(Contd.)

Table 6.5   (Continued)

Itaconic acid by *Aspergillus terreus*

Penicillin by *Penicillium notatum, Penicillium chrysogenum*

Pentonic acids by *Pseudomonas*

Riboflavin by *Ashbya gossypii*

Subtilin by *Bacillus subtilis*

Amylase by *Aspergillus niger*

Soy Bean Meal

Soy bean meal is the residue after extraction of oil from soy bean. It has 50% protein content, 30% carbohydrate, 1% fat and 1.8% lecithin. Soy products stand out by their large diversity of critical nutrients. This is especially true where it is used in concentrate feeds. It has the same protein levels, low fibre concentrations and high metabolizable energy value. Lysine concentration of soy products is high and other amino acids are fairly constant across soy products. Cooking or boiling does not affect the amino acid profile in soy bean. Soy meal is frequently used in the production of enzymes and antibiotics. During production, catabolic repression will not occur because of slow catabolism of this complex mixture. The contents of soy bean meal is given in Table 6.6.

Table 6.6   Contents of the soy bean meal

Nutrients	%
Crude protein	48.0%
Fat	1.0%
Crude fibre	3.0%
Neutral detergent fibre	7.1%
Acid detergent fibre	5.3%
Calcium	0.2%
Phosphorus	0.65%
Total digestible nutrients	78.0%
Dry matter	89%
Net energy—Lactation	81.1 Mcal/100 g

Malt Extract

Malt extract is the crush of cereals like barley and wheat. It contains several organic compounds. Malt extract is mainly used in brewery industries. It is added as an active ingredient in the broth for cultivation of microorganisms. Commercially available wheat germ (moisture-free) malt extract has the following composition: sugar 7.71%, dextrin 7.50%, starch 18.21%, ash 4.91%, fibre 2.31%, protein 31.00%, fat 10.46% and pentosans 8.29%. The nutrition value of malt extract is given in Table 6.7.

Table 6.7   Nutritive value of malt extract (per 100 g)

Energy	1559 KJ
Carbohydrate	92.8 g
Sugars	65.0 g
Fat	less than 0.1 g
Sodium	30.5 mg
Potassium	383 mg

STERILIZATION

Sterilization is an important step in fermentation technology because any contamination may retard the achievement of production. The equipment, media and other supplementary components should be sterile for better cultivation. If a system has only those organisms that are supposed to be present, the situation is aseptic. In the past, industrial bioprocesses could tolerate a very few foreign organisms relative to the desired organisms. The cost of achieving absolute sterility was not justified for processes such as penicillin fermentation where the production culture hardly competes with a few foreign organisms and will usually develop a high titre of product. Today the situation is different as cell cultures grow slowly and are very easily contaminated.

In industrial fermentation, the medium and other additives are sterilized by autoclaving. The vessel is sterilized separately by passing steam, and air that is passing through the medium during aeration is sterilized by air filter. In some batch fermenters, the medium along with the fermenter vessel is sterilized by passing steam. This kind of sterilization is known as in-built sterilization.

Usually, a fermenter and its medium are sterilized together, and all connecting lines are flushed with steam. A fermenter may be sterilized and then filled with sterile medium. It is a common practice to sterilize water and insensitive materials in the fermenter so as to minimize the volume that must be sterilized by other methods such as filtration or passage through a continuous sterilizer. Several factors such as temperature, pH, osmotic pressure, shear, mass transport, and concentration of the substrate influence sterilization. These factors operate synergistically, and affect the kinetics of a reaction.

Batch Sterilization

Batch sterilization is reduction of contaminant organisms through heating of the vessel. The entire volume of the medium is sterilized through thermal or radiation techniques. In thermal batch sterilization, a system undergoes three steps: heating, holding and cooling (Figure 6.3). Heating requires the addition of energy throughout the entire medium. This can be done by adding hot water through the water jacket around the vessel. The temperature is raised to reach the sterilization temperature and held for a set period of time. During this phase, most unwanted microorganisms are destroyed. Finally, the system is cooled to bring the sterile medium back to the desired temperature. In radiation batch sterilization, the process is similar except it uses radiation intensity instead of heat.

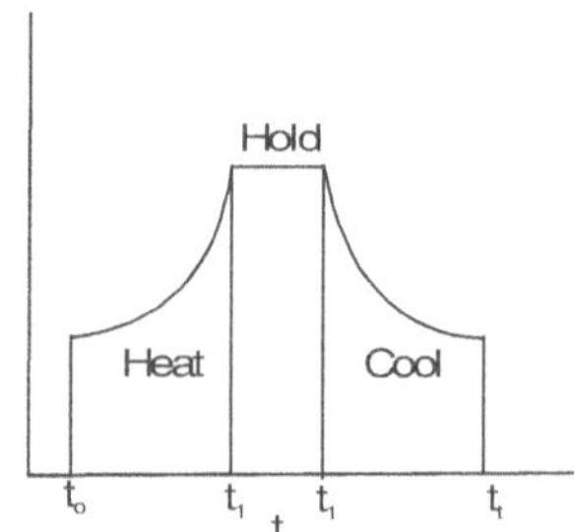

Figure 6.3   Thermal batch sterilization process

$$\nabla_{total} = \nabla_{heat} + \nabla_{hold} + \nabla_{cool}$$

$$\nabla_{heat} = \ln\frac{N_0 V_0}{N_1 V_1} = \int_{t_0}^{t_1} k\,dt$$

$$\nabla_{hold} = \ln\frac{N_1 V_1}{N_2 V_2} = k(t_2 - t_1) \qquad\qquad \text{(At constant room temperature)}$$

$$\nabla_{cool} = \ln\frac{N_2 V_2}{N_f V_f} = \int_{t_2}^{t_f} k\,dt$$

where,

t	=	time,
N_o	=	initial spore concentration,
N_f	=	final spore concentration,
V_o	=	initial batch volume,
V_f	=	final batch volume and
k	=	death rate constant.

By Arrhenius equation, the death kinetics is $k = Ae^{-e/RT}$.

In order to sterilize a batch, calculate the total area underneath the curve. Therefore, model death using first order kinetics and integrate as seen above. This will yield a temperature and the corresponding duration of time needed to sterilize the media.

Advantages in batch sterilization

- Most widely used

- Simple operation

- No additional materials are added

Disadvantages in batch sterilization

- More expensive heat requirements than continuous sterilization

- Best results occur in well-mixed closed vessels

Wet Sterilization

In moist conditions, microorganisms germinate and grow vegetatively. However, it is relatively easy to kill them in wet environment. A laboratory autoclave is set up for thirty minutes at 1 atmosphere of pressure or 121°C. When a solid, e.g. soy meal, is present in the culture medium, the exposure time should be increased to allow heat to penetrate into solid clumps.

Dry Sterilization

The absence of moisture encourages the formation of bacterial endospores, which are a means of preserving the organism. These spores are 1000 times more resistant to heat than that of vegetative cells. Sterilization of such materials such as glass pipettes uses not a steam autoclave but an oven. Typical conditions of sterilization are several hours at 200°C. Anti-foam oil is essentially dry and should be sterilized either at higher temperatures or for very prolonged periods in a steam autoclave.

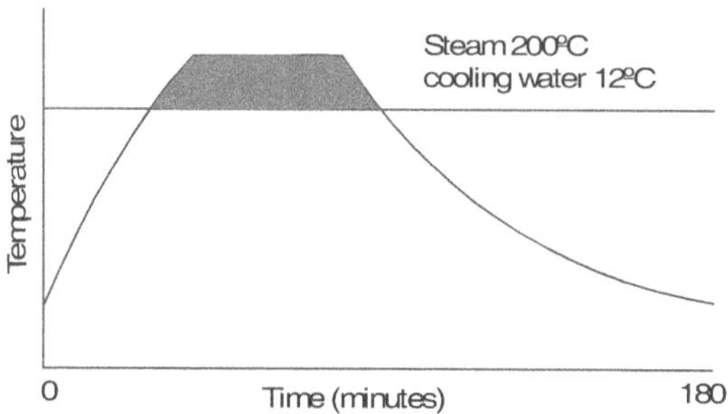

Figure 6.4   Relationship between time and temperature in sterilization

Figure 6.4 shows temperature versus time. The rising and descending portions of this curve can be changed by the design of the fermenter, the specifications of steam supply, and supply of cooling water. The fermenter contents are heated by coils or injection of steam. The region of heat transfer and sizing of steam lines determine the rate of heating. The cooling depends on heat transfer area and force of water. Additional cooling can come from venting the steam from the fermenter. This provides evaporative cooling, but it requires care as foaming may carry the nutrient medium away with steam.

For sterilization, temperature is applied by two ways. One is applying high temperature for short time and other is applying low temperature for long time. In fermentation sterilization, to kill the contaminants, both the medium and nutrients should be cooked well. Though the medium is cooked well complete killing of microbes is not achieved when low temperature is applied for long time. But in the other process cooking and killing are equally achieved. The rectangular area (dark shaded) in Figure 6.4 shows the temperature and duration of time required in industrial sterilization.

Continuous Sterilization

Continuous sterilization is rapid transfer of heat to the medium through steam condensate without the use of a heat exchanger. Once the media is in a holding loop, steam is injected to the system via a nozzle. The medium stays in this loop for a predetermined holding time or until the entire medium is sterile. This is more efficient than batch sterilization, because small portions of inlet streams are heated at a time. By looping sterile media tubes (which are at higher temperatures) past the inlet tubes, the difference in temperature can be used to heat the unsterile medium. So instead of having a cold-water stream to cool the sterile media, the unsterile media stream at lower temperature absorbs heat from the warm stream and cools the sterile media. Finally, the sterile medium is flash-cooled through an expansion valve to adjust the temperature to meet process parameters (Figure 6.5).

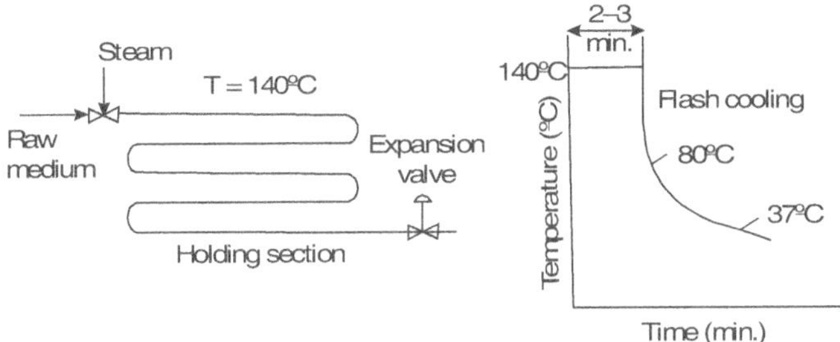

Figure 6.5   Continuous sterilization curve

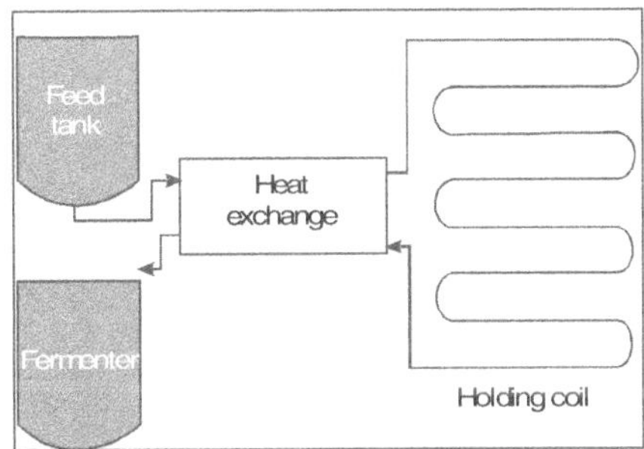

Figure 6.6   Schematic representation of continuous sterilization

Advantages in continuous sterilization

- Uniform steam requirements throughout the duration of sterilization
- Simplified process control
- Short sterilization time means less thermal degradation of medium

Disadvantages in continuous sterilization

- High demand for steam
- Concentration of medium becomes dilute due to steam condensation
- Since steam disperses in media, steam must be clean to avoid contamination

Heat Economy in Batch and Continuous Sterilization

Batch sterilization uses steam or direct firing to elevate the temperature, and then cool water brings the material back to room temperature. Both hot and cool water are spent with no opportunity for energy recovery. Large volumes should be passed continuously through heat exchangers for energy economy with the hot, treated fluid heating the cold, incoming feed. One method of continuous sterilization injects steam into the medium (no heat exchanger). The medium stays in a loop for a predetermined holding time until the entire medium is sterilized.

Better heat economy comes from substituting heat exchangers for direct steam injection. Instead of having a cold water stream to cool the sterile media, the lower temperature unsterile media stream absorbs heat from the warm stream, cooling the sterile media. A system for continuous sterilization has a holding coil for detention long enough to kill all microorganisms. The medium from a make-up vessel flows through the exchanger, is held in the coil, and passes back through the heat exchanger, heating more unsterile medium while becoming cool itself, as it is collected in a sterile fermenter.

This design would work only with an exchanger with infinite heat transfer area because there is no driving force for heat transfer as the temperatures for the two streams approach closely. A real design would have another small exchanger to raise the temperature to the set point after the main exchanger has done all it can do. There is no need for a cooler before entering the fermenter because it has a jacket or coils to control temperature.

Heat economy is not important for continuous sterilization in a small pilot plant, so direct steam injection is simpler. A heat exchanger is then needed with cooling water to bring the medium back quickly to a temperature at which it is not overcooked. High temperature for short time is preferable in preparing nutrient media for industrial fermentations and in pasteurizing milk, as it causes less damage to biochemicals than prolonged time at lower temperature. This exploits the temperature effects on activation energies because bacterial killing is affected by a temperature change more than is heat destruction of biochemicals.

HEAT EXCHANGERS IN BIOPROCESSES

The main purpose of heat exchangers in bioprocesses is sterilization.

A heat exchanger works on the principle that heat transfer occurs when there is a difference in temperature. In a heat exchanger, there exists both a cold stream and a hot stream. The two streams are separated by a solid wall. The wall must be thin and conductive in order for heat exchange to occur. Yet, the wall must be strong enough to withstand any pressure from the fluid. Copper is a common choice for construction.

Figure 6.7 shows how heat is transferred in a heat exchanger.

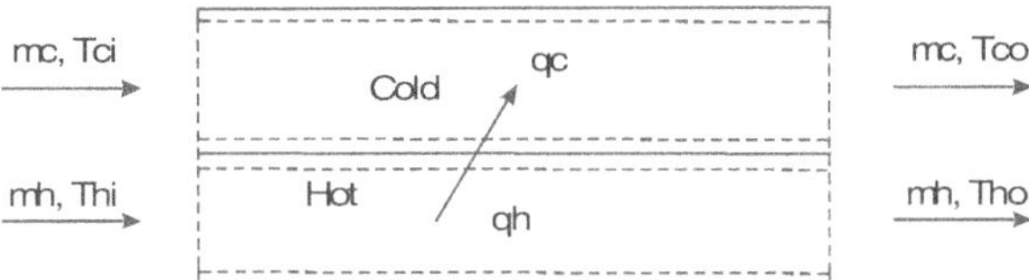

Figure 6.7   Heat exchanger

This arrangement is called co-current flow. If the direction of one of the streams is reversed, the arrangement can be called counter-current flow.

Note that the temperature profiles are different for co-current flow and counter-current flow along the heat exchanger.

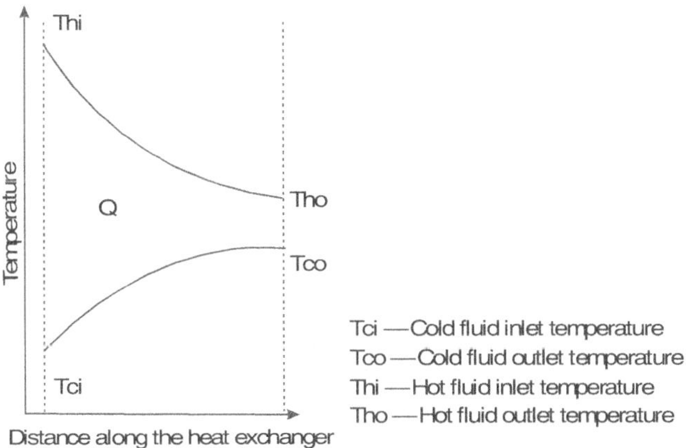

Figure 6.8   Heat exchange curve

The area between the curves is the heat transfer rate (Q). We can see that the heat transfer rate for counter-current flow is larger than the rate for co-current flow. In Figure 6.8, the initial temperature of the heat exchanger fluid (Thi) is high and the temperature of the medium (Tci) is low. But the temperature of the medium while passing through the heat exchanger, is increased because it receives heat from the heat exchanger fluid whereas the temperature of the heat exchanger fluid is gradually reduced because it transfer heat to the medium. When the medium passes through the heat exchanger for a long distance, at a particular time, the temperature of the medium (Tco) and heat exchanger fluid (Tho) become equal. Most industrial heat exchangers are counter-current flow heat exchangers.

Here are some ways to improve the performance of a heat exchanger:

- Increase heat transfer area

- Decrease velocity of the fluid

- Increase temperature gradient

These suggested ways of improvements are based on the equation for heat transfer rate, which is:

$$Q = U \times A \times dTlm$$

where,

Q =   Heat transfer rate between the fluids,

U =   Overall heat transfer coefficient,

A =   Heat transfer area and

$dTlm$ =   Log mean temperature difference of the system.

Heat Transfer Area

The heat transfer area (or contact area) is directly proportional to the heat transfer rate. If heat transfer area increases, heat transfer rate increases as well. A simple way to increase heat transfer area is adding fins (Figure 6.9) to the surface. It is inexpensive to add fins to the heat transfer area, but fins may increase fouling, especially in bioprocesses.

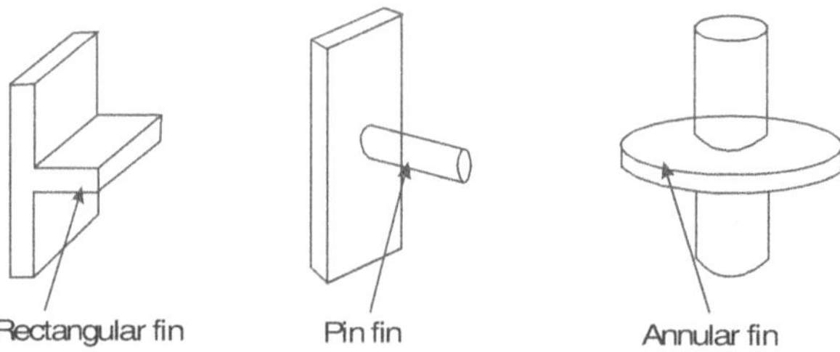

Figure 6.9   Different types of fins

Fluid Flow Rate

The importance of fluid flow in a heat exchanger is that it changes the overall heat transfer coefficient, U. The data obtained from heat transfer experiments show that the velocity of the cooling fluid is directly proportional to the overall transfer coefficient. The following is a plot of 1/U vs. $1/V^{0.8}$ during one of the runs in a lab experiment (Figure 6.10).

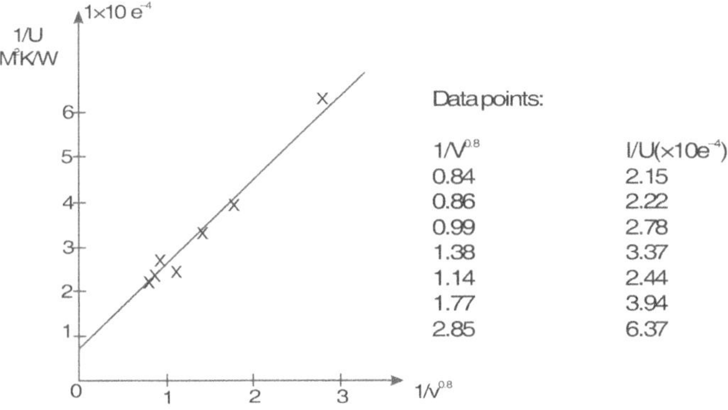

Figure 6.10   Wilson plot for hot water counter-current flow

As the velocity of cooling fluid increases, it dissipates more heat effectively. However, it may not be good in some bioprocesses as high velocity may create high shear stress in flow. Some proteins or cells are very delicate. They cannot withhold such force and they will be destroyed.

Temperature Gradient

Temperature gradient is certainly an important part of heat transfer. If we can introduce fluids with greater temperature difference into the heat exchanger, the heat transfer rate (Q) will be greater. If we go back to the temperature profiles of co-current and counter-current flows, we can see that the driving force is great for co-current at the beginning but decreases drastically as it moves along the heat exchanger. The counter-current flow provides relatively consistent driving force and therefore performs better than co-current flow.

Fouling is a major problem in mass or heat transfer. Materials are concentrated at the transfer surface and they decrease the flux flow. In bioprocesses, the problem of fouling is more apparent. Many biomaterials (such as proteins and bacteria) can easily stick to the wall and cause fouling. Cleaning the tubes can be a major task.

TYPES OF HEAT EXCHANGERS

Shell and Tube Exchangers

Shell and tube exchangers (Figure 6.11) are common in chemical process industries for heat economy. Many tubes go from the header on one side to the header on the other. The other fluid is in the space outside the tubes. Hot streams exchange energy with cold streams so that thermal energy of hot streams is not wasted. However, discharge of hot streams into the environment can be harmful. Safe removal of hot stream is important.

In Figure 6.11, the hot fluid is passed through the inlet of the shell and the cold medium which is to be sterilized is passed through the tube. When the hot fluid reaches the outlet of the shell, the temperature would have been reduced due to transferring from hot to cold medium. But when the medium reaches the tube outlet, the temperature would have been increased since it gets heated by the hot fluid.

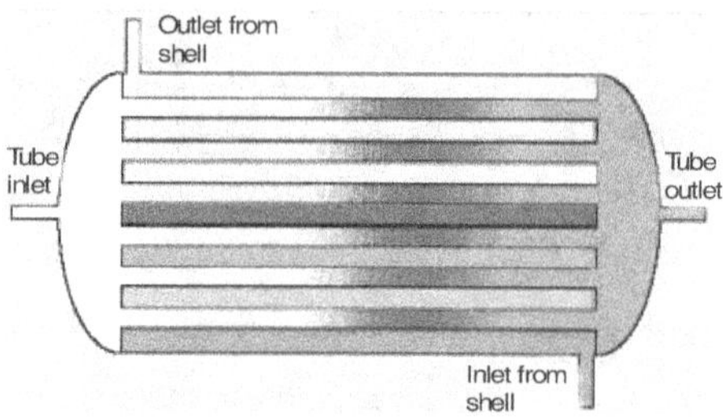

Figure 6.11   Shell and tube exchanger

Plates in a Heat Exchanger

The plates in an exchanger encourage more turbulent flow to make films thinner to promote better heat transfer. The headers for feeding and removing the fluids pass through all the plates, but inappropriate headers are blocked off on a plate. Figure 6.12 shows plates that are supported by rods that pass through holes in the top and bottom, but plates are often designed with ears on the sides that slide on rods that are external to the plates. Between each plate is a gasket. Plates for the two fluids alternate.

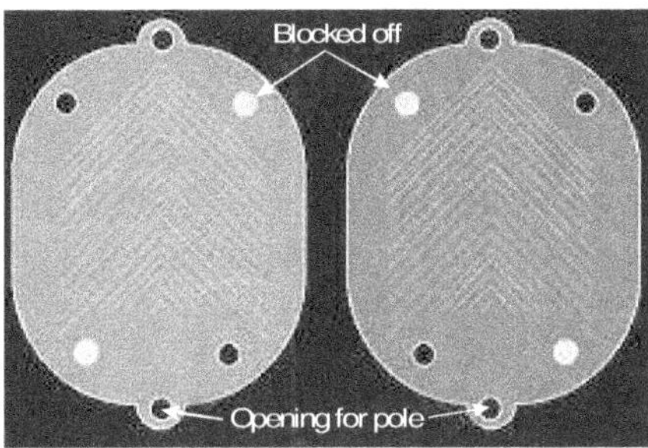

Figure 6.12   Arrangement of plates in heat exchanger

Air Sterilization

The magnitude of air sterilization can be seen from a highly aerobic fermentation where roughly 1 volume of air per volume of medium per minute may be used. For a factory with 20 fermenters of 100,000 L each, 2 million L/m (70,000 ft³/m) of air is handled. At least two large compressors are used. In the past, air filters were made of columns that could approach diameters of only one-fourth of the fermenter diameter. The packing material was slag wool that lumped up with repeated use; fibreglass broke down due to repeated thermal expansion and contraction, and beads of carbon sometimes underwent spontaneous combustion and melted the column. Carbon packing works fairly well but is too bulky.

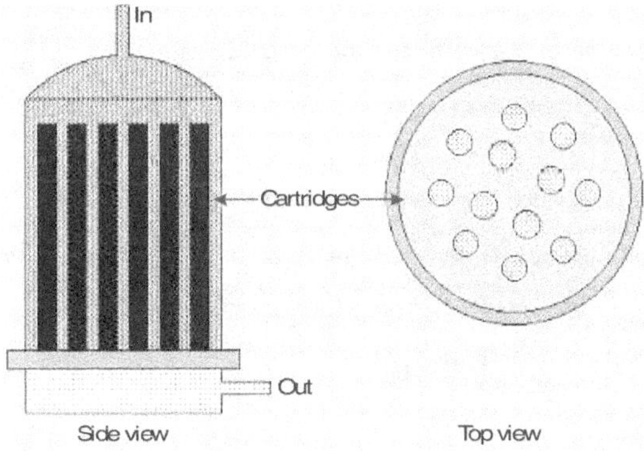

Figure 6.13   Cross sectional view of air sterilization filters

Currently, there is a pronounced trend to use membrane filters in a cartridge configuration to obtain excellent performance with units of relatively small size. Moisture is bad for all methods of air sterilization and may help microorganisms to pass. A membrane pore size of 0.2 to 0.3 μm is recommended. Hydrophilic membranes should not be used because moisture held tightly in the pores does not dislodge unless there is quite high pressure drop across the membrane. Moisture tends to drain from hydrophobic membranes and collect in a sump. The units are housed in a shell with a manifold. Sizing is based on the number of cartridges needed.

Air leaving a vessel in which pathogenic organisms are cultured is sterilized by heating. Air inside the room for culturing microorganisms may be exposed to ultraviolet light to reduce the potential contaminants. Ultraviolet light penetrates poorly through glass, so organisms in shake flasks are not killed. Ultraviolet lights are also mounted on flow devices for water sterilization, but quartz bulbs are needed to circumvent the attenuation of UV wavelengths. Such devices may be plagued by turbidity in water or by dirt forming on the transparent surfaces. There have been some attempts to commercialize enzymatic sterilization of air. The basic concept is to bring microorganisms and viruses into contact with enzymes that attack nucleic acids. Viruses are destroyed by passage through a labyrinth of surfaces coated with deoxyribonuclease enzymes.

REVIEW QUESTIONS

1. What is the difference between growth medium and production medium?

2. Give the characteristics of fermentation medium.

3. What are the different raw materials used in industrial fermentation? Explain their characters.

4. What is sterilization? Explain about different types of industrial sterilization.

5. Write about heat exchangers used in industrial sterilization? Give their properties.

6. Give an account on types of heat exchangers.

7. Give an account on air sterilization.

PROCESS CONTROL IN FERMENTATION

INTRODUCTION

Biological factors include the characteristics of the cells, their maximum specific growth rate, Monod constant, yield coefficient, pH range and temperature range.

We have seen how the productivity of fermentation is determined by the mode of operation of the fermentation process. The advantages of fed-batch and continuous fermentations over batch fermentations were discussed. Likewise mass transfer, in particular, oxygen transfer was an important factor which determines how a reactor must be designed and operated.

Cost is also an important factor. The larger the reactor or the faster the stirrer, cost involved is more.

In this chapter, we shall look into how bioreactors can be designed to meet cost, biological and engineering needs.

AERATION

Bioreactors vary in size and complexity from a 10 ml volume in a test tube to a computer-controlled fermenter with volume greater than 100 m^3. Similarly they vary in their costs.

Let us discuss methods of aeration in the following reactors:

- ➲ Standing cultures
- ➲ Shake flasks

- ⮑ Stirred tank reactors

- ⮑ Bubble column and airlift reactors

Standing Cultures

In standing cultures, little or no power is used for aeration. Aeration is dependent on the transfer of oxygen through the surface of the culture (Figure 7.1).

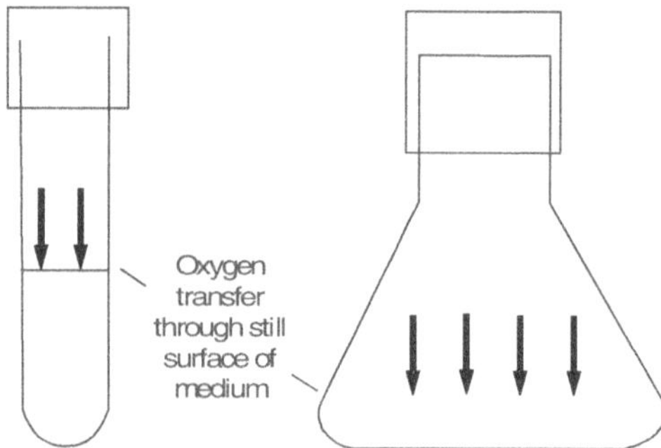

Figure 7.1   Aeration in standing culture

The rate of oxygen transfer will be poor due to small surface area. Standing cultures are commonly used in small-scale laboratory systems in which oxygen supply is not critical. For example, biochemical tests for the identification of bacteria are often performed in test tubes containing 5–10 ml of biochemical medium.

Standing cultures—surface cultures   Standing culture aeration may not be restricted to the laboratory. In regions where electricity is unreliable, citric acid can be produced using surface culture techniques (Figure 7.2). In these cultures, *Aspergillus niger* mycelia are grown on the surface of liquid media in large shallow trays. The medium is neither gassed nor agitated.

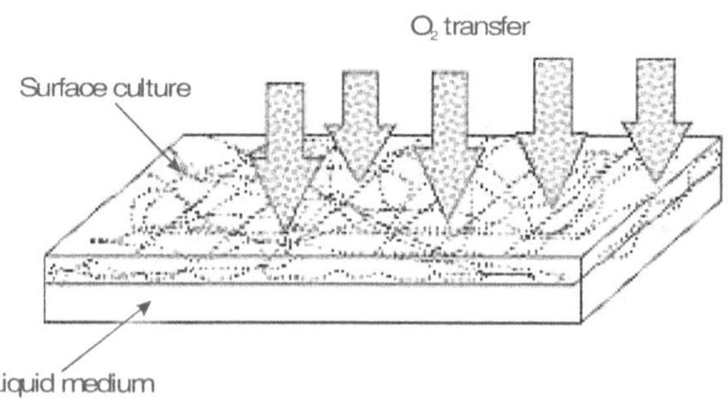

Figure 7.2   Surface culture techniques

Aerobic solid-substrate fermentation is another example of standing cultures. In this type of fermentation, the biomass is grown on solid biodegradable substrates such as water-softened bran, rice or barley.

The solid may be continuously or periodically turned over to improve aeration and to regulate culture temperature. One example of commercial, solid-substrate fermentation is the production of koji with *Aspergillus oryzae* on soy beans. Another example is mushroom cultivation. Considerable research is currently being invested into the feasibility of solid substrate fermentation in producing biochemicals.

Shake Flasks

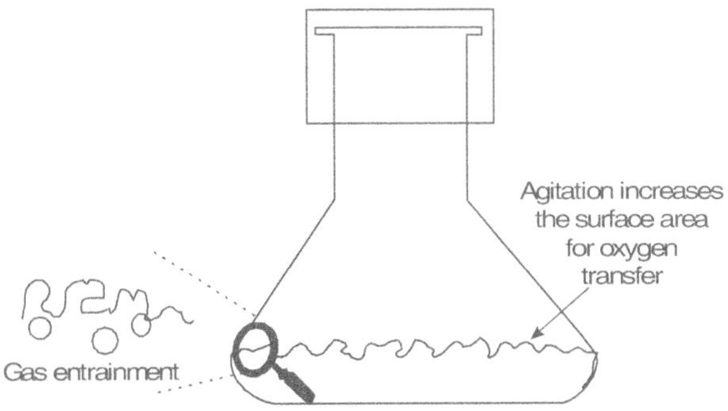

Figure 7.3   Shake flask

Shake flasks are commonly used for small-scale cell cultivation. Through continuous shaking of the culture fluid, higher oxygen transfer rate can be achieved. Continuous shaking helps break the liquid surface and thus provides a greater surface area for oxygen transfer. Figure 7.3 shows increased rates of oxygen transfer may also be achieved by entrainment of oxygen bubbles at the surface of the liquid.

Although higher oxygen transfer rates can be achieved with shake flasks, oxygen transfer limitations will still be unavoidable especially when trying to achieve high cell densities.

Mechanically Stirred Tank Bioreactors

For aeration of liquid volume greater than 200 ml, various options are available. Non-sparged, mechanically agitated bioreactors can supply sufficient aeration for microbial fermentation with liquid volume up to 3 litres. However, stirring speeds of up to 600 rpm may be required to distribute the dissolved oxygen that is already present throughout the culture. In non-sparged reactors, oxygen is transferred from the headspace above the fermenter liquid. Continuous agitation breaks the liquid surface and increases the surface area for oxygen transfer (Figure 7.4).

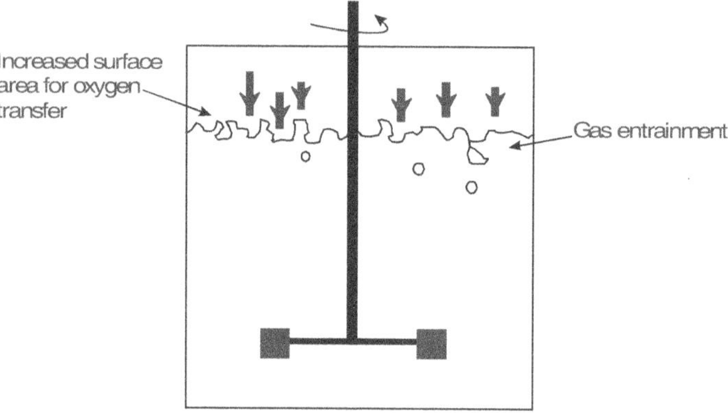

Figure 7.4 Mechanically stirred tank reactors

Sparged Stirred Tank Bioreactors

For liquid volume greater than 3 litres, air sparging is required for effective oxygen transfer. The introduction of bubbles into the culture fluid leads to a dramatic increase in oxygen transfer area.

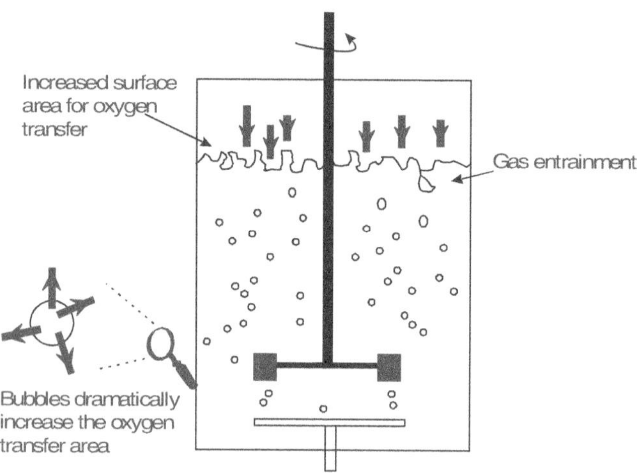

Figure 7.5 Sparged stirred tank reactors

Agitation is used to break up bubbles and further increase k_La. Sparged fermenters require significantly lower agitation speeds for aeration efficiencies comparable to those achieved in non-sparged fermenters (Figure 7.5). Air-sparged fermenters can have liquid volumes greater than 500,000 litres.

HEADSPACE AND WORKING VOLUME

A bioreactor is divided into working volume and headspace volume (Figure 7.6). The working volume is the fraction of total volume taken up by the medium, microbes and gas bubbles. The remaining volume is called headspace.

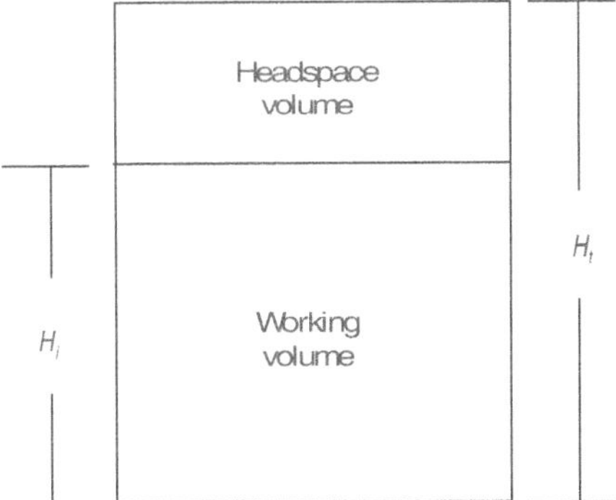

Figure 7.6   Headspace and working volume

Typically, working volume will be 70–80% of the total fermenter volume. This value will, however, depend on the rate of foam formation during the reaction. If the medium has a tendency to foam, a larger headspace and smaller working volume is required.

OXYGEN DELIVERY SYSTEM

The oxygen delivery system consists of

- a compressor

- inlet air sterilization system

- an air sparger

- exit air sterilization system

Compressor

A compressor forces air into the reactor (Figure 7.7). It should generate sufficient pressure to force the air through the filter, sparger holes and into the liquid. Compressors used in large-scale bioreactors typically produce air at 250 kPa. The air should be dry and oil free so as not to block the inlet air filter or contaminate the medium. It is very important that an 'instrument air' compressor is not used. Instrument air is usually generated at higher pressure but is aspirated with oil. Instrument air compressors are commonly used for pneumatic control.

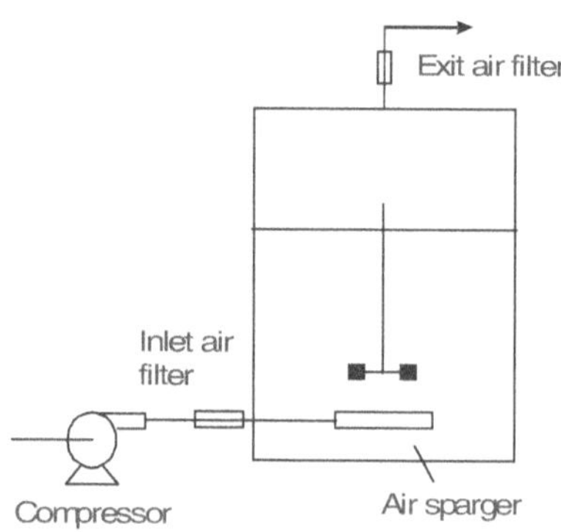

Figure 7.7   Compressor

Air Sterilization System

The inlet air is sterilized to prevent contaminating organisms from entering the reactor. On the other hand, the exit air is also sterilized to prevent organisms in the reactor from contaminating the air. A common method for sterilizing the inlet and exit air is filtration. For reactors with volume less than 5 litres, disc shaped hydrophobic teflon membrane housed in polypropylene is used (Figure 7.8). Teflon is tough, reusable and does not readily block.

By pleating the membrane, it will be possible to create a compact filter with a very large surface area. Increasing the filtration area would decrease the pressure required to pass a given volume of air through the filter. However, sterilization of the inlet and exit air in large bioreactors (> 10,000 litres) can present a major problem. Large-scale membrane filtration is very expensive and the energy required to pass air through the filter can be quite considerable.

Heat sterilization may be an alternative. Steam can be used to sterilize air. In older style compressors, it was possible to use heat generated by the compressors to sterilize air. However, modern compressors (Figure 7.9) are multi-stage devices which are cooled at each stage and disinfecting temperatures are never reached.

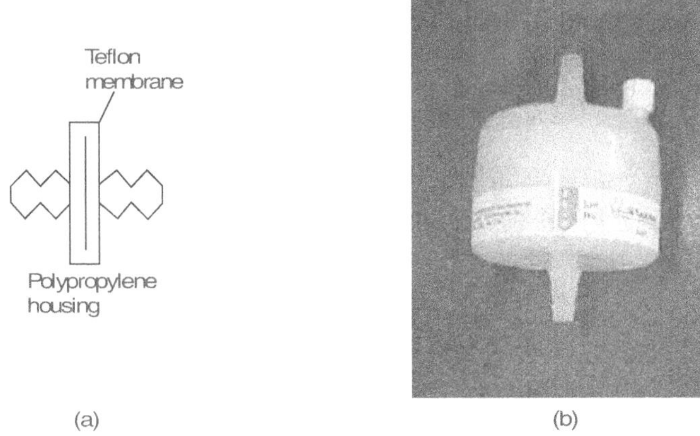

Figure 7.8   Sterilizing air using teflon

In small reactors, the exit air system will typically include a condenser.

Figure 7.9   Mini compressor

The condenser is a simple heat exchanger through which cool water is passed. Volatile materials and water vapour condense on the inner surface of the condenser. This minimizes evaporation and loss of volatiles. Drying the air also prevents blocking of the exit air filter.

Positive pressure   In sterilization, the concept of 'positive pressure' will often be used.

Maintaining positive pressure means that during sterilization, cooling and filling, air must be pumped into the reactor.

In this way, the reactor always maintains the pressure, and thus aerial contaminants will not be sucked into the reactor. It is very important that positive pressure is maintained when the bioreactor is cooled after sterilization. Without air being continuously pumped into the reactor, vacuum will form and contaminants will be drawn into the reactor. Figure 7.10 (a and b) shows the positive pressure under presence and absence of aeration. Figure 7.10a shows that without aeration, a vacuum forms when the reactor cools and 7.10b shows that with aeration, positive pressure is always maintained and contaminants are pushed away from the reactor.

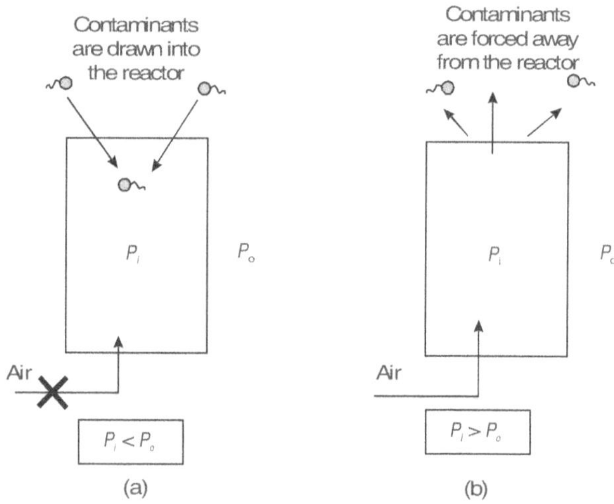

Figure 7.10   Positive pressure

Sparger

An air sparger breaks the incoming air into small bubbles. It may be made of glass or metal. The most commonly used type of sparger in modern bioreactors is the sparge ring.

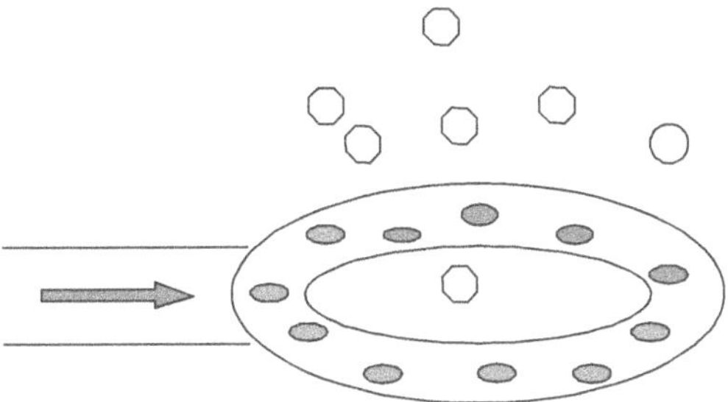

Figure 7.11   Sparge ring

A sparge ring (Figure 7.11) consists of a hollow tube in which small holes are drilled. It is easier to clean and is less likely to block during fermentation.

The sparge ring must be located below the agitator at approximately the same diameter as the impeller. Thus, the bubbles raise directly into the impeller blades, facilitating easy break up of bubbles (Figure 7.12).

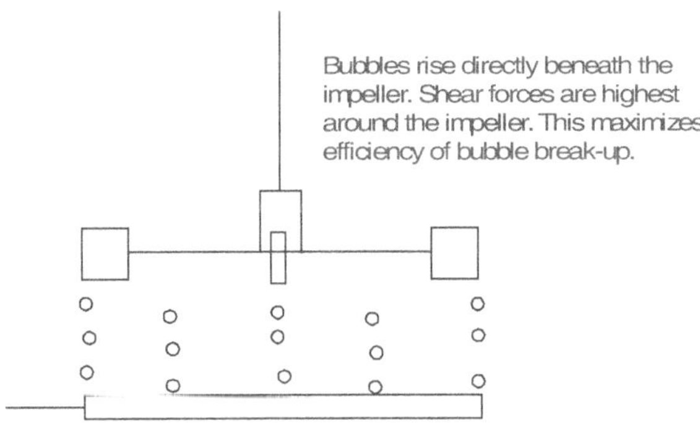

Figure 7.12   Bubbles breaking up in a sparge ring

When the fermenter is emptied, the air feed valve should be closed. This will minimize contamination of inlet air.

Effect of impeller speed   The shear forces that an impeller generates play a major role in determining bubble size. If the impeller is too slow, the bubbles will break down. In addition, the bubbles will tend to rise directly to the surface due to buoyancy (Figure 7.13a).

Another consequence of slow impeller speed is flooded impeller. Under these conditions, the bubbles will accumulate and coalesce under the impeller, leading to the formation of large bubbles and poor oxygen transfer rates. A similar phenomenon will happen when aeration rate is too high. In this case, the oxygen transfer efficiency will be low. In the fast impeller speed (Figure 7.13b) Smaller bubbles will be generated, which will move throughout the reactor increasing the gas hold up and bubble residence time.

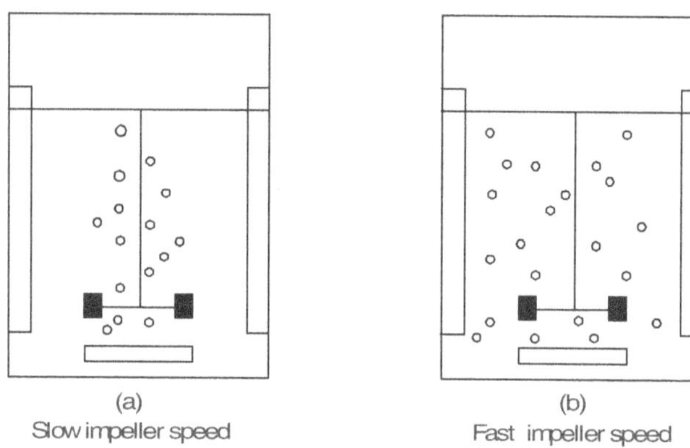

Figure 7.13   Effect of impeller speed

Air flow rate   Air flow rate is typically reported in terms of volume per volume per minute or VVM.

$$VVM = \frac{\text{Volumetric air flow rate}}{\text{Liquid volume}}$$

$$VVM = \frac{F_A(\text{litres.min}^{-1})}{V_L(\text{litres})}$$

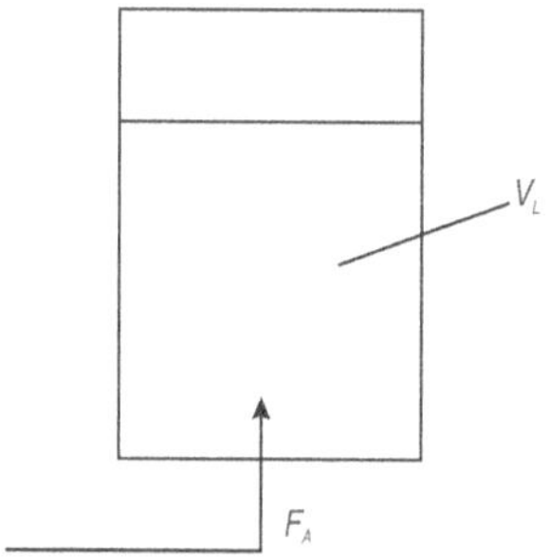

Figure 7.14   Air flow rate

Air flow rate is expressed in terms of volume per minute. Air flow rate and liquid volume must have the same basal unit (Figure 7.14).

FOAM CONTROL SYSTEM

Foam is a critical obstacle to the fermentation process. Severe foaming may cause contamination or loss of material, which will decrease the yield. Foam can be treated with defoaming agents, which usually gives satisfactory results. In addition to the control of foam by adding separate anti-foaming agents, some times the oil substrates that are used to carryout the fermentation may act as an anti-foaming agent. In another case, the metabolic product produced during fermentation may have anti-foaming properties and control the foam without the addition of separate anti-foaming agents.

Foam control is an essential operation in sparged bioreactor. Figure 7.15 shows the accumulation of foam in a 2-litre laboratory reactor.

Figure 7.15   Accumulation of foam in a laboratory bioreactor

Excessive foam can block exit air filters, and pressure can build up in the reactor, which may cause damage to the reactor and even injury to the operator.

Anti-foaming agents are usually based on silicone or vegetable oils. However, excessive anti-foaming agents may result in poor oxygen transfer rates.

Detecting Foam

There are several methods for detecting foam. Manual inspection or valve turning may be waste of manpower, and errors may lead to contamination or loss of contents. There is also a need for reliable oil detection system in the vessel. Mechanical devices such as paddle or float could sense the foam level. However, electrical methods are better because they are sterile. One method could be to use a conducting probe, insulated from the wall of the tank, so that the maximum allowable rise of foam completes a circuit and current flow is detected. Relays can be actuated to control the valves or pumps in the oil system. One drawback with probes is that they may sometimes be short-circuited due to the collection of materials across the insulation at the point of entry into the tank. This results in addition of excess anti-foam oil which may spoil the batch. Another type of device could sense the change in capacitance as foam becomes dielectric in the danger zone in the tank. Simple relay circuitry may be used to signal anti-foam oil control. Unlike the probes, the capacitance detector cannot be fouled.

Foam can be detected using two conductivity or level probes (Figure 7.16 a and b). In one process, when the upper level probe is below the foam level, no current will pass between the level probes, and the antifoam pump remains turned off (Figure 7.16a). In the other process, when the upper level probe is immersed in the foam layer, current is carried in the foam. This turns the antifoam pump on (Figure 7.16b).

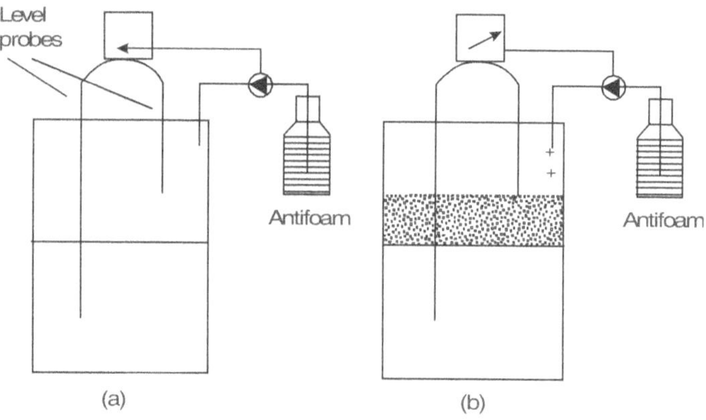

Figure 7.16   Foam controlling set-up

Foam Control

Foam can be controlled with the help of mechanical foam breakers or antifoams. Mechanical methods are a crude approach. Therefore, antifoams are extensively used commercially. Oil is of much importance to the fermentation industry in the control of foam. Excessive foaming causes contamination or loss of material, while excessive oil additions may decrease product formation. Anti-foam oils may be natural, e.g. lard, soybean, linseed, tall and rape seed oils, or synthetic—silicones or polyglycols. Either can substantially change the physical structure of foam, principally by reducing surface elasticity. Some alcohols are suitable as antifoams.

Action of antifoams   The prime action of an anti-foaming agent is to reduce surface elasticity. Industrial antifoam systems usually operate automatically. Methods for analysing oil under aseptic conditions are:

- timed delivery through a solenoid

- using two solenoids with an expansion chamber in between

- using hypodermic syringe

- using certain industrial pumps

Analysing different sterile oil-based anti-foaming agent is rather specialized operation. In most of the cases, the formation of foam is detected by the sensor and subsequent addition of anti-foaming agent takes place. But in large fermentor, the anti-foaming agents are added at particular time intervals without anti-foaming sensors set-up. In such cases, whether the foam is formed or not but the addition of anti-foaming agent will happen continuously at particular time intervals. Some times, the oil-based metabolic products formed during fermentation may reduce the foam formation automatically. In such cases, the fermentor with anti-foaming sensor system detect the reduction of foam and automatically the addition of anti-foaming agents will be stopped. But in the case of adding, continuously the anti-foaming agents at particular time interval leads excess and waste of anti-foaming agents.

TEMPERATURE CONTROL SYSTEM

The temperature control system consists of:

- temperature probes

- a heat transfer system

Typically, the heat transfer system uses a jacket to transfer heat in or out of the reactor. The jacket is a shell that surrounds part of the reactor Figure 7.17. The liquid in the jacket does not come in direct contact with fermentation fluid.

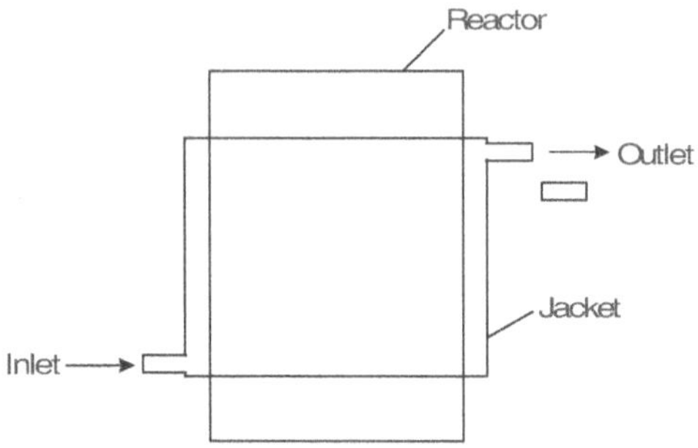

Figure 7.17   Temperature control system

The jacket will usually be dimpled to encourage turbulence and thus increase heat transfer efficiency. An alternative to jackets is coils. Coils have much higher heat transfer efficiency than that of jackets. However, coils take up valuable reactor volume and can be difficult to clean and sterilize. The heating/cooling requirements differ between laboratory scale and production scale reactors as follows:

	Laboratory scale reactors	Production scale reactors
Heating	Electric heaters	Steam generated in boilers
Cooling	Tap water or refrigerated water bath	Cooling water produced by cooling towers or refrigerants such as ammonia

In production scale reactors, heating is required only during the initial and final stages of fermentation as most processes that occur during fermentation are exothermic, which includes the following:

- ⮞ biological reactions (e.g. growth)
- ⮞ chemical reactions
- ⮞ mixing

pH CONTROL SYSTEM

The main function of the pH control systems in the fermentor is to control the pH at particular set point.

The pH control system consists of:

- a pH probe
- alkali delivery system
- acid delivery system

The pH probe is a sensor which can detect the current pH of the fermentation medium. This sensor is connected with pH controlling unit which has link with acid and alkali system (Figure 7.18). When the level of pH is either increased or decreased over the set pH (set point), the pH controller automatically switch on either acid controller or alkali controller depending on the pH. As a result the addition of alkali or acid is made and the pH is brought to the set point level. In fermenter, the sterile acid and alkali is prepared and filled up in the respective controller. The pH probe are heat proof and can be sterilized by autoclave.

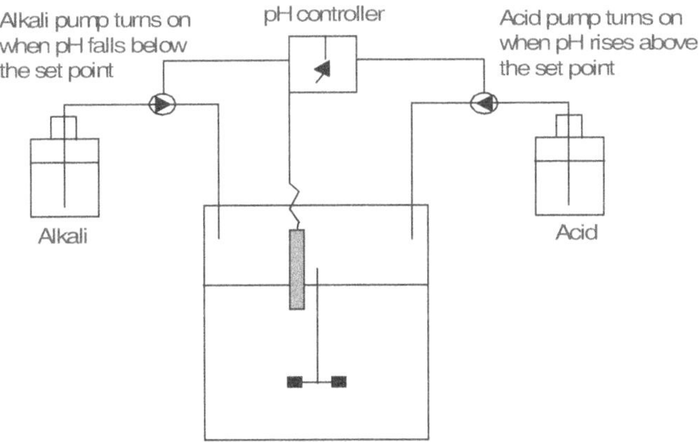

Figure 7.18   pH controller

Neutralizing agents   The neutralizing agents used should be non-corrosive and, in the medium, should be non-toxic to cells. KOH is preferred to NaOH, as potassium ions tend to be lesser toxic to cells. However, KOH is expensive. Sodium carbonate is also commonly used in small-scale bioreactor systems. Hydrochloric acid should never be used because it is corrosive. Similarly, the sulphuric acid concentration should not be beyond 10% since it is more corrosive.

In fermentation of organisms that produce large amounts of acids (e.g. lactic acid), high concentrations of alkali (4 M and above) are used; this prevents dilution of the medium. In laboratory fermenters, a peristaltic pump is used to add the neutralizing agents. Silicone tubing is often used. However, silicone tubing may decay in the presence of high alkali concentrations. Hence, a thick-walled silicone or, alternatively, tygon or neoprene tubing can be used. Tygon is not autoclavable but can be sterilized by passing NaOH through the tubing for 1 hour. Neoprene is autoclavable but is not transparent or translucent like tygon and silicone.

Set point and dead band

A pH control system (and indeed all other fermenter control systems) may be designed to have a dead band. The dead band prevents excessive alkali or acid.

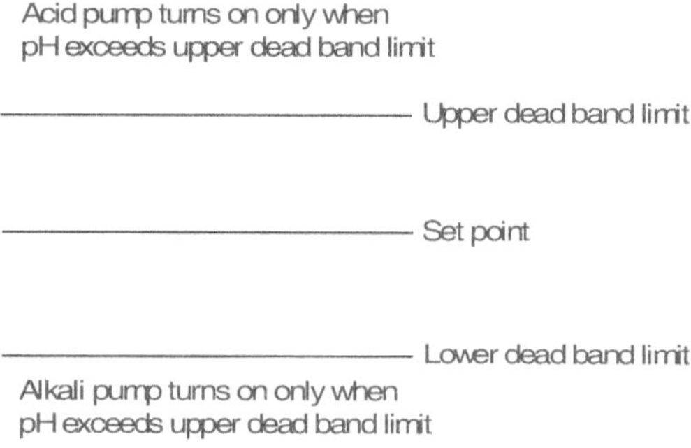

Figure 7.19   pH control dead band

The set point is the pH at which the fermenter is being attempted to be controlled (Figure 7.19). For example, if fermentation is to run at a constant pH of 6.5, the set point is 6.50. If dead band is set to 5%, the upper dead band limit will be

$$1.05 \times 6.5 = 6.83$$

and lower dead band limit will be

$$0.95 \times 6.5 = 6.18$$

If dead band is too small, it may lead to excessive alkali or acid addition. The trade off is that a wide dead band will lead to less precise pH control.

Indeed most fermentations produce acids. Hence acid addition is often not required.

AGITATOR DESIGN AND OPERATION

Agitators may have radial flow or axial flow characteristics. With radial flow mixing, the liquid flow from the impeller is initially directed towards the wall of the reactor; along the radius of the tank. With axial flow mixing, the liquid flow is directed downwards to the base of the reactor, in the direction of the axis of the tank.

Radial flow impellers are primarily used for gas–liquid contacting (like the mixing in sparged bioreactors) and blending processes. Axial flow impellers provide gentle and efficient mixing and are used for shear sensitive cells and particles.

Radial Flow Impellers

The radial flow impeller is shown in Figure 7.20 contain two or more impeller blades that are set at vertical pitch.

The flat-blade radial turbine has blades in the vertical plane, parallel to the mixer shaft having a minimum of two blades. Four blades are most commonly used although radials are available with as few as two blades to as many as eight. The flow is discharged radially and splits into two equal flows after leaving the blade tips. Suction is equal from top to bottom. Two types of radial turbines are generally in use. They are the open type with blades fastened to the impeller axis and the Rushton turbine (also known as disc type) with blades fastened to a disc which is fastened to the axis. Flat-blade turbines are used when radial flow and high shear is desirable. This impeller is ideal for immiscible liquid emulsion applications. The working of radial flow impeller is shown in Figure 7.21.

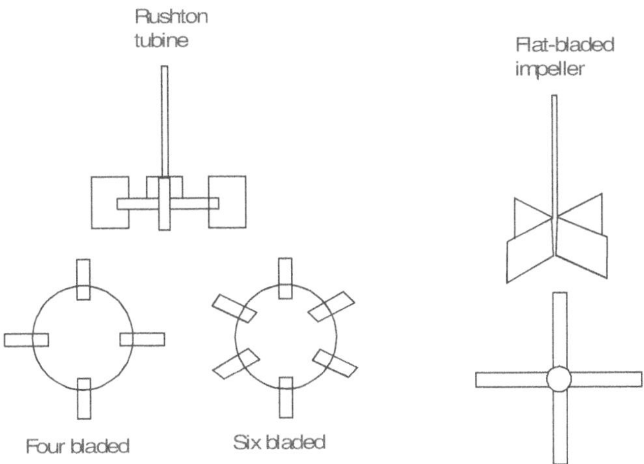

Figure 7.20   Radial flow impellers

With radial flow impellers, vertical (or axial) mixing is achieved with the use of baffles. Radial flow mixing is not as efficient as axial flow mixing. In radial flow impellers, a much higher input of energy is required to generate a given level of flow.

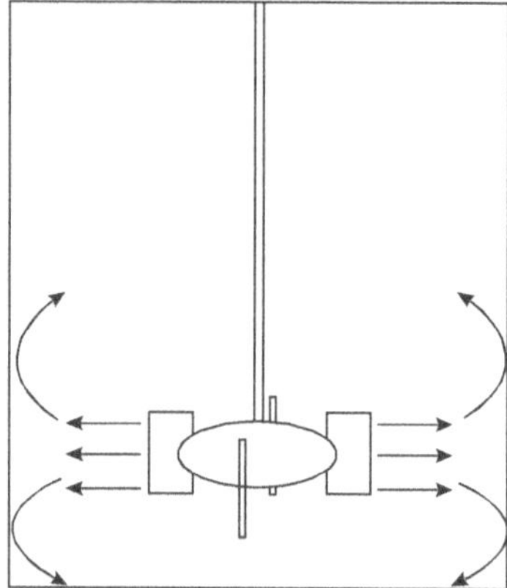

Figure 7.21   Working of radial flow impellers

Shear characteristics in radial flow impellers   Radial flow impellers generate high shear conditions. This is achieved by the formation of vortices in raising the level of the impeller.

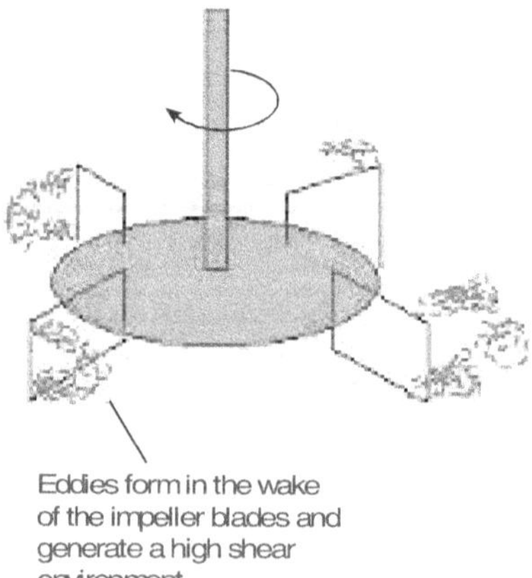

Figure 7.22   Shearing way of radial flow impeller

High shear is effective at breaking up bubbles. It is for this reason that radial flow impellers are used to culture aerobic bacteria (Figure 7.22). However, high shear may damage shear sensitive materials (e.g. crystals), cells (e.g. filamentous fungi and animal cells).

Rushton turbine   The most commonly used agitator in microbial fermentation is Rushton turbine (Figure 7.23). Like any radial flow impeller, the Rushton turbine is designed to provide high shear conditions. The Rushton turbine has 4 or 6 blades fixed onto a disc.

The diameter of Rushton turbine should be one-third of that of the tank diameter. A Rushton turbine is often referred to as a disc turbine.

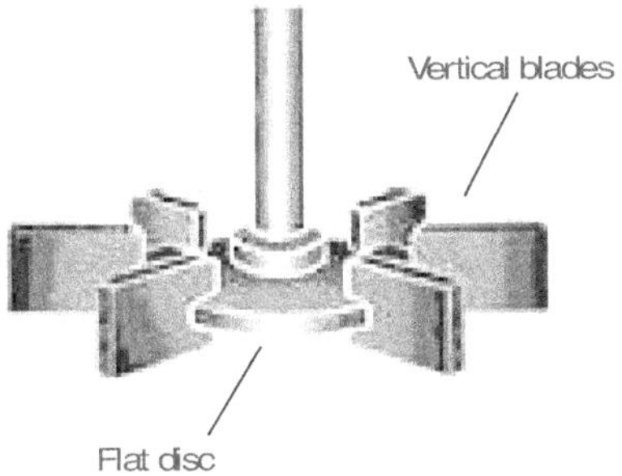

Figure 7.23   Rushton turbine

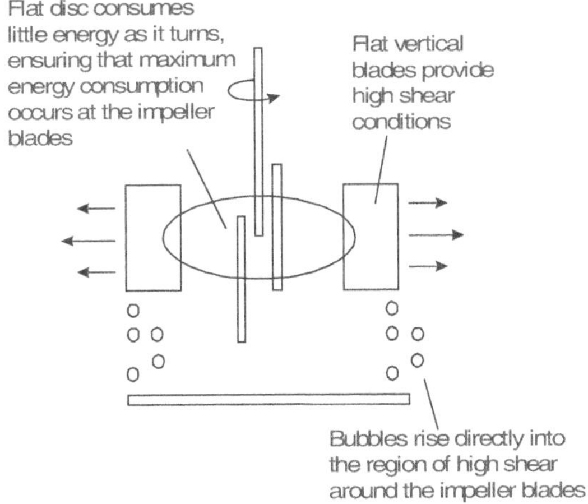

Figure 7.24   Working principle of Ruston turbine

The design of the disc ensures that most of the motor power is consumed at the tips of the agitator, thus maximizing the energy used for bubble shearing Figure 7.24.

Axial Flow Impellers

An axial flow impeller along with its working principle is shown in Figures 7.25, 7.26 and 7.27.

The blades of an axial flow impeller are pitched to direct the liquid flow towards the base of the tank. Examples of axial flow impellers include marine impellers and hydrofoil impellers.

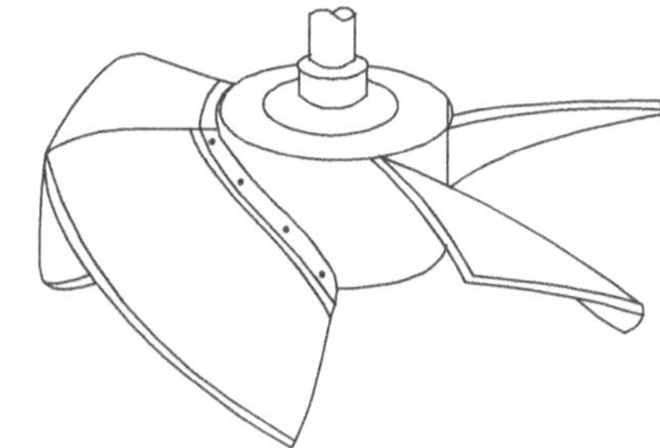

Figure 7.25   Axial flow of impellers

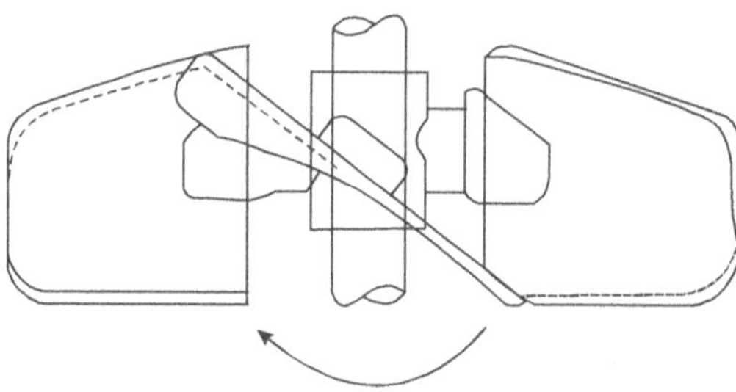

Figure 7.26   Working principle of an axial flow impeller

Axial flow impellers have low shear properties. They are more effective at lifting solids from the base of the tank. The angled pitch coupled with thin trailing edges of the impeller blades reduce the formation of eddies in the wake of the moving blades.

Axial flow impellers are widely used for mixing shear sensitive materials and animal cells. However, their low shear characteristics make them ineffective at breaking up bubbles and thus unsuitable for use in aeration of bacterial fermentation. Low shear conditions are achieved by pitching the impeller blades and by making the edges of the impeller blades thin and smooth.

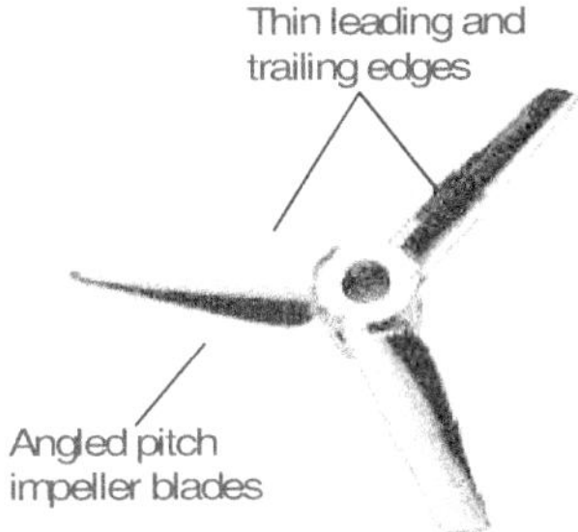

Figure 7.27   Structure of long blade axial cylinder

Intermig impeller   An intermig impeller is an axial flow impeller used for microbial fermentation (Figure 7.28).

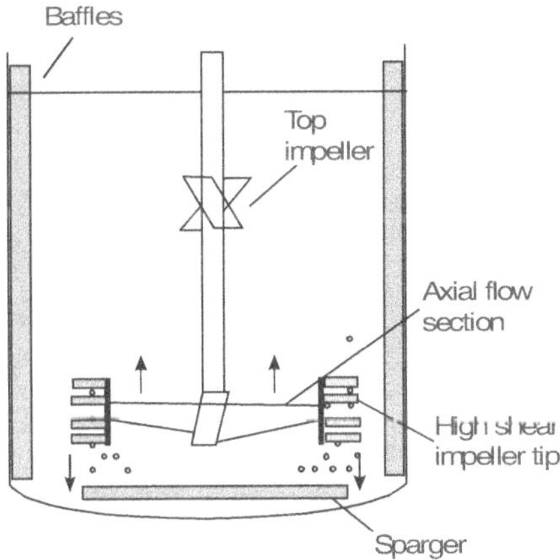

Figure 7.28   Intermig impeller

Intermig impeller has two impellers (Figure 7.28). The bottom one has a large axial flow section. The tips of the impeller contain finger-like extensions for breaking bubbles. Since high shear region exists only at the tip, the overall shear conditions in the reactor are lower than that generated by other radial flow impellers like Rushton turbine. Intermig impellers are widely used for agitation and aeration in fungal fermentation.

REVIEW QUESTIONS

1. Give a brief account on parameter optimization of fermentation.

2. Explain about different types of aeration and their principles.

3. What is the necessity of foam control in fermentation? How is it achieved?

4. Discuss temperature and pH control system.

5. Explain about medium agitation and their principles.

DOWNSTREAM PROCESSING

Downstream processing in fermentation refers to the process of product recovery from the culture. The methods and techniques involved in downstream processing differ from one product to another. However, there are three steps involved in downstream processing, such as separation of cells from the culture, cell fractionation and extraction, and purification of product. All these steps are necessary to recover intracellular products. For extracellular products, cell fractionation may be skipped because they are produced in exhausted medium and extracted from the culture filtrate.

CELL SEPARATION

Cell separation is the first step in downstream processing. Whether the product is extracellular or intracellular, the cells need to be separated from the cultured medium. After separation of cells, the cell-free culture medium is taken for further extraction of extracellular products while the separated cells are taken for further extraction of intracellular products. The general methods employed in cell separation are as follows.

Centrifugation

Centrifugation is a common method used to separate cells from cultured broth. It employs centrifugal force to promote accelerated settling of particles in a solid–liquid mixture. Centrifuges achieve separation by means of accelerated gravitational force achieved by rapid rotation. This may either replace normal gravity in the suspension or provide the driving force in filtration through a filter medium.

In a cylindrical vessel that rotates at an angular speed ω(rad/s) or N (rpm) and contains a liquid ring of mean radius R(m), the centrifugal acceleration Fc (m/s) to which the particles are subjected is:

$$F_c = \omega^2 R = 0.011 \ N^2 R$$

The force exerted on a particle per unit of weight may be expressed by:

$$F_c = 0.011 \ N^2 R \ (rs - rl) \times 1/g = G \ (rs - rl),$$

with $G = \omega^2 R/g = 0.11 \ N^2 R/9.81 = 11.2 \times 10^{-4} \ N^2 R$

where,

   rs = density of particle and

   rl = density of interstitial liquid.

Types of Centrifuges

A variety of centrifuges are used in bioprocessing industries. The type or size of the centrifuge is chosen based on the nature of production.

Tubular bowl centrifuge   The tubular bowl centrifuge (Figure 8.1) has been widely used than most other designs of centrifuge. It is a simple machine made of a tube, of length several times its diameter, rotating between bearings at each end. The suspension enters at the bottom of the centrifuge and high centrifugal forces act to separate the solids and liquids. The bulk of the solids will adhere on the walls of the bowl, while the liquids exit at the top of the centrifuge.

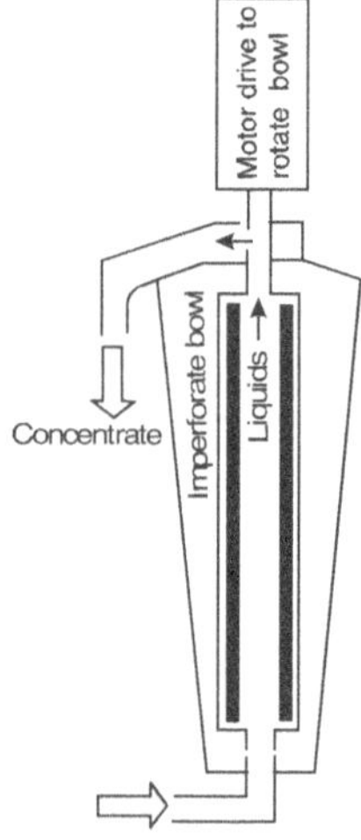

Figure 8.1   Tubular bowl centrifuge

Since the system lacks a provision for solid rejection, the solids can only be removed by dismantling the system and scraping the solids out manually. Tubular bowl centrifuges have dewatering capacity, but limited solid capacity. Foaming can be a problem unless the system includes special skimming or centripetal pumps.

Chamber bowl centrifuge   A chamber bowl centrifuge (Figure 8.2) has a number of tubular bowls arranged co-axially. The main bowl contains cylindrical inserts that divide the volume of the bowl into a series of annular chambers, which operate in sequence. The feed enters the centre of the bowl and the suspension passes through each chamber in turn at increasing distances from the axis. The solids settle on the outer wall of each chamber and the clarified liquid overflows from the largest diameter chamber. An advantage of the device is that it provides a classification of the suspended solids: coarse particles deposit on the inner chamber and finer particles deposit on the subsequent chambers. As said earlier, the removal of sediment solids requires halting the system.

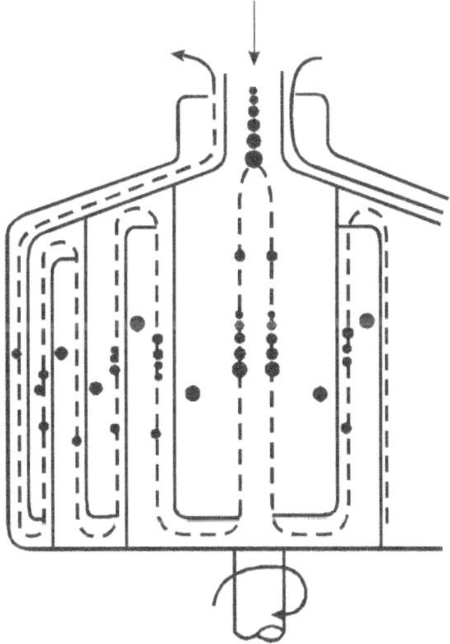

Figure 8.2   Chamber bowl centrifuge

Imperforate basket centrifuge   An imperforate basket centrifuge is used when the solid content of the suspension is higher. It consists of a simple drum-shaped basket or bowl, usually rotating around vertical axis. The solids accumulate and are compressed by the effect of centrifugal force, but they are not dewatered. The residual liquid drains out when the rotation of the bowl is stopped. The solids are removed manually by scraping or shovelling. However, unloading could be done without switching off the machine. To achieve this, a skimmer pipe can be used to remove the residual liquid, and subsequently lowering a knife blade into the solid for cutting it out from the bowl.

Disc stack separator   The simplest design of disc stack separator is a closed bowl, containing the disc stack, with an arrangement to collect residual solids at the outer part of the bowl, from where they have to be removed manually after stopping rotation. The solids are discharged from the bowl through nozzles, which are always open. In a more complicated design, valved nozzles could open automatically when the solid depth in the bowl reaches a certain value, and then close when much of the solids have been discharged.

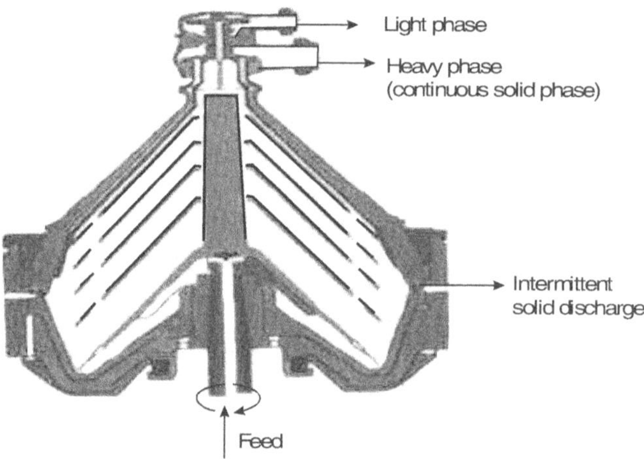

Figure 8.3   Disc stack separator

Decanter centrifuge   Decanter centrifuge is the only sedimentation centrifuge designed to handle significant solid concentration in feed suspension. At the same time, it can achieve a good degree of clarification of the liquid concentrate. Although a complicated piece of machinery, it embodies a simple principle. It basically consists of a horizontal cylindrical bowl rotating at a high speed, with a helical extraction screw placed co-axially. The screw perfectly fits the internal contour of the bowl, only allowing clearance between the bowl and scroll. The differential speed between the screw and scroll provides the conveying motion to collect and remove solids that accumulate at the bowl wall.

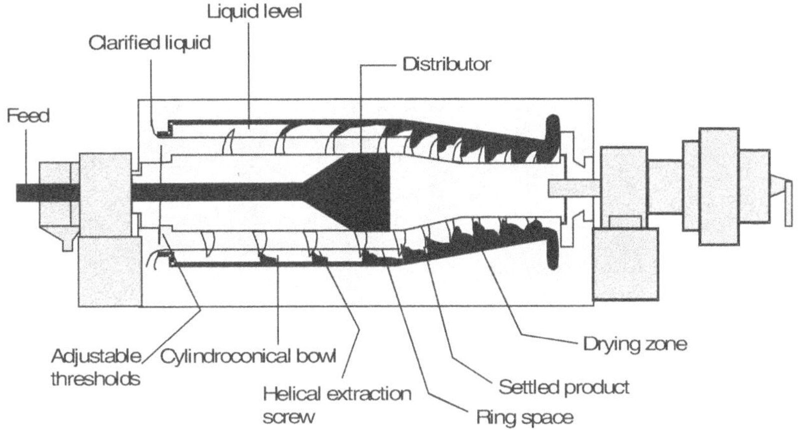

Figure 8.4   Decanter centrifuge

FILTRATION

In mechanical separations, a screen is placed against the flow to impose total restraint on particles above a given size. The fluid is subject to a force that moves it past the retained particles. This process is called filtration. The particles suspended in the fluid, which will not pass through the apertures are retained and build up into what is called a filter cake.

Fine apertures used in filtration comprise fabric cloths, meshes and screens of plastics or metals. Sometimes, a thin preliminary coat of cake or fine particles is put on the aperture prior to filtration. This preliminary coating, known as precoat, gives sufficiently fine pores on the filter.

The analysis of filtration is largely a question of studying the flow system. When the fluid passes through the filter, the normal flow rate will be less. The flow rate can be increased by giving pressure to the fluid. While applying pressure, the pressure of the fluid will be higher than the outer side of the filter. The filtration rate is greatly influenced by the given fluid pressure and the resistance of the membrane. Thus, we can use the following equation:

rate of filtration = driving force/resistance

Resistance usually arises from the filter cloth, mesh or bed as well as from the filter cake that accumulates. Resistance from filter cake can be calculated by multiplying specific resistance of the filter cake (resistance per unit thickness) by thickness of the cake. The resistances of the filter material and precoat are combined into a single resistance called filter resistance. It is convenient to express filter resistance in terms of fictitious thickness of filter cake. This thickness may be multiplied by specific resistance of the filter cake to give filter resistance. The overall equation to find the volumetric rate of flow dV/dt is

$$\frac{dV}{dt} = \frac{(ADP)}{R}$$

where,

A   is the filter area and

DP is the pressure drop across the filter.

As total resistance is proportional to the viscosity of the fluid, we can write

$$R = mr\,(L_c + L)$$

where,

R is the resistance to flow through the filter,

m is the viscosity of the fluid,

r is the specific resistance of filter cake,

L_c is the thickness of the filter cake, and

L is the fictitious thickness of the filter cloth and precoat.

If the rate of flow of the liquid and its solid content are known (and assuming that all solids are retained on the filter), the thickness of the filter cake may be expressed by

$$L_c = \frac{wV}{A}$$

where,

w is the fractional solid content per unit volume of liquid,

V is the volume of fluid that has passed through the filter, and

A is the area of filter surface on which the cake forms.

Total resistance may now be expressed as

$$R = mr\,[w(V/A) + L]$$

and the equation for flow through the filter, under the driving force of the pressure drop is

$$\frac{dV}{dt} = \frac{ADP}{mr}\left[w\left(\frac{V}{A}\right) + L\right]$$

This is the fundamental equation for filtration. It expresses the rate of filtration in terms of quantities. It may be used to predict the performance of large-scale filters on the basis of laboratory tests. Two applications of this equation are filtration at constant flow rate and filtration under constant pressure.

Constant-rate Filtration

In early stages of a filtration cycle, the filter resistance may be larger than the resistance of the filter cake because the cake would be thin. Under this circumstance, the resistance offered to the flow is virtually constant, so filtration proceeds at a constant rate. The equation (dV/dt) can be integrated to give the quantity of liquid passed through the filter in a given time. The terms on the right-hand side of the equation (dV/dt) are constant, and hence integration is very simple

$$\int \frac{dV}{Adt} = \frac{V}{At} = \frac{DP}{mr}\left[w\left(\frac{V}{A}\right) + L\right]$$

$$\text{or} \quad DP = \frac{V}{At} \times mr\left[w\left(\frac{V}{A}\right) + L\right]$$

Using this equation, the pressure drop required for a desired flow rate can be found. If a series of runs is carried out under different pressures, the results may be used to determine the resistance of the filter cake.

Constant-pressure Filtration

Once the initial cake has been built up, the flow occurs under a constant-pressure differential. This could be true for many practical filtration operations. Under this condition, the term DP would be constant and so

$$mr\left[w\left(\frac{V}{A}\right) + L\right]dV = ADPdt$$

and integration from $V = 0$ at $t = 0$ to $V = V$ at $t = t$

$$mr\left[w\left(\frac{V^2}{2A}\right) + LV\right] = ADPt$$

$$\text{i.e.,} \quad \frac{tA}{V} = \left[\frac{mrw}{2DP}\right] \times \left(\frac{V}{A}\right) + \frac{mrL}{DP}$$

$$\frac{t}{(V/A)} = \left[\frac{mrw}{2DP}\right] \times \left(\frac{V}{A}\right) + \frac{mrL}{DP} \tag{1}$$

Equation (1) would be useful as it covers a practical situation frequently experienced in filtration plants. If a test is carried out with constant pressure, and collecting and measuring the filtrates at fixed time intervals, a filtration graph could be plotted of t/(V/A) against (V/A). It may be seen that this graph would be a straight line. The slope of the graph will correspond to mrw/2DP and the intercept on t/(V/A) axis will give the value of mrL/DP. Since m, w, DP and A are known, the values of the slope and intercept will help to calculate L and r.

A filtration test was carried out with a particular product slurry on a laboratory filter press under a constant pressure of 340 kPa. The volumes of filtrate were collected as follows:

Filtrate volume (kg)	20	40	60	80
Time (min.)	8	26	54.5	93

The area of the filter was 0.186 m². Consider a plant-scale filter with an area of 9.3 m². It is desired to filter the same material at 50% greater concentration and under a pressure of 270 kPa. Let us estimate the quantity of filtrate that would pass through in 1 hour.

From the experimental data:

V (kg)	20	40	60	80
t (s)	480	1560	3270	5580
V/A (kg/m²)	107.5	215	323	430
$t/(V/A)$ (s/m² kg⁻¹)	4.47	7.26	10.12	12.98

The values of t/(V/A) are plotted against the corresponding values of V/A in Figure 8.5. From the graph, we will find that the slope of the line is 0.0265 and intercept is 1.6.

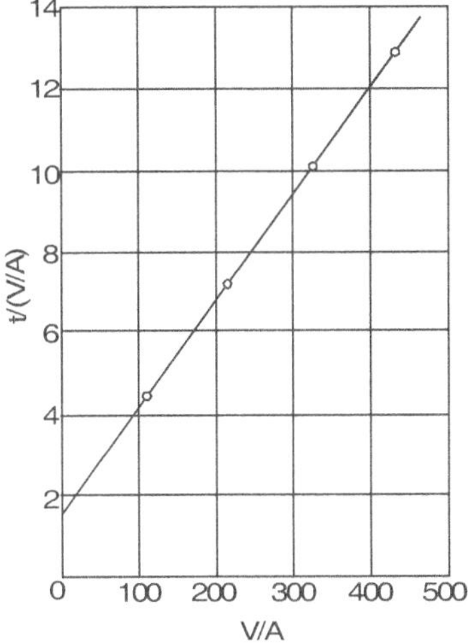

Figure 8.5 Filtration graph

Substituting the values in equation (1), we get:

$$\frac{t}{(V/A)} = 0.0265 \left(\frac{V}{A} \right) + 1.6$$

To match the desired conditions for a plant filter, the constants in this equation will have to be modified. If all the factors in equation (1) except those which are varied in the problem are combined into constants K and K′, we can write

$$\frac{t}{(V/A)} = \left(\frac{w}{DP} \right) K \times \left(\frac{V}{A} \right) + \frac{K'}{DP} \tag{2}$$

In a laboratory experiment, $w = w_1$ and $DP = DP_1$

$$K = \left(\frac{0.0265 \, DP_1}{w_1} \right) \text{ and } K' = 1.6 DP_1$$

For the new plant condition, $w = w_2$ and $P = P_2$, so that, substituting in the equation (2) we then have for the plant filter, under the given conditions:

$$\frac{t}{(V/A)} = \left(0.0265 \, \frac{DP_1}{w_1} \right) \left(\frac{w_2}{DP_2} \right) \left(\frac{V}{A} \right) + (1.6 DP_1) \left(\frac{1}{DP_2} \right)$$

Since, from these conditions,

$$\frac{DP_1}{DP_2} = \frac{340}{270}$$

and

$$\frac{W_2}{W_1} = \frac{150}{100}$$

$$\frac{t}{(V/A)} = 0.0265\left(\frac{340}{270}\right)\left(\frac{150}{100}\right)100\left(\frac{V}{A}\right) + 1.6\left(\frac{340}{270}\right)$$

$$= 0.05\left(\frac{V}{A}\right) + 2.0$$

$$t = 0.05\left(\frac{V}{A}\right)^2 + 2.0\left(\frac{V}{A}\right)$$

To find the volume (V) that passes the filter in 1 hour (i.e., $t = 3600$ s),

$$3600 = 0.05\left(\frac{V}{A}\right)^2 + 2.0\left(\frac{V}{A}\right)$$

Solving this quadratic equation, we will find that $V/A = 250$ kg m^{-2}

and so the slurry passing through 9.3 m^2 in 1 hour would be:

$$= 250 \times 9.3$$

$$= 2325 \text{ kg}$$

Filter-cake compressibility   The specific resistance of a filter cake varies with the pressure drop across it. This is because the cake becomes denser under higher pressure, which provides fewer and smaller passages for flow. This effect is termed as compressibility of the cake. Soft and flocculent materials provide highly compressible filter cakes, whereas hard granular materials, (e.g. sugar and salt crystals), are least affected by pressure. To allow for cake compressibility, the empirical relationship has been proposed.

$$r = r'DPs$$

where,

r is the specific resistance of the cake under pressure P,

DP is the pressure drop across the filter,

r' is the specific resistance of the cake under pressure drop of 1 atm. and

s is constant compressibility of the material.

The expression for r may be inserted into the filtration equation (equation 1) and values for r' and s can be determined by carrying out experimental runs under various pressures.

Types of Filtration Equipment

The basic requirements for equipment in filtration are:

- ➲ mechanical support for filter medium

- ➲ flow accesses to and from the filter medium

- ➲ provision for removing excess filter cake

Sometimes, washing the filter cake may be necessary to remove traces of the solution. Pressure can be provided on the upstream side of the filter, or a vacuum can be drawn downstream, or both can be used to drive the wash fluid through.

Plate and frame filter press   In plate and frame filter press (Figure 8.6), a cloth or mesh is spread over the plates which support the cloth along ridges and, at the same time, allow a large area free for the filtrate to flow. The plates along with filter cloths are usually hung vertically with a number of plates operated in parallel to give sufficient area.

The filter cake builds up on the upstream side of the cloth, which is away from the plate. In early stages of the filtration cycle, the pressure drop across the cloth is small and filtration proceeds at a constant rate. As the cake increases, the pressure becomes more, and this is the case throughout the cycle. When the space between successive frames is filled with cake, the press has to be dismantled and the cake scraped off and cleaned, after which a new cycle can be initiated.

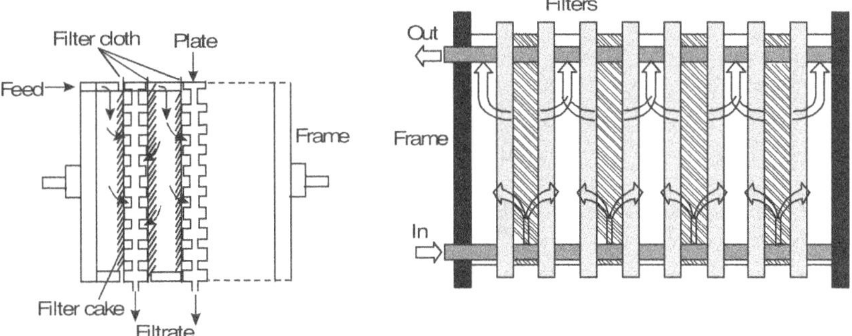

Figure 8.6   Plate and frame filter press

The plate and frame filter press is economical, but it is difficult to mechanize to a great extent. The variants of the plate and frame press have been developed, which allow easier discharging of filter cake. In an ideal design, the plates, which may be rectangular or circular, are supported on a central hollow shaft and the whole assembly enclosed in a pressure tank contains the slurry. Filtration may be done with pressure or vacuum. The advantage of vacuum filtration is that the pressure drop can be maintained whilst the cake can be removed easily. The disadvantages are that the cost of maintaining a given pressure drop by applying vacuum is higher and the maximum force delivered is 80 kPa. On the other hand, in pressure filtration, the driving force is limited only by the economics of generating the pressure and by the mechanical strength of the equipment.

Rotary filters   In rotary filters (Figure 8.7), the flow passes through a rotating cylinder. Either pressure or vacuum can provide the driving force, but vacuum is more useful. In rotary filters, the cloth is supported on the periphery of the horizontal cylindrical drum that dips into a bath of slurry. Vacuum is drawn into those segments of the drum surface where the cake is building up. A suitable bearing applies the vacuum at the stage when actual filtration commences and breaks the vacuum when the cake is being scraped off after filtration. The filtrate is removed through trunnion bearings. Rotary vacuum filters are expensive, but they provide a considerable degree of mechanization and convenience.

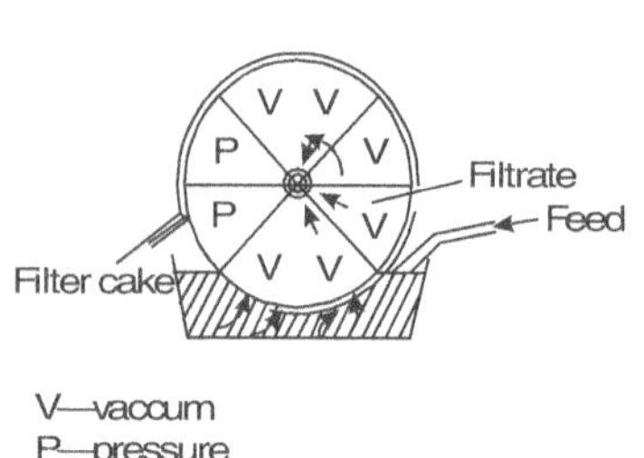

Figure 8.7   Rotary filters

Centrifugal filters   Centrifugal force may be used to provide the driving force in filters (Figure 8.8). In centrifugal filters, centrifuges are fitted with a perforated bowl that has filter cloth on it. The liquid is fed into the bowl, which passes out under centrifugal forces through the filter material.

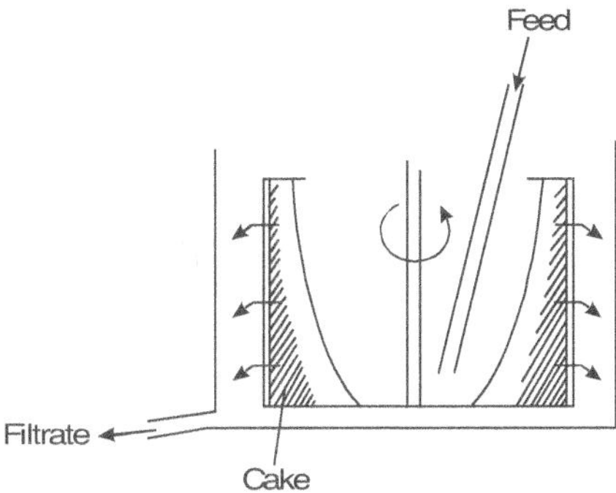

Figure 8.8   Centrifugal filters

Cross-flow filtration (CFF)   In cross-flow filtration, the feed flow is driven within the filter in downward spiralling direction along the filter surface. While the flow is parallel to the filter surface, it is being pressurized against the filter surface at the same time. Schematic diagram of a cross-flow filter is shown in Figure 8.9.

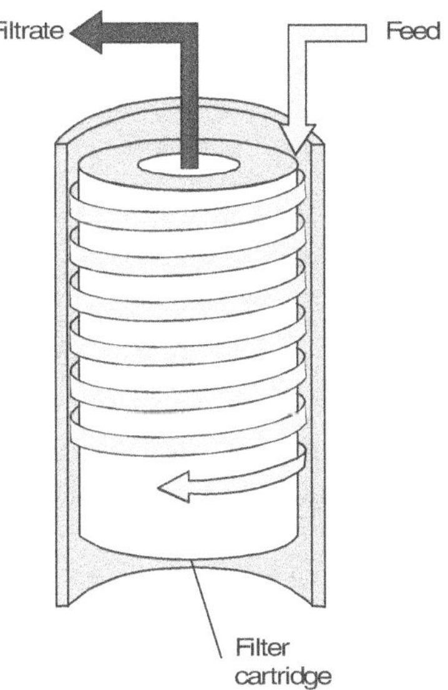

Figure 8.9   Cross-flow filter

Flow velocity is of fundamental importance in cross-flow filters. If flow velocity falls to zero, the cross-flow filtration (Figure 8.10) stops and dead-end filtration begins (Figure 8.11). The cake layer that forms on the filter media at zero velocity becomes thinner when the flow velocity (parallel to the medium) increases. The thickness of the cake layer in a flow channel can be determined by the shear force on the filter media surface, which is roughly in direct proportion to feed viscosity and feed flow velocity. Therefore, higher flow velocity results in thinner deposit layer, lower hydraulic resistance, and therefore higher filtrate flux rate.

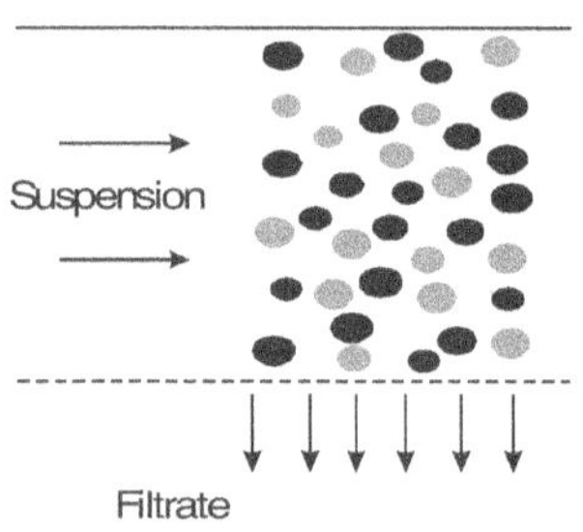

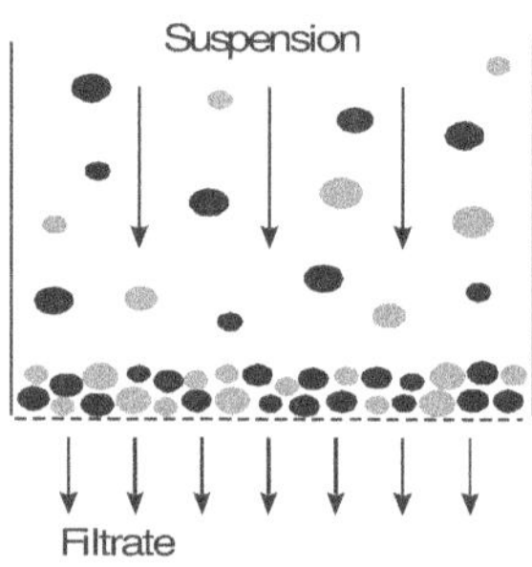

Figure 8.10   Cross-flow filtration   **Figure 8.11**   Dead-end filtration

Most filter cartridges for micro-filtration have good depth. These filters operate with dead-end pressure feed and, therefore, are prone to plugging due to cake formation. Further problems may arise when dead-end filters are allowed to dry. For example, particles are usually cemented inside flow passages, which may greatly reduce the flux rate during subsequent operations. Therefore, these filters are used only in light-duty applications.

Air filters   Filters are extensively used to remove suspended dust particles from airstreams. The air or gas moves through a fabric and the dust is collected in it. Air filters are more useful in removing fine particles. One type of air filter is bag filter, which consists of a number of vertical cylindrical cloth bags 15–30 cm in diameter. Air enters the bags, usually from the bottom, and passes out. Some bag filters are provided for mechanical removal of accumulated dust. A familiar example of a bag filter is the domestic vacuum cleaner. For removal of particles that range in the sizes of bacterial cells and spores in modern air sterilization units, paper filters and packed tubular filters are used.

CELL DISRUPTION

As far as intracellular products are concerned, they must be released from the cells. This is usually achieved by cell disruption (lysis). Cell disruption is a sensitive process because the cell wall is poorly resistant to high osmotic pressure. Furthermore, complexities may arise from non-controlled cell disruption, which results from an unhindered release of all intracellular products (proteins, nucleic acids, cell debris) as well as cell disruption without denaturation of the desired products.

Cell disruption may be mechanical or non-mechanical. The results of these methods are often evaluated in terms of the activity level of a cellular enzyme released to the disrupted suspension, combining the efficiency of the disrupting process with an estimate of the degree of cell disruption.

Mechanical Methods

Usually, intracellular products are released from microorganisms by mechanical disruption of cells. In this process, the cell envelope is physically broken, releasing all intracellular components into the surrounding medium.

Equipment used for mechanical cell disruption

Homogenizers   A homogenizer (Figure 8.12) pumps the slurry through a restricted orifice valve. It uses high pressure (up to 1500 bar) followed by an instant expansion through a special exiting nozzle. Cell disruption is accomplished by three different mechanisms: impingement on the valve, high liquid shear in the orifice, and sudden pressure drop upon discharge, causing an explosion of the cell. The method is mainly applied for the release of intracellular molecules. According to Hetherington *et al.*, cell disruption (and consequently the rate of protein release) is a first-order process, described by the relation

$$\log \left[\frac{Rm}{(Rm - R)} \right] = K\ N\ P$$

where,

R = the amount of soluble protein,

Rm = maximum amount of soluble protein,

K = temperature dependent rate constant,

N = number of passes through the homogenizer (represents the residence time) and

P = operating pressure.

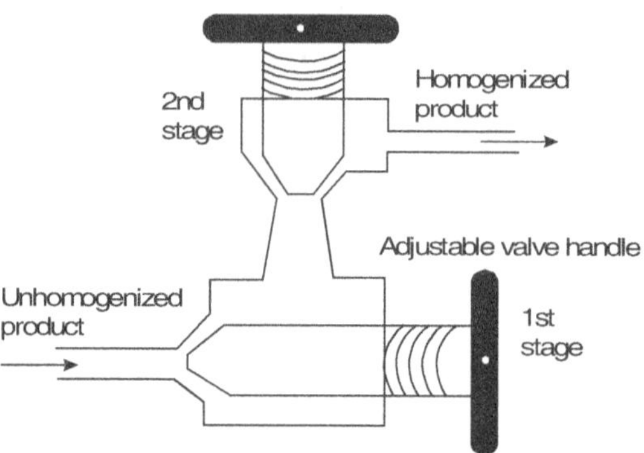

Figure 8.12   Homogenizer

Ball mills   Ball mills, also known as centrifugal or planetary mills, are devices used to rapidly grind materials to colloidal fineness (1 micron or below) by developing high grinding energy *via* centrifugal and/or planetary action. The grinding is carried out by pounding and rolling a charge of steel or ceramic balls within the cylinder. The cylinder rotates at a relatively slow speed, allowing the balls to cascade through the mill base, thus grinding or dispersing the materials.

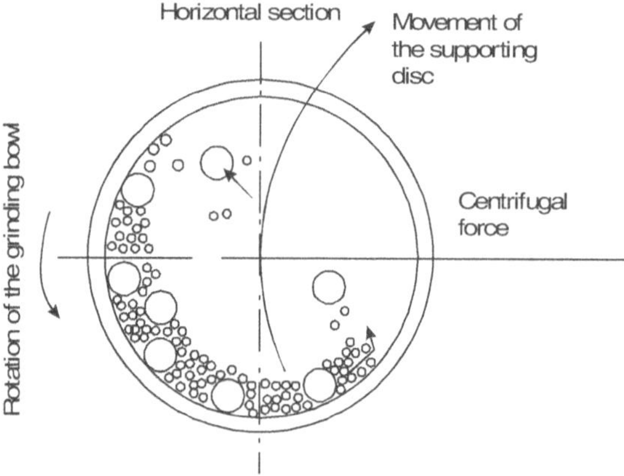

Figure 8.13   The principle of a ball mill

In a ball mill (Figure 8.13), the cells are agitated in suspension with small abrasive particles. Cells break due to shear forces, grinding between beads and colliding with beads. The beads disrupt the cells to release biomolecules. The kinetics of biomolecule released in this method is of first-order process. The simplest method of ball milling is grinding the cells with beads using mortar and pestle. However, mortar and pestle grinding is a manual process and cannot be applicable in industrial process.

The grinding media available include agate, sintered corundum, tungsten carbide, tempered chrome steel, stainless steel, zirconium oxide and polyamide plastic. The type of bowl and balls depends on the type of material being ground. For example, very hard samples require tungsten carbide balls and steel bowls. Typically, agate is a good choice. In this method of grinding, cross-contamination of the sample with the grinding media can be a complication.

Ultrasonic Disruption

Another widely applied method of cell lysis uses high-frequency sound produced electronically and transported through a metallic tip to an appropriately concentrated cellular suspension. This method is costly, and it can only be used at laboratory scale for lysis of less-resistant cell walls, such as bacteria and fungi. The concept of ultrasonic disruption is based on the creation of cavities in cell suspension. Sonicator (Figure 8.14) is an ultrasonic disruption device used commonly.

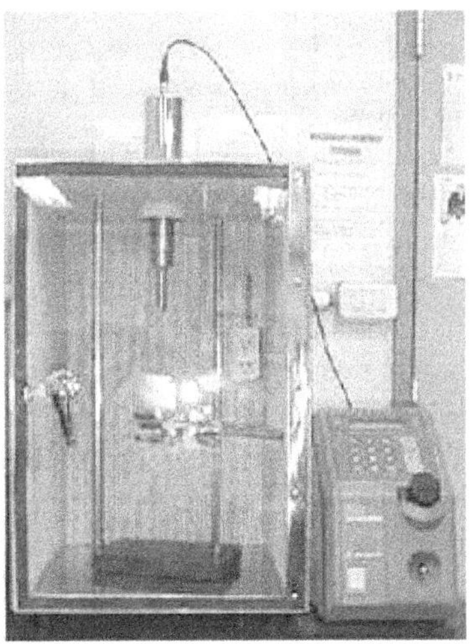

Figure 8.14   Sonicator

Blenders (High speed or Waring)

The french press, or even centrifugation in case of weak cell walls, also disrupt the cells by using the same concepts.

Disadvantages in Mechanical Disruption

Mechanical disruption has several drawbacks. Because cells break completely, all intracellular materials are released together. Therefore, the product of interest must be separated from a complex mixture of proteins, nucleic acids, and cell wall fragments. Nucleic acids may increase the viscosity of the solution and may complicate subsequent processing, especially chromatography.

The cell debris often consists of cell fragments, which makes the solution difficult to clarify. Moreover, for the complete product to release, it requires more than one pass through the disruption device, which further reduces the size of fragments. These are difficult to remove by continuous centrifugation, because the throughput of the device is inversely related to the square of the particle diameter. Filtration may be complicated by the gelatinous nature of the homogenate and by its tendency to foul the membranes. Furthermore, mechanical methods expose the cells and the extracted product to harsh conditions. While most proteins can tolerate the high pressure inside a homogenizer or a ball mill, others will be denatured by the heat generated, unless the device is sufficiently cooled.

Non-mechanical Methods

Another way to disrupt the cells is permeabilization. This can be achieved by numerous methods. The most important of them are:

Chemical permeabilization   Many chemical methods have been employed to extract intracellular components from microorganisms by permeabilizing the outer-wall barriers. Chemical permeabilization is achieved with organic solvents that can create canals through the cell membrane: toluene, ether, phenyl ethyl alcohol, dimethyl sulphoxide, benzene, methanol and chloroform. Chemical permeabilization can also be achieved with antibiotics, thionines, surfactants (Triton, Brij, Duponal), chaotropic agents, and chelates. EDTA (chelating agent) is widely used for permeabilization of gram-negative microorganisms. It can effectively bind the divalent cations of Ca^{++}, Mg^{++}. The latter ones stabilize the structure of outer membranes by binding lipopolysaccharides to each other. Once these cations are removed from EDTA, the lipopolysaccharides are separated, resulting in increased permeability of the outer walls.

Chaotropic agents   Chaotropic agents such as urea and guanidine, are capable of bringing some hydrophobic compounds into aqueous solutions. They accomplish this by disrupting the structure of water, making it less hydrophilic, and weakening the hydrophobic interactions among solute molecules.

Mechanical permeabilization   One method is osmotic shock. While cells exposed to slowly varying extracellular osmotic pressure are usually able to adapt to such changes, cells exposed to rapid changes in external osmolarity can be mechanically injured. This procedure may be conducted by first allowing the cells to equilibrate internal and external osmotic pressure in a high sucrose medium and then rapidly diluting away the sucrose. The resulting overpressure of the cytosol is assumed to damage the cell membrane. Enzymes released by this method are believed to be periplasmic, or at least located near the surface of the cell.

Enzymatic permeabilization   Enzymes can also be employed to permeabilize cells, but this method is often limited to release periplasmic or surface enzymes. In this method, EDTA is often used to destabilize the outer membrane of gram-negative cells, making the peptidoglycan layer accessible to the enzyme used. Enzymes used for enzymatic permeabilization include beta(1-6) and beta (1-3) glucanases, proteases, and mannase. The main drawbacks in recovering intracellular products in large-scale enzymatic permeabilization include huge investment and the necessity to remove lytic enzyme from the product.

Other techniques   Basic proteins (such as protamine) or cationic polysaccharides (e.g. chitosan) can permeabilize yeast cells. Similarly, mammalian cells can be permeabilized by exposure to several natural substances such as streptolysin and even viruses. Electrical discharges have also been demonstrated to permeabilize mammalian cells in order to study secretion by exocytosis.

SEPARATION OF PRODUCTS

The separation of products is an important step in bioprocessing. Separation of extracellular products is quite easier than that of intracellular products. In extracellular products, the cells are separated from the culture medium and processed to recover products from cell-free culture medium. In case of intracellular products, the cells are fractionated, cell debris is removed, filtration is carried out, and then products are recovered from debris-free medium. The general recovery methods are as follows.

Liquid–Solid Extraction

In liquid–solid extraction, a solvent (hydrophilic, hydrophobic, acidic, neutral or basic) is added to the solid. Insoluble material can be separated by gravity or vacuum filtration, and soluble material is extracted into a solvent. A sequence of solvents, of varying polarity or pH, can be used to separate complex mixtures into groups. The filtered solution can have many practical uses, and the solvent can be evaporated to recover the solute in powder or crystalline form.

Liquid–liquid Extraction

Extraction is a process to separate components based on chemical differences rather than differences in physical properties. The basic principle of extraction involves the contacting of a solution with an immiscible solvent. The solvent is also soluble with a specific solute contained in the solution. Two phases are formed after the addition of the solvent, due to the differences in densities. A solvent should be so chosen that the solute in the solution has more affinity towards the added solvent. Therefore, mass transfer of the solute from the solution to the solvent occurs. Further separation of the extracted solute and solvent will be necessary. However, these separation costs may be desirable in contrast to distillation and other separation processes for situations where extraction is applicable.

Partitioning   The partitioning of solute(s) between two phases is an equilibrium process and is characterized by an equilibrium constant called partition coefficient:

$$\kappa \text{ partition} = \frac{\text{Hydrophobic phase}}{\text{Hydrophilic phase}}$$

where, the notation κ refers to concentration. If a characteristic (such as pH or polarity) of one of the solvents is changed, κ partition likewise changes. One important thing to note is that it is more efficient to extract with multiple small volumes than once with an equivalent large volume.

The use of expensive anhydrous organic solvents is not necessary. The solubility of water in many common solvent is not zero, so upon mixing the phases water dissolves in the organic solvent. If an anhydrous organic solvent is used, it would not be anhydrous once the extraction begins. If solvated water needs to be removed (it usually does, whether to aid crystallization of the solute or to prevent damage to the GC column), the organic solvent can be passed through an anhydrous salt capable of forming a solid hydrate after separation from the aqueous phase. Anhydrous sodium sulphate is commonly used.

Acid–base extraction   As mentioned earlier, the pH of the system has an effect on κ partition. Therefore, the analyst can move some compounds between the phases.

➲ Neutral organic solutes can be extracted from water, aqueous base or aqueous acid using an appropriate organic solvent.

➲ For organic bases, such as amines,

   ◑ the "free base" is often more soluble in non-polar organic solvents than in polar organic solvents, water or aqueous base.

   ◑ the quaternary ammonium salts of amines (formed by reaction with acids) are usually more soluble in polar or aqueous media than in non-polar organic solvents. Note that, if the aqueous solvent is basic, amine compounds may be in free base form.

➲ For organic acids,

   ◑ the "free acid" may be more soluble in an organic solvent than in water (or aqueous acid).

   ◑ salts formed by the reaction with bases are usually more soluble in aqueous media than in organic solvents. If the aqueous medium is acidic, however, the compound may be in free acid form and hence can be partitioned into the organic phase.

Applying acid–base extraction   By manipulating the pH of the aqueous layer, the partitioning of a solute can be changed. Consider a mixture of caffeine (neutral) and procaine (basic) that you wish to separate. Caffeine, being neutral, does not form a salt. Procaine, on the other hand, can form a quaternary ammonium salt, and this salt will partition into an aqueous phase if that phase is acidic; if the aqueous layer is basic, procaine will be partitioned into the organic layer. The procedure to separate these common drugs is as follows:

➲ White powder (caffeine + procaine) is added to CH_2Cl_2 in a separator funnel.

➲ Aqueous acid (e.g. 0.1 M H_2SO_4) is added.

   ◑ The neutral caffeine is partitioned into the methylene chloride (bottom) layer.

   ◑ The sulphate salt of procaine is partitioned into the aqueous (top) layer.

➲ The bottom layer is drained; methylene chloride can be evaporated to obtain nearly pure caffeine.

➲ Another portion of methylene chloride is added to the separator funnel, which still contains the aqueous acid and dissolved procaine.

⮑ Solid or aqueous base is added to the aqueous phase until that phase turns basic. Procaine, converted to free base, is partitioned into the methylene chloride (bottom) layer.

⮑ The bottom layer is drained; methylene chloride can be evaporated to obtain nearly pure procaine. However, procaine base will not crystallize.

Process operation   A general extraction column (Figure 8.15) has two input streams and two output streams. The input streams consist of a solution feed at the top containing the solute to be extracted and a solvent feed at the bottom which extracts the solute from the solution. The solvent containing the extracted solute leaves the top of the column and is referred to as the extract stream. The solution exits through bottom of the column which contains small amounts of solute called raffinate. Further separation of output streams may be required using other separation processes.

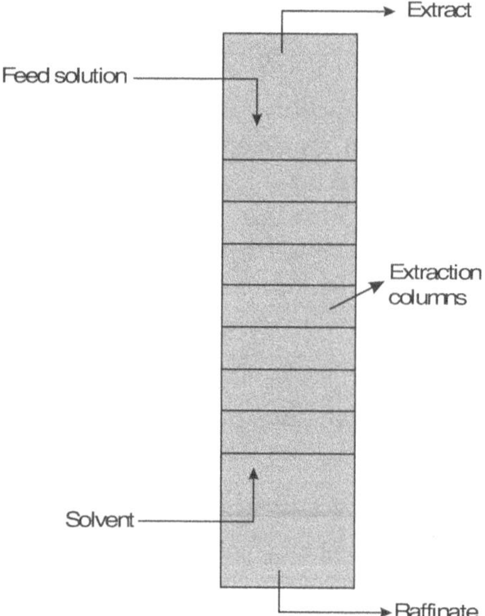

Figure 8.15   Extractor column

There are certain design variables in an extraction process.

⮑ Operating temperature

⮑ Operating pressure

➲ Feed flow rate

　　ʊ Composition

　　ʊ Temperature of entering stream

　　ʊ Pressure of entering stream

In most separation processes, the pressure and temperature conditions play an important role. In order to have a good split of the feed, the pressure and temperature conditions should ensure that all components remain in the liquid phase. The process will be adversely affected if one or more of the components are allowed to become vapour, or extraction may not occur at all if a large portion of a component is allowed to vaporize. In addition, the temperature should be high enough that the components are soluble in one another. If extremes in temperature are present, finding a suitable solvent for extraction can be a problem. This is, however, generally not the case since extraction can be done at ambient pressures and temperatures. In many applications, such as the pharmaceutical industry, an extreme temperature will destroy the desired product. For these applications, extraction is ideally suited, since the only temperature requirement is that dictated by the solubility. At this point, the biggest challenge would be to find a suitable solvent for extraction.

Another example of benefits of extraction in pharmaceutical industry is the volumes involved for effective extraction. However, extraction can become very expensive if the solvent used is expensive. The cost can be contained if a batch process is being used as in the case of medicines. In non-batch processing, the solvent would need to be constantly supplied.

Limitations   We must consider under what extremes extraction can be used as separation process.

1. **Suitable solvent**

　　➲ Solvent partially soluble with the carrier

　　➲ Feed components immiscible with the solvent

　　➲ Solute is soluble in the carrier and at the same time completely or partially soluble in the solvent

　　➲ Different densities than the feed components for a phase separation to facilitate and maintain the capacity of the extractor high

　　➲ Extremely high selectivity for the solute for the solvent to dissolve the maximum amount of solute and the minimum amount of the carrier

- ⇒ Large distribution coefficient to reduce the theoretical number of stages

- ⇒ Contributing to a greater efficiency

- ⇒ Low viscosity increases the capacity of the extraction column and does not allow for the settling rate of dispersion to be slow

- ⇒ Chemically stable and inert toward other components of the system

- ⇒ Low cost, non-toxic, and non-flammable

2. **Equipment**

- ⇒ Interfacial tension and viscosity

- ⇒ High interfacial tension and viscosity leads to more power being supplied to maintain rapid mass transfer throughout the extraction process

- ⇒ Low interfacial tension and viscosity leads to the formation of an emulsion

3. Temperature preferred to be higher since solubility increases, but temperature not more than the critical solution temperature.

4. Pressure for condensed system must be maintained below the vapour pressure of the solutions such that a vapour phase will not appear and interrupt liquid equilibrium.

5. Separation may only occur for compositions in the region between the feed composition and that apex of the carrier.

Applicability   With all the key components in the design of an extractor system to be discussed, the equipment selection can be evaluated. We must choose an extractor that would be relevant for the situation at hand. The specifications for each of these different systems are relatively the same. The following design constraints should be placed on each system in order to optimize the individual process: (1) maximize surface area of mass transfer, and (2) adjust flow feeds for maximum solute recovery.

In general, there are three main types of extractors:

Mixer-settlers   Mixer-settlers are used when there will be only one equilibrium stage in the process. For such a system, the two liquid phases are added and mixed. Due to their density differences, one phase will settle out and the mixture will be separated. The drawback to this type of extractor is that it requires a large volume vessel and has a high liquid demand.

Contacting columns   Contacting columns are practical for most liquid—liquid extraction systems. The packing, trays or sprays increase the surface area in which the two liquid phases can mix together. This also allows for a longer flow path that the solution can travel through. In the selection of a packing, it is necessary to select a material that is wetted by the continuous phase. Lastly, the flow in a column should always be counter-current.

Centrifugal contractors   Centrifugal contractors are ideal for systems in which the density difference is less than 4%. In addition, this type of system should be utilized if the process requires many equilibrium stages. In these systems, mechanical devices are used to agitate the mixture to increase the interfacial area and decrease mass transfer resistance.

Table 8.1   Advantages and disadvantages in liquid–liquid extractor types

Unit of operation	Advantages	Disadvantages
Mixer-settler	Efficient	Large floor
	Low head room	High set-up costs
	Induces good contacting	High operation costs
	Can handle any number of stages	
Columns (without agitation)	Small investment	High head room
	Low operating costs	Difficult to scale-up from lab
		Less efficient than mixer-settler
Columns (with agitation)	Good dispersion	Difficult to separate small density differences
	Low investment	
	Can handle any number of stages	Does not tolerate high flow ratios
Centrifugal extractors	Can separate small density differences	High set-up cost
	Short holding time	High operating and maintenance costs
	Small liquid inventory	Cannot handle many stages

Properties   The following is a partial list of physical properties desirable in liquid–liquid extraction separations. However, other properties will be needed for some calculations, especially those needed to size the diameter of the column.

Temperature plays a little role in this type of extraction than in other separation processes. However, the temperature of the streams fed into the column is significant. There is no heating requirement in the process and pH of mixing is generally insignificant. Hence, this type of extraction can be regarded as an isothermal process.

Pressure too plays a little role in extraction. The fact that extraction processes can be run at isothermal and isobaric conditions is quite beneficial to the phase stability of the system. Phase stability, from thermodynamic standpoint, is temperature- and pressure-dependent; since these are not changing, the stability of the phases will not change.

Activity coefficients are the most important physical property in the extraction process. These are used to determine the miscibility of the solute in both the solvents involved. There are many different equations available to determine a particular activity. When working with liquid–liquid systems, the NRTL and UNIFAC models are the most accurate in predicting the activities of the liquids involved. Although Van Laar or Margules are better than such predictive models, they still fall short of perfection. Once a predictive model has been plotted, it will most likely be necessary to fix the exact equilibrium line for the most accurate data. The activity coefficients also determine the partition factor which will determine whether or not a good separation is possible.

Viscosity is a property that cannot be overlooked. Its presence appears in two different areas: flooding and choice of equipment. Flooding can occur in extraction just as it can for other unit operations. Viscosity is also valuable in the determination of what type of system to use for extraction.

Components that have high viscosity cannot be used in spray or packed columns.

PRECIPITATION

Precipitation is widely used for product recovery from biomolecules, especially proteins. Precipitation is usually induced by the addition of a salt or an organic solvent, or by changing the pH of the solution. The most common type of precipitation for proteins is induced by salt. Protein solubility depends on several factors. At low concentration of the salt, solubility of proteins usually increases. This is termed as salting in. However, at high concentrations of salt, the solubility of proteins drops sharply. This is termed as salting out, and the proteins precipitate out.

Another method is the addition of an organic solvent. If there is an average decrease in dielectric constant with the addition of an organic solvent, the solubility should decrease. Here we can expect precipitation. The third method is precipitation by changing the pH of the protein solution. This effect is due to different functional groups of a protein. At some point of pH, known as the isoelectric point, the net charge on the protein is zero. This varies for different proteins. Precipitation of a protein can also be achieved by the addition of a non-ionic polymer or metal ions.

Direct Precipitation

This is an uncommon method. The early method for purification of the antibiotic cycloserine was to add silver nitrate to the clarified fermentation broth to precipitate its insoluble silver salt. This was too expensive, and alternate methods were soon discovered. Furthermore, dry silver salt could explode.

Another instance is the recovery of proteolytic enzymes for use in detergents. Acetone is added directly to the fermentation broth. Some yield could be lost due to enzyme instability in acetone, and the volumes of solvent are enormous. However, other methods of recovery tend to be more expensive.

Solvent Precipitation

Many substances that are soluble in water are much less soluble in liquids that are miscible with water. Examples include methanol, ethanol, propanol and acetone. A good method of precipitating the substances is to add the solvent to an aqueous solution. The yield would be better if the starting aqueous solution is concentrate. Otherwise, the decreased solubility after adding the solvent may not produce an acceptable yield.

Salt Precipitation

Ammonium sulphate precipitation is often used as the first purification and concentrating procedure. The proteins can be stored in ammonium sulphate solution, which reduces bacterial contamination, denaturation and proteolysis. Ammonium sulphate takes up the water molecules around the protein exposing hydrophobic sites on the protein. Because hydrophobic groups tend to prefer to be together, the proteins will aggregate and thus come out of the solution. Many other salts can be used for this purpose; however, ammonium sulphate is the cheapest.

The aim is to ensure that only the target protein precipitates. This can be achieved by using an appropriate ammonium sulphate concentration. The concentration to be used must be determined by experiments.

The precipitation reaction should be performed at low temperatures to prevent denaturation of the target protein. Ammonium sulphate must be added slowly, and mixing is important to distribute ammonium sulphate and to prevent unwanted proteins in the precipitate. Ammonium sulphate should be prepared as a stock solution at 4°C. The required final concentration is often reported as a volume of saturated ammonium sulphate added.

After precipitation, the pellet is separated by centrifugation and washed in ammonium sulphate solution. The pellet is then dissolved in buffer for further purification or analysis. Ammonium sulphate is preferable for the precipitation of protein because

- most proteins precipitate in saturated solution (4 M)
- low heat of solubilization
- low density of saturated solution (1.25 g/cm^3); proteins can be collected in pellets by centrifugation
- concentrated solutions prevent microbial growth
- protects most proteins from denaturation
- economical

Another way of precipitation is done using polyethylene glycol and protamine sulphate. Polyethylene glycol is a neutral, non-denaturating compound, which binds water with steric exclusion mechanism. Protamine is a basic protein from sperm (many Arg and Lys residues), which precipitates large protein complexes (ribosomes), DNA and RNA.

Dialysis

Dialysis is the movement of molecules by diffusion from high concentration to low concentration through a semi-permeable membrane (Figure 8.16). Only those molecules that are small enough to pass through the membrane pores move through the membrane and reach equilibrium with the entire volume of solution in the system. Once equilibrium is reached, there is no further net movement of the substance because molecules keep moving through the pores in and out of the dialysis unit at the same rate. In contrast, large molecules that cannot pass through the membrane pores will remain on the same side of the membrane. To remove unwanted substances, it is necessary to replace the dialysis buffer so that a new concentration gradient can be established. Once the buffer is changed, movement of particles from high (inside the membrane) to low (outside the membrane) concentration will resume until the equilibrium is once again reached. With each change of dialysis buffer, substances inside the membrane are further purified by a factor equal to the volume difference of the two compartments. For example, if one compartment is dialysing 1 ml of sample against 200 ml of dialysis buffer, the concentration of the dialysable substance at equilibrium will be diluted 200 times less than at the start. Each new exchange against 200 ml of new dialysis buffer will dilute the sample 200 times more. For example, for three exchanges of 200 ml, the sample will be diluted 8,000,000 times (i.e., $200 \times 200 \times 200$), assuming complete equilibrium was reached each time before changing the dialysis buffer.

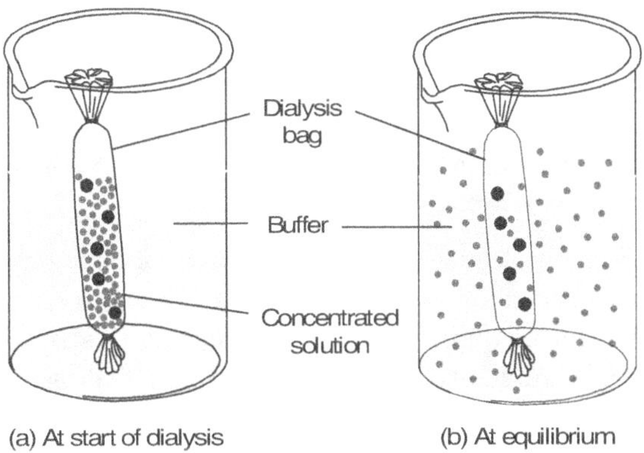

Figure 8.16   Dialysis

Factors affecting dialysis rate   Factors that affect dialysis include (1) dialysis buffer volume, (2) buffer composition, (3) the number of buffer changes, (4) time, (5) temperature and (6) particle size vs. pore size. Substances that are very much smaller than the pore size will reach equilibrium faster than substances that are only slightly smaller than the pores.

The dialysis equipment uses membranes that contain convoluted pores, not the tube-like pores often found in traditional dialysis tubing. Convoluted pores allow small molecules to pass through the twists and turns of the pores with greater ease. The greater the difference in molecular weight of unwanted molecule vs. molecular weight cut-off of the pore size, the greater the rate of dialysis.

Because dialysis involves molecules passing through the semi-permeable membrane in both directions and each substance reaches its own equilibrium independent of other substances, sample dilution can occur. If a substance's concentration is higher on the outside of the membrane and is small enough to pass through the pores, it will have a net movement from the dialysis buffer into the sample. Water is such a small molecule that it is capable of passing through virtually all dialysis membranes. When a high solute concentration is dialysed against a dilute dialysis buffer, there will be a net movement of water (and possibly salts) into the dialysis unit through the membrane. Glycerol and some sugars are especially hygroscopic, and as they diffuse rapidly across the membrane to reach equilibrium, they also significantly affect the osmosis of water across the membrane and cause a change in volume of the sample. Care should be employed when dialysing with large differences in glycerol or sugar concentration between sample and dialysis membrane. We can prevent the movement of water and consequent change in sample volume by dialysing in "stepwise" fashion. For example, when we process a sample with very high solute concentration against a buffer with very low solute concentration, we should first dialyse against a fairly concentrated dialysis buffer. With each subsequent replacement of dialysis buffer, we should use a less concentrated buffer until the desired final buffer concentration is reached.

Finally, some molecules may stick to the dialysis membrane, which is made of regenerated cellulose. This non-specific binding can result in sample loss. The per cent of total protein lost is partially dependent on protein concentration. Protein loss caused by non-specific binding to the membrane is negligible for concentrated samples (>0.5 mg/ml) but may be significant for dilute protein samples (<0.1 mg/ml). Adding a "carrier" protein such as BSA to dilute the protein sample before dialysis will prevent this loss.

Reverse Osmosis

Reverse osmosis (RO) is a pressure-driven membrane separation process that separates dissolved and suspended substances from water. The membrane acts as a selective barrier, removing unwanted substances such as salts.

Adsorption

Adsorption refers to the binding of molecules or particles to a surface. The term must be distinguished from absorption, which means the filling of pores in a solid. The binding to the surface is usually weak and reversible. Just about anything, including the fluid that dissolves or suspends the material of interest, would bind, but compounds that have colour, taste or odour tend to bind strongly. Compounds that contain chromogenic groups (atomic arrangements that vibrate at frequencies in visible spectrum) are strongly adsorbed on activated carbon. Decolorization can be achieved efficiently by adsorption and with negligible loss of other materials.

The most common industrial adsorbents are activated carbon, silica gel and alumina, because they present enormous surface areas per unit weight. Activated carbon is produced by roasting organic material to decompose into granules of carbon; coconut shell, wood and bone are common sources. Silica gel is a matrix of hydrated silicon dioxide. The alumina is derived from mining or derived by precipitating either aluminium oxide or aluminium hydroxide. Although activated carbon is a magnificent material for adsorption, its black colour persists and adds a grey tinge even after treatment; however filter materials with fine pores remove carbon quite well.

Freshly prepared activated carbon has a clean surface. Charcoal differs from activated carbon in that its surface is contaminated by other products; however further heating will drive off these compounds to produce a surface with high adsorptive capacity. A surface already heavily contaminated by adsorbates is not likely to have much capacity for additional binding. Although carbon atoms and linked carbons are most important for adsorption, the mineral structure contributes to shape and mechanical strength. Activated carbon can be regenerated after use by roasting, but the thermal expansion and contraction eventually disintegrate the structure, so some carbon is lost or oxidized. Temperature affects adsorption profoundly, and measurements are usually at a constant temperature. Most operations using adsorbents have little variation in temperature.

Electrophoresis

Electrophoresis is a process used to monitor the complexity of the protein mixture by separating the proteins on acrylamide gels. This thin gel allows exquisite separations, but unfortunately it cannot be scaled up to production levels without major losses in efficiency. Electrophoresis separates on the same basis as ion exchange chromatography. The first proteins to elute from anion exchange gels will be the slowest moving on a normal acrylamide electrophoresis gel. When the gel is increased in thickness, the heat buildup causes disturbances in the flow, and only small sample volumes can be applied to electrophoresis gels. These have so far made electrophoresis impractical for commercial separations. We can use electrophoresis to 1) estimate where a protein will elute on a salt gradient from an anion exchange column; 2) determine the number of proteins in a mixture; and 3) separate proteins on the basis of size by the use of SDS gels. SDS is sodium dodecylsulphate, a detergent. SDS coats a protein and, when it is heated and denatured, the polypeptide becomes totally coated with SDS such that it has no native charge. The proteins move at the same rate, but the gel is made of small pore size that it restricts the flow of these non-linear molecules and they separate based on size and not on charge. SDS electrophoresis is extremely common in biochemistry laboratories and is used to monitor proteins during purification.

Chromatography   Chromatography involves a broad range of physical methods used to separate and analyse complex mixtures. The components to be separated are distributed between two phases: a stationary phase (bed) and a mobile phase, which percolates through the stationary bed. A mixture of components is flushed through the system at different rates. The differential rates at which the mixture moves over adsorptive materials provide separation. Repeated sorption/desorption during the movement of the sample over the stationary bed determine the rates. The smaller the affinity a molecule has for the stationary phase, the shorter the time spent in a column.

In a chemical or bioprocessing industry, the need to separate and purify a product from a complex mixture is a necessary step. Today, there exist a number of separation methods. Chromatography (Figure 8.17) is a special separation process. It can separate complex mixtures with great precision.

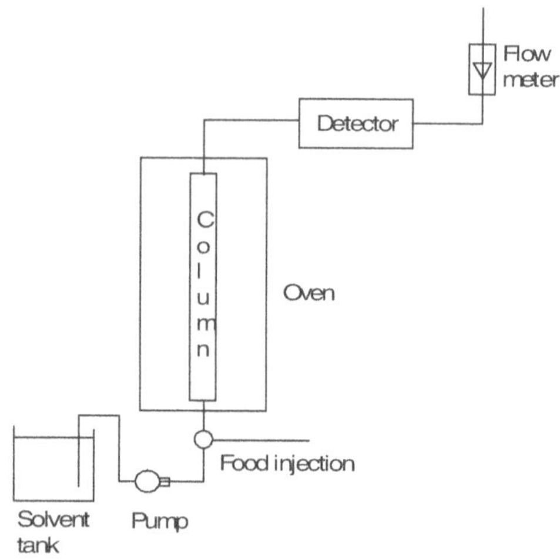

Figure 8.17   Chromatography

Even identical components, such as the proteins that may vary by a single amino acid, can be separated using chromatography. In fact, chromatography can purify any soluble or volatile substance if the right adsorbent material, carrier fluid and operating conditions are employed. Moreover, chromatography can be used to separate delicate products.

Column   Although there are many types of chromatography available, most modern applications of chromatography employ a column. The column is where actual separation takes place. It is usually a glass or metal tube of sufficient strength to withstand the pressures applied across it. The column contains the stationary phase. The mobile phase runs through the column and is adsorbed into the stationary phase. The column may either be a packed bed or open tubular column. A packed bed column comprises a stationary phase, which is in granular form and packed into the column as a homogeneous bed. The stationary phase completely fills the column. An open tubular column has a stationary phase made of a thin film or layer on the column wall. There is a passage through the centre of the column.

Mobile and stationary phases   The mobile phase comprises a solvent into which the sample is injected. The solvent and sample flow through the column together; thus the mobile phase is often referred to as the "carrier fluid". The stationary phase is a material in the column for which the components to be separated have varying affinities. The materials which comprise the mobile and stationary phases vary depending on the type of chromatographic process being performed. The mobile phase in gas chromatography is generally an inert gas. The stationary phase is generally an adsorbent or liquid distributed over the surface of a porous, inert support. The mobile phase in liquid chromatography is a liquid of low viscosity, which flows through the stationary phase bed. This bed may comprise an immiscible liquid coated on a porous support, a thin film of liquid phase bonded to the surface of a sorbent, or a sorbent of controlled pore size.

Basic operation   The process of chromatographic separation takes place within a chromatography column (Figure 8.18). This column, made of glass or metal, is either a packed bed or open tubular column. A packed bed column contains particles that make up the stationary phase. Open tubular columns are lined with a thin film stationary phase. The centre of the column is hollow. The mobile phase is typically a solvent moving through the column, which carries the mixture to be separated. This may be a liquid or a gas, depending on the type of process. The stationary phase is usually a viscous liquid coated on the surface of solid particles which are packed into the column. However, solid particles may also be taken as the stationary phase. In any case, the partitioning of solutes between the stationary phase and mobile phase leads to the desired separation.

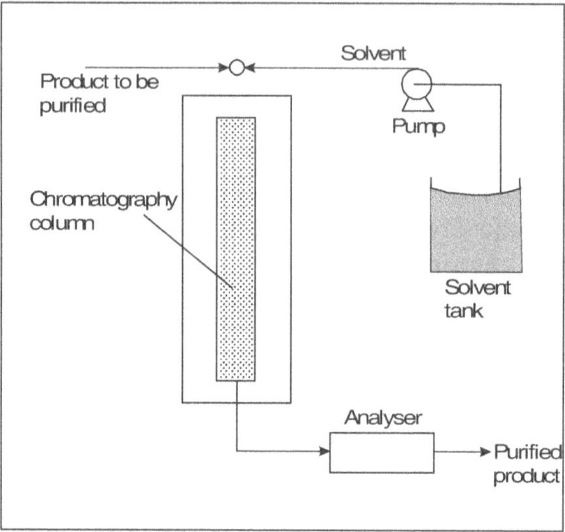

Figure 8.18   Basic operation of column chromatography

1. *Feed injection*   The feed is injected into the mobile phase. The mobile phase flows through the system by the action of a pump. In olden analytical chromatography, capillary action or gravity is employed to move the mobile phase.

2. *Separation in the column*   As the sample flows through the column, its different components will be adsorbed to the stationary phase at varying degrees. Those with strong attraction to the support move more slowly than those with weak attraction. This is how the components are separated.

3. *Elution from the column*   After the sample is flushed or displaced from the stationary phase, the different components will elute from the column at different time. Components with least affinity for the stationary phase (the most weakly adsorbed) will elute first, while those with greatest affinity (the most strongly adsorbed) will elute last.

4. *Detection*   The components are collected as they emerge from the column. A detector analyses the emerging stream with respect to concentration and characteristic of chemical composition by measuring the refractive index or ultra-violet absorbance.

Types of Chromatography

Gel filtration   Gel filtration chromatography is a process of separation based on size. It is also called molecular exclusion or gel permeation chromatography. In gel filtration chromatography, the stationary phase consists of porous beads with a well-defined range of pore sizes. The stationary phase for gel filtration is said to have a fractionation range, which means that molecules within that molecular weight range can be separated. Proteins that are smaller can fit inside all the pores in the beads. The smaller proteins have access to the mobile phase inside the beads as well as between beads, and they elute last in gel filtration separation.

Proteins that are too large to fit inside the pores are excluded. They have access only to the mobile phase between the beads and, therefore, elute first.

Proteins of intermediate size are partially included; they can fit inside some pores but not all in the beads. These proteins will elute between the large (excluded) and small (totally included) proteins.

Consider a mixture of glutamate dehydrogenase (molecular weight (MW 290,000), lactate dehydrogenase (MW 140,000), serum albumin (MW 67,000), ovalbumin (MW 43,000), and cytochrome c (MW 12,400) on a gel filtration column packed with Bio-Gel P-150 (fractionation range 15,000–150,000). When the protein mixture is applied to the column, glutamate dehydrogenase would elute first because it is above the upper fractionation limit. Therefore, it is totally excluded from inside the porous stationary phase and would elute with void volume. Cytochrome c is below the lower fractionation limit and would be completely included, eluting last. Other proteins would be partially included and elute in the order of decreasing molecular weight.

Gas chromatography   Gas chromatography makes use of a pressurized gas cylinder and a carrier gas, such as helium, to carry the solute through the column. The most common detectors used in this type of chromatography are thermal conductivity detectors and flame ionization detectors. There are three types of gas chromatography: gas adsorption, gas–liquid and capillary gas chromatography.

Gas adsorption chromatography involves a packed bed comprising an adsorbent used as the stationary phase. This method is commonly used to separate mixtures of gases. Some common adsorbents are zeolite, silica gel and activated alumina. Gas–liquid chromatography is a more common type of analytical gas chromatography. In this type of column, an inert porous solid is coated with a viscous liquid that acts as the stationary phase. Diatomaceous earth is commonly used. Solutes in the feed stream dissolve into the liquid phase and eventually vaporize. The separation is thus based on relative volatilities.

Capillary gas chromatography is the most common analytical method. Glass or fused silica makes up the capillary walls which are coated with an absorbent or a solvent. Because of the small amount of stationary phase, the column can contain only a limited capacity. However, this method can yield rapid separation of mixtures.

Liquid chromatography   There is a variety of types in liquid chromatography. In liquid adsorption chromatography, an adsorbent is used. This method can be used in large-scale applications since adsorbents are relatively inexpensive. Another method called liquid–liquid chromatography is analogous to gas–liquid chromatography. The modern liquid chromatography includes three main types, namely reverse phase, high performance and size exclusion liquid chromatography, along with supercritical fluid chromatography.

Reverse phase chromatography is a powerful analytical tool and involves a hydrophobic, low polarity stationary phase which is chemically bonded to an inert solid such as silica. Separation is essentially an extraction operation. This method is useful for separating non-volatile components.

High performance liquid chromatography (HPLC) is similar to reverse phase. In this method, the process is conducted at a high velocity and pressure drop. The column is shorter and has a small diameter, but it is equivalent to possessing a large number of equilibrium stages.

Size exclusion chromatography also known as gel permeation or filtration chromatography, does not involve adsorption and is extremely fast. The packing is a porous gel and is capable of separating large molecules from smaller ones. The larger molecules elute first since they cannot penetrate the pores. This method is common in protein separation and purification.

Supercritical fluid chromatography is a relatively new analytical tool. In this method, the carrier is a supercritical fluid, such as carbon dioxide mixed with a modifier. Supercritical fluids have larger solubilities and densities than liquids, and they have larger diffusivities and viscosities. This method has not yet been implemented on a large scale.

Ion exchange chromatography   Ion exchange chromatography is commonly used in the purification of biological materials. There are two types of exchange: cation exchange, in which the stationary phase carries a negative charge, and anion exchange, in which the stationary phase carries a positive charge. Charged molecules in the liquid phase pass through the column until a binding site in the stationary phase appears. The molecule will not elute from the column until a solution of varying pH or ionic strength is passed through it. Separation by this method is highly selective. Since the resins are fairly inexpensive and high capacities can be used, this method of separation is applied early in the overall process.

Affinity chromatography   Affinity chromatography involves the use of packing that has been chemically modified by attaching a compound with a specific affinity for the desired molecules, especially biological compounds. The packing material used, called affinity matrix, must be inert and easily modified. Agarose is the most common substance used. The ligands, or "affinity tails", that are inserted into the matrix can be genetically engineered to possess a specific affinity. The desired molecules adsorb to the ligands on the matrix until a solution of high salt concentration is passed through the column. This causes desorption of the molecules from ligands, and they elute from the column. Fouling of the matrix may occur when a large number of impurities are present; therefore, this type of chromatography is usually implemented late in the process.

Scaling-up of chromatography columns   In many instances, it may be necessary to increase the size of a chromatography column in order to increase the production capacity. This is scaling-up of the column to a suitable size. The simplest way to scale-up a column is maintaining the same column length and increasing the cross-sectional area to maintain a constant ratio of unknown column volume. We may use the following equation to calculate the scaled diameter

$$\frac{\text{load}_2}{\text{load}_1} = \left(\frac{\text{radius}_2}{\text{radius}_1} \right)^2$$

To recreate identical operating conditions on the scaled column as the unscaled one, the linear flow rate must be kept constant. The volumetric flow rate required to keep a constant linear flow rate can be known from the equation

$$\frac{\text{volume flow}_2}{\text{volume flow}_1} = \frac{\text{load}_2}{\text{load}_1}$$

FINISHING THE PRODUCT

Finishing is the final step in downstream processing. After purification, the products are dried and crystallized.

Drying and Evaporation

The removal of solvents from purified wet products is usually carried out by evaporation and drying. The purpose and principle of evaporation and drying are the same. Based on the nature of product, drying or evaporation is employed to remove solvents from the product.

The basic factors that affect the rate of evaporation and drying are:

- rate at which heat can be transferred to the liquid

- quantity of heat required to evaporate each kg of water

- maximum allowable temperature of the liquid

- pressure at which evaporation takes place

- changes to the food materials during the course of evaporation

Evaporators

An evaporator has two principal functions: to exchange heat and to separate vapour that is formed from liquid.

Some important practical considerations in evaporators are as follows:

- maximum allowable temperature , which may be substantially below 100°C

- promotion of circulation of liquid across heat transfer surfaces, to attain reasonably high heat transfer coefficients and to prevent any local overheating

- viscosity of the fluid will often increase substantially as the concentration of the dissolved materials increases

- tendency to foam makes separation of liquid and vapour difficult

A typical evaporator (Figure 8.19) is made up of three functional sections: the heat exchanger, the evaporating section, where the liquid boils and evaporates, and the separator in which the vapour leaves the liquid and passes to the condenser. In many evaporators, all three sections are set up in a single vertical cylinder. At the centre of the cylinder, there is a steam heating section, with pipes passing through it in which the evaporation of liquors takes place. At the top of the cylinder, there are baffles that allow the vapours to escape but check liquid droplets that may accompany the vapours. A conventional evaporator is shown in Figure 8.19.

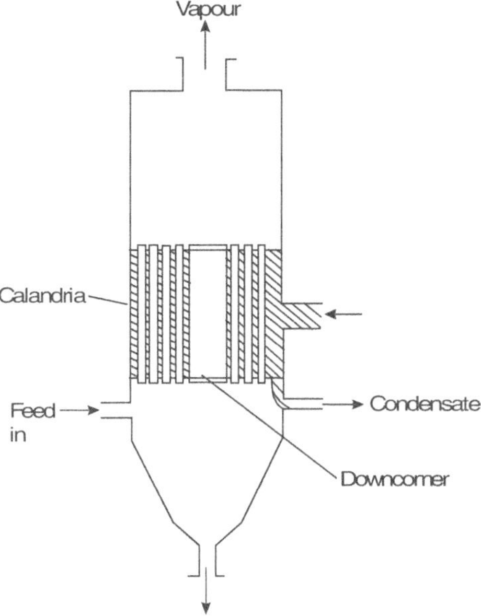

Figure 8.19   Evaporator

In the heat exchanger section, called calandria, steam condenses in the outer jacket and the liquid being evaporated boils inside the tubes and in the space above the upper tube plate. The resistance to heat flow is imposed by the steam, liquid film coefficients and by the material of the tube walls. The circulation of liquid greatly affects evaporation rates, but circulation rates and patterns are very difficult to predict. The values of overall heat transfer coefficients that have been reported for evaporators are of the order 1800–5000 J m^{-2} s^{-1} °C^{-1} for evaporation of distilled water in a vertical-tube evaporator with heat supplied by condensing steam. However, with dissolved solids in increasing quantities as evaporation proceeds leading to increased viscosity and poorer circulation, heat transfer coefficients in practice may be much lower than this.

As evaporation progresses, the remaining liquors become more concentrated and the boiling temperatures rise. The rise in the temperature of boiling reduces the available temperature drop, assuming no change in the heat source. So the total rate of heat transfer will drop correspondingly. Also, with increasing solute concentration, the viscosity of the liquid will increase, often quite substantially, and this affects circulation and heat transfer coefficients, leading to lower rates of boiling again. Another complication is that the overall heat transfer coefficients have been found to vary with the actual temperature drop. Hence the design of an evaporator is inevitably subject to wide margins of uncertainty.

Types of evaporators

Open pans   The basic evaporator consists of an open pan in which the liquid is boiled. Heat may be supplied through steam jacket or coils, and scrapers or paddles may be fitted to provide agitation. Such evaporators are simple and economical, but they are expensive in their running cost.

Horizontal-tube evaporators   The horizontal-tube evaporator is a development of open pan evaporator, in which the pan is fixed in a vertical cylinder. The heating tubes are arranged in a horizontal bundle immersed in the liquid at the bottom of the cylinder. Liquid circulation is poor in this type of evaporator.

Vertical-tube evaporators   Vertical-tube evaporators can give good heat transfer. The standard evaporator is an example of this type. Recirculation of the liquid is through a large "downcomer". The liquid rise through the vertical tubes 5–8 cm in diameter, boil in the space just above the upper tube plate, and recirculate through the downcomers. The hydrostatic head reduces boiling on the lower tubes, which are covered by the circulating liquid. The length to diameter ratio of the tubes is 15 : 1. The basket evaporator is a variant of calandria evaporator in which the steam chest is contained in a basket suspended in the lower part of the evaporator, and recirculation occurs through the annular space round the basket.

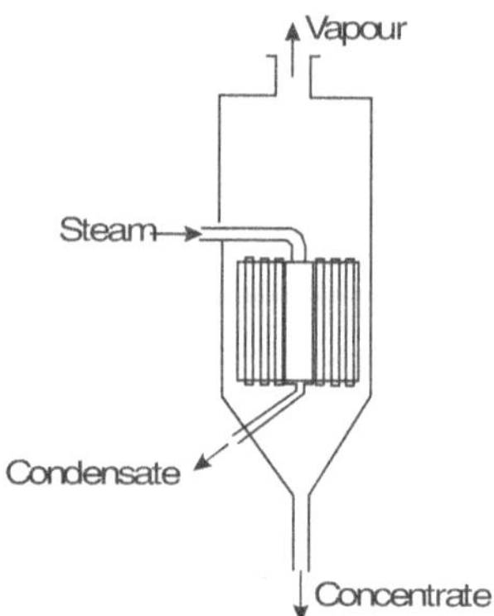

Figure 8.20   Vertical tube evaporator (Basket type)

Plate evaporators   The plate heat exchanger may be adapted for use as an evaporator. The spacing between the plates can be increased and appropriate passages provided so that much larger volume of vapours can be accommodated. Plate evaporators can provide good heat transfer, and they are easy to clean.

Long-tube evaporators Tall, slender, vertical tubes may be used in evaporators (Figure 8.21). The tubes, which have a length to diameter ratio of 100:1, pass vertically upward inside the steam chest. The liquid may either pass down through the tubes, called a falling-film evaporator, or carried up by the evaporating liquor, in which case it is called a climbing-film evaporator. Evaporation occurs on the walls of the tubes. Because circulation rates are high and the surface films are thin, good concentration of heat sensitive liquids can be obtained due to high heat transfer rates and short heating times.

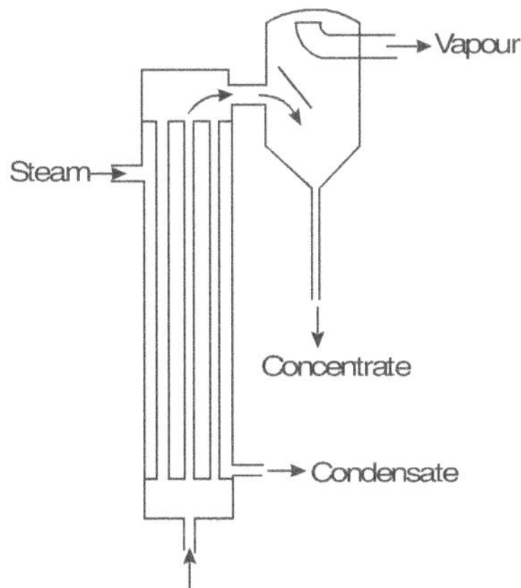

Figure 8.21   Long-tube evaporators

Generally, the liquid is not recirculated, and if sufficient evaporation does not occur in one pass, the liquid is fed to another pass. In the climbing-film evaporator, the liquid boils inside the tube slugs of vapour form and this vapour carries up the remaining liquid that continues to boil. Tube diameters may be 2.5 to 5 cm, and contact time may be as low as 5–10 seconds. The overall heat-transfer coefficients may be up to five times as great from a heated surface immersed in a boiling liquid. In the falling-film type, the tube diameters are rather greater, about 8 cm, and these are especially suitable for viscous liquids.

Forced circulation evaporators Heat transfer coefficients from condensing steam are usually high, so the major resistance to heat flow in an evaporator is in the liquid film. Tubes are generally made of metals with high thermal conductivity, though scale formation may occur on the tubes which may reduce tube conductance.

Liquid film coefficients can be increased by improving the circulation of the liquid and by increasing its velocity of flow across the heating surfaces. Pumps or impellers can be fitted in the liquid circuit to achieve this. Using pump circulation, the heat exchange surface can be divorced from the boiling and separating sections of the evaporator. Alternatively, impeller blades may be inserted into flow passages such as the downcomer of the calandria-type evaporator. Forced circulation is used particularly with viscous liquids; this may also be worth considering in expensive heat exchange surfaces for corrosion or hygiene requirements. In this case, it obtains the greatest possible heat flow through each square metre of heat exchange surface. Figure 8.22 shows the schematic diagram of a forced circulation evaporators.

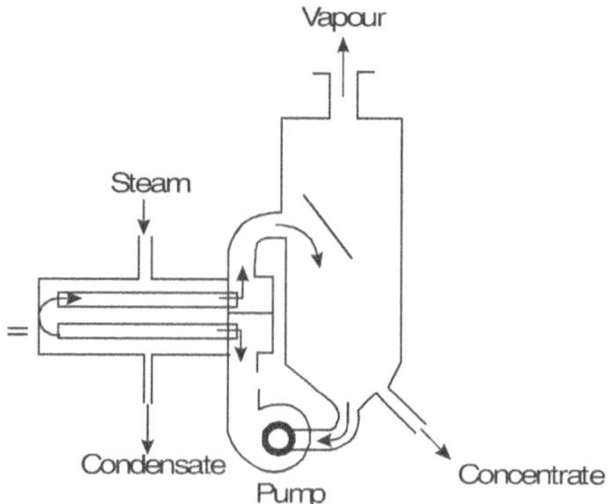

Figure 8.22   Forced-circulation evaporators

The forced-circulation evaporators are of various models. In one type the material to be evaporated passes down over the interior walls of a heated cylinder and it is scraped by rotating scraper blades to maintain a thin film, high heat transfer and a short and controlled residence time exposed to heat.

Dryers

In an industry as diversified and extensive as the bioprocessing industry, a number of different types of dryers would be used. The total range of equipment is much too wide to be described in any introductory book such as this. However, the principles of drying may be applied to any type of dryer.

Tray dryers   In tray dryers (Figure 8.23), the food material is spread out quite thinly on trays, and heat is supplied by air current sweeping across the trays, or by conduction from heated trays or heated shelves or by radiation from heated surfaces. Usually tray dryers are heated by air, which also removes moist vapour.

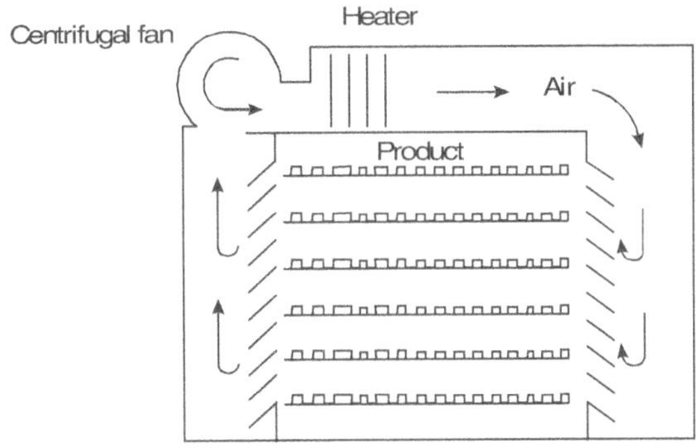

Figure 8.23   Tray dryer

Tunnel dryers   These have developed from tray dryers. In tunnel dryers, the trays move through a tunnel on trolleys where heat is applied and vapours removed. Usually, air current is used in tunnel drying, and the material moves through the dryer either parallel or counter-current to the air flow. Sometimes, the dryers are compartmentalized, and cross-flow may also be used.

Roller or drum dryers   In drum dryers (Figure 8.24), the food material is spread over the surface of a heated drum. The food material remains on the drum surface for a greater part of rotation, during which time drying takes place, and the material is then scraped off. Drum drying is an example of conduction drying.

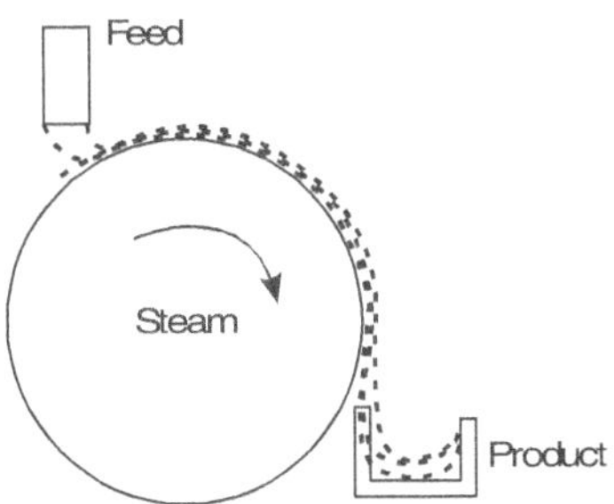

Figure 8.24   Roller dryer

Fluidized bed dryers   In a fluidized bed dryer (Figure 8.25), the food material is suspended against gravity in an upward-flowing air stream. The horizontal air flow helps to convey the food material through the dryer. Heat is transferred from air to the food material, mostly by convection.

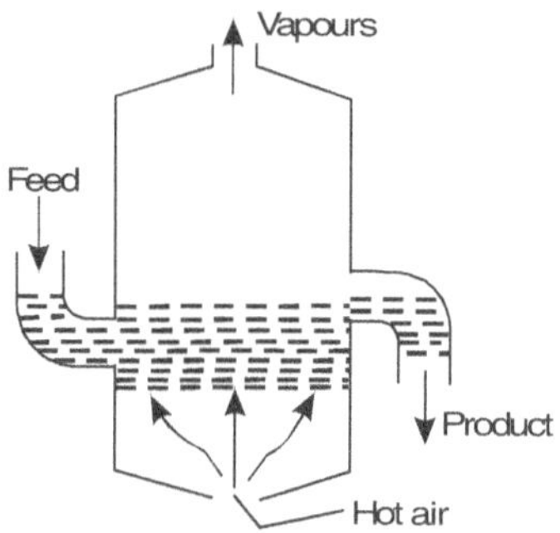

Figure 8.25   Fluidized dryer

Spray dryers   In a spray dryer (Figure 8.26), liquid or fine solid particles are sprayed into a current of heated air. The air current and the materials may move in parallel or in counter-current. Drying occurs very rapidly. Hence this process is very useful for materials that delicate to heat. The dryer body is large so that the particles can settle, as they dry, without touching the walls. Commercial dryers can be very large of dimension 10 m diameter and 20 m height.

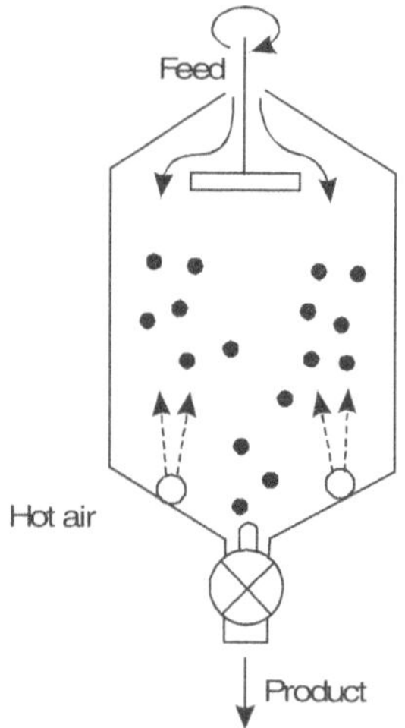

Figure 8.26   Spray dryer

Pneumatic dryers   In a pneumatic dryer (Figure 8.27), the solid food particles are conveyed rapidly in an airstream. The velocity and turbulence of the stream maintain the particles in suspension. Heated air accomplishes drying. Often some form of a classifying device is included in the equipment. In the classifier, the dried material is separated and the moist remainder is recirculated for further drying.

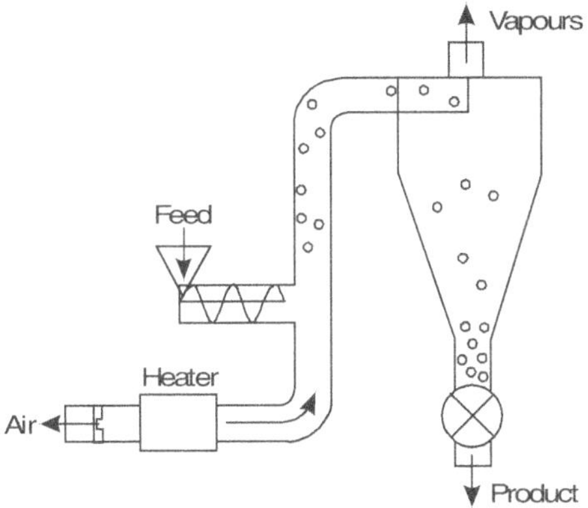

Figure 8.27   Pneumatic dryer

Rotary dryers   In rotary dryers (Figure 8.28), foodstuff is filled in a horizontally inclined cylinder and heated either by air flow through the cylinder or by conduction of heat from the cylinder wall. In some cases, the cylinder rotates, and in others a paddle rotates within the cylinder conveying the material through.

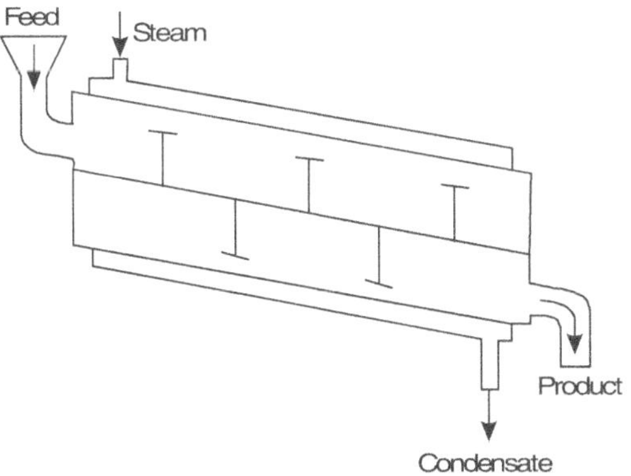

Figure 8.28   Rotary dryer

Trough dryers   The materials are placed in a trough-shaped conveyor belt, made out of mesh, and air is blown through the bed of materials. The movement of the conveyor continuously turns over the material, exposing fresh surfaces to hot air.

Bin dryers   In bin dryers, the food material is placed in a bin with perforated bottom through which warm air is blown upwards, which passes through the material and dries it up.

Belt dryers   The food material is spread as a thin layer over a mesh or solid belt and air is passed through or over the material. In some dryers, the belt would move, and in others, the material is transported by scrapers.

Batch vacuum dryers   Batch vacuum dryers are to a great extent the same as tray dryers, except that they operate under vacuum, and heat transfer is largely by conduction or radiation. The trays are enclosed in a large cabinet, which is evacuated. Water vapour produced is generally condensed, so that the vacuum pumps have to deal with only non-condensable gases. Another type consists of an evacuated chamber containing a roller dryer.

Lyophilization

The word lyophilized was derived from the Greek "made solvent-loving". Freeze-drying, technically known as lyophilization, is a process of sublimation, in which water molecules in a solid phase specimen are directly converted to free water molecules in vapour phase. The free water molecules are then trapped and removed. Porous dried specimens can be easily rehydrated. The purpose of freeze-drying is to preserve a specimen, be it food or microbial organisms, and, in some circumstances, to decrease the size of the product. Depending on the nature of specimen, the temperature and drying time used to freeze and sublime differs from one specimen to another. Since lyophilization is the most complex and expensive form of drying, its use is restricted to delicate, heat-sensitive materials.

Substances that are not usually damaged by freezing can be lyophilized (exceptions are mammalian cells, nearly all of which can be destroyed when lyophilized). Many microorganisms and proteins survive lyophilization. Hence, lyophilization is a favoured method for drying vaccines, pharmaceuticals, blood fractions and diagnostics. Some food products are also lyophilized. They rehydrate quickly because of the porous structure left after the ice has sublimed.

Often, materials are lyophilized to achieve a porous, friable structure rather than for preservation. Lyophilizers are sometimes used for achieving concentration of delicate materials. The form of the product and the type of container to be used for freeze-drying influence the type of lyophilizer needed and how it should be operated.

Principle of lyophilization   First the material is frozen and transferred to a drying chamber. During the drying stage, the material is subjected to high vacuum. Heat is applied gradually to the material, and a condenser is used to collect the water. When water is leaving rapidly, heat is taken out of the material, which helps to keep the material cool and safe. As the material dries, this cooling diminishes so that it is possible to overheat and damage the material.

Heat supplies the energy necessary for sublimation of water. An ice crystal is composed of pure water rigidly confined in a crystal lattice. The molecules have natural vibrations so that extra thermal energy increases the probability of breaking freely. When the water molecule breaks freely, it diffuses through the already dried surface of the solid and sublimes. As water molecules diffuse and sublime, the thickness of the dry outer surface of the specimen increases, and thus more energy is required to transport the molecules through the dry shell. The actual force driving the water vapour from drying boundary to the specimen surface through dry shell is based on the concentration gradient of the sample and not due to the vacuum.

As the molecules sublime and use up latent thermal energy, the thermal reserves in the specimen are depleted and thus the probability of further sublimation decreases. The rate of transfer through dried solids is low. The rate of drying of the specimen decreases until such time that so much external thermal energy would have to be supplied that the specimen may be harmed. Sublimation may continue safely if no heat is supplied, but drying time is greatly extended.

The removal of water vapour that reached the specimen surface is critical to the drying process. Water molecules that have successfully sublimed must be removed from the free space of the vacuum. The molecules move through vacuum-induced free space and are trapped by condensation. Most condensers are usually plates, but a device known as "cold finger" is common. Cold finger is a long, thin condenser.

Wet samples are frozen by placing them in vacuum. The more energetic molecules escape, and the temperature of the sample falls by evaporative cooling, and eventually it freezes. About 15% of water in the wet material is lost.

A simplest form of lyophilizer (Figure 8.29) would consist of a vacuum chamber, into which wet sample could be placed, together with an arrangement to remove water vapour so as to freeze the sample by evaporative cooling and then maintain the water-vapour pressure below triple-point pressure. The temperature of the sample would continue to fall below the freezing point and sublimation would slow down until the rate of heat gain in the sample by conduction, convection and radiation would equal the rate of heat loss as the more energetic molecules sublimed away.

However, this simple approach creates numerous difficulties. When a material is frozen by vacuum cooling, it froths like boiling. Frothing can be suppressed by low-speed centrifugation. Centrifugation helps the material to dry faster by reducing the thickness and exposing a greater surface area.

An alternative would be to freeze the material before it is placed under vacuum. This is commonly done with laboratory lyophilizers where material is frozen inside a flask. The flask is attached to a manifold connected to the ice condenser. To speed up the process, the material may be shell-frozen by rotating the flask in a low-temperature bath, giving a large surface area and small thickness of material.

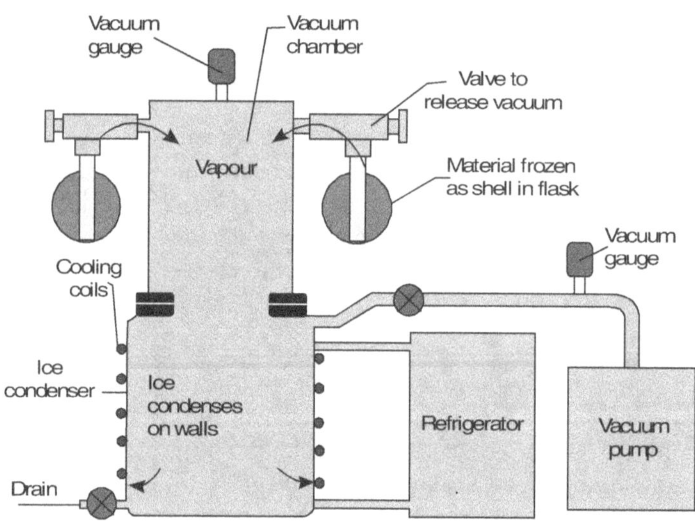

Figure 8.29   Principle of a lyophilizer

For large-scale equipment, the material is usually placed on product-support shelves inside the drying chamber, which will be cooled so that the material is frozen at atmospheric pressure before vacuum is created. Without a controlled heat input to the sample, its temperature would fall until drying was virtually at standstill. It is for this reason that heat supply is arranged at product-support shelves so that, after their initial use for freezing the product, they can be used to replace the energy lost with subliming water vapour and maintain the product at a constant low temperature.

One milliliter of ice produces more than 1,000,000 mL of water vapour at typical lyophilization cycle pressures. Even the more energy-efficient vacuum pumps cannot handle large quantities of water vapour. Hence, a refrigerated trap (called ice condenser) is fixed between lyophilization chamber and vacuum pump. Modern lyophilizers incorporate refinements to include the following:

- Separated drying chamber and ice condenser to reduce cross-contamination

- Provision of an isolation valve between the chamber and ice condenser to allow for end point determination and simultaneous loading and defrosting

- Construction of a chamber and an ice condenser as pressure valves to allow for steam sterilization at 121°C or higher.

- Cooling and heating of product-support shelves by a circulating intermediate heat-exchange fluid to give even and accurate temperature

- Instruments to control, monitor and record process variables

- Movable product-support shelves to close the slotted bungs used in vials and to facilitate cleaning and loading

- Automatic control system with safety interlocks and alarms, duplicated vacuum pumps, refrigeration systems, and other moving parts to enable drying to proceed without endangering the product in the event of mechanical breakdown

REVIEW QUESTIONS

1. What is downstream process in bioprocessing?

2. Give a detailed account on cell separation techniques.

3. Write about centrifugation and discuss their types.

4. What is the principle behind filtration? Give their types.

5. What is cell disruption? Give the different methods employed for cell disruption.

6. What is ultrasonic disruption? Explain the principle and applications.

7. Give a detailed account on separation of product.

8. What is the principle behind liquid–liquid extraction?

9. What is chromatography? Explain in detail the types of chromatography.

10. Explain about precipitation and dialysis.

12. What is reverse osmosis?

13. What is adsorption?

14. What is electrophoresis?

15. Give a detailed account on dryers and evaporators.

16. What is lyophilization? Explain the principle.

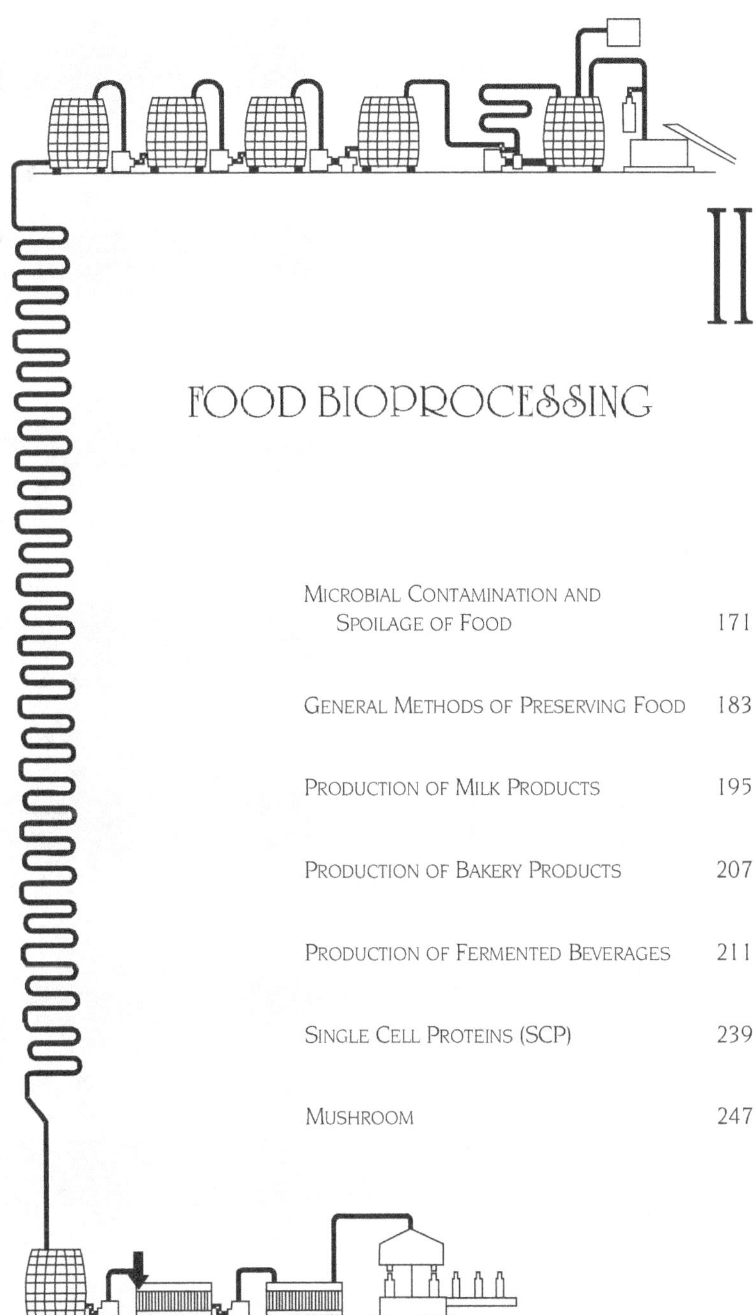

II

FOOD BIOPROCESSING

MICROBIAL CONTAMINATION AND SPOILAGE OF FOOD

Food is a substance that yields energy and growth to the living beings. The presence of microorganisms in food may lead to spoilage and contamination of food. The source of contamination may be indigenous or other sources from environment such as water and land. Further, handling of food materials by human beings adds to contamination. Generally, sanitation is employed rather than sterilization to avoid or reduce contamination. Complete sterilization of food may be quite difficult because it may change the nature of the food.

9

CONTAMINATION AND SPOILAGE OF DIFFERENT KINDS OF FOOD

Cereals and Cereal Products

Grains may be contaminated by the soil or from insects and other sources. Freshly harvested grains will have a high load of contaminants belonging to the families of Pseudomonadaceae, Micrococcaceae and Bacillaceae. Washing, milling and bleaching can reduce the load of microbes considerably. However, the wheat flour contains bacterial populations including the spores of *Bacillus* and coliforms. The mould spores of *Aspergillus*, *Alternaria* and *Penicillium* can be identified in wheat flour. The corn flour generally contains *Fusarium* and *Penicillium* dominantly.

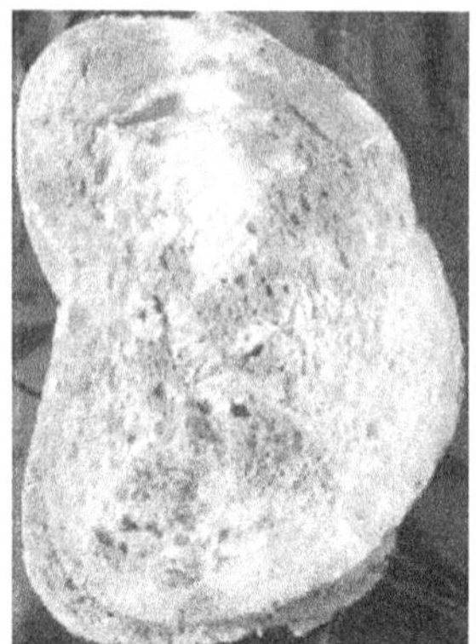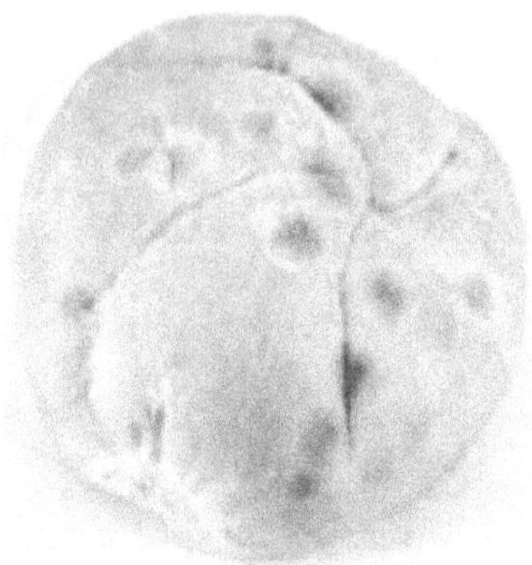

Figure 9.1   Mould growth in bread

Cereal grains and flours do not usually suffer microbial spoilage when stored properly. However, the presence of moisture facilitates the growth of contaminants. In flours, the spoilage may be carried out sequentially or non-sequentially based on the contaminants. Bread made from cereal flours can be greatly contaminated by several moulds (Figure 9.1) and bacteria. The ropiness of bread is caused by various mucoid *Bacillus* species, which hydrolyse the flour proteins and starch by producing extracellular hydrolysing enzymes that form sticky, yellow layer together with bad odour. Another kind of spoilage is Red bread, caused by *Serratia marcescens*. The growth of this pigmented organism over the surface of bread shows the appearance of bloody red. Moulds, e.g. *Monilia sitophila*, impart the pink or red colour over the surface of bread. Chalky bread is another spoilage which results from white chalk like spots on the bread. It is usually caused by the yeast like fungi *Endomycopsis* and *Trichosporon*.

Sugar and Sugar Products

Sugar and sugar products are the main energy sources for the growth of microorganisms. However, sugar or concentrated sugar products restrict the growth of most organisms through osmotic imbalance caused due to their hypertonic state. The osmophilic bacteria like *Bacillus* sp. and *Leuconostoc* are common contaminants in sugar products. The raw sugar cane juices contain accessory nutrients for microorganisms. Therefore, it can be readily deteriorated by numerous organisms. The contaminants cause clarification of gum and the formation of slime in the juices. Microbial spoilage of molasses is not common, although it is difficult to sterilize by heat because of the protective effect of sugar. Canned molasses or syrups may be spoiled by some osmophilic yeasts.

Vegetables and Fruits

Vegetables and fruits undergo spoilage easily due to their moisture and nutrient content. The diseases in vegetables and fruits may be either due to the growth of organisms that obtain food and eventually damage their host or from adverse environmental conditions that cause abnormalities in the functions and structures of vegetables and fruits. Deterioration of raw vegetables and fruits may result from both intrinsic and extrinsic factors. The diseases caused by pathogens and decomposition of nutrients facilitate the growth of saprophytic organisms, which cause further spoilage. Contamination in vegetables usually occurs in a series of events such as harvesting, grading, packing, transportation, storage and handling by the seller.

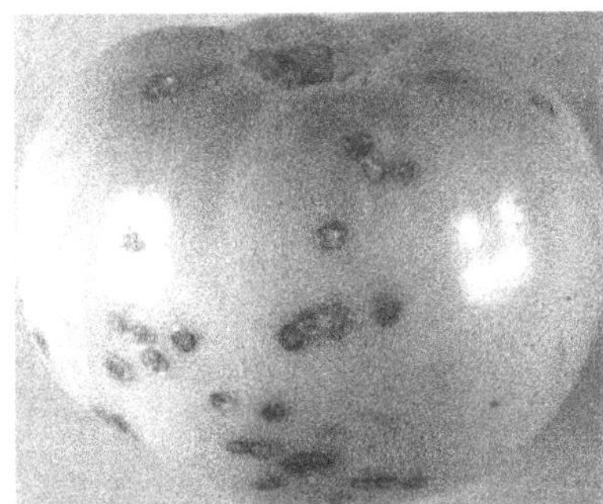

Figure 9.2   Bacterial spot on tomato

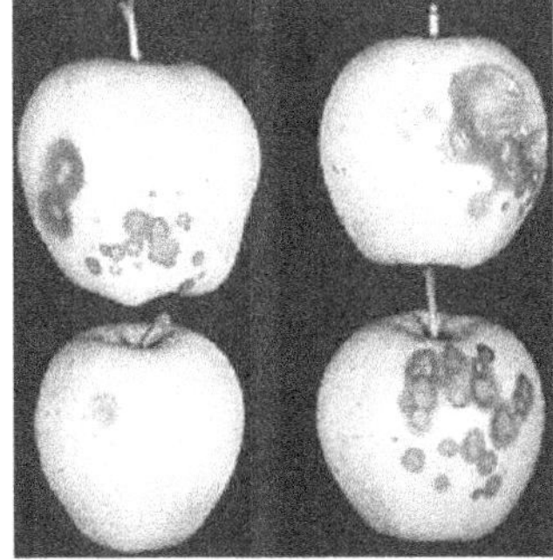

Figure 9.3   Mould rot on apple

Pathogens can spoil different parts of vegetables and fruits. Saprophytic organisms are secondary invaders of fruits and vegetable (Figure 9.2 and 9.3). Rot is a common cause of spoilage in vegetables and fruits, which is caused by various bacteria and fungi. Bacterial rot, known as soft rot, is usually caused by *Erwinia carotovora*, *Pseudomonas marginalis* and *Clostridium*. Fungal rot, known by many names as grey mould rot, anthracnose, blue mould rot, black mould rot, pink mould rot and mildew, is caused by *Botrytis*, *Colletotrichum*, *Penicillium*, *Aspergillus* and *Trichothecium*. Bacterial rot often appears as spots of different colours whereas fungal rots appear water-soaked mush areas.

Meat and Meat Products

Healthy flesh has been reported to contain few or no microorganisms. *Staphylococcus*, *Streptococcus*, *Clostridium* and *Salmonella* have been reported as main contaminants of meat. Usually contamination in meat occurs from a series of events like skinning and slaughtering. Other than bacteria, moulds of many genera may reach the surface of meat and grow there. *Cladosporium, Sporotrichum, Geotrichum, Mucor, Penicillium, Alternaria* and *Monilia* are surface contaminants of a variety of meat.

Meat spoilage shows in two different conditions, namely aerobic and anaerobic conditions. In aerobic conditions, spoilage may be detected by the formation of surface slime and change in the meat colour. Surface slime and colour change is generally caused by *Pseudomonas, Acinetobacter, Moraxella, Alcaligenes, Bacillus* and *Micrococcus.* Often, the oxidation of unsaturated fatty acids together with lipolysis causes several changes to meat such as stickiness, spots, patches and bad taste. In anaerobic conditions, facultative and anaerobic bacteria grow within the meat and spoil it. Anaerobic bacteria like *Clostridium* and facultative organisms like *Pseudomonas, Proteus* and *Alcaligenes* are common contaminants of meat. These organisms decompose the protein in anaerobic conditions and cause putrefaction, taints and souring of meat.

Milk and Milk Products

Usually milk contains few bacteria when released from the udder of the cattle, but they do not grow under usual conditions. The main sources of contamination are the utensils used to collect milk, hands of dairy workers, air, etc. Micrococci, streptococci and staphylococci have been reported as primary contaminants in fresh milk. Coliforms, *Lactobacillus*, brevibacteria, *Bacillus* and enterococci have been reported as secondary contaminants of milk.

The contaminants cause several changes to milk like gas formation, proteolysis and ropiness. Changes in milk fat, colour and odour are other important events. The gas production is due to the formation of acid from milk sugar. During fermentation of milk, organisms like *Clostridium* and some gas-forming aerobic organisms like *Bacillus* liberate gas as the by-product of the acid. Proteolysis is due to *Pseudomonas, Alcaligenes, Proteus*, etc. which produce acid or alkali based on the nature of proteolysis. Bacterial ropiness is caused by slimy capsular material from the cells, usually gums or mucins. In milk and milk creams, ropiness is mainly caused by *Alcaligenes.* Ropiness is usually less when the acidity of the milk is increased.

The decomposition of milk fat is caused by bacteria, moulds and yeast. Unsaturated fatty acids are oxidized into acids, ketones and aldehydes, and buffered fats are hydrolysed into fatty acids and glycerol by lipase activity. *Pseudomonas, Proteus, Alcaligenes, Bacillus, Clostridium*, etc. hydrolyse fat by producing lipase. Flavour and colour changes are always coupled with spoilage of milk. This is mainly due to the production of microbial metabolites and growth of pigmented organisms. The sour flavour is due to production of acid, and the bitter flavour is due to production of alkali by Streptococci, *Leuconostoc* and *Clostridium*. The change in colour of the milk to blue, yellow, red and brown are caused by different species of *Pseudomonas*.

BACTERIAL FOOD-BORNE ILLNESS

Food-borne infection is usually caused by bacteria. If the food we eat contains numerous bacteria, they may continue to grow in the intestines and cause illness. *Salmonella, Campylobacter,* haemorrhagic *E. coli* and *Listeria* cause infections.

Food intoxication results from toxins (or poisons) produced by bacterial growth. It is the toxin, not bacteria, that causes illness. Toxins may not alter the appearance, odour or flavour of food. The bacteria commonly involved are *Staphylococcus aureus* and *Clostridium botulinum*. In the case of *Clostridium perfringens*, illness may be caused by toxins released during sporulation in the gut. Some of the facts relating to food-borne illness are listed below.

- ➲ Food-borne illness is usually the result of mishandling food.

- ➲ Bacterial food-borne illnesses include food infection and food intoxication.

- ➲ *Salmonella, Campylobacter* and *Listeria* cause food infection.

- ➲ *Staphylococcus* and *Clostridium botulinum* produce a toxin (or poison), which causes food intoxication.

- ➲ *Clostridium perfringens* can multiply in food to sufficient numbers resulting in food poisoning.

- ➲ Sanitation, and proper heating and refrigeration practices will help prevent food-borne illness.

Salmonellosis

Salmonellosis is a form of food infection that may result when food containing *Salmonella* is consumed. Once ingested into the body, the bacteria may continue to grow in the intestine and develop an infection. The severity of the illness depends on the resistance of the host and type of organism causing the illness.

The bacteria may spread through direct or indirect contact with the intestinal contents or excrement of animals and humans. They may also be spread to raw meat when handled by an infected person. It is important to make sure that hands and working surfaces are thoroughly washed after contact with raw meat, fish and poultry. *Salmonella* survive at temperatures between (4.4° and 60°C). They can be readily destroyed by cooking to (73.9°C). However, they can survive refrigeration and freezing.

Symptoms of salmonellosis include headache, diarrhoea, abdominal pain, nausea, chills, fever and vomiting. These usually shown within 12 to 36 hours after eating contaminated food and may last for two to seven days. Infants, elderly persons and persons already ill have least resistance to disease effects. Food commonly involved are eggs or egg-based food, salads (of tuna, chicken or potato), poultry, pork, processed meat, meat pies, fish, cream desserts and fillings, sandwich fillings, and milk products.

Campylobacteriosis

Campylobacteriosis or *campylobacter enteritis* is caused by ingesting food or water contaminated by *Campylobacter jejuni*. A pathogen principally of veterinary significance until recent years, these bacteria are now thought to be responsible for 2.5 times more food poisoning outbreaks per year than *Salmonella*. *C. jejuni* is commonly found in the intestinal tracts of healthy animals (especially chickens) and in untreated surface water. Raw and inadequately cooked food of animal origin and non-chlorinated water are the most common sources of human infection. The organism grows best in reduced oxygen environment, and it can easily be killed by heating (48°C). It can be inhibited by acid, salt and drying and will not multiply at temperature below (30°C).

Campylobacter infections affect both small and large intestines and produce diarrhoeal illness. Diarrhoea, nausea, abdominal cramps, muscle pain, headache and fever are common symptoms. The onset usually occurs two to five days after eating contaminated food. The duration of illness is usually two to seven days but may go for several weeks with complications such as meningitis, urinary tract infections and reactive arthritis. Deaths, although rare, have been reported. Preventive measures include pasteurizing milk, controlling post-pasteurization contamination, consuming cooked meat and fish, and avoiding cross-contamination between raw and cooked or ready-to-eat food.

Listeriosis

Prior to 1980s, listeriosis, a disease caused by *Listeria monocytogenes*, was primarily associated with abortions and encephalitis in sheep and cattle. As a result of its wide distribution in the environment and its ability to survive under adverse conditions and refrigeration temperatures, *Listeria* has been recognized as an important food-borne pathogen. *L. monocytogenes* is widely distributed in nature and frequently carried by humans and animals. The organism grows in the pH range of 5.0 to 9.5 and at temperature between (1° and 45°C). It is salt-tolerant and relatively resistant to drying, but can easily be destroyed by heating.

Listeriosis primarily affects newborn infants, pregnant women, the elderly and those with compromised immune system. In a healthy person, listeriosis may occur as a mild illness with symptoms such as fever, headache, nausea and vomiting. In pregnant women, the infection affects the foetus, resulting in abortion. If born alive, the infant may develop meningitis. The mortality rate is 20–35 per cent. The incubation period may range a few days up to three weeks. The most recent cases are like raw milk, cheese made from raw milk, and post-pasteurization contamination of milk. Preventive measures include maintaining good sanitation, pasteurizing milk, avoiding post-pasteurization contamination, and cooking foods thoroughly.

Staphylococcal Intoxication

Staphylococci are commonly found on skin and in nose and throat of most people; persons with cold and sinus infections are special carriers. Infected wounds, pimples, boils and acne are generally rich sources. Staphylococci are widespread in untreated water, raw milk and sewage. When staphylococci get into warm food and multiply, they produce a toxin that causes illness. This toxin may not be detectable by taste or smell. While the bacteria can be killed at a temperature of 48°C, its toxin is heat resistant; therefore, it is important to prevent the organism from growing in food. This can be achieved by keeping the food either hot (above 60°C) or cold (below 4.4°C), refrigerating or freezing the leftover food, and serving uncovered food as quickly as possible.

Symptoms of staphylococcal intoxication include abdominal cramps, vomiting and severe diarrhoea. These usually appear within one to eight hours after eating the infected food, and the illness lasts for one or two days. The illness is seldom fatal. Food commonly involved are ham, processed meat, tuna, chicken, sandwich fillings, cream fillings, potato and meat salads, custards, milk products and creamed potatoes. Foods that are handled frequently during preparation are prime targets for staphylococci contamination.

Botulism

Clostridium botulinum causes botulism, the deadliest kind of food poisoning. The organism lives on dead and decaying organic matter and is found in soil, fruits and vegetables. The greatest danger of botulism is in home-canned foods, such as meat and low-acid vegetables. *Clostridium botulinum*, spore-forming bacteria, can grow in places without oxygen such as canned foods and produces toxin that can cause death. The toxin causes nausea, vomiting, general weakness, constipation and headache. It attacks the nervous system, progressively causing double vision, impaired speech, muscle paralysis and difficulty in breathing. If not treated, death may result in three to seven days.

The treatment includes administration of botulinal antitoxin and appropriate supportive care, particularly respiratory assistance. Recovery may take several weeks or months. The death rate was quite high in the past. Today, however, with improvements in detection and treatment procedures, mortality rate has come down to 10 per cent. Foods commonly involved are canned or air-tight packaged, low-acid foods (e.g. processed or smoked meat and fish), raw meat, beans, peas, corn, beets and olives. Low-acid foods can be packaged in a pressure canner at 115.6°C to destroy botulism spores. Home-canned vegetables and meat should be boiled 10 minutes plus 1 minute per 1,000 feet elevation to destroy any toxin that may have developed if the spores were not adequately destroyed during processing.

Clostridium perfringens   *Clostridium perfringens* belong to the same genus as botulinum organism. However, the disease produced by perfringens is not as severe as botulism and very few deaths have occurred. Spores of *Clostridium perfringens* may be found in soil, non-potable water, unprocessed food and the intestinal tract of animals and humans. Meat and poultry are frequently contaminated with these spores.

Spores of some strains of *Clostridium perfringens* are heat-resistant and they survive boiling for four or more hours. Furthermore, cooking drives off oxygen and kills competitive organisms, which promote their germination. Once the spores have germinated, a warm, moist, protein-rich environment with little or no oxygen is sufficient for the growth of vegetative cells.

Holding meat at warm room temperature for several hours or cooling large pots of gravy of meat too slowly in the refrigerator encourages sufficient numbers of vegetative cells to produce and cause illness when ingested. Symptoms of poisoning occur within 8 to 24 hours after ingesting the contaminated food. They include acute abdominal pain and diarrhoea. Nausea, vomiting and fever are less common. Recovery is usually within a day or two. Foods commonly involved are cooked, cooled or re-heated meat, poultry, stews, meat pies, casseroles and gravies. Holding foods at warm (43°C) temperatures and cooling foods too slowly are primary causes of perfringens contamination.

E. coli   *Escherichia coli* belongs to a family of microorganisms called coliforms. Many strains of *E. coli* live in the gut, helping the growth of more harmful microorganisms. *E. coli* 0157:H7 has recently been identified as the cause of a distinctive form of gastroenteritis characterized by severe abdominal cramps and bloody diarrhoea. Symptoms develop in three to five days after eating contaminated food and may last 10 days or more. They may include fever, nausea or vomiting. The disease may sometimes be severe and requires hospitalization.

Further, serious complications are seen in children, the elderly and in people with compromised immune system. In children under 10 and the elderly, blood vessels in the kidney may be damaged, leading to kidney failure. Blood clots in brain may also result, causing strokes and seizures. The sources of infection include raw and undercooked foods of animal origin, especially ground beef. Infected workers in food processing are likely to spread the bacteria. Prevention is a matter of thorough cooking and avoiding cross-contamination of cooked meat with raw meat.

PREVENTION OF FOOD-BORNE ILLNESS

Food-borne illness can be prevented. The following practices have been identified by the Food Safety Inspection Service of USDA as essential to prevent bacterial food-borne illnesses.

Purchase and storage

- Keep packages of raw meat and poultry separate from other foods, particularly foods to be eaten without further cooking. Use proper packing to prevent raw juices from dripping on other foods or refrigerator surfaces.

- Accept products labelled "keep refrigerated" only from a refrigerated case. Refrigerate promptly.

- Ensure that the product you buy has not expired and packaging is sound.

⮑ Buy unpackaged meat or poultry products only if it is not in contact with other foods, especially raw foods.

Preparation

⮑ After handling raw meat or poultry, wash the hands with soap water .

⮑ Thaw the refrigerated items using cold water and change the water every 30 minutes till the materials get complete thawing.

⮑ Wash the utensils with hot soap water before keeping the cooked materials and use different utensils to keep raw food and cooked food materials.

⮑ Stuff raw products just before cooking, never the night before.

⮑ Never taste meat, egg, fish or shellfish raw or during cooking.

⮑ Use pasteurized milk and milk products.

⮑ Avoid interrupted cooking. Never cook products partially and refrigerate. Never put food in the oven with a timer set to begin cooking later in the day.

⮑ Boil all home-canned vegetables and meat 10 minutes plus 1 minute per 1,000 feet.

Serving

⮑ Wash hands with soap and water before serving or eating food. Serve cooked food on clean utensils and with clean hands.

⮑ Keep foods to be served hot above 60°C and foods to be served cold below 4°C.

⮑ At 32°C or warmer temperature, leave cooked food out no longer than one hour before re-heating, refrigerating or freezing. At temperatures below 32°C, leave out no more than two hours.

Handling leftovers

⮑ Wash hands with soap and water before handling leftovers and use clean utensils and surfaces.

⮑ Remove stuffing before cooling or freezing.

⮑ Refrigerate or freeze cooked leftovers in small, covered containers within two hours after cooking. Leave space around containers to help assure rapid and even cooling.

⮕ Avoid tasting leftovers.

⮕ Cover and re-heat leftovers to appropriate temperature before serving.

⮕ If in doubt, throw it out. Discard outdated, unsafe leftovers that cannot be consumed.

REVIEW QUESTIONS

1. Write a detailed note on contamination and spoilage of different food products.

2. What is food-borne illness? Explain in detail.

GENERAL METHODS OF PRESERVING FOOD

10

Food preservation is a process of treating and handling food in such a way to stop or greatly slow down microbial growth while maintaining the nutritional value, texture and flavour of food. Unless food is preserved in some manner, it soon begins to spoil.

Physical methods of preservation such as canning and freezing kill the microorganisms present or at least prevent their growth for long enough to allow the food to be safely consumed. Other physical methods include drying, gamma irradiation, ultraviolet or high intensity white light, ultra high pressure and filtration.

Chemical preservatives work either by directly attacking the microbes or reducing the pH to a level of acidity that prevents the growth of microorganisms. Acetic acid, better known as vinegar, has been used as a food preservative since ancient times. Salted, pickled or dried foods were the only nourishment that were offered on long sea voyages to sailors.

PHYSICAL METHODS OF PRESERVATION

Freezing

Freezing is the simplest and least time-consuming way to preserve food. Freezing does not sterilize food, but the extreme cold retards growth of microorganisms and slows down changes that affect quality or cause spoilage in food. Freezing cannot improve the flavour or texture of food, but it can preserve the quality of fresh product.

Foods must be properly packed and kept in the freezer. Packaging protects their flavour, colour, moisture content and nutritive value from dry climate in the freezer. The packaging materials must

- ⮑ be moisture- and vapour-resistant
- ⮑ be durable and leak-proof
- ⮑ be unbreakable at low temperature
- ⮑ be resistant to oil, grease or water
- ⮑ protect foods from loss of flavours or odours

Two types of packaging materials are used for freezing, such as rigid containers and flexible bags or wrappings.

Rigid containers   Rigid containers made of plastic or glass are suitable for liquid food. They are often reusable and make the stacking of foods in the freezer easier. Cardboard cartons are not sufficiently moisture- and vapour-resistant for long-term freezer storage, unless they are lined with a freezer bag or wrap. Covers for rigid containers should fit tightly or the seal should be reinforced with freezer tape, which is especially designed to stick at freezing temperature.

Glass jars break easily at freezer temperatures. However, wide-mouth dual-purpose jars used for freezing and canning have been tempered to withstand extremes temperatures. The wide mouth allows easy removal of partially thawed foods. If standard canning jars with narrow mouths are used for freezing, expansion of the liquid could cause the jars to break at the neck. Hence, it would be safe to leave extra headspace to allow for expansion of foods during freezing. Some foods will need to be thawed completely before removal from the jar.

Flexible bags or wrappings   Bags and sheets of moisture- and vapour-resistant materials, heavy-duty aluminum foil, and laminated papers are suitable for dry-packing vegetables and fruits, meat, fish and poultry. Bags can also be used for liquid foods. Laminated papers are sometimes used as protective over-wraps. Cardboard cartons may be used to protect bags and sheets against tearing and to make stacking easier. Light-weight (household) aluminum foil, wax paper and bread wrappers are not sufficiently moisture- and vapour-resistant to be suitable for long-term freezer storage.

Preservation by Heating

Pasteurization   Pasteurization is a process that uses the application of heat to destroy pathogens in foods. In dairy industry, the terms "pasteurization", "pasteurized" and similar terms shall mean the process of heating every particle of milk or milk product, in properly designed equipment, to one of the temperatures given in Table 10.1 and held for at least the specified time.

Table 10.1   Temperature range of pasteurization

Temperature	Time	Pasteurization type
63°C (145°F)	30 minutes	Vat pasteurization
72°C (161°F)	15 seconds	High temperature short time pasteurization (HTST)
89°C (191°F)	1.0 second	Ultra pasteurization (UP)
90°C (194°F)	0.5 second	Ultra pasteurization (UP)
94°C (201°F)	0.1 second	Ultra pasteurization (UP)
96°C (204°F)	0.05 second	Ultra pasteurization (UP)
100°C (212°F)	0.01 second	Ultra pasteurization (UP)
138°C (280°F)	2.0 seconds	Ultra-high temperature (UHT) sterilization

The original method of pasteurization was vat pasteurization, in which milk or other liquid ingredients are heated in a large tank for at least 30 minutes. It is now used for preparing starter cultures in the processing of cheese, yoghurt, buttermilk and for pasteurizing ice cream mixes.

In United States, the most common method of pasteurization is HTST pasteurization, which uses metal plates and hot water to raise milk temperature to at least 71°C for not less than 15 seconds, followed by rapid cooling. Ultra pasteurization (UP) is similar to HTST pasteurization, but uses slightly different equipment, higher temperatures and longer time. UP offers longer shelf life to the product but still requires refrigeration. UHT sterilization raises the temperature of milk to at least 137°C for two seconds, followed by rapid cooling. UHT-pasteurized milk packaged aseptically is "shelf stable" and does not require refrigeration until opened.

Methods of Pasteurization

There are two basic methods, namely batch and continuous.

Batch method   The batch method uses a vat pasteurizer, which consists of a jacketed vat surrounded by circulating water, steam or heating coils. In the vat, the milk is heated and held throughout the holding period while being agitated. Milk may be cooled within the vat or removed hot after the holding time. As a modification, milk may be partially heated in tubular or plate heater before entering the vat. This method is more useful for milk by-products (e.g. creams, chocolate).

Continuous method   The continuous method has several advantages over the vat method, the most important being time and energy saving. In continuous processing, a high temperature short time (HTST) pasteurizer is used. Heat treatment is accomplished by a plate heat exchanger, which consists of a stack of corrugated stainless steel plates clamped together in a frame. Several flow patterns can be used. Gaskets are used to define the boundaries of the channels and to prevent leakage. The heating medium may be vacuum steam or hot water.

Preservation by Drying

Drying is one of the oldest methods of preserving food. It may be an alternative to canning or freezing. Drying is simple and safe. With modern food dehydrators, fruit leathers, banana chips, pumpkin seeds and beef jerky can be dried year-round at home. Dried foods are ideal for backpacking and camping. They are light-weight and do not require refrigeration.

Drying removes moisture from the food so that bacteria, yeasts and moulds cannot grow and spoil the food. It slows down the action of enzymes but does not inactivate them. By adding water to the food, it is ready to use. Foods may be dried in the sun, in an oven or in a food dehydrator with the right combination of warm temperatures, low humidity and air current.

The optimum temperature for drying food is 60°C. If higher temperatures are used, the food may get cooked. If moisture in food cannot escape while drying, "case hardening" may occur. The food will eventually mould. Thus, drying should never be hurried by raising the temperature.

Low humidity aids the drying process. If the surrounding air is humid, drying will be slowed down.

Air current speeds up drying by moving the surrounding moist air away from the food.

At home, foods can be dried using modern food dehydrators, counter-top convection ovens or conventional ovens. Microwave ovens are recommended only for drying herbs, because there is no way to create enough air flow to dry denser foods.

Types of Drying

Sun drying   Vegetables and meat are not recommended for outdoor drying. Vegetables are low in sugar and acid, which increases the risk for food spoilage. Meats are high in protein, making them ideal for microbial growth. It is best to dry meat and vegetables indoor using controlled conditions of an oven or food dehydrator.

The high sugar and acid content in fruits makes them safe to dry outdoors at favourable conditions. It takes several days to dry foods outdoor, and because weather is uncontrollable, this method can be risky. Hot, breezy days with humidity below 60 per cent are best. A temperature of 29°C or higher is needed. Fruits dried outdoor must be covered at night. At night, cool air condenses and could add moisture back to the food, slowing down the drying process.

Racks or screens placed on blocks allow for better air movement around the food. It is best to place the racks or screens on a concrete ground or over a sheet of aluminum or tin. The reflection of the sun on the metal increases the drying temperature.

The screens may be made of stainless steel, teflon-coated fibreglass or plastic. However, screens made from galvanized metal coated with cadmium or zinc can oxidize and leave harmful residues on food. So also, copper destroys vitamin C and increases oxidation and aluminium tends to discolour and corrode.

Solar drying   Solar drying is an improvement of sun drying. It uses sun as the heat source, but a specially designed dehydrator increases the temperature and air current to speed up drying. Shorter drying time reduces the risk of food spoilage or moulding.

Vine drying   Another method of drying outdoor is vine drying. To dry beans (e.g. soybeans), leave bean pods on the vine in the garden until the beans rattle. When the vines and pods are shrivelled, pick the beans and shell them. No pretreatment is necessary. The beans will mould if not thoroughly dried. If needed, drying can be completed in the oven or a dehydrator.

Food Dehydrators

A food dehydrator is an electrical appliance for drying foods. It has an electric element for heat and a fan and vents for air circulation. Dehydrators are designed to dry foods quickly at 60°C.

There are two basic designs for dehydrators. In horizontal air flow dehydrators, the heating element and fan are located on the side, whereas vertical air flow dehydrators have the heating element and fan located at the base. The major advantages in horizontal flow are: (1) it reduces flavour mixture so that different foods can be dried at one time; (2) all trays receive equal heat penetration; and (3) liquids do not drip down into the heating element.

Oven drying An oven works as a food dehydrator too. By combining the factors of heat, low humidity and air current, an oven can be used as a dehydrator. An oven is ideal for drying meat jerkies, fruit leathers, banana chips or for preserving celery or mushrooms. However, an oven may not be satisfactory for preserving abundant garden produce. Oven drying is usually slower than dehydrators because it does not have a built-in fan. (Some convection ovens do have a fan.) It takes twice as long to dry food in an oven than in a dehydrator, and it uses more energy.

Dehydrofreezing

Dehydrofreezing is a new method of food preservation that combines the techniques of drying and freezing. Fruits dried at home normally have 80 per cent of their moisture removed; vegetables, 90 per cent. Subsequently, the dried fruits or vegetables may be stored in a freezer. The low temperature in the freezer inhibits microbial growth. Moreover, the dried food takes up less room in the freezer. Dehydrofrozen fruits and vegetables retain flavour and colour. They reconstitute in about one-half the time it takes for traditionally dried foods.

Dehydrofreezing is different from freeze-drying. Freeze-drying is a commercial technique that forms a vacuum while the food is freezing. The effect of various control temperatures are shown in Table 10.2.

Table 10.2 Temperature suitable to control bacteria in food

Temperature		Effect
°C	°F	
121.14	250	Canning temperatures for low-acid vegetables, meat and poultry in pressure canner.
115.6	240	Canning temperature for fruits, tomatoes and pickles in water-bath canner.
100	212	Cooking temperatures destroy most bacteria.
73.9	165	Warning temperatures prevent growth but allow survival of some bacteria.
60	140	Some bacterial growth may occur. Many bacteria survive.

(Contd.)

Table 10.2   (Continued)

Temperature		Effect
°C	°F	
51.7	125	**Danger zone** Temperatures in this zone allow rapid growth of bacteria and production of toxins. Do not hold food in this temperature zone for more than 2 hours.
15.6	60	Growth of food-poisoning bacteria may occur.
4.4	40	Cold temperatures permit slower growth of bacteria that cause spoilage.
0	32	Freezing temperatures stop growth of bacteria, but may allow bacteria to survive.

PACKAGING AND STORING DRIED FOODS

Dried foods are susceptible to insect contamination and moisture reabsorption and must be properly packaged and stored immediately. First, cool it completely. Warm food causes sweating, which could provide enough moisture for mould to grow. Then, pack the food in clean, dry, insect-proof containers without crushing.

Glass jars, metal cans or boxes with tightly fitted lids or moisture-vapour resistant freezer cartons are suitable for storing dried foods. Heavy-duty plastic bags are acceptable but are not usually insect and rodent proof.

Pack food in amounts that will be fully used in a recipe. Every time a package is opened, the food is exposed to air and moisture, which may affect the quality of the food.

Fruits that have been sulphured should not touch any metal. Place the fruit in a plastic bag before storing it in a metal can. Sulphur fumes will react with the metal and cause colour change in the fruit.

Dried foods should be stored in cool, dry, dark places. Dried foods can be stored and used for four months up to one year. Heat can affect the quality of dried foods; the higher the temperature, the shorter the storage time. Most dried fruits can be stored for one year at 15°C, six months at 25°C. Vegetables have about half the shelf life of fruits.

Foods that are packaged "bone dry" may spoil if moisture is reabsorbed during storage. Check dried foods frequently during storage to see if they are still dry. Glass containers are excellent for storage because any moisture that collects on the inside can be seen easily. Foods affected by moisture should be used immediately or redried and repackaged. Mouldy foods should be discarded.

Canning

Canning is a safe method of food preservation if practiced properly. The process of canning involves placing foods in jars and heating them to a temperature that destroys microorganisms that could be a health hazard or cause the food to spoil. Canning inactivates enzymes that could cause the food to spoil. Air is driven out from the jar during heating, and as it cools, a vacuum seal is formed. The vacuum seal prevents air from getting into contact with the product.

Clostridium botulinum is the main reason why pressure canning is necessary. Though the bacterial cells are killed at boiling temperatures, they can form spores that can withstand these temperatures. The spores grow well in low-acid foods or in the absence of air. When the spores begin to grow, they produce the deadly botulinum toxins.

These spores can be destroyed by canning the food at a temperature of 115°C or above for specified time. This temperature is above the boiling point of water, so it can only be reached in a pressure canner.

Canning methods   Two ways of canning are available, depending on the type of food being canned. They are boiling water bath method and pressure canner method.

Boiling water bath   This method is safe for fruits, tomatoes and pickles as well as jam, jellies and other preserves. In this method, jars of food are completely covered with boiling water (100°C).

In high-acid foods (pH of 4.6 or less), *Clostridium botulinum* spores cannot grow and produce toxins. High-acid foods include fruits and pickled vegetables. These foods can be safely canned at boiling temperatures in a boiling water bath.

Tomatoes and figs have pH close to 4.6. Lemon juice or citric acid must be added to them before canning.

Pressure canner   Pressure canning is the only safe method of canning low-acid foods (pH of more than 4.6). Low-acid foods include vegetables, meat, poultry and seafood. These foods must be canned in a pressure canner. Jars of food are immersed in 2 to 3 inches of water in a pressure canner and then heated to a temperature of 115°C.

Methods of Packing

Fruits and vegetables may be packed raw or they may be preheated and then packed into cans or jars. The hot pack yields better colour and flavour, especially when foods are processed in boiling water bath. For both raw pack and hot pack, there should be enough syrup, water or juice to fill around the solid food in the jar and to cover the food. If not covered by a liquid, food at the top tends to darken and develop abnormal flavours. It takes from ½ to 1½ cups of liquid for a quart jar.

Raw pack   In this method, raw, unheated food is directly introduced in jars. Boiling hot water, juice or syrup is poured over the food to obtain proper headspace. Fruits and vegetables should be packed tightly because they may shrink during processing; however, corn, lima beans, potatoes and peas should be packed loosely because they expand during canning.

Hot pack   In this method, the food is cooked for a specified time and then packed in jars. Hot foods should be packed fairly loosely.

Preservation by Radiation

Non-ionized rays like UV and ionized rays like gamma, beta and X-rays are commonly used for food preservation. Food is irradiated to give the same benefits as when it is processed by heat, refrigeration, freezing or treated with chemicals. Irradiation destroys disease-causing bacteria and reduces the incidence of food-borne illness. Hence hospitals use irradiation to sterilize food for immunocompromised patients.

Irradiated foods are usually wholesome and nutritious. All known methods of food processing and even storing food at room temperature can lower the content of nutrients, especially vitamins. At low doses of radiation, nutrient losses are not significant. However, at higher doses, which are used to extend shelf life or control harmful bacteria, nutritional losses are less than or about the same as cooking or freezing.

CHEMICAL PRESERVATIVES

Two commonly used preservative chemicals are

- ➲ nitrates and nitrites used to preserve meats;

- ➲ sulphites used to prevent fungal spoilage and the browning of fruits and vegetables after they have been peeled.

Preservatives, however, have developed a bad name in many countries. Salt is now widely shunned because of its effects on blood pressure. Nitrites and sulphites can cause asthma, nausea, vomiting and headaches.

Consumers demand foods containing lower levels of chemical preservatives. A number of food-processing techniques have been developed to prolong the shelf life of foods and permit a reduction in preservative levels.

Natural Preservatives

Bacteriocins or small proteins are a new breed of protective compounds. They are used in a wide variety of foods. For example, someone who uses yoghurt gets the benefits of bacteriocins. Bacteriocins are produced by good bacteria to kill competing organisms such as *Listeria monocytogenes*. The whole bacteria or purified bacteriocin can be added to foods such as soft cheese. Another example of bacteriocins is the use of nisin in crumpets to restrict the growth of *Bacillus cereus*.

Table 10.3 shows the list of some food additives and their hazards.

Table 10.3   The list of some food additives and their hazards

Preservatives	Side effects
Sorbic acid	Obtained from berries or synthesized from ketone; possible skin irritant
Potassium sorbate	No known adverse effects
Calcium sorbate	No known adverse effects
Benzoic acid	Added to alcoholic beverages, baked goods, cheese, gum, condiments, frozen dairy, relishes, soft sweets, cordials and sugar substitutes; used in cosmetics and as an antiseptic in many cough medications and an antifungal in ointments; can cause asthma, especially in those dependant on steroid asthma medications; is also known to cause neurological disorders and to react with sulphur bisulphite, provokes hyperactivity in children.
Sodium benzoate	Used as an antiseptic, as a food preservative and to disguise taste; orange soft drinks contain a high amount of it, up to 25 mg per 250 ml; also in milk and meat products, relishes and condiments, baked goods and lollies; used in many oral medications including Actifed, Phenergan and Tylenol; known to cause nettle rash and aggravate asthma.
Potassium benzoate	People with a history of allergies may show allergic reactions.

(Contd.)

Table 10.3   (Continued)

Preservatives	Side effects
Calcium benzoate	People with a history of allergies may show allergic reactions.
Propyl *p*-hydroxybenzoate	Possible contact allergen.
Methyl *p*-hydroxybenzoate	Allergic reactions possible, mainly affecting skin.
Sulphur dioxide (derived from coal tar or by combustion of sulphur or gypsum)	All sulphur drugs are toxic; known to provoke asthma attacks and cause difficulty to metabolize in those with impaired kidney function; destroys vitamin B_1; typical products are beer, soft drinks, dried fruit, juices, cordials, wine, vinegar and potato products.
Nisin (antibiotic derived from bacteria)	Found in beer, processed cheese products and tomato paste.
Natamycin (mould inhibitor derived from bacteria)	Used to treat candidiasis; can cause nausea, vomiting, anorexia, diarrhoea and skin irritation; typical products are meat and cheese.
Potassium nitrite	Colour fixative and curing agent for meat; nitrites can effect the body's ability to carry oxygen, resulting in shortness of breath, dizziness and headache; potential carcinogen; not permitted in foods for infant and young children.
Sodium nitrite	May provoke hyperactivity and other adverse reactions; potentially carcinogenic; restricted in many countries; can combine with chemicals in stomach to form nitrosamine.
Sodium nitrate	Used in the manufacturing of nitric acid; as a fertilizer and in fermented meat products.
Potassium nitrate (derived from dead animal or vegetable matter)	Used in gunpowder, explosives and fertilizers, and in the preservation of meat; may provoke hyperactivity and other adverse reactions; potentially carcinogenic; restricted in many countries.
Sodium propionate	May be linked to migraine; typical products are flour products.
Sorbitol	Artificial sweetener and humectants; obtained from glucose or berries; used in lollies, dried fruit, pastries, confectionary, low calorie foods, pharmaceutical syrups and ophthalmic preparations, and as a preservative in cosmetics; not permitted in foods for infants and young children, can cause gastric disturbance.
Mannitol	Artificial sweetener and humectants; derived from seaweed or the manna ash tree; possible allergen; not permitted in infant foods; can cause diarrhoea, nausea, vomiting and kidney dysfunction; typical products are low calorie foods.

REVIEW QUESTIONS

1.   Explain in detail about different preservation methods of food products.

2.   Explain cold preservation.

3.   What is pasteurization? Explain its types with principle.

4.   Explain how food is preserved by drying? Give the different methods of food drying.

5.   What is canning?

6.   Give a detailed account on preservation by radiation and chemical preservations.

PRODUCTION OF MILK PRODUCTS

Milk has an excellent nutrient profile, providing significant amounts of high-quality protein, calcium, riboflavin, magnesium, phosphorus, niacin equivalents, vitamin B_{12}, vitamin B_6, vitamin A, and when added, vitamin D, as well as several other essential nutrients. In fact, milk is a nutrient-dense food, providing a high nutrient content in relation to its calories. All types of milk contain 2% reduced fat, 1% low fat content with the exception of calories and fat. The nutrient content of flavoured milks such as chocolate milk is similar to that of the corresponding unflavoured milk. The major difference is the higher content of carbohydrate and calories in chocolate milk due to the addition of sucrose and other nutritive sweeteners. In general, chocolate-flavoured milks have about 60 calories more than their unflavoured counterparts.

11

CHEESE

Cheese is made from milk, and milk comes from animals as diverse as cows, sheep, goats, horses, camels, water buffalo, and reindeer. In the past, cheeses were not fermented. They consisted solely of salted white curds drained of whey, similar to cottage cheese. The next step was to develop a natural separation process. This was achieved by adding rennet to the milk. Rennet is an enzyme from the stomachs of young ruminants (e.g. cow). Rennet is still the most popular way of starting cheese, though other starting agents such as lactic acid and various plant extracts are used.

Most cheese makers expedite the curdling process with rennet, lactic acid, or plant or vegetable extracts of wild artichokes, fig leaves, safflower, or melon.

In addition to curdling agents, cheese may contain various ingredients that enhance flavour and colour. Cheese may also be salted or dyed, usually with annatto, an orange colouring made from the pulp of a tropical tree, or carrot juice. They may be washed in brine or covered with ashes. Cheese makers who wish to avoid rennet may encourage the bacterial growth necessary for curdling by other methods. Cheese made from unpasteurized milk naturally possesses these bacteria. Some reports say that cheese is made using the native organisms isolated from cow dung and other relative natural environment.

The texture and flavour of processed cheese are obtained by adding salt, milk fat, cream, whey, water, vegetable oil and other fillers. Processed cheese will have preservatives, emulsifiers, gums, gelatin, thickeners and sweeteners as ingredients. Most processed cheese and some natural cheese are flavoured with ingredients like paprika, pepper, chives, onions, cumin, caraway seeds, jalapeño peppers, hazelnuts, raisin, mushroom, sage and bacon. Cheese may also be smoked to give it a distinctive flavour.

CHEESE MAKING

Although cheese making is a linear process, it involves many factors. Several grades of cheese can be produced by varying additives or changing the procedure at different points. The various processes involved in cheese making are shown in Figure 11.1.

Preparing Milk

Cheese factories may accept morning milk (which is richer), evening milk, or both. The milk would contain the bacteria necessary to produce lactic acid, one of the agents that triggers curdling. The milk is monitored until enough lactic acid has formed to begin producing a particular type of cheese. Depending on the type of cheese being produced, the cheese makers may heat the ripening milk. However, large cheese factories, which usually purchase pasteurized milk, consequently add a culture of bacteria to produce lactic acid.

Separating Curd from the Whey

The next step is to add animal or vegetable rennet to milk, for separating curd and whey. When the curd forms, it is cut vertically and horizontally with knives. In large factories, huge vats of curdled milk are cut vertically with a machine that consists of sharp, multi-bladed, wire knives. The machine then agitates the curd and slices them horizontally. Soft cheeses are cut into big chunks, while hard cheeses are cut into tiny chunks. After cutting, the curds may be heated to hasten the separation.

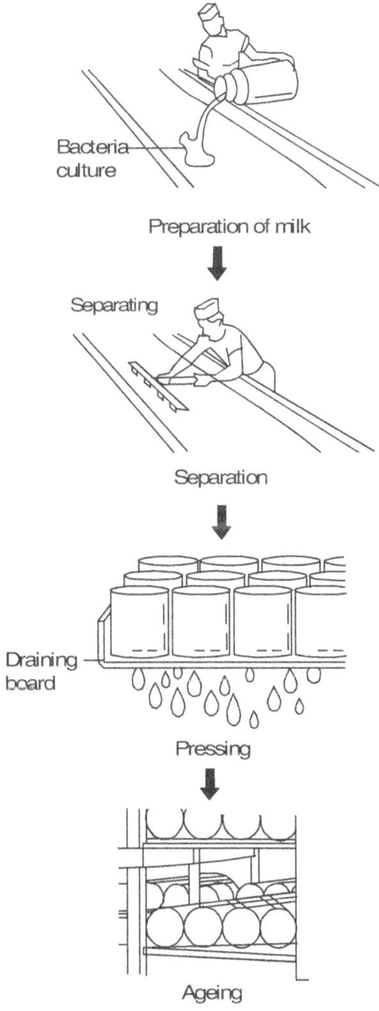

Figure 11.1   Process of cheese making

Pressing the Curd

The amount of moisture removed depends on the type of cheese. For some types with high moisture content, the whey-draining process removes sufficient moisture. Other types require the curds to be cut, heated and/or filtered to remove excess moisture. To make Cheddar cheese, for example, cheese makers finely chop the curd. To make hard, dry cheeses such as Parmesan, cheese makers cook the curd. The curds are put into moulds for ageing. Here, curds are pressed to give shape and size. Soft cheeses such as cottage cheese are not aged.

Ageing the Cheese

The cheese may be inoculated with a flavouring mould, bathed in brine and wrapped in cloth or hay before depositing it in a place of proper temperature and humidity to age. Some cheeses are aged for a month, some for several years. Ageing sharpens the flavour of cheese; for example, Cheddar aged more than two years may be labelled extra sharp.

Wrapping Natural Cheese

Some cheeses may develop a rind naturally, as their surfaces dry. Sometimes, rinds may form from the growth of bacteria that has been sprayed on the surface of the cheese. In place of rinds, cheeses may be sealed in cloth or wax. For domestic use, this may be the packaging that is necessary. However, for exporting to distant countries, cheeses may be heavily salted or sealed in impermeable plastic or foil.

Wrapping Processed Cheese

Inferior cheese (yet edible) can be turned into processed cheese. Cheeses are cut and very finely ground. The powder is mixed with water to form a paste. Salt, fillers, emulsifiers, preservatives, and flavourings are added to the paste. The mixture is heated under controlled conditions. While still warm and soft, the cheese paste is extruded into long ribbons and sliced. The small sheets of cheese are wrapped in foil by a machine.

Quality Control

Cheese making has never been an easily regulated, scientific process. Analytical tests of cheese characteristics may yield a good cheese. Developing a set of standards for cheese is difficult because each variety of cheese has its own range of characteristics. For example, good soft blue cheese will have high moisture and a high pH; Cheddar will have neither.

Many cheese makers think whether it is necessary to pasteurize milk that goes into cheese. Pasteurization was promoted because *Mycobacterium tuberculosis*, disease-causing bacteria, persists in milk products. However, it is possible to eat cheese made from unpasteurized milk to no ill effect. In fact, cheese connoisseurs insist that pasteurizing destroys the natural bacteria necessary for quality cheese manufacture. They claim that modern cheese factories are so clean and hygienic that pasteurization is unnecessary.

Because cheeses possess such disparate characteristics, different types of cheese are required to meet different compositional standards. Based on moisture and fat content, cheese may be labelled soft, semi-soft, hard, or very hard. For example, Cheddar, a hard cheese, contains no more than 39 per cent water and no less than 50 per cent fat. In addition to meeting compositional standards, cheese must also meet standards for flavour, aroma, texture, colour, appearance, and finish. To test a batch of cheese, inspectors examine a sample to detect any inconsistencies in it. Cheese is usually assigned points for each of these characteristics, with flavour and texture weighing more than colour and appearance.

Processed cheese is also subject to restrictions and standards. It must contain at least 90 per cent real cheese. Products labelled "cheese food" must be 51 per cent cheese, and others are 65 per cent. Products labelled "cheese spread" must be 51 per cent cheese; it contains more water and gums to make it spreadable. A "cheese product" usually refers to diet cheese that has more water and less cheese than cheese food or cheese spread. Similarly, cheese is usually not a main ingredient in "imitation cheese". In general, quality processed cheese should be smooth and evenly coloured. It should easily melt in the mouth and should not be rubbery.

TYPES OF CHEESE

Fresh Soft Cheese

Some cheeses are made by souring the milk and straining the curd by passing it through a separator to remove moisture, giving a white, crumbly, soft product. The flavour is mildly acidic. This type of cheese lends itself to mixing with herbs, spices, fruits, etc.

Another form of soft cheese is made by cutting the curd and placing it in a mould to drain. This gives the cheese a soft silky texture. Salt is sprinkled or rubbed onto the surface. The flavour is mild for a few days. The cheese can mature under optimal conditions and a distinctive mouldy rind develops.

The cheeses are light in texture and lower in fat than many of the hard cheeses. Soft cheese is much used in cooking.

Soft Mould Ripened Cheese

Mould ripened cheeses are suspicious. Moulds would grow, and it may sometimes be undesirable. However, the white and blue are desirable, which not only assist in the process of maturing the cheese but also provide another wonderful array of textures and flavour.

Soft mould ripened cheeses are cylindrical in shape. At their best, they are creamy in texture and have mushroom flavour.

Specially selected white moulds of *Geotrichum candidum* and *Penicillium camemberti* may be added with the starter culture or sprayed on the surface of the drained cheese. After a few days, a bloom appears and develops into a distinct white fluffy rind.

When young, the cheese is firm and has a very mild taste. As it matures or ripens, it becomes soft. The cheese has a limited shelf life.

Hard and Semi-hard Cheese

Cheddar is the most popular hard cheese. Caerphilly is soft cheese, sometimes classified as semi-hard cheese. Hard cheese of the Cheddar type can be defined as a firm cheese with a close texture. The Cheshire type can be a little more open, almost granular in texture, whereas Caerphilly has a close but flaky texture. Traditional cheese usually has a degree of openness, adding to the character of the cheese. However, none of these types of cheese have holes typical of some continental cheese such as Emmenthal. The holes are the result of the gas produced by bacteria added to the starter culture. The colour of hard cheese will vary according to season and milk used. The milk of cow is creamy yellow colour whereas milk of sheep or goat is much white. Carotene or annatto may be added to the milk to produce coloured cheese such as Double Gloucester and Leicester. These are natural plant colourants. They have no effect on the texture or flavour of cheese.

The process of making hard and semi-hard cheese is basically the same, but the result will vary considerably.

Washed Rind Cheese

This may be a semi-soft or a hard type of cheese. A distinguishing feature is the colour of the rind, caused by washing or wiping the surface of cheese with a cloth soaked in brine and containing the bacteria *Brevibacterium linens* often with other select microorganisms. Some cheeses have a dry orange brown rind, whereas others have a distinctive sticky rind with a rather pungent aroma. The texture of this type of cheese may be soft, pliable or creamy. The flavour would be mildly aromatic.

Blue Cheese

Blue cheese is a mould ripened type, with blue green veins throughout the cheese. This kind of cheese may not be available throughout the year. Granston is a type of blue cheese made only when the milk is deemed to be at its best. Landsker cheeses are pressed whereas others are soft creamy type.

Some types of cheese have a white mould on the surface and blue mould within the body. This combination of the two *Penicillium* moulds provides a mild soft cheese usually classified within the soft ripened cheese category. The *Penicillium roqueforti* mould is now normally added with milk. However, this type of cheese does not create any antibiotic resistance or affect the usefulness of the antibiotic penicillin.

Blue cheese is not pressed as it is essential for the curds to be loosely packed, leaving space for the mould to grow. The mould requires air to grow and turn it a blue colour. This process is aided by piercing the cheese with stainless steel needles. The tunnels created by the needles allow air to enter into the body of the cheese and quickly develop the attractive blue mould.

The flavour of blue cheese varies according to the type of milk and season. It would be mellow, with a slightly peppery taste and mushroom overtone. There should not be a bitter after taste. The cheese is left unwrapped during maturing and the rind can be eaten.

Smoked Cheese

Smoking of foods has been a means of preservation. It has also been a tradition to smoke certain types of cheese. Smoked cheese provides another variant of cheese.

Cheese is smoked using oak or other local wood. Smoking is done with low temperatures so that the cheese is not cooked, but it does bring some fat to the surface. This, together with the deposits from smoke vapours, provides an attractive and characteristic colour to cheese.

Table 11.1   Bacteria used in milk products

Species	Major function	Product
Propionibacterium shermanii	Flavour and eye formation	Swiss cheese family.
Lactobacillus bulgaricus, *Lactobacillus lactis,* *Lactobacillus helveticus*	Acid and flavour	Bulgarian buttermilk, yogurt, kefir, koumiss, Swiss, Emmenthal and Italian cheese.
Lactobacillus acidophilus	Acid	Acidophilus buttermilk.
Streptococcus thermophilus	Acid	Emmenthal, Cheddar and Italian cheese, and yogurt.
Streptococcus diacetilactis	Acid	Sour cream, ripe cream, butter, cheese, buttermilk and starter cultures.
Streptococcus lactis, *Streptococcus cremoris*	Acid	Cultured buttermilk, sour cream, cottage cheese, all types of foreign and domestic cheese, and starter cultures.
Streptococcus durans, *Streptococcus faecalis*	Acid and flavour	Soft Italian, Cheddar, and Swiss cheese.
Leuconostoc citrovorum, *Leuconostoc dextranicum*	Flavour	Cultured buttermilk, sour cream, cottage cheese, ripened cream butter, and starter cultures.

YOGURT AND OTHER MILK PRODUCTS

Yogurt is a semi-solid, fermented milk product, which originated centuries ago in Bulgaria. It is now consumed in most parts of the world. Although the consistency, flavour and aroma may vary from one region to another, the basic ingredients and manufacturing processes are essentially consistent.

In general, yogurt is made with several ingredients including milk, sugars, stabilizers, fruits and flavours, and a bacterial culture (*Lactobacillus bulgaricus*). During fermentation, these organisms interact with milk and convert it into curd. They also change the flavour of milk to a characteristic yogurt flavour, of which acetaldehyde is one of the important contributors.

The primary by-product in fermentation is lactic acid. The yogurt fermentation completes in three to four hours.

Various ingredients are added to modify certain properties of yogurt. To make yogurt sweeter, sucrose (sugar) may be added at approximately 7%. For reduced calorie yogurts, artificial sweeteners such as aspartame or saccharin are used. Cream may be added to provide a smooth texture. Consistency and shelf stability of yogurt can be improved with stabilizers such as food starch, gelatin, locust-bean gum, guar gum and pectin. These materials are used because they do not have a significant impact on the final flavour.

To improve taste and flavours, fruits are added to yogurt, which include strawberries, blueberries, bananas and peaches, but almost any fruit can be added. Besides fruits, other flavouring agents are added such as vanilla, chocolate, coffee and mint.

MANUFACTURING METHOD

The general process of making yogurt includes modifying the composition of milk and pasteurizing it; fermenting at warm temperatures; cooling it; and finally adding fruit, sugar and other materials.

Modifying Milk Composition

The composition of milk is modified before it is used to make yogurt. This process involves reducing the fat content and increasing the total solids. Fat content may be reduced by using a standardizing clarifier and a separator (a device that applies centrifugation to separate fat from milk). In the clarifier, the milk is placed in a storage tank and tested for fat and solids. The solids in the milk is increased to 16% with 1–5% being fat and 11–14% being solids-not-fat (SNF). This could be accomplished either by evaporating some water or adding concentrated milk or milk powder. Increasing the solids improves the nutritional value of yogurt and makes it easier to produce a firmer yogurt.

Pasteurization and Homogenization

 After the solids' composition is adjusted, stabilizers are added and the milk is pasteurized. This step has many benefits. First, it will destroy the microorganisms in milk that may interfere with fermentation. Second, it will denature the whey proteins in milk which will give yogurt better appearance. Third, it will not alter the flavour of milk greatly. Finally, it helps to release the compounds in milk that will stimulate the growth of the starter culture. Pasteurization may be continuous or batch process. Both these processes involve heating milk to a relatively high temperature for a set amount of time. While the milk is being heat treated, it is also homogenized. Homogenization is a process in which the fat globules in milk are broken up into smaller, more consistently dispersed particles. In commercial yogurt making, homogenization has the benefits of giving a uniform end product, which will not separate. Homogenization is conducted with a homogenizer or viscolizer, in which the milk is forced through small openings at a high pressure and fat globules are broken up due to shearing forces.

Fermentation

When pasteurization and homogenization are complete, the milk is cooled to 43–46°C, and the fermentation culture is added in a concentration of 2%. The mixture is held at this temperature for three to four hours while the incubation process takes place. During this time, the bacteria metabolize certain compounds in milk, producing the characteristic yogurt flavour. An important by-product of this process is lactic acid.

Depending on the type of yogurt, the incubation process may be done in a large tank or individual containers. Stirred yogurt is fermented in bulk and then poured into the final containers, whereas set yogurt, also known as French style, is allowed to ferment in the container. In both instances, the level of lactic acid completes the production of yogurt. When a desired acid level is reached, the yogurt is cooled, modified as necessary and dispensed into containers.

Adding Other Ingredients

Fruits, flavours and other additives may be added to yogurt at various stages in manufacturing. In non-fruit yogurts, flavours are added as the milk is processed. Fruits and flavours may also be added into the containers, to create a bottom layer. The inoculated milk is added on top and the carton is sealed and incubated. Otherwise, pasteurized fruits may be added to bulk yogurt, which is then dispensed into containers.

Quality Control

Many milk products including yogurt are subject to a variety of safety tests for microbial quality, degree of pasteurization, and various forms of contaminants. The microbial quality of raw milk may be determined with dye reaction test. If the microbial count is too high, the milk will be rejected. On the other hand, the degree of pasteurization is determined by measuring the level of an enzyme called phosphatase in milk. Besides microbial contamination, raw milk is subject to other kinds of contaminants such as antibiotics, pesticides and even radioactivity.

In addition to safety tests, the final yogurt is also evaluated for characteristics such as pH, rheology, taste, colour and odour. These factors may be tested with equipment such as pH meters and viscometers.

Other Fermented Milk Beverages

Cultured buttermilk   This product was originally the by-product of butter manufacturing, but today it is more common to produce cultured buttermilk from skim or whole milk. The cultures most frequently used are *S. lactis* and *S. cremoris*. Milk is heated to 95°C and cooled to 20–25°C before adding the starter culture. The starter is added at 1–2%, and fermentation is allowed to proceed for 16–20 hours, to an acidity of 0.9% lactic acid. This product is frequently used as an ingredient in bakeries.

Acidophilus milk   Milk fermented with *Lactobacillus acidophilus* (LA) has been thought to have therapeutic benefits in the gastrointestinal tract. Skim or whole milk may be used. Milk is heated to 95°C for 1 hour to favour the slow growing LA culture. Then milk is inoculated at 2–5% and incubated at 37°C to coagulation. Acidophilus milk may have acidity as high as 1% lactic acid, but for therapeutic purposes 0.6–0.7% is more common.

A variation of this is sweet acidophilus milk, in which the LA culture is added but no incubation is done. It is thought that the culture would reach the GI tract where its therapeutic effects will be realized, but the milk has no fermented qualities, which is considered undesirable by some people.

Sour cream   Cultured cream has a fat content between 12–30%. The starter is similar to that used in cultured buttermilk. The cream is heated to 75–80°C and is homogenized at >13 MPa to improve appearance. Inoculation and fermentation are similar to those for cultured buttermilk, except that fermentation is stopped at an acidity of 0.6%.

Others   There are several other fermented dairy products, including kefir, koumiss, beverages based on bulgaricus or bifidus strains and labneh. Many of these products vary in flavour, texture and producing gas and ethanol as by-products since different organisms are used to carryout the production.

REVIEW QUESTIONS

1. Give the importance of milk and milk products.

2. Describe the process of cheese making and add notes on types of cheese.

3. What is yogurt? How is it prepared?

4. Explain in detail about other milk products.

PRODUCTION OF BAKERY PRODUCTS

BREAD

Wheat and barley were the earliest known crops cultivated by primitive people living around 5000 BC. Eventually people experimented with cooking the grain. In this manner, porridge and flat breads were developed.

Bread is made with three basic ingredients: grain, water and baker's yeast. The grains are composed of three parts: bran (the hard outer layer), germ (the reproductive component) and endosperm (the soft inner core). All the parts are ground together to make whole wheat and rye breads. To make white flour, the bran and germ must be removed. Since bran and germ contain much of nutrients, the white flour is often "enriched" with vitamins, minerals, fibre and calcium.

Grinding takes place at grain mills. In the baking factory, water and yeast are mixed with the flour to make dough. Other ingredients such as salt, fat, sugar, honey, raisins and nuts are also added.

Process

The sifted flour is poured into an industrial mixer. Temperature-controlled water is piped into the mixer. This mixture is called gluten, which gives bread its elasticity. Yeast is added to the mixture. As the yeast grows, it produces gas bubbles, which leaven the bread. Depending on the type of bread to be made, other ingredients are added into the mixer. Modern mixers can process up to 908 kg of dough per minute.

The mixer is enclosed in a drum that rotates at speeds between 35 and 75 revolutions per minute. Inside the drum, mechanical arms knead the dough to the desired consistency. Mixing takes about 12 minutes.

Fermentation   A simple method would be to fix high-speed machinery in the plant to manipulate dough at extreme speeds, which force the yeast cells to multiply rapidly. Fermentation may also be induced by the addition of chemical additives such as 1-cysteine (naturally occurring amino acid) and vitamin C. Sometimes, breads could ferment naturally when the dough is placed in covered metal bowls and stored in a temperature-controlled room until it rises.

Division and gas reproduction   After the dough has fermented, it is loaded into a divider with rotating blades that cuts the dough. A conveyer belt moves the pieces of dough to a moulding machine. The moulding machine shapes the dough into balls and drops them into a layered conveyer belt that is enclosed in a warm, humid cabinet called prover. The dough moves through the prover slowly and the gas reproduction progresses.

Moulding and baking   When the dough emerges from the prover, it is conveyed to the second moulding machine that shapes the dough into loaves and drops them into pans. The pans travel to another prover, which is set at high temperature and humidity. The dough regains the elasticity lost during fermentation and resting period.

From the prover, the pans enter a tunnel oven. The temperature and speed are carefully decided so that when the loaves emerge from the tunnel, they are completely baked and partially cooled. While inside the tunnel, the loaves are mechanically dumped from the pans into shelves. The baking and cooling process lasts for 30 minutes. Figure 12.1 shows the process of bread making.

Slicing and packaging   As the bread moves from the oven to the slicing machine, it cools down. In the slicing machine, vertically serrated blades move up and down at great speeds, slicing the bread into consistent sizes.

Metal plates hold the slices together while picking up each loaf and passing it to the wrapping machine. Pre-printed plastic bags are mechanically slipped over each loaf. The bags are tied with wire twists or sealed.

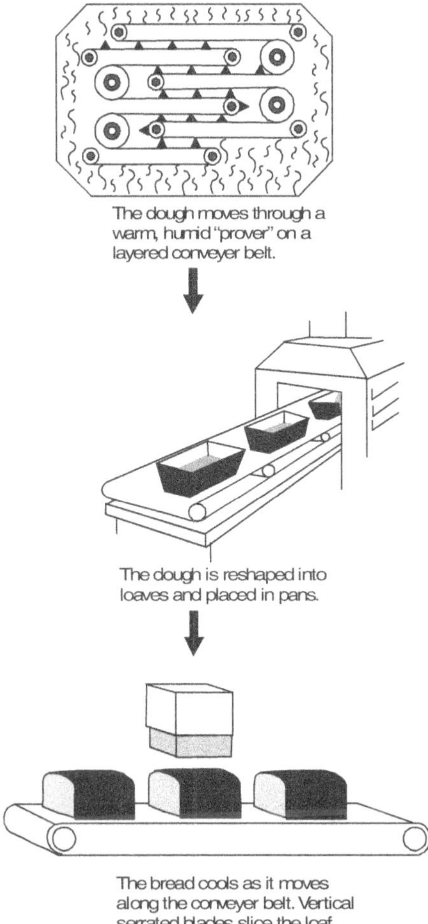

Figure 12.1   Overview of bread making

Quality control   Commercial bread-making (Figure 12.2) maintains high quality standards for the appearance, texture and flavour of bread. Therefore, quality checks are performed at each step of the production process. Producers employ a variety of taste tests, chemical analyses, and visual observation to ensure quality.

Moisture content is particularly critical. A ratio of 12 to 14% is ideal for the prevention of bacterial growth. However, freshly baked breads have moisture content as high as 40%. Therefore, it is imperative that the bakery plants be kept scrupulously clean. The use of fungicides and irradiation is popular.

Figure 12.2 Industrial production of bread

Types of Bread

White bread White bread is the most common type of bread. It is made from wheat flour and in many different sizes, shapes and textures. Different ingredients are added to enhance the flavour of bread.

Whole meal or whole wheat bread It is made from whole meal flour. It has become more popular for its nutritional benefits.

Mixed grain bread It may be made from any combination of flours, grains and seeds.

Kibbled wheat or cracked wheat bread It is a type of bread rolled in wheat grains.

Fibre increased white bread It is made with additional fibre materials.

Rye bread It is made from a combination of rye flour and wheat flour.

Pumpernickel It is heavy and dark and made from rye flour, rye meal and cracked grain.

Sourdough bread It has a slight sour flavour and has a dense texture.

REVIEW QUESTIONS

1. What is bakers' yeast? What are the basic ingredients used in the preparation of bread.

2. Describe in detail about bread preparation and types of bread.

PRODUCTION OF FERMENTED BEVERAGES

YEAST

INTRODUCTION

In brewing, alcoholic fermentation comprises the conversion of sugar into carbon dioxide (CO_2) and ethyl alcohol. This process is carried out by yeast enzymes. There is a complex series of conversions in alcoholic fermentation. In brewing, we use the sugar fungi form of yeast. These cells gain energy from breaking down the sugar. The by-product, CO_2, bubbles through the liquid and dissipates into air. Alcohol remains in the liquid. The yeast cells die when the level of alcohol exceeds beyond tolerance. Brewer's yeast tolerates about 5% alcohol. Beyond this level, the yeast cannot continue fermentation. Wine yeast, on the other hand, tolerates about 12% alcohol. The level of alcohol tolerance by yeast varies between 5% and 21% depending on yeast strain.

Fermentation of alcohol is also controlled by temperature. Temperatures above 27°C can kill the yeast, and below 15°C, yeast activity is too slow.

An excess amount of sugar in the solution can prevent fermentation. Some recipes suggest adding sugar in parts throughout fermentation rather than in the beginning. This is especially true if the brew is aimed at producing high level of alcohol.

The overall process of fermentation aims to convert glucose sugar ($C_6H_{12}O_6$) into alcohol (CH_3CH_2OH) and carbon dioxide. The reactions within yeast make this happen; the overall process is as follows:

$$C_6H_{12}O_6 \longrightarrow 2(CH_3CH_2OH) + 2(CO_2)$$

Sugar $\longrightarrow$ Alcohol + Carbon dioxide

(Glucose)   (Ethyl alcohol)

YEAST IN BREWERY INDUSTRY

Brewer's Yeast

Brewer's yeast is dried, pulverized cells of *Saccharomyces cerevisiae*, a type of yeast that reproduces by budding. They are biologically classified as fungi and are responsible for converting fermentable sugars into alcohol and other by-products. There are hundreds of varieties and strains of this yeast. In the past, there were two types of beer yeast: ale yeast (the "top-fermenting" type, *Saccharomyces cerevisiae*) and lager yeast (the "bottom-fermenting" type, *Saccharomyces uvarum*, formerly known as *Saccharomyces carlsbergensis*). As a result of recent reclassification of *Saccharomyces* species, both ale and lager yeast strains are considered to be members of *S. cerevisiae*. However, throughout this book, both ale and lager yeasts will be referred, as previously, as members of *S. cerevisiae* and *S. carlsbergensis*, respectively. Top-fermenting yeasts are used for brewing ales, porters, stouts, Altbier, Kölsch, and wheat beers. Some of the lager styles made from bottom-fermenting yeasts are Pilsners, Dortmunders, Märzen, Bocks and American malt liquours.

Ale Yeast

Ale yeast strains are best used at temperatures between 10 and 25°C, though some strains will not actively ferment below 12°C. Ale yeast is generally regarded as top-fermenting yeast since it rises to the surface during fermentation, creating a very thick, rich yeast head. That is why the term "top-fermenting" is associated with ale yeast. Ale yeasts produce beer high in esters.

Lager Yeast

Lager yeast strains are best used at temperatures between 7 and 15°C. Lager yeasts grow less rapidly than ale yeasts, and with less surface foam they tend to settle to the bottom of the fermenter at the end of fermentation. That is why they are often referred as "bottom fermenting" yeasts. The final flavour of beer will, however, depend greatly on the strain of yeast and the temperatures at which it is fermented.

YEAST LIFE CYCLE

Yeast is activated from dormancy as soon as it is added (pitched) to the wort. Yeast growth follows four phases; however the phases may sometimes overlap: (1) the lag period, (2) the growth phase, (3) the fermentation phase, and (4) the sedimentation phase.

Lag phase   Reproduction is the priority in pitching, and the yeast will not show any other activity until food reserves are built up. This stage is marked by a drop in pH due to utilization of phosphate and reduction in oxygen. Glycogen, an intracellular carbohydrate reserve, is a prime energy source for cell activity since wort sugars are not assimilated early in the lag phase. Glycogen is broken down into glucose, which is utilized by the yeast cell for reproduction. Low glycogen levels produce abnormal levels of vicinal diketones (especially diacetyl) and may result in longer fermentations.

Growth phase   The growth phase, often referred to as the respiration phase, follows the lag phase once sufficient reserves are built up within the yeast. This phase is evident from the covering of foam on the wort surface due to the liberation of carbon dioxide. In this phase, the yeast cells use oxygen in the wort to oxidize a variety of acid compounds, resulting in a significant drop in pH. In this connection, some yeast strains will result in a much greater fall in pH than others within the same fermenting wort.

Fermentation phase   The fermentation phase quickly follows the growth phase when the oxygen supply has been depleted. Fermentation is an anaerobic process. In fact, any oxygen remaining in the wort is stripped out of the solution by carbon dioxide bubbles produced by yeast. This phase is characterized by the reduction of wort gravity and the production of carbon dioxide, ethanol and beer flavours. During this period, yeast is mostly in suspension, allowing itself dispersal and maximum contact with the beer wort to quickly convert fermentables. Most beer yeasts will remain in suspension for 3 or 7 days, after which flocculation and sedimentation will commence.

Sedimentation phase   The sedimentation phase is when the yeast flocculates and settles to the bottom of the fermenter. The yeast produces a substance called glycogen, which will preserve its life as it prepares itself for dormancy. Glycogen is necessary for cell maintenance during dormancy and is an energy source during the lag phase.

Selection of Yeast

Selection of yeast is vital from both product quality and economic standpoint. The choice for yeast will vary according to the brewing equipment and beer style; however, the characteristics for selection of yeast include the following:

Rapid fermentation   Rapid fermentation without excessive yeast growth is important, as the objective is to produce beer with maximum attainable ethanol content consistent with the overall flavour balance of the product.

Yeast stress tolerance   The yeast strain should be tolerant to alcohol, osmotic shock and temperature. Other instances of stress can be the collection, separation (centrifuging/pressing), and transfer (pumping) throughout the process.

Flocculation   Flocculation of yeast is of great importance. The term "flocculation" refers to the tendency to form clumps of yeast called flocs. The flocs (yeast cells) settle at the bottom in the case of bottom-fermenting yeasts or rise with carbon dioxide bubbles to the surface in case of top-fermenting yeasts. The flocculation characteristics need to be matched to the type of fermentation vessel used. A strongly cropping strain will be ideal for skimming from an open fermenter but unsuitable for a cylindroconical fermenter.

Attenuation   Attenuation refers to the percentage of sugars converted into alcohol and carbon dioxide, as measured by specific gravity. Most yeasts ferment sugar glucose, sucrose, maltose and fructose. To achieve efficient conversion (good attenuation) of sugars into ethanol requires the yeast to be capable of completely utilizing the maltose and maltotriose. Brewing yeasts vary significantly in the rate and extent to which they use sugars. Lager strains are often better than their ale counterparts at utilizing maltotriose. The degree of attenuation obtainable exerts a great influence on the organoleptic properties of the resultant beer and, consequently, is one of the determinant factors in the process of yeast selection.

Flavour component   The selection of yeast strain is perhaps the most important contributor to beer flavour. Different strains will vary markedly in the by-products they produce: esters, higher alcohols, fatty acids, hydrogen sulphide, and dimethyl sulphide. The yeast strain must also be capable of reproducible flavour production.

Storage characteristics   The storage characteristics of a yeast are very important for maintaining viability during storage between fermentations and rapid attenuation when repitched.

Yeast mutation   Yeast mutation is common in breweries, but their presence may never be detected. Usually, the mutant has no adverse effect since it cannot compete with yeast and generally disappears rapidly. However, in some cases, the mutant yeast will overcome the normal brewing yeast and may express itself in many different ways. For example, mutation could affect the fermentation of maltotriose, or there could be a continuous variation in the fermentation rate. Reportedly, lager yeast mutates more rapidly than ale yeast. Most commonly, mutations are due to poor handling of brewing yeast.

Yeast degeneration   Yeast degeneration refers to the gradual deterioration in the performance of brewing yeast. Yeast degeneration may be harmful to brewing fermentation. It may have the following symptoms: sluggish fermentation, premature cessation of fermentation (resulting in high residual fermentable levels in beer), gradual lengthening of fermentation time, and poor foam or yeast head formation. Sometimes, the flavour of beer becomes increasingly "dry" due to yeast degeneration.

BEER

Beer is a fermented beverage produced from barley, which is rich in starch. The production of beer involves processing of enzymes and fermentation. Initially, the barley malt starch is converted into simple sugar by enzymatic process. Later, the simple sugar is converted into beer through fermentation by the action of yeast.

$$\text{Barley malt (wort)} \xrightarrow{\text{(Enzyme)}} \text{Simple sugar} \xrightarrow{\text{(Yeast)}} \text{Beer}$$

BEER PRODUCTION

Preparation of Malt

Malting   The object of milling is to split the husk in order to expose the starchy endosperm and allow for efficient extraction and subsequent filtration of the wort. However, it is necessary to compromise between the requirements for extraction and filtration, because although a fine grind potentially yields more extract, it may lead to subsequent filtration problems and loss of extract in grains. In addition, excessive milling may lead to increased wort viscosity due to enhanced beta-glucan extraction. Furthermore, it may cause "balling" of the grist, which is the formation of clumps of malt that are wet outside but dry inside. This will result in a loss of extract recovery since the grist is not converted during the mashing process.

Mashing   Mashing is the process of converting starch from malt and solid adjuncts into fermentable and unfermentable sugars to produce wort of desired composition. The cooked malt is mixed with malt adjunct (uncooked malt as a enzyme source) to convert the complex sugar into simple sugar. The mixer is kept at 60°C in which saccharification takes place. The formation of clear saccarified solution is called wort. The composition of wort will vary according to the style of beer. Mashing involves mixing milled malt and solid adjuncts with water at a set temperature and volume to continue the biochemical changes initiated during the malting process.

Wort separation and boiling with hops Hops (*Humulus lupulus*) (Figure 13.1), a minor ingredient in beer, are used for bitter taste and enhancing flavour. They contain resins, oils and polyphenol. Hops have pronounced bacteriostatic activity, which inhibits the growth of gram-positive bacteria in finished beer and, in high concentrations, aids in precipitation of proteins.

Figrue 13.1   Hop flowers

When the starch has been broken down by mashing, it is necessary to separate the liquid extract (wort) from the solids (grain particles and adjuncts). Wort separation is important because the solids contain large amounts of protein, poorly modified starch, fatty materials, silicates, and polyphenols (tannins). The objectives of wort separation (lautering) include the following:

- ➲ to produce clear wort

- ➲ to obtain good extract recovery

- ➲ to operate within the acceptable cycle time

It should be emphasized that the quality of grist from the mill can greatly affect wort clarity, extraction recovery, and overall processing time.

Following the extraction of carbohydrates, proteins and yeast nutrients from the mash, the clear wort must be conditioned by boiling in the kettle. The purpose of wort boiling is to stabilize the wort and extract the desirable components from the hops. The principal biochemical changes that occur during wort boiling are as follows:

- sterilization

- destruction of enzymes

- protein precipitation

- colour development

- isomerization

- dissipation of volatile constituents

- concentration

- oxidation

After boiling and clarification, the wort is cooled for the addition of yeast and subsequent fermentation. The principal changes that occur during wort cooling are as follows:

i. cooling the wort to yeast pitching temperature

ii. formation and separation of cold break

iii. oxygenation of wort to support yeast growth

Fermentation   Fermentation is the process in which fermentable carbohydrates are converted by yeast into alcohol, carbon dioxide, and numerous by-products. The by-products have a considerable effect on the taste, aroma, and other characteristics of beer. Fermentation is dependent on the composition of wort, yeast and fermentation conditions.

Wort composition influences fermentation by the presence and concentration of various nutrients, pH, and degree of aeration and temperature. These factors may affect the rate of fermentation, the extent of fermentation, the amount of yeast produced, and the quality of beer.

Fermentation time–temperature profiles vary widely for lagers. Traditional lager brewing involves pitching the yeast between 5 and 6°C and allowing the temperature to rise between 8 and 9°C. This generally results in a better quality beer because low fermentation temperature retards the development of by-products—esters, fusel alcohols and diacetyl, all of which are inappropriate in lagers. However, the lag period is generally longer at lower fermentation temperatures. At the end of primary fermentation, the temperature is reduced by 1 to 1.5°C per day and transferred to the lager cellar between 4 and 5°C. Figure 13.2 shows the tower fermenter used for industrial fermentation process.

Figure 13.2   Tower fermenter

Beer Conditioning

Many undesirable flavours may be present in the "green" or immature beer. Conditioning reduces the levels of these undesirable compounds to produce a more finished product. The processes of conditioning include the following.

Maturation   Maturation of beer involves four schemes: traditional lagering, bottle conditioning, casking, and accelerated lagering.

Lagering   Lagering involves secondary fermentation of the remaining fermentable extract at low temperatures and low yeast count. Low temperatures aid in settling the remaining yeast and precipitating haze-forming materials (protein/polyphenol complexes). The evolution of carbon dioxide during secondary fermentation carbonates the beer and reduces by-products (including sulphur compounds and other volatiles). Bottle conditioning involves secondary fermentation and clarification in the bottle induced by adding yeast and sugar to the beer. Cask-conditioning involves clarification in the cask induced by adding yeast, sugars, hops and finings. In accelerated lagering, the beer is fully attenuated, virtually free of yeast, and stored at higher temperatures. Unlike beers that have undergone traditional or accelerated lagering, beers conditioned in a bottle or cask are neither filtered nor pasteurized.

Beer bottle conditioning   The practice of using priming sugars for beer bottle conditioning has been refined by British brewers. Belgian brewers use this method to add unique flavours. Bottle conditioning usually involves a short time in the conditioning tanks to improve the overall stability and flavour before adding priming sugars. Some brewers allow yeast to pass through for secondary fermentation, while others prefer to completely remove the primary fermented yeast and repitch with ale or lager yeast. Some brewers use lager yeast because it generally has smaller cell mass, is less likely to leave an autolysed flavour, and flocculates and settles better than ale yeast. Thompson reports pitching rates between 0.75 and 3.0 million cells/ml. After a brief conditioning, the beer may be filtered with stirring agitators, where priming sugar (e.g. glucose, dextrose or invert sugar) is added. The beer is then bottled for secondary fermentation, which takes 10–14 days. After secondary fermentation, the bottled beer is moved to cold storage (5–15°C) to protect the flavour and to expedite yeast sedimentation.

Beer casking   Casking beer has its origin in the British Isles and is widely used to make pale ales (bitters), porters and stouts. Beer is cased when fermentation is sufficiently complete or when the correct charge of yeast is present (0.25 to 4.00 million cells/ml). If too little yeast is present in beer, secondary fermentation will be too slow and insufficient carbon dioxide is dissolved in beer. However, if too much yeast is suspended in beer, secondary fermentation may be violent.

Traditional beer lagering   Lagering beer was developed in Germany, and it involves a long, cold storage between 1 and 3 months in conditioning tanks. Today, the trend is shorter storage time between 2 and 3 weeks at higher temperatures. However, shorter lagering may result in beer with higher levels of yeast and haze, requiring subsequent clarification steps.

Accelerated beer lagering   Generally, lagering serves four purposes: extract reduction, clarification, carbonation, and flavour maturation. However, many brewers cannot afford longer storage time. Today, with better equipment the wort is usually fully attenuated during primary fermentation, eliminating the need for traditional lagering. The beer is usually conditioned for 2 or 5 days at temperatures between −1 and 2°C before filtration.

Clarification   Fermented beer may be quite turbid due to the presence of yeast, protein/polyphenol complexes, and other insoluble materials, which are responsible for haze formation in beer. Extended lagering at low temperatures, addition of finings to beer, and centrifugation are some techniques that brewers use to reduce these substances.

Lagering   Lagering involves holding the beer in shallow vessels at temperatures between −1 and 2°C. The low temperatures reduce the solubility of by-products and encourage precipitation of yeasts and other haze-loading materials. Natural sedimentation yields beer that is slightly turbid. Although sedimentation is a simple technique employed by many brewers, it requires a long storage period; however, tanks may be under counter-pressure to expedite sedimentation. Another drawback is that the sludge in the tank may warm up, causing yeast autolysis and producing off-flavours in beer. The sludge may contain a considerable amount of beer, which could be recovered with filtration or centrifugation. Moreover, removing sludge manually from the tank is expensive. In reference to ales, this process is often referred to as cold conditioning.

Fining   Although simple sedimentation can give good clarity, better results would be obtained in a short time by adding fining agents during transfer from the fermenter or during storage. Finings improve clarity and physical stability. Isinglass and wood chips may be used to improve clarity by precipitating negatively charged yeast cells. The use of finings is not universal. They are widely used in the United Kingdom, while there is renewed interest in North America.

Centrifugation   Centrifuging is a popular method for reducing the solid content of beers. It may replace lagering or fining. Lagering requires a high investment for setting up storage tanks. Finings can produce relatively bright beer, but some yeasts do not respond to fining action, and finings themselves tend to create tank residues.

Chill proofing   Sometimes, fining alone cannot provide satisfactory clarity due to the presence of proteins and polyphenols (tannins). During conditioning, positively-charged proteins combine with polyphenols to form a colloidal haze, which is soluble at warmer temperatures but insoluble at cooler temperatures. These reactions may continue after filtration if sufficient quantities of soluble protein and polyphenols remain in the beer. To remove proteins or polyphenols and to improve physical stability, the beer undergoes a process called "chill proofing".

Proteolytic enzymes, tannic acid, hydrolysable tannins, silica gels, polyvinylpolypyrrolidone (PVPP) and bentonite are used in chill proofing.

Carbonation   The next major process after filtration is carbonation. Carbon dioxide enhances foaming, acts as a flavour enhancer, and plays an important role in extending the shelf life of beer.

The level of dissolved carbon dioxide in beer following primary fermentation varies with temperature, pressure, yeast, type of fermentation vessel, and initial wort clarity. Typically, carbon dioxide levels range from 1.2–1.7 volumes per volume of beer (v/v) for non-pressurized fermentations. Consequently, carbon dioxide levels are adjusted, unless the beer has undergone traditional lagering. The common practice is to raise the carbon dioxide level between 2.2 and 2.8 v/v prior to packaging. Secondary fermentation and mechanical carbonation are the two modes of carbonation.

1. *Secondary fermentation* The traditional method involves carbonating beer during secondary fermentation at low temperatures and under counter-pressure. Beer with 0.5 to 1.0% fermentable extract is placed under 12–15 psi in conditioning tanks. If the pressure is too high, yeast growth may be affected and flavour characteristics of beer may change. In the tank, the remaining extract ferments and creates sufficient carbon dioxide to saturate the beer to equilibrium.

2. *Mechanical carbonation* Mechanical carbonation is accomplished by in-line or in-tank technique. This technique involves addition of carbon dioxide from an external source. The source of carbon dioxide is from two ways. In one way, the commercially available carbon dioxide is purchased and used for mechanical carbonation. In other way, the carbon dioxide released during fermentation of wort by yeast is collected and used for mechanical carbonation. Some brewers report that use of mechanical carbonation actually has a greater influence on reducing acetaldehyde levels than does kraeusening the beer.

Bottling and pasteurization Once the final quality of beer has been achieved, it is ready for packaging. Packaging (Figure 13.3) of beer is the most labour intensive of the entire production process.

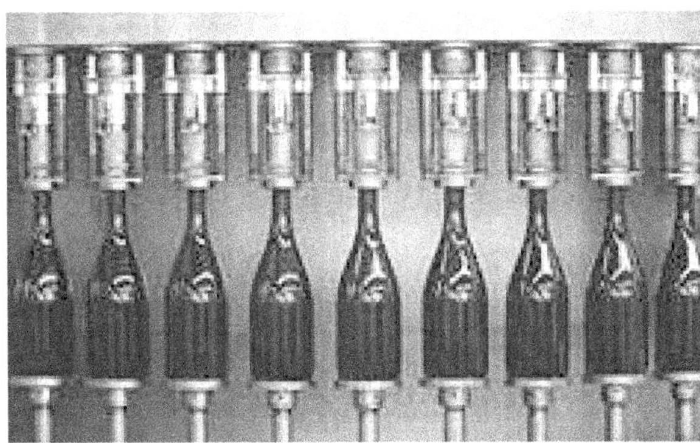

Figure 13.3 Bottling of beer

Pasteurization is an alternative to sterile filtration. A pasteurization unit (PU) is defined as a one-minute exposure to a temperature of 60°C. A PU is a measure of the lethal effect of the heat treatment on microorganisms. The aim is to reach a minimum degree of pasteurization necessary to inactivate beer-spoiling organisms. The two main types of beer pasteurization are flash and tunnel. Flash pasteurization is used for continuous treatment of bulk beer prior to bottling. It is carried out in a plate heat-exchanger. Tunnel pasteurization is used mainly for in-pack treatment following the crowning of bottles.

Flash pasteurization is widely adopted in Europe and Asia. In flash pasteurization, the beer is heated to 71.5–74°C and held at this temperature for 15 or 30 seconds. Flash pasteurization typically uses a two- or three-stage plate heat exchanger with hot water as the medium. The heat exchanger is designed to achieve maximum efficiency at a particular flow rate. Consequently, the flow rate—not the temperature-must be adjusted to alter the number of PUs for the given beer.

BRANDY

The name brandy comes from the Dutch word brandewijn, meaning "burnt wine". Brandy is usually made by applying heat, originally from open flames, on wine. The heat concentrates the alcohol naturally present in wine. Because alcohol has a lower boiling point (78°C) than water, it evaporates quickly leaving the water portion of wine behind. Heating a liquid to separate components of different boiling points is called heat distillation. While brandy is usually made from wine or other fermented fruit juices, it can be distilled from any liquid that contains sugar. The liquid is allowed to ferment and the low level of heating is made. The heating temperature is generally below the boiling point of the water. Low-boiling point liquids distilled from wine include alcohol, a small amount of water, and many of wine's organic chemicals. These chemicals give brandy its taste and aroma.

RAW MATERIALS

French brandy is made from the wine of grapes. However, anything that will ferment can be distilled and turned into brandy. Grapes, apples, blackberries, sugar cane, honey, milk, rice, wheat, corn, potatoes and rye are commonly fermented and distilled. During World War II, people in London made wine out of cabbage leaves and carrot peels, which they subsequently distilled to produce a truly vile form of brandy.

Heat is used to warm the stills for brandy production. In France, the stills are usually heated with natural gas. It would require about 20 ft^4 of wood (0.6 m^4) to produce 25 gal (100 l) of brandy.

PROCESS OF BRANDY PRODUCTION

Fine brandies retain the concentrated flavour of the underlying fruit. However, mass-produced brandies can be made out of anything.

The first step in brandy production is to allow the fruit juice (typically grape) to ferment, by placing it in a large vat at 68–77°F (20–25°C) and leaving it for five days. During this period, natural yeast present in the distillery environment will ferment the sugar into alcohol and carbon dioxide. The white wine grapes usually ferment to an alcohol content of 10%.

Fine brandies are always made in small batches using pot stills. A pot still is a large pot, usually made out of copper, with a bulbous top.

The pot still is heated with the fermented liquid inside to reach the boiling point of alcohol. Alcohol vapours, which contain a large amount of water vapour, rise into the bulbous top.

The vapours are collected in a condenser through a bent pipe, where the vapours are chilled to condense into a liquid with much higher alcohol content. The purpose of the bulbous top and bent pipe is to allow undesirable compounds to condense and fall back into the still. Thus, these elements do not mix with the final product.

Sometimes, fine brandy may be double-distilled. It takes about 34 L of wine to make 3.8 L of brandy. After first distillation for eight hours, 13,249 L of wine may have been converted into 4,542 L of concentrated liquid (not yet brandy) with an alcohol content of 26–32%. The product of second distillation has an alcohol content of 72%. The higher the alcohol content, the more neutral (tasteless) the brandy will be. The lower the alcohol content, the more the underlying flavours remain in the brandy. However, there is a greater chance that off-flavours make their way into the final product.

After second distillation, the brandy must be placed in oak casks and allowed to age. As the brandy ages, it absorbs flavours from the oak while its own structure softens, becoming less astringent. Through evaporation, brandy loses about 1% of alcohol every year.

MASS PRODUCTION

Mass-produced brandy, other than having the same alcohol content, has very little in common with fine brandy. Both start with wine, though mass-produced brandies are likely to be made from table grape varieties rather than wine grapes. Mass-produced brandies are made with fractional distillation in column stills. In column stills (or continuous stills), the raw material is continuously poured from the top, while the final product and wastes come out of the sides and bottom.

A column still is about 30 ft (9 m) high and contains a series of horizontal, hollow baffles that are interconnected. Hot wine is poured from the top of the column while steam is run through the hollow baffles; the steam and wine do not mix directly. Alcohol and other low-boiling point liquids in wine evaporate. The vapours rise while the non-alcoholic liquids fall. As the still is cooler at the top, the rising vapours eventually condense, each at a temperature just above its own boiling point.

Once they condense, the liquids fall into the still and boil again. This process of boiling and condensing happens over and over again in the column. The various components of the wine fraction and collect in the column where the temperature is just below the boiling point of that component. This allows the alcohol condensate to be separated out of the column at the point where it is collected. The resulting product is pure spirit, colourless, odourless, and tasteless, with an alcohol content of 96.5%. Pure spirit can be used to fuel automobiles. It may be diluted to make vodka or diluted and flavoured with juniper berries to make gin.

Mass-produced brandies are aged in oak casks and finally blended, diluted to 40% alcohol, and bottled.

QUALITY CONTROL

Usually brandies that pick up off-flavours during distillation are discarded. The quality of brandy is much dependent on its alcohol content. Alcohol is less dense than water, and a hydrometer can help check the alcohol content in brandy. A hydrometer is a glass float with a rod sticking out at the top. The rod is calibrated and a line on the rod will be exactly at the liquid surface if the hydrometer is floating in water. As alcohol is less dense than water, the hydrometer will sink deeper in alcohol than it will in water. By calibrating the rod scale with different blends of known alcohol content, it can be used to determine the percentage of alcohol in a water–alcohol mixture.

VODKA

Vodka is an alcoholic beverage distilled at a high proof from a fermented vegetable or grain mash. Proof is a measurement of alcoholic content. Each degree of proof equals a half per cent of alcohol. For example, 100 proof equals 50% alcohol. Because distilled vodka can have proof as high as 145, it is a neutral spirit, having no taste and odour. Water is added to bring the proof down to between 80 and 100.

RAW MATERIALS

Grains   Vodka may be distilled from virtually any fermentable material. Originally, it was made from potatoes. Although vodkas are still made from potatoes and corn, most high quality vodkas are distilled from cereal grains, such as wheat.

Water   Water is added at the end of distillation process to bring down the alcohol content.

Malt meal   Because vegetables and grains contain high starches, an active ingredient must be added to the mash to facilitate the conversion of starch to sugar. These converted sugars, maltose and dextrin respond effectively to diastase, an enzyme found in malt. Therefore, malt grains are soaked in water and allowed to germinate. Then, they are coarsely ground into a meal and added to the mash.

Yeast   Yeast contains enzymes that allow food cells to extract oxygen from starches or sugars, producing alcohol. In the manufacture of alcoholic beverages, the yeast species *Saccharomyces cereviseae* is commonly used.

Flavourings   Herbs, grasses, spices and fruit essences may be added to vodka after distillation.

PRODUCTION PROCESS

Mash preparation   The grain or vegetables are loaded into an automatic mash tub. The tub is fitted with agitators that break down the grain. Malt meal is added to promote the conversion of starches into sugar.

Sterilization and inoculation   The mash is sterilized by heating it to the boiling point. Then, it is injected with lactic-acid bacteria to raise the acidity level needed for fermentation. When the desired acidity level is reached, the mash is inoculated again.

Fermentation   The mash is poured into large stainless-steel vats and yeast is added. Over the next two to four days, the enzymes in yeast convert the sugars in the mash into ethyl alcohol.

Distillation and rectification   The liquid ethyl alcohol is pumped into stills, made up of stainless-steel columns with vaporization chambers stacked on top of one another. The alcohol is continuously cycled and heated with steam, until the vapours are released and condensed. This process helps remove impurities. The vapours rise into the upper chambers (still heads), where they are concentrated. The residue materials flow into the lower chambers and excreted. Some grain residues may be used as livestock feed.

Water   Concentrated vapours, or fine spirits, contain 95–100% alcohol. This translates to 190 proof. In order to make the spirit consumable, water is added to it to decrease the alcohol percentage to 40, and the proof to 80.

Bottling   Alcoholic beverages are stored in glass bottles because glass is non-reactive. Other receptacles, such as plastic, would cause a chemical change in beverage. The bottles are cleaned, filled, capped, sealed, labelled, and loaded into cartons with highly mechanized equipment.

Quality control   There are some strict guidelines for production, labelling, import and advertising of vodka. Flavoured vodkas must list the predominant flavour (pepper, lemon, peach, etc.) on the label.

WINE

Wine is an alcoholic beverage produced through partial or total fermentation of grapes. However, berries, apples, cherries, dandelions, elder-berries, palm and rice, can also be fermented to produce wine.

Grapes belong to the botanical family Vitaceae. Many species of grapes are available. The most widely used species in wine production are *Vitis labrusca* and *Vitis vinifera*.

WINE PRODUCTION

The wine grape itself contains all the necessary ingredients for wine: pulp, juice, sugars, acids, tannins and minerals. However, some manufacturers add yeast to increase strength and cane or beet sugar to increase alcoholic content. During fermentation, winemakers may add sulphur dioxide to control the growth of wild yeast.

The process of wine production has remained much the same for ages, but new sophisticated equipment have helped streamline and increase the production of wine. The equipment include mechanical harvesters, grape crushers, temperature-controlled tanks, and centrifuges.

While the manufacturing process is highly automated in medium- and large-sized wineries, small wineries still use hand operated presses and store wine in musty wine cellars.

A universal factor in the production of fine wine is timing. It includes picking grapes at the right time, removing must at the right time, monitoring and regulating fermentation, and storing wine.

The wine-making process may be divided into four distinct steps: harvesting and crushing grapes; fermenting must; ageing the wine; and packaging (Figure 13.4).

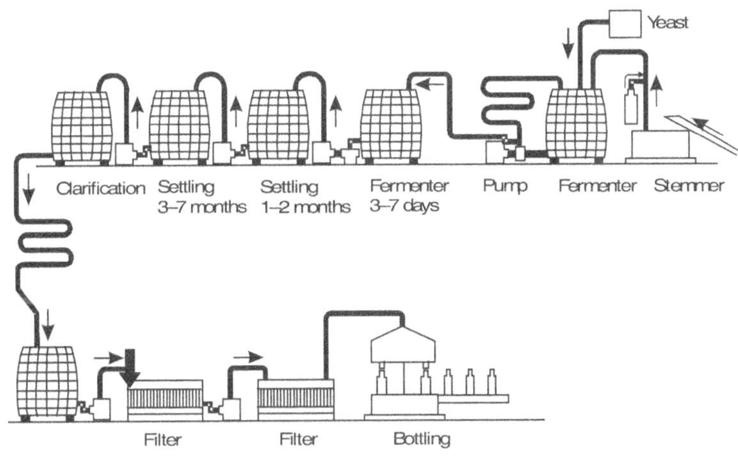

Figure 13.4   Process of wine-making

Harvesting and Crushing Grapes

At the vineyard, sample clusters of wine grapes are inspected with a refractometer to check if the grapes are ready to be picked. The refractometer is a small, hand-held device that accurately checks the amount of sugar in grapes.

A mechanical harvester (usually a suction picker) is used to gather and funnel the grapes into a field hopper or mobile storage container. Some mechanical harvesters (Figure 13.5) have grape crushers mounted on them, allowing vineyard workers to gather grapes and press them at the same time. The result is that vineyards can deliver newly crushed grapes, called must, to wineries, eliminating the need for crushing at the winery. This can prevent oxidization of the juice through tears or splits in grape skin.

Mechanical harvesters eliminate the need for hand-picking. Mechanical harvesters have been first used in California vineyards in 1968. Grapes may be gathered at night when they are cool, fresh and ripe.

The grapes are transported to the winery where they are unloaded into a crusher-stemmer machine. Some crusher-stemmer machines are hydraulic while others are driven by air pressure.

The grapes are crushed into must and the stems are removed. The must is fermented, settled, clarified and filtered. After filtering, the wine is aged in stainless steel tanks or wooden vats. White and rose wine may age for a year to four. Red wine may age for seven to ten years. Large wineries age their wine in large temperature-controlled stainless steel tanks, while smaller wineries store their wine in wooden barrels in damp wine cellars.

Figure 13.5   Harvesting and crushing of grapes

Fermentation

For white wine, the grape skins are separated from must by filters or centrifuges before fermentation. For red wine, the whole crushed grape goes into the fermentation tank (Figure 13.6) or vat. The pigment in the grape skin gives red wine its colour. The more the time the skins are left in the tank, the darker the colour of wine.

During fermentation, wild yeast is fed into the tank or vat to turn the sugar in must into alcohol. In addition, cane or beet sugar may be added. Adding sugar is called chaptalization. Usually chaptalization is done if the grapes have not received enough sunlight prior to harvesting. The winemaker uses a handheld hydrometer to measure sugar content in the must. The wine ferments in the tank for seven to fourteen days.

Figure 13.6 Fermentation tank

Ageing

After the wine is fermented, it is stored, filtered and properly aged. In some instances, the wine is blend with other alcohol. Small wineries store wine in damp, subterranean wine cellars, but larger wineries store wine above ground in epoxylined and stainless steel tanks. Water is circulated inside the lining of the tank shell to control temperature.

Certain wines (usually red wine) are crushed again and allowed to ferment for a second time for three to seven days. This is done to extend the shelf life and ensure clarity and colour stability of wine. The wine is then pumped into settling tanks or vats (racking)and allowed to remain in it for one or two months. Typically, racking is done at 10 to 15°C for red wine and 0°C for white wine.

During the settling (racking) process, the debris (stem pieces, etc.) settle to the bottom of the tank and are removed. Racking or settling produces smoother wine. Additional settling may be necessary for certain wines. After the settling process, the wine passes through a number of filters or centrifuges where clarifying substances trickle through wine. Subsequent to filtering, the wine is aged in stainless steel tanks or wooden vats (Figure 13.7). The wine is then filtered the last time to remove sediments.

Packaging

Medium- and large-sized wineries use automated bottling machines. Expensive wine bottles have corks made of a special oak. Cheaper wines may have an aluminum screw-off cap or plastic stopper. Wine is usually shipped in wooden crates, though cheaper wines may be packaged in cardboards.

Figure 13.7   Wine production in wooden vats

Quality Control

All facets of wine production must be carefully controlled to create a quality wine. Such variables as the speed with which harvested grapes are crushed; the temperature and timing during both fermentation and ageing; the per cent of sugar and acid in the harvested grapes; and the amount of sulphur dioxide added during fermentation all have a tremendous impact on the quality of the finished wine.

WHISKY

Whisky is spirit produced from fermented grain and aged in wood. Spirit is an alcoholic beverage in which the alcohol content has been increased by distillation. Spirits include brandy (distilled from wine), rum (distilled from sugar cane juice or molasses), vodka (distilled from grain but not aged), and gin (distilled from grain and flavoured with juniper berries and other ingredients but not aged.)

Whisky is made with water, yeast and grain. Water is considered most important in making good whisky. It should be clean, clear, and free from impurities such as iron. In the United States, water that contains carbonates, found in areas rich in limestone, is often used. Scottish water is best suited for making fine whisky.

In the production of whisky, yeast is grown on barley malt and kept free from bacterial contamination. Whisky makers use several kinds of yeast.

The type of grain used varies with the kind of whisky being made, but all whisky contain at least a small amount of malted barley, which is needed to start the fermentation process. Scotch malt whisky contains only barley. Other types of whisky contain barley in combination with corn, wheat, oats and rye. Corn whisky must contain at least 80% corn, while Bourbon whisky and Tennessee whisky must contain at least 51% corn. Rye whisky must contain at least 51% rye, and wheat whisky must contain at least 51% wheat. Straight whisky contains no additives, but blended whisky may contain a small amount of additives such as caramel colour and sherry.

WHISKY PRODUCTION

Preparing the Grain

The grain is inspected and cleaned to remove dust and other foreign particles. All grains except barley are first ground into a meal in a gristmill. The meal is then mixed with water and cooked to break down the cellulose walls that contain starch granules. Cooking is usually done in a pressure cooker at 155°C, or more slowly in an open cooker at 100°C.

Barley is malted instead of being cooked. The first step in malting barley is soaking it in water until it is thoroughly saturated. It is then spread out and sprinkled with water for about three weeks, at which time it begins to sprout. During germination, the enzyme amylase is produced, which converts the starch in the barley into sugars. Sprouting is halted and the barley is dried and then heated with hot air from a kiln. For Scotch whisky, the fuel used in the kiln is peat. It gives Scotch whisky a characteristic smoky taste. The malted barley is then ground.

Mashing

Mashing is mixing cooked grain with malted barley and warm water. Amylase in malted barley converts the starch in other grains into sugars. After several hours, the mixture is converted into a turbid, sugar-rich liquid known as mash. In Scotch malt whisky making, the mixture consists only of malted barley and water. After mashing, the mixture is filtered to produce a sugar-rich liquid known as wort.

Fermenting

The mash or wort is transferred to a fermentation vessel made of wood or stainless steel. Yeast is added to begin fermentation; the single-celled yeast converts the sugars in the mash or wort to alcohol. The yeast used may be a portion of a previous batch of fermentation (sour mash process) or new yeast cells (sweet mash process). The sour mash method is often used because it is effective at room temperature and its low pH (high acidity) promotes yeast growth and inhibits the growth of bacteria. The sweet mash method is difficult to control, and it must be used at temperatures above 27°C to speed up fermentation and to avoid bacterial contamination. After three or four days of fermentation, the end product is a liquid containing 10% alcohol, which is known as distiller's beer (wash in Scotland).

Distilling

Scottish whisky makers often distill wash in traditional copper pot stills. The wash is heated to 78°C so that most part of alcohol is transformed into vapour. This vapour is collected and transferred into liquid alcohol in a water-cooled condenser. Most modern distilleries use a continuous still. It consists of a tall cylindrical column filled with a series of perforated plates. Steam enters the still from the bottom, and distiller's beer is piped from the top. As the beer slowly drips through the plates, it is distilled , and alcohol is condensed into a liquid. The product of initial distillation is known as low wine, which is distilled a second time to produce a product known as high wine or new whisky, which contains about 70% alcohol.

The temperature of distillation and other factors determine the proportions of water, alcohol and other substances (called congeners) in the final product. If it contains more than 95% alcohol, it may have no flavour because it has no congeners. This product is known as grain neutral spirit. If the final product has too many congeners, it will taste bad. Bad-tasting congeners (usually aldehydes, acids, esters, and higher alcohols) could be removed in several ways. Some congeners boil at a lower temperature than alcohol and can be boiled off. Some are lighter than alcohol and will float, and they can be skimmed off. Tennessee whisky is unique in that the high wine is filtered through charcoal before it is aged. This filtration removes unwanted congeners to produce smooth whisky. Premium Tennessee whisky may be filtered a second time after it is aged to produce an even smoother product.

Ageing

Water is added to high wine to reduce its alcohol content to 50 or 60% for American whisky and 65% or higher for Scotch whisky. Whisky is aged in wooden barrels, usually made from charred white oak. White oak allows the water in whisky to move back and forth within the pores of wood, which helps to add flavour. In the United States the barrels are not reused. In most other countries, it is common to reuse barrels. However, new barrels add more flavour than used barrels. Scotch whisky is aged in cool, wet conditions, so it absorbs water and becomes less alcoholic. American whisky is aged in warmer, drier conditions, so it loses water and becomes more alcoholic.

Ageing is a complex process, where at least three factors are involved. First, the original mixture of water, alcohol and congeners react with each other over time. Second, these ingredients react with oxygen in the outside air and oxidize. Third, water absorbs substances from wood as it moves within it. (Charring wood makes these substances more soluble in water.) All these factors can change the flavour of whisky. Whisky generally takes at least three or four years to mature, and some may take ten or fifteen years.

Blending

Straight whisky and single malt Scotch whisky are not blended; that is, they are produced from single batches and bottled straight from the barrel. Other whisky types are blended. Different batches of whisky may be mixed to produce better flavour. Often neutral grain spirit is added to lighten the flavour; caramel is added to standardize the colour; and a small amount of sherry or port wine is added to help the flavours blend. Blended Scotch whisky usually consists of several batches of strongly flavoured malt whisky mixed with less strongly flavoured grain whisky. However, a few blends contain only malt whisky. Blending is often considered the most difficult and critical process in producing premium Scotch whisky. A premium blended Scotch whisky may contain more than 60 individual malt whisky blended in proper proportions.

Bottling

Glass is always used to store mature whisky because it does not react with whisky to change the flavour. Modern distilleries can produce as many as 400 bottles of whisky per minute. The glass bottles are cleaned, filled, capped, sealed, labelled, and packed in cardboard boxes with mechanized equipment.

Quality Control

Although the making of good whisky is more of an art, there are certain precautions that whisky makers take to ensure quality. The water used must be filtered to make it free from organic matter. The grain used must be clean. Grains are passed through screens to remove small grains. The yeast is carefully grown to avoid contamination by other microorganisms. The temperature of distillation is monitored with thermometers in the boiling liquid, which are visible through glass windows in the still. During aging, samples of whisky are evaluated continuously. The blending process is also monitored.

VINEGAR

Vinegar is an alcoholic liquid that has been allowed to sour. It is primarily used to flavour and preserve foods. Vinegar is also used as a cleaning agent. The word originates from the French vin (wine) and aigre (sour).

Vinegar may be made from a variety of diluted alcoholic products, the most common being wine and beer. Some distilled vinegars are made from wood products such as beech.

The process of vinegar making relies on the presence of acetobacters, bacteria that live on oxygen bubbles. Acetobacters grow freely in natural processes. In the vinegar factory, the growth of acetobacters is induced by feeding acetozym nutrients into the tanks of alcohol.

The gooey film that appears on the surface of alcohol is turned into vinegar. It is a natural carbohydrate called cellulose. This film holds the highest concentration of acetobacters. It may be skimmed off and added to subsequent batches of alcohol to speed up the formation of vinegar. Herbs and fruits are often used to flavour vinegar. Herbs commonly used include tarragon, garlic and basil. Raspberries, cherries and lemons are also used. Figure 13.8 provides the flowchart of vinegar production.

PROCESS OF VINEGAR PRODUCTION

Orleans Method

On the wooden barrels, bungholes are drilled at the top side and plugged with stoppers. Holes are also drilled at the ends of the barrels.

Alcohol is poured into the barrel through long-necked funnels inserted into the bungholes. Acetozym nutrients are added at this point. The barrel is filled to a level just below the holes on the ends. Nets or screens are placed over the holes to prevent insects from getting into the barrels.

The barrels are kept at rest for several months. The temperature is maintained around 29°C. Samples are taken periodically by inserting a spigot into the side holes and drawing liquid off. When alcohol is completely turned into vinegar, it is drawn off through the spigot. About 15% of the liquid is left in the barrel for the next batch.

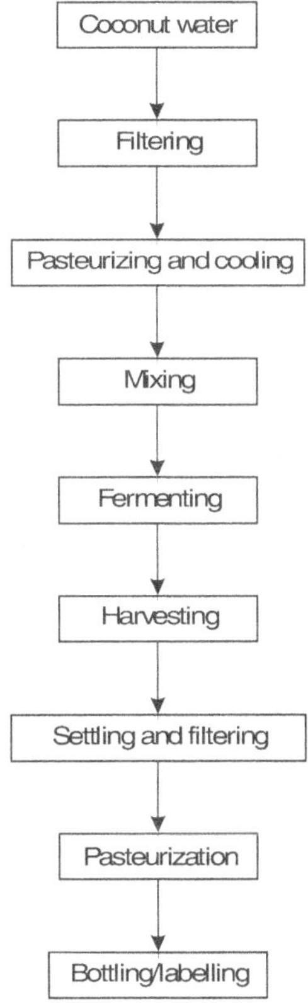

Figure 13.8   Vinegar production using *Acetobacter*

Submerged Fermentation Method

Submerged fermentation is commonly used in the production of vinegar from wine. Large stainless steel tanks called acetators are used in production. The acetators are fitted with centrifugal pumps in the bottom that pump air bubbles into the tank much like what an aquarium pump does.

As the pump stirs the alcohol, acetozym nutrients are piped into the tank. The temperature is maintained between 26–38°C.

Within several hours, alcohol in the acetators is converted into vinegar, which is piped into a filtering machine. The stainless steel plates press the alcohol through paper filters to remove sediments, usually about 3% of the total product. The sediment is flushed out while the filtered vinegar moves to the dilution station.

Generator Method

Distilled and industrial vinegars are often produced with the generator method. Tall oak vats are filled with vinegar-moistened beechwood shavings, charcoal, or grape pulp. Alcohol is poured from the top of the vat, which slowly drips through the fillings. An air compressor blows air through the holes in the vat.

When alcohol is converted into vinegar, it is drawn off from the vat into storage tanks. Vinegar produced in this manner has very high acetic acid content (14%) and must be diluted with water to bring its acetic acid content to 5 or 6%. To produce distilled vinegar, the diluted vinegar is poured into a boiler and heated to its boiling point. The vapour is collected in a condenser, which is cooled and condensed to distilled vinegar.

BALSAMIC VINEGAR

The production of balsamic vinegar is much similar to the production of fine wine. Balsamic vinegar is made from the juices of Trebbiano and Lambrusco grapes. The juice is boiled and poured into barrels (Figure 13.9) containing oak, chestnut, cherry, mulberry and ash.

Figure 13.9   Balsamic vinegar

The juice is allowed to age, ferment, and condense for five years. Every year, the aged liquid is mixed with younger vinegar and placed in a series of smaller barrels. The finished product absorbs aroma from oak and colour from chestnut.

Quality Control

The growing of acetobacters requires vigilance. In the Orleans method, bungholes must be protected with nets to prevent insects. In the generator method, the temperature inside the tank should be maintained at 26–38°C. Workers routinely check the thermostats on the tanks. Failure of electricity may kill acetobacters within seconds, and vinegar plants should have electrical backup systems.

REVIEW QUESTIONS

1. What is yeast? What are the types of yeast used in brewery industry?

2. Give a brief account on yeast life cycle.

3. Explain in detail about beer production process and types of beer.

4. What is hops? What is its function in beer production?

5. What is wort? How is it prepared?

6. What is carbonation in beer production? How is it achieved?

7. Explain about brandy production.

8. Explain in detail about vodka production and its quality control.

9. What is wine? How is it made?

10. Write about whisky production.

11. Describe in detail about vinegar production.

SINGLE CELL PROTEINS (SCP)

A single cell protein may be defined as microbial protein used as food or livestock feed. In other words, dead and dry cells of microorganisms, such as yeast, bacteria, fungi and algae, used as food are called single cell proteins. Single cell proteins have the potential to be developed into a very large source of supplemental protein, which could be used in livestock feed. In some regions, single cell proteins could become the principal protein source for domestic livestock. This could be possible because microbes can be used to ferment vast amounts of waste materials, such as straws; wood and wood processing wastes; food, cannery and food processing wastes; and residues from alcohol production; and human and animal excreta. Producing and harvesting microbial proteins is not without costs. In nearly all instances, the single cell protein will be found in dilute solutions, usually less than 5% solids. Methods available for concentrating include filtration, precipitation, coagulation, centrifugation, and the use of semi-permeable membranes. The equipment required for these de-watering methods are quite expensive and would not be suitable for small-scale operations. Removal of water to stabilize the material for storage, in most instances, is not currently economical. Single cell protein must be dried to about 10% moisture, or condensed and acidified to prevent spoilage.

HISTORY OF SCP

During 1934 and 1938, Asia, Africa and South America were the main exporters of grains to the developed world. Since 1948, the food flow has reversed, due to the growth of population which is much higher in the less developed countries. It was projected that there will be 8 billion people living on earth by 2015 and 10.5 billion by the year 2110. This means that we must produce as much food.

Death from starvation, malnutrition and related diseases is a reality in many countries today. The World Health Organization (WHO) estimates that 12,000,000 people die of hunger and starvation related diseases every year. Half are children under the age of 5. Man could not depend only on agriculture, animal husbandry and fisheries for food. Energy crisis has an immediate impact on agriculture and hence man should search for other types of food for his survival.

During the sixties, the idea of single cell protein was gaining research interest among scientists in universities and industry, particularly in oil. The result was the development of SCP technology. Large-scale production of SCP began after the First World War in Berlin with the growth of *S. cerevisiae* to replace as much as 60% of foodstuff. Germany had been importing after the war. *Candida arborea* and *C. utilis* made an important contribution to human diet in Germany in World War II. After World War II, the application of fungi for the production of antibiotics led to an investigation of the potential microfungi as flavour additives to replace mushrooms.

In 1950's, some oil industries became interested in growing micro-organisms on alkanes. Many companies that produce SCP appeared on the scene, including BP (UK), *Kanegafuichi* (Japan) and *Liquichimica* (Italy). Many SCP plants rely on hydrocarbons as the source of energy for microorganisms. Other potential substrates for SCP include bagasse, citrus wastes, sulphite waste liquor, molasses, animal manure, whey, starch and sewage. The nutritional value of SCPs and doubling time of microbes are given in Table 14.1.

Table 14.1   Nutritional value of SCPs and doubling time of microbes

Organism	Mass doubling	Nutritional value (% of dry weight)			
		Protein	Fat	Ash	Nucleic acid
Algae	2–6 hours	40–60	7–20	8–10	3–8
Bacteria	10–120 minutes	50–65	1–3	3–7	8–12
Fungi	2–8 hours	30–45	2–8	9–14	7–10
Yeast	10–120 minutes	45–55	2–6	5–10	6–12

Advantages of SCP from microbes

1. Microbes possess relatively high content of protein and vitamins.

2. They are independent of land and climate, the bio mass production is associated with waste utilization and the production of other products like alcohols, acids, etc.

3. They can be grown constantly and used continuously in production.

4. It can be genetically controlled.

5. It causes less pollution and emission of hazardous materials.

Drawbacks in SCP from microbes

1. Most algae have non-digestible cell wall, which cannot be digested by mammals.

2. Microbial protein has a high nucleic acid content and different profile of amino acids and vitamins, which may cause indigestion and allergies, and hence the levels should be limited in the diet of monogastric animals.

3. Some algae may contain certain pigments that cause unacceptable colouration and disagreeable flavour.

4. Presence of endotoxins of microbes may cause harmful effects.

PRODUCTION OF SCP FROM BACTERIA

Bacterial Population

Based on nutrient requirements, bacterial populations have been categorized into two types: (i) hydrocarbon utilizing bacteria and (ii) hydrogen utilizing bacteria.

Hydrocarbon utilizing bacteria   The requirement of hydrocarbon varies from one strain to another.

Polyhydric alcohol utilizing bacteria   These utilize simple sugars for its energy source. For example, the milk fermenting *Lactobacillus* and streptococci utilize sugar lactose for its growth.

Aliphatic hydrocarbon utilizing bacteria   *Aliphatic hydrocarbon* utilizing bacteria include *Methanomonas methanica, M. methanooxidans, Methylococcus capsulatus, Pseudomonas methanica, Bacillus* sp., *Xanthomonas* sp. and *Vibrio* sp. They mainly use aliphatic hydrocarbons such as methane and methanol as substrates for its cultivation. However, the production of this organism is little risky since the substrates are inflammable as well as they give problems in mass transfers across cell membranes. In spite of this, the production can be carried out by keeping the oxygen level below 12% and reducing the methanol concentration.

Aromatic hydrocarbon utilizing bacteria   Some bacterial population utilizes the aromatic hydrocarbons as substrate, and those strains are cultivated by the medium amended with paraffin as a substrate. Strains of *Brevibacterium insectivorum, Corynebacterium* spp., *Pseudomonas ligustri, Cellulomonas* sp. and *Micrococcus* sp. are used for the production of SCP using paraffin as substrate. The production medium comprises paraffin as a carbon source and ammonium salts as effective nitrogen source.

Hydrogen utilizing bacteria   In these bacteria, hydrogen serves as energy source for the strains and carbon dioxide serves as carbon source. Strains of *Pseudomonas facilis, P. saccharophila, P. flava, Alcaligenes eutrophus* and *A.paradoxus* are widely used for the production of SCP. The supplements of trace elements like potassium, magnesium, calcium, nickel and sulphur are needed for better growth of these bacterial strains.

Production of Bacterial Biomass

As the upstream process of production, strains are isolated from the natural resources; they are standardized and pure cultures are maintained in appropriate solid medium. The inoculum is prepared using appropriate broth cultures and introduced into bioreactors with optimizing growth parameters. The composition of production medium, supplements and duration of cultivation may vary from strain to strain based on cultural condition. The downstream process for bacterial SCP production involves centrifugation and flocculation followed by killing and drying. Continuous production of SCP is schematically represented in Figure 14.1.

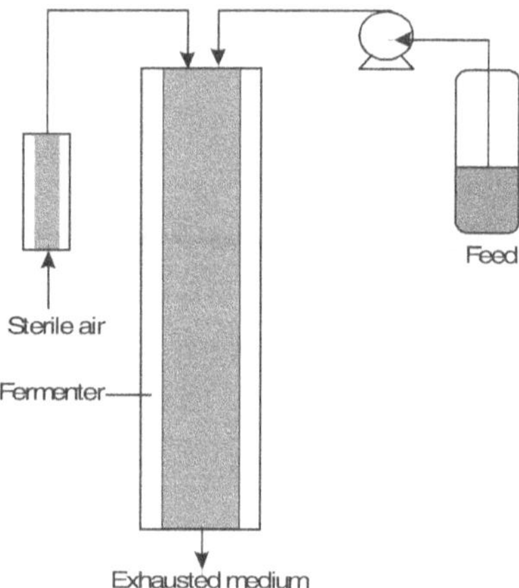

Figure 14.1   Schematic diagram of continuous production of SCP

PRODUCTION OF SCP FROM ALGAE

Algal Population

The algal SCP has high content of protein with extra β-carotene, vitamin K, fatty acids (12%–14%), carbohydrates (10%–17%) and little quantity of vitamin C. The content of protein from algal SCP is approximately 50 per 100 g of dry weight with low level of sulphur containing amino acids. A significant drawback in algal SCP is that it shows low digestibility and unpleasant taste. Algal strains widely used for the production of SCP include *Scenedesmus, Chlorella, Chlamydomonas, Spirulina, Anabaena* and *Nostoc.*

Unlike bacterial cultivation, algae are cultivated in shallow ponds with continuous recirculation of medium since sunlight is an important factor for the growth of photosynthetic system of algae. The shallow pond is filled with the medium containing ammonia and some necessary trace elements with alkaline pH. The medium is circulated by a circulatory pump, and continuous agitation is made by a petal wheel roller with CO_2 supplement. The unicellular algae are harvested by flocculation using aluminium sulphate and calcium hydroxide as effective flocculants, whereas the multi-cellular filamentous algae are harvested manually using nylon nets. Once harvested, the algae are de-watered and dried in open sand beds or using dryers. Figure 14.2 shows the shallow pond cultivation of algae.

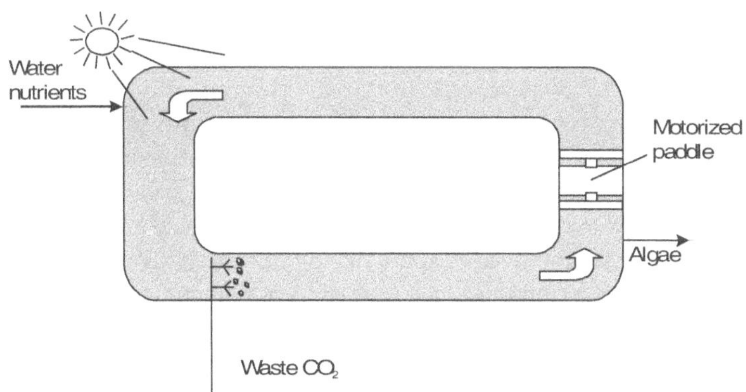

Figure 14.2   Shallow pond cultivation of algae

However, the shallow, open-air ponds are vulnerable to contamination. Serious problems of contamination have not been encountered, although occasional infestation of the culture with *Brachionus* and *Chironomus* larvae develop during certain seasons. In waste stabilization ponds, contamination with other algae is frequently seen, but this could be overcome by a gradual population build-up of the desired organism.

Production of *Spirulina* Biomass

Recently, the production of SCP from *Spirulina* has received a great attention. Among the algal population, *Spirulina* is widely used as a nutrient supplement since it possesses a broad range of nutrients.

Other than the major protein and carbohydrates, it contains vitamins, minerals and essential amino acids.

Table 14.2 Nutrient contents of *Spirulina*

Vitamins	Minerals	Amino acids
β-Carotene	Calcium	Lysine
Biotin	Phosphate	Cysteine
Vitamin B$_{12}$	Iron	Methionine
Folic acid	Sodium	Phenyl alanine
Riboflavin	Potassium	Threonine
Thiamine		
Tocopherol		

Moreover, the cultivation of *Spirulina* has several advantages. The growth of *Spirulina* will be high at alkaline pH, and contamination by other organisms is less possible. Since it has gas vacuoles, it floats over the medium, which facilitates easy harvesting. It has a thin cell wall, and so it requires less heat during drying, and no need of spray drying like other algal strains. It shows high digestibility as it has less cell wall content and nucleic acid.

Spirulina may be cultivated in semi-natural lake system or artificial ponds. However, semi-natural lake system may yield less quality product due to contamination by other organisms. On the other hand, an artificial pond system uses fresh water and provides necessary light, pH and temperature, and even a waste water system can be set up. In a waste water system, the solid particles settle through settle tanks and the effluent is allowed to flow into the artificial ponds. *Spirulina* is seeded over the effluent and allowed to grow. It is harvested manually using nets, and dried with heat dryers.

PRODUCTION OF SCP FROM FUNGI

Fungal SCP is of much less importance because it has least quantity of amino acid methionine and high content of nucleic acid. The chemical composition and amino acid composition of SCP as well as soybean meal is given in Tables 14.3 and 14.4 respectively. However, the production has some significance because fungal strains can be grown in organic wastes, and some fungi can grow using cellulose as a carbon source which is available in abundance. Hence, cellulose is used as a main substrate for the production of fungal SCP. Fungal strains that grow over cellulosic wastes are *Trichoderma* spp., *Aspergillus* spp., *Rhizopus* sp. and *Fusarium* spp.

The production is carried out by solid-state fermentation of semi-solid cellulosic wastes and organic wastes. Usually, cellulosic materials are treated with alkali prior to inoculating the organisms. This process helps break the glucosidic bonds and facilitate better fungal growth. The treated medium is then acetified to bring the pH to acetic condition. Further, the medium is amended with ammonium salts as nitrogen source and minerals. The pure fungal cultures are seeded over the medium and the room temperature is applied. The recovery of fungal mass is by decanting and filtration through porous plate filters.

Table 14.3   Comparison of chemical composition (%) of SCP with soybean meal

	Yeast	Bacteria	Fungi	Algae	Soybean meal
Dry matter	96	90	86	94	88
Ash	6	8	2	7	6
Organic matter	90	81	84	87	82
Protein	54	63	27	49	42
Crude fat	9	8	5	15	1
Crude fibre	-	-	28	11	6
Nitrogen-free extract	20	-	20	12	30

Table 14.4   Amino acid composition of SCPs and soybean meal

	Yeast	Bacteria	Fungi	Algae	Soybean meal
Lysine	7.0	5.5	4.8	4.6	6.2
Methionine + cysteine	2.9	3.1	2.5	3.2	2.9
Arginine	4.8	4.7	5.2	-	7.2
Histidine	2.0	1.9	2.0	-	2.5
Isoleucine	4.5	3.9	4.1	3.1	4.9
Leucine	7.0	6.3	6.4	7.0	7.6
Phenylalanine + tyrosine	7.9	6.2	8.1	6.0	8.4
Threonine	4.9	4.2	4.4	4.9	4.2
Tryptophan	1.4	0.8	1.4	1.7	1.3
Valine	5.4	4.8	5.6	4.7	5.0

REVIEW QUESTIONS

1. What is SCP? Give its uses.

2. Mention the different types of bacteria used as SCP.

3. Write about algal SCP production method.

4. Describe in detail about *Spirulina* cultivation.

5. Give an account on fungal SCP.

MUSHROOM

Mushrooms are members of macro-fungi and belong to the group Ascomycetes and Basidiomycetes. They show heterotrophic mode of nutrition. The fruiting body (carp) of fungi is usually edible and rich in proteins. The Basidiomycetes members like *Morchella* and Tuber and the Ascomycetes members like *Agaricus, Auricularia* and *Tremella* are widely used in the commercial production of mushrooms. Some mushrooms (e.g. *Amanita verna* and *A. virosa*) are highly poisonous. The nutritional values of *Volvariella* mushroom is given in Table 15.1.

MUSHROOM CULTIVATION

Mushroom cultivation involves mainly four steps: 1) raising pure cultures, 2) preparation of spawn, 3) preparation of substrates, and 4) spawn running and cropping.

Raising Pure Cultures

A pure fungal culture would be raised and maintained in a laboratory using potato dextrose agar or malt extract agar. Sugar and starch present in the medium will enhance the growth of fungi. To raise a pure culture, a piece of fruiting body from previous running bed will be used as an inoculum. Before inoculation, the tissue will be washed with distilled water to remove soil particles followed by washing with 70 per cent ethyl alcohol as surface sterilization. The piece will then be cut into two halves and placed asceptically into agar medium for further growth.

Table 15.1   Nutrient value of *Volvariella* mushroom

Nutrient per 100 g of edible portion	Fresh	Dried
Moisture (%)	87.7	14.9
Food energy (calories)	39.0	274.0
Protein (g)	3.8	16.0
Fat (g)	0.6	0.9
Total carbohydrate (g)	6.9	64.6
Fibre (g)	1.2	4.0
Ash (g)	1.0	3.6
Calcium (mg)	3.0	51.0
Phosphorus (mg)	94.0	223.0
Iron (mg)	1.7	6.7
Thiamine (mg)	0.11	0.09
Riboflavin (mg)	0.17	1.06
Niacin (mg)	8.3	19.7
Ascorbic acid (mg)	5.0	-

Preparation of Spawn

Spawn is the growth medium with mycelia, which serves as the inoculum for cultivation. Care should be taken to avoid contamination by bacteria, mycophages and other fungi. A variety of substrates may be used to prepare spawn (e.g. paddy straw, wheat straw, grains of rice, wheat, sorghum and rye) (Figures 15.1 and 15.2). A substrate may be used either alone or combined with other substrates. Sterile substrates are mixed with the mycelium of fungi and incubated at suitable temperature. In case straw is used, it is chopped into small pieces of 6 to 10 cm and autoclaved. The pure fungal mycelium will be seeded with sterile straw and incubated.

Figure 15.1 Spawn preparation in jarn

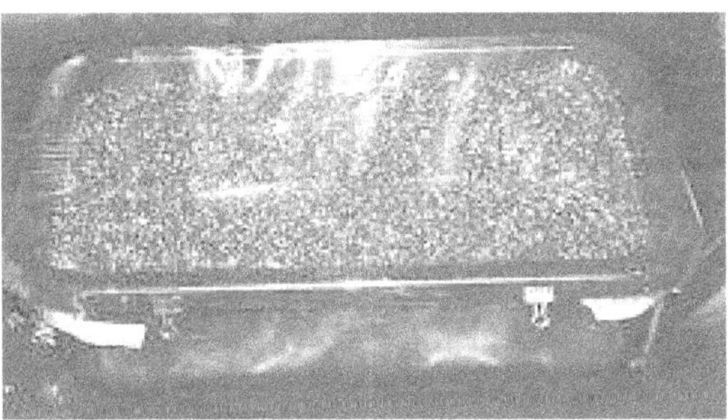

Figure 15.2. Spawn preparation in trays

Preparation of Substrates for Cultivation

We saw how substrates are prepared for the cultivation of mushroom. When several substrates are used in combination, some nitrogen sources are also added. The preparation of combined substrates along with nitrogen and mineral sources is called compost. Usually, composts are prepared in wooden boxes or polythene bags and placed in the cultivation room (Figure 15.3). The compost is pasteurized by passing steam at 50–60°C for five or six hours. The compost bed is usually covered with a thin layer of soil to maintain temperature during cultivation.

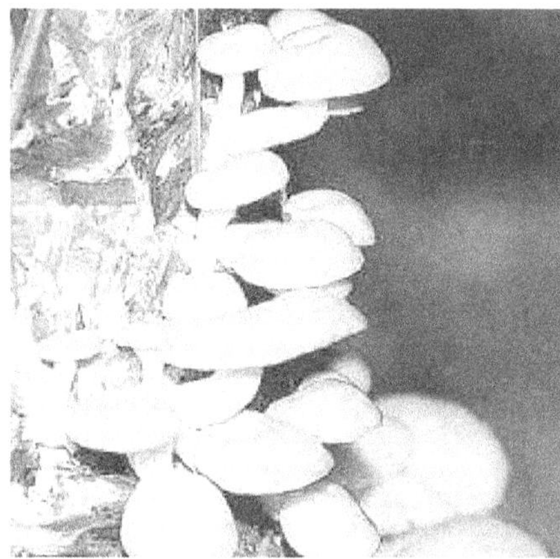

Figure 15.3   Mushroom cultivation in bags

Rice straw as bedding material   Long rice straw, thoroughly dried, is preferable. Properly prepared straw produces a better yield of mushrooms. Straws are cut to a uniform length of 45 cm, bundled to a size of 8 cm in diameter, tied at the middle with twine, and soaked in water for three hours. The straw bundles should cover the bed foundation to its whole length. The butt ends are placed in an alternating fashion. If the butt ends of the first layer are on the left side, the butt ends of the second layer must be on the right side. This manner of arrangement is continued until four to six layers are made. About 240 bundles are needed for a six-layer, 4-metre-long bed. Each layer must be pressed firmly to make the surface flat, and then the bed is watered.

Simultaneously, several newspapers are soaked in a container with aqueous urea. The mushroom spawn and soaked paper are distributed on the straw bed in thumb-sized pieces. The plantings should be 5 to 8 cm from the edge of the bed and 5 cm apart. For every six-layer bed, three bottles of spawn are used. One-half bottle of spawn is apportioned to plant one layer. The spawn is buried with the paper 4 cm deep in the layer. The same procedure is repeated for the remaining layers. Left-over straws are spread to a thickness of 10 cm at the centre.

The straw bed is protected with an elevated, transparent plastic sheet attached to a bamboo frame. During dry season, a four-layer bed is recommended because of lower relative humidity. Beds of six layers or more are possible in wet season.

Banana leaves as bedding material Dry banana leaves hanging on the plant are gathered and cut to a uniform length of 45 cm, bundled to a diameter of 8 cm, and soaked in water for three or four hours. The leaf bed is made in the same way as straw bed. Four- or six-layer beds may be constructed, depending on the season. The leaf bed should also be protected with elevated plastic sheet on a bamboo frame.

For both rice-straw and banana-leaf beds, no water should be added in the first five days after planting. During dry season, the bed may be watered gently on the sixth or seventh day after planting and repeated daily until the mushroom pinheads have developed. During rainy season, the bed may not need watering, or at least not as much as during dry months. When the mushrooms are at pin-head stage, the bed should not be watered. Water should be applied again when the mushrooms reach the size of corn seeds and the bed has become dry.

Spawn Running and Cropping

Spawn running is the main part in mushroom cultivation. Spawing is a seedling method in which spawn is spread over the compost (Figure 15.4) and the bed is covered with polythene to maintain temperature and moisture. Temperature and moisture are important factors that govern the growth and production of mushrooms. The main drawback in spawn running is pests and nematodes that destroy the fruiting bodies in mushrooms. They can be controlled with extracts of castor leaf, neem leaf, chrysanthemum and marigold. Either the extracts are sprinkled over the bed or the dry leaves are mixed with the compost during preparation.

Figure 15.4   White button mushroom in compost bed

The maturation of mushrooms happens at intervals. When mushrooms mature, they are harvested without disturbing nearby immature beds. Under normal conditions, the first harvest of mushrooms happens 10–14 days after planting. The harvest usually lasts for three days. This is called first flush. The average daily production is 1.2 kg. The bed is allowed to rest for five or seven days, and the second harvest continues. The average production for second harvest is 0.42 kg. This manner of production may continue for a month or even longer.

Growing Mushrooms under Semi-controlled Conditions

Recently, it has been demonstrated that *Volvariella* mushrooms can be grown on a commercial scale in special growing houses. Compared to the open-bed method, the yield of box-grown mushrooms has improved from 10.5–25 per cent. There is greater assurance of consistent production because pests and diseases are more easily controlled. The only possible drawback is the higher cost of constructing the growing houses and boxes.

Growing houses   A growing house consists of a concrete floor set on a gravel base, four layers of hollow blocks set on the floor, wooden frames, plastic walls, plain galvanized sheets for roofing, and a plastic screen for ventilation. The roof is convex for better air circulation, and there are two ventilators with hinged covers, properly screened to keep away insects. The room has larger width and there are several rows of one-foot-high cement blocks where mushroom boxes are accommodated. In this way, a series of growing houses can be constructed.

Making the box   The box is made of wood. The wooden frame is 60 cm long, 45 cm wide, and 20 cm thick. The bedding materials (rice straw or dry banana leaves) are cut to the length of the box. The box is filled to the brim in anticipation of loosening when the pack is soaked in water. Protruding materials are trimmed to avoid obstruction during watering. Then the box is soaked for at least three hours in water or until the straw becomes dark-brown or the banana leaves exhibit a certain degree of transparency. The box is then removed from the soaking tank and excess water is drained off (Figure 15.5).

Planting the spawn   Young spawn of thumb-sized pieces are distributed on the surface of the bedding material. Starting on one side of the box, four pieces are distributed along the width and five pieces along the length at approximately 5 cm from the edge of the box. The spawn is buried deep in the bedding material, and the surface is massaged to close the spaces. The procedure is repeated on the other side of the box.

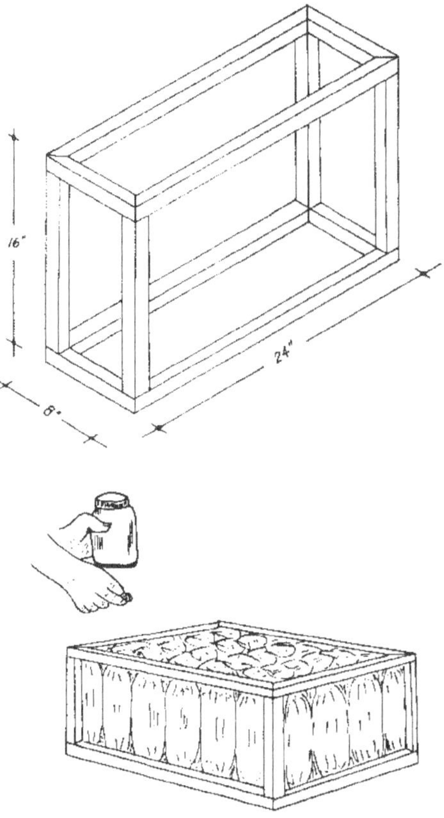

Figure 15.5   Bed preparation

Incubation of boxes   The boxes are placed in a special incubation room, where the temperature is maintained between 35–38°C with high relative humidity. The boxes can be covered with plastic sheets. The boxes are removed from the incubation room after three days, or once a good spawn run has been obtained. While the box is covered with plastic, it may take five days to attain good mycelial growth.

The incubated boxes should not be exposed to conditions that may cause serious stress on mycelial growth. Twenty-four hours after removing the boxes from the incubation room, the temperature should be lowered gradually by opening the ventilators or by allowing fresh air to circulate in the growing room. The temperature should be maintained between 26 and 28°C with a relative humidity of 75–85 per cent. Spraying superfine mist will maintain the desired relative humidity, and the preferable temperature can be achieved by manipulating ambient air through aeration or closing the ventilators. The bedding material must be watered with fine mist to avoid destroying the delicate mycelial threads of growing mushroom moulds.

REVIEW QUESTIONS

1. What are mushrooms? Mention the organisms used as edible mushrooms.

2. Explain in detail about mushroom production.

3. What is spawn? Give a detailed account on spawn preparation.

4. What are the different substrates used for mushroom cultivation? Write about their preparations.

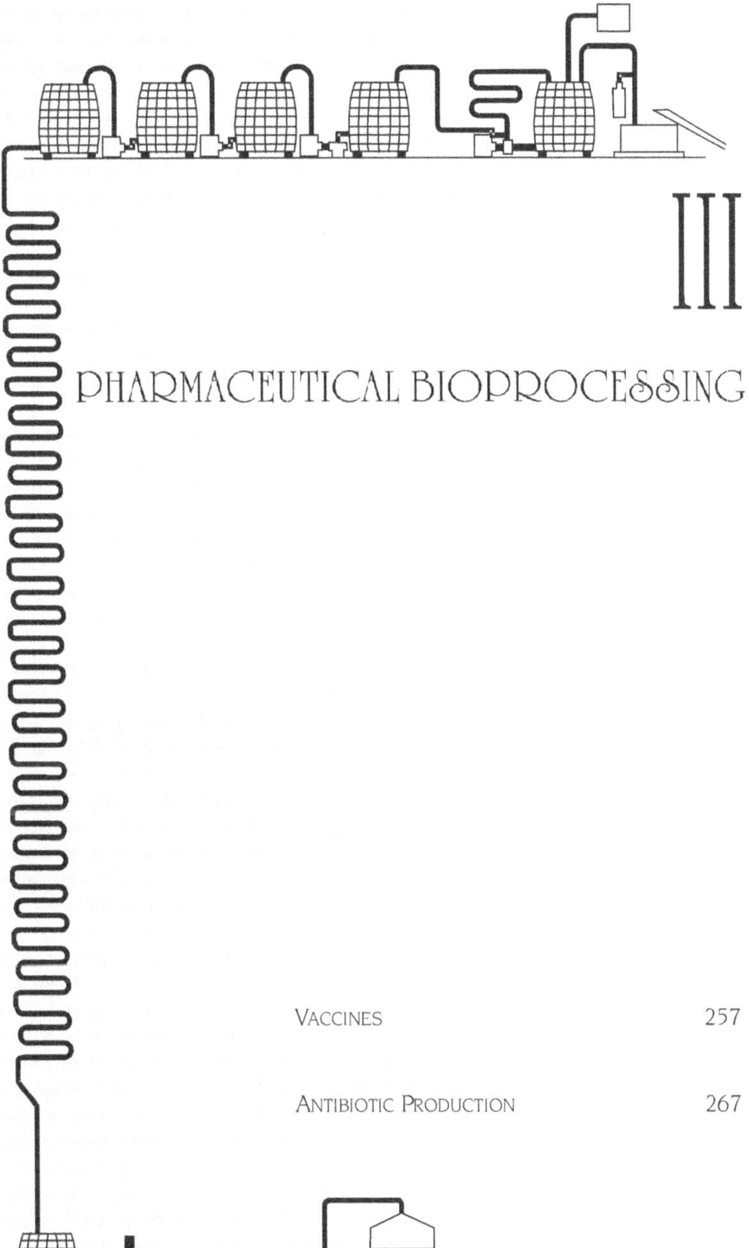

III

PHARMACEUTICAL BIOPROCESSING

VACCINES

INTRODUCTION

Vaccines are agents that give immunity or induce immunity against pathogens. The prevention of a disease with vaccines is called vaccination. Vaccines either inactivate or attenuate the whole cell of pathogens or the immunologically active compounds produced from pathogens. Such a vaccine is generally called active vaccine because it will not interact with pathogen directly, and it gives protection by provoking the immune system against the pathogen. Sometimes, antibodies are administrated as vaccines. These kinds of vaccines are known as passive vaccines because they interact with specific pathogens without stimulating the immune system.

TYPES OF VACCINES

Killed or Inactivated Vaccines

The whole cell of killed pathogenic organisms is given as vaccine. Heat or chemicals are used to kill these organisms. These vaccines are less effective.

Attenuated Vaccines

The ability of a pathogen to cause a disease in the host is called pathogenicity. Pathogenicity can even be removed by growing the pathogen in specific environment or by chemical treatment. The removal of pathogenicity is generally called attenuation. Attenuated vaccines show high efficiency over inactivated vaccines.

16

Toxoid Vaccines

Some pathogens cause diseases through exotoxins, so the vaccine should inhibit the toxin. Toxins are used as vaccines after detoxification with heat or chemicals.

Macromolecular Vaccines

The purified, immunologically-active compounds extracted from pathogens are called macromolecular vaccines.

Recombinant Vaccines

Recombinant vaccines are employed when other types of vaccines do not succeed. The gene coding for the antigenic molecule is cloned into the compatible organisms to produce vaccines. Sometimes, the whole cloned cell is used as vaccine where the expression of macromolecules on the surface of the clone acts as immune inducer.

Vaccines may be administrated into the body either alone or adsorbed with adjuvants. Adjuvants are compounds adsorbed with vaccine, which enhance the prolonged exposure of vaccine into the immune system.

PRODUCTION OF VACCINES

The steps involved in the production of vaccines include the following.

1. Selection of Materials

The selection of appropriate organisms is known as seed lotting. The organisms are derived from master cell seed. The master cell seed (MCS) is a collection of aliquots of cells for use in the preparation of a product, distributed into containers, processed together in such a manner as to ensure uniformity, and stored in such a manner as to ensure stability. The master cell seeds are prepared by collecting and processing the pathogens from the host tissues and stored.

Bacterial seed

General requirements   The genus and species of bacteria used in vaccine shall be stated. The origin, date of isolation and designation of bacterial strains and details of media used at each stage shall also be given. The bacteria used in production shall be derived from a seed lot system wherever possible. Each master seed lot shall be tested. A record of the origin, passage history (including purification and characterization procedures), and storage conditions shall be maintained for each seed lot. Each seed lot shall be assigned a specific code for identification purposes.

Identity and purity   Each seed lot shall be shown to contain only the species and strain of bacterium stated. A brief description of the method of identifying each strain by biochemical, serological and morphological characteristics shall be provided, as shall also the methods of determining the purity of the strain. If the seed lot is shown to contain living organisms of any kind other than the species and strain stated, it is unsuitable for vaccine production.

Seed lot requirements   The minimum and maximum number of subcultures of each seed lot shall be specified. The methods used for the preparation of seed cultures, preparation of suspensions for seeding, techniques for inoculation of seeds, titre and concentration of inocula and the media used shall be described. It shall be demonstrated that the characteristics of the seed material (e.g. dissociation or antigenicity) are not changed by these subcultures.

The conditions under which each seed lot is stored shall be described.

Virus seed

General requirements   Viruses used in production shall be derived from a seed lot system. Each master seed virus (MSV) shall be tested. A record of the origin, passage history (including purification and characterization procedures), and storage conditions shall be maintained for each seed lot. Each MSV shall be assigned a specific code for identification purposes. The MSV shall normally be stored in aliquots at –70°C or lower if it is in liquid form or at –20°C or lower if in a lyophilized form. The production of vaccine shall not normally be undertaken using virus more than five passages from MSV. The nature of virus seed is maintained by cultivating them in the specific host like embryonated eggs and cell lines. Periodical batch tests will be conducted for sterility of the material and mycoplasma contaminations.

2.   Large-scale Cultivation of Organisms

Cultivation of bacterial strains   Bacteria are generally cultivated by fermentation techniques. The fermentation conditions like formulation of medium composition and process optimization differ from one organism to another. Aerated batch cultivation is widely used for obtaining biomass. For anaerobic cultivation, aeration is achieved by passing suitable sterile gases other than oxygen. Generally, the freeze-dried seeds are pre-propagated in inoculum medium until it reaches the log phase of growth and subsequently transferred to the production medium for better propagation and production. In most cases, the ingredients of the production medium have nutrients that bring high rates of bacterial biomass since either live or killed cells are needed for the preparation of vaccines except toxoid vaccines. In case of toxoid vaccines, the production medium should have some inducible materials and conditions for the production of toxins. Besides, care should be taken over the composition not to suppress the production of valuable toxins.

Cultivation of viral strains   Unlike bacterial strains, virus propagation needs specific biological systems. As far as vaccine production is concerned, the chick embryo and monolayer cell cultures are the biological systems widely employed. Even in the biological system, viruses show specific locations for its maximum growth. The embryonated eggs are conveniently used to grow the influenza virus and yellow fever virus. The influenza virus grows and accumulates specifically in allantoic fluid, and the yellow fever virus accumulates in the nervous system of chick embryo. Hence, embryonated eggs are rich media for most organisms. However, care should be taken to avoid contamination of various avian viruses and other organisms. During viral cultivation, sterility checks are made to avoid general contaminants like mycoplasma and phages.

When monolayers are used in the propagation of virus, the cells must be tested for contamination by viruses.

The tested monolayers must be at least 70 cm², prepared and maintained using medium and additives, and grown under conditions similar to those used in the preparation of the biological product. The monolayers must be maintained in the culture for at least 28 days. The subcultures should be made at 7-day interval, unless the cells do not survive for this length of time, in which case subcultures should be made on the latest day possible. Monolayers must be examined regularly throughout the incubation period for the presence of cytopatheic effects (CPE) and at the end of the observation period for CPE, haem adsorbent viruses and specific viruses by immunofluorescence and other appropriate tests.

3.   Processing

Harvesting of cells   The method of harvesting cells differs from one type of vaccine to other. In live attenuated vaccines, there is no need to take much effort because the attenuated pathogen is cultivated and it is harmless. In case of toxoid and killed vaccines, the pathogens should be cultivated and processed in safety conditions. Generally, the harvesting of cells is by means of centrifugation in aseptic manner. Ultra filtration and precipitation of cells by reducing the pH are other techniques used to harvest cells. In toxoid vaccines, after harvesting, the supernatant is preserved and processed to purify the toxins. In other cases, the cells are preserved and processed further. In the preparation of influenza vaccine, the allantoic fluid of the embryonated egg along with virus is centrifuged to concentrate the particles and re-suspended in sterile water. The yellow fever virus accumulates at the nervous system of the chick embryo. The nervous system will be crushed to recover the viral particles and re-suspended in sterile distilled water and centrifuged to separate the debris. The supernatant along with viral particles are preserved for further processing. When those viruses are cultivated in cell culture, the infected fluid is clarified satisfactorily by filtration.

Killing   The killing process is mainly employed for killed vaccines and macromolecular vaccines where non-viable cells are utilized. Usually, heat and disinfectants are used to carryout this process. For example, heat or chemicals like thiomersal and formaldehyde are used to kill *Bordetella pertussis* for the preparation of pertussis vaccine. Phenol is used to kill *Vibrio cholerae* and *Salmonella typhi* for the preparation of Vibrio and Salmonella vaccines.

Fractionation   Cell destruction is employed for the production of macromolecule vaccines. In this process, the harvested cells are purified and immunogenic macromolecules are removed by cell destruction using physical or chemical methods. The polysaccharide antigens of *Neisseria meningitidis* are separated by treating the cells with hexadecyl trimethyl ammonium bromide followed by extraction using calcium chloride and ethanol. The *Streptococcus pneumoniae* are extracted with sodium deoxycholate followed by ethanol precipitation. After processing, the molecules are purified by ultra filtration and dialysis, and the purity is improved by washing with suitable solvents.

Detoxification   Detoxification is a process employed in the preparation of toxoid vaccines. After harvesting the cells, the extracellular toxins present in the exhaust medium are separated by salt precipitation, dialysis or solvent extraction methods. Usually, chemicals are used to carry out detoxification. Detoxification may be carried out in the whole culture before harvest or after separating the toxins. However, the whole culture detoxification is advisable even though the product quality will be comparatively less. Formaldehyde is used to detoxify the toxins of *Corynebacterium diphtheriae*, *Clostridium botulinum* and *Clostridium tetani*. The toxins of *Bordetella pertussis* are detoxified by treatment with tetranitromethane and hydrogen peroxide.

Adsorption   Adsorption is a process employed only for adsorbed vaccines. Vaccines that are combined with adjuvants are called adsorbed vaccines. Adjuvants are agents that prolong the exposure and release of vaccines into the biological systems. Most often used mineral adjuvants include aluminium hydroxide and aluminium phosphate, and rarely calcium phosphate. Vaccines for diphtheria, tetanus and pertussis are prepared as adsorbed vaccines.

Conjugation   Conjugation is a process mainly used in macromolecular and toxoid vaccines to improve immunogenicity. Some biological molecules are adsorbed with vaccines to increase immunogenicity. For example, the immunogenicity of influenza vaccine may be greatly increased when it is combined with tetanus toxoid.

4. Blending

Blending is a process in which the final bulk product is made by adding appropriate diluents and recommended additives. Blending of vaccines is done in stir tank batch fermenters. For blending bacterial vaccines, the required volume of diluents is taken into the fermenter and sterilized. Additives like sterile preservatives and the required concentration of the vaccine constantly flow into the vessel with continuous stirring sequentially. Thiomersal is a widely used preservative in most vaccines, and phenoxy ethanol is an alternative. For blending viral vaccines, it is necessary to maintain their immunogenicity and infectivity because dilution may preclude the nature of the vaccine. So mostly it may be remain undiluted or it will be with less dilution.

5. Filling and Drying

The bulk vaccines are dispensed into necessary volume based on the size of dosage. Usually, the single dose is packed in ampoules and multi dose is packed in vials. Powder vaccines are subject to freeze drying.

6. Quality Assessment

The quality assessment of vaccines is carried out in two steps—first at the starting point of production and next on the final product. Further, general safety tests are carried out over the product before distribution.

In-process control   The in-process control is mainly carried out over the seeds and substrates to check sterility and natural characters.

Identity and purity checks for bacteria   Each seed lot shall be shown to contain only the species and strain of the bacterium stated. A brief description of the method of identifying each strain by biochemical, serological and morphological characteristics and distinguishing it as far as possible from related strains shall be provided, as shall also the methods of determining the purity of the strain. If the seed lot is shown to contain living organisms of any kind other than the species and strain stated, it is unsuitable for vaccine production.

Identity and purity checks for virus   The virus seed shall be shown to contain only the virus stated. The virus is identified using serological methods like ELISA and EIA. The MSV shall pass the tests for sterility and freedom from mycoplasma contaminations. The virus titre will be checked using appropriate host. The animal experiments are carried out for the retainment of attenuation.

Tests for cell lines   After preparation of the vaccines, the growth ability of the live viral particle and other characters are checked up with animal cell lines.

Detection of cytopatheic viruses   Two monolayers of at least 6 cm^2 each must be stained with an appropriate cytological stain. The entire area of each stained monolayer is examined for any inclusion bodies, abnormal numbers of giant cells or any other lesion indicative of a cellular abnormality, which might be attributable to a contaminant.

Detection of haemadsorbent viruses   Monolayers totalling at least 70 cm^2 must be washed several times with an appropriate buffer and a sufficient volume of a suspension of appropriate red blood cells is added to cover the surface of the monolayer evenly. After different incubation times, the cells are examined for the presence of haemadsorption.

Detection of specified viruses   Tests should be carried out for contaminants specific to the species or origin of the cell line and for the species for which the product is intended. Sufficient cells on appropriate supports must be prepared to carry out tests for the agents specified. Appropriate positive controls must be included in each test. The cells are subject to appropriate tests using fluorescein-conjugated antibodies or similar reagents.

Final product control   Final product control is the quality assessment over the final product. Samples are taken from each batch of the product and the following tests are done.

Tests for efficacy   This test checks the potency of the vaccines produced. The potency of live cell vaccines is estimated based on the viable cells present in them and multiplication capacity in appropriate medium or host. The live bacterial vaccines are suspended completely in sterile distilled water and subject to total viable count check in a suitable medium. The potency is estimated by comparing the total viable count of the product with standard count. In case of viral vaccines, the viral particles are made to infect the monolayer of the cell culture and the number of cells infected is calculated. The potency is assayed by comparing the number of cells infected by virus particles with standard count. On the other hand, the potency of killed and toxoid vaccines is estimated by administrating these vaccines. Based on time taken by the animal to recover, the potency is estimated. The other kind of potency test is checking the immunogenicity of the product. Healthy test animals are vaccinated with appropriate dose. Periodically, the serum is collected from the animals and the antibody titre is observed. The potency of vaccine is estimated by comparing the antibody titre with standard titre.

Safety test Safety test over the final product is important because vaccines are basically prepared from pathogens. Inadequate processing may cause harmful effects. Generally, killed bacterial vaccines are inoculated with suitable medium and observation is made for the growth. The absence of growth provides the assurance of safety. Likewise, vaccines composed of killed viral particles are inoculated into the suitable cell culture and the cells are examined for the growth of viral particles. In the same manner, toxoid vaccines are tested on animals. Observations are made for any manifestation caused by the toxins.

Live attenuated vaccines are tested by administering them on susceptible animals. Periodical observations are made over the animals for infection. Besides, live attenuated viruses are made to infect through cell lines, and observations are made for any cytopatheic effects of the cell and other abnormalities.

General tests for application

Sterility test In general, vaccines should be free from contaminants. The sterility of bacterial vaccines may be tested either by inoculating it into a bacteriological medium or by passing it through a filter.

General toxicity and abnormality Usually, the components of a vaccine may have some sort of toxicity, which may not cause considerable effects. However, toxicity in vaccines should be tested on animals. Generally, two or four healthy Guinea pigs are vaccinated with a single human dose, and the entire animal should survive four days. This indicates sustainable toxicity in the product.

Pyrogenicity test   Apart from other tests, it is necessary to monitor the pyrogenicity of the prescribed dose of the vaccine. It can be assayed with pharmacopoeial method using rabbit as test animal.

Test for chemicals   Many chemicals are used as adjuvants and preservatives in the production of vaccines. The concentrations of chemicals should be as advised in pharmacopoeia. Generally, the concentration of minerals like aluminium and calcium used as adjuvants in adsorbed vaccines should not exceed 1.25 mg per dose. Likewise, the concentrations of phenol and formaldehyde used as preservatives should not exceed 0.25 w/v and 0.02% per dose, respectively.

REVIEW QUESTIONS

1. Define vaccine.

2. What is attenuation?

3. Give an account on different types of vaccines.

4. Describe the steps involved in bacterial vaccines production.

5. Explain in detail about viral vaccine production.

6. Explain in detail about quality control of the vaccines.

ANTIBIOTIC PRODUCTION

INTRODUCTION

Although Fleming discovered penicillin in 1929, its large-scale production was not achieved until early 1940s when Florey realized the potential of this drug in the treatment of wartime casualties. Initially, penicillin was produced in milk bottles. Over time, it became apparent that deep tank fermenters be used to achieve efficiency and more productivity.

The early success in isolating a medically useful compound from a microorganism leads to a massive hunt for antibiotics that continues to the present day with over 5000 compounds. By far, the largest proportion has been isolated from *Streptomyces* cultures. However, only a small number of compounds have been commercially successful in therapy. Penicillins and cephalosporins account for approximately 70% of market share. These are used either in their natural form or as semi-synthetic derivatives.

Antibiotics are produced as a defence mechanism generally, as a result of some environmental stress that triggers a metabolic response.

The organism will incur significant metabolic overhead when synthesizing these compounds under environmental stress. The extent of production may be exaggerated by subjecting the original microorganisms to mutagenic influences.

Antibiotics have widespread clinical applications as well as agricultural purposes. Antibiotics presently used in agriculture include the antifungal blasticidin S and cycloheximide; the antibacterial kasugamycin; and the peptides nosiheptin, siomycin and thiopeptin.

Antibiotics are also useful as growth promoting substances in poultry and animals.

CLASSES OF ANTIBIOTICS

There are several classification schemes for antibiotics based on bacterial spectrum (broad versus narrow), route of administration (injectable, oral or topical), or type of activity (bactericidal or bacteriostatic). The most useful classification is based on chemical structure. Antibiotics within a structural class generally show similar patterns of effectiveness, toxicity and allergic potential.

Based on chemical nature, antibiotics are divided into groups as β-lactam groups, amino glycosides, glycopeptide antibiotics, macrolides, quinolones and tetra cyclones.

Beta-lactam Antibiotics

β-lactam antibiotics are a broad class of antibiotics, which include penicillin derivatives, cephalosporins, monobactams, carbapenems, β-lactamase inhibitors and any antibiotic agent that contains a β-lactam nucleus (Figure 17.1). They are the most widely used group of antibiotics (Figure 17.2).

Penicillins   The narrow spectrum of activity of penicillins, along with poor activity of the orally-active phenoxymethyl penicillin, led to the search for derivatives of penicillin which could treat a wider range of infections. The first major development was ampicillin, which offered a broader spectrum of activity than the original penicillins to treat a number of both gram-positive and gram-negative infections. Amoxicillin was a subsequent development with improved duration of action.

Further, beta-lactamase-resistant penicillins were created, which include flucloxacillin, dicloxacillin and methicillin. These were effective against beta-lactamase-producing *Staphylococcus* species but ineffective against methicillin-resistant *Staphylococcus aureus* strains that subsequently emerged.

The last in the line of true penicillins were the antipseudomonal penicillins, such as ticarcillin and piperacillin. They are useful against gram-negative bacteria. However, the usefulness of the beta-lactam ring was such that related antibiotics, including the mecillinams, the carbapenems and, most importantly, the cephalosporins, have it at the centre of their structures.

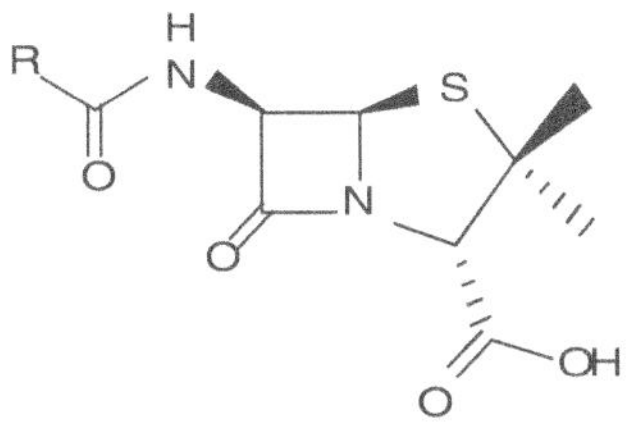

Figure 17.1   Penicillin nucleus 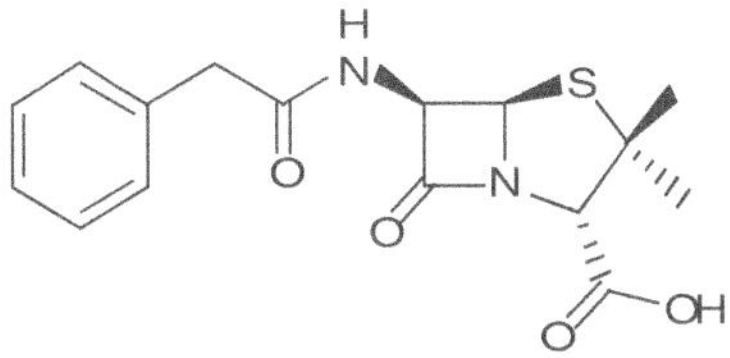   Figure 17.2   Penicillin G

Cephalosporins   Cephalosporins are a class of β-lactam antibiotics. Together with cephamycins, they belong to a sub-group called cephems. The cephalosporin nucleus (Figure 17.3) can be modified to achieve different properties. Cephalosporins may be grouped into generations. For example, the first cephalosporins were designated first generation while the latter, spectrum extended cephalosporins were classified as second generation cephalosporins. Each newer generation of cephalosporins has significantly greater gram-negative antimicrobial properties and, in most cases, decreased activity against gram-positive organisms. Fourth generation cephalosporins, however, have true broad spectrum activity.

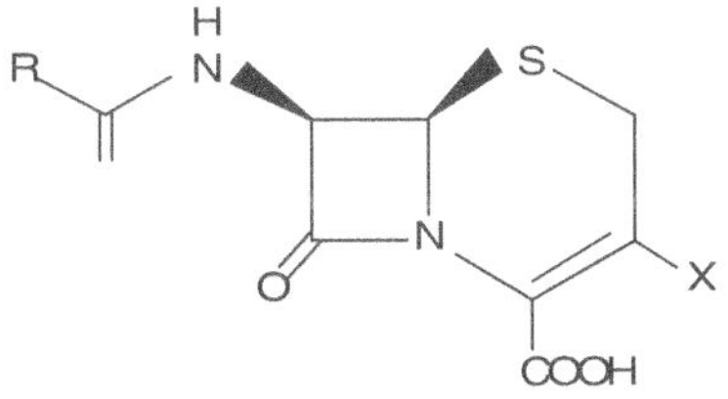

Figure 17.3   Cephalosporin nucleus

However, the classification of cephalosporins into generations is often imprecise. For example, the fourth generation of cephalosporins is not yet recognized in Japan. In Japan, cefaclor is a first generation cephalosporin; and cefbuperazone, cefminox and cefotetan are second generation cephalosporins. Cefbuperazone, cefminox and cefotetan are second generation cephems. Cefmetazole and cefoxitin are third generation cephems. Flomoxef and latamoxef are in a new class called oxacephems.

Carbapenems   Carbapenems (Figure 17.4) are a class of β-lactam antibiotics. Imipenem, meropenem, ertapenem, faropenem and doripenem belong to the carbapenem class.

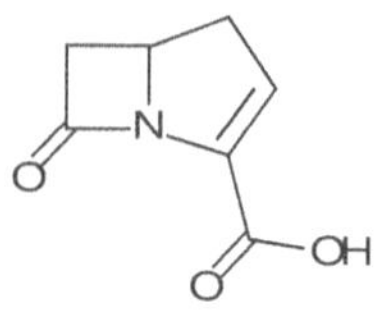

Figure 17.4   Carbapenem

Glycopeptide Antibiotics

Glycopeptide antibiotics consist of glycosylated, cyclic or polycyclic nonribosomal peptide. Vancomycin, teicoplanin, ramoplanin and decaplanin are common glycopeptide antibiotics. This class of antibiotics prevents the synthesis of cell walls in susceptible microbes by inhibiting peptidoglycan synthesis. Due to toxicity, their use is restricted in patients who are critically ill or who have demonstrated hypersensitivity to β-lactams. Principally effective against gram-positive cocci, they exhibit a narrow spectrum of action. They are the last line of defense for cases of methicillin-resistant *Staphylococcus aureus.*

Vancomycin is usually given intravenously. It may cause tissue necrosis and phlebitis at the injection site if given too rapidly. Pain at site of injection is a common adverse event.

Macrolides

Macrolides are a group of antibiotics whose activity stems from the presence of a macrolide ring (a large lactone ring to which one or more deoxy sugars, usually cladinose and desosamine, are attached). The lactone ring may be 14-, 15- or 16-membered. Macrolides belong to the polyketide class of natural products.

Commonly prescribed macrolides include erythromycin, azithromycin, clarithromycin (Figure 17.5), dirithromycin and roxithromycin (Figure 17.6).

Developmental macrolides includes: carbomycin A, josamycin, kitasamycin, oleandomycin, spiramycin, troleandomycin and tylosin.

Ketolides are a new class of antibiotics structurally related to macrolides. They are used to treat respiratory tract infections caused by macrolide-resistant bacteria.

Macrolides are used to treat infections of respiratory tract and soft tissue. The antimicrobial spectrum of macrolides is slightly wider than that of penicillin, and therefore macrolides are common substitutes for penicillin.

Beta-haemolytic streptococci, pneumococci, staphylococci and enterococci are usually susceptible to macrolides. Unlike penicillin, macrolides have been shown to be effective against mycoplasma, mycobacteria, some rickettsia and chlamydia.

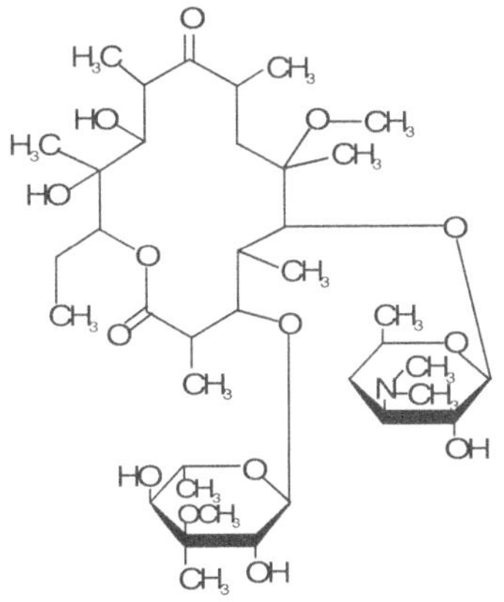

Figure 17.5   Clarithromycin

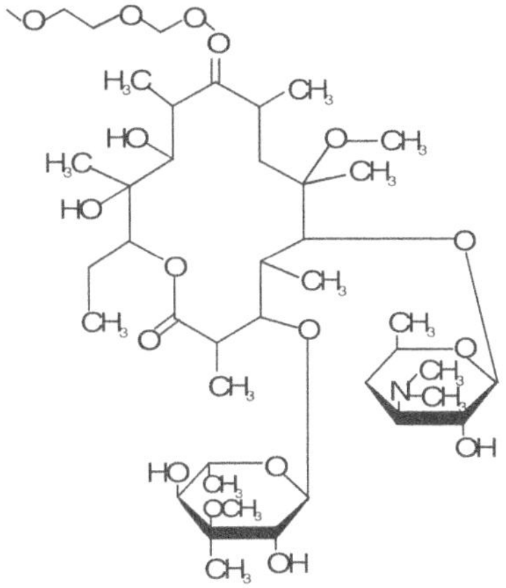

Figure 17.6   Roxithromycin

The mechanism of action of macrolide includes inhibition of bacterial protein synthesis by binding reversibly to the subunit 50S of the bacterial ribosome, thereby inhibiting translocation of peptidyl-tRNA. This action is mainly bacteriostatic but can also be bactericidal in high concentrations. Macrolides accumulate within leucocytes and are transported into the site of infection.

Quinolones

Quinolones are a family of broad-spectrum antibiotics. The parent of the group is nalidixic acid. A majority of quinolones in clinical use belong to the subset of **fluoroquinolones**, which have a fluoro group attached to the central ring system.

Quinolones and fluoroquinolones actively kill bacteria. Quinolones inhibit the bacterial DNA gyrase or topoisomerase IV enzyme, thereby inhibiting DNA replication and transcription. They can enter cells easily and therefore are often used to treat intracellular pathogens such as *Legionella pneumophila* and *Bacillus anthracis*. Quinolones are divided into generations based on their antibacterial spectrum. The earlier generation agents have generally more narrow spectrum than the latter ones.

Tetracyclines

Tetracyclines are a group of broad-spectrum antibiotics whose general usefulness has been reduced with the onset of bacterial resistance. However, they continue to remain the treatment of choice for some specific indications. Tetracyclines are so named for their four (tetra) hydrocarbon ring (cycl) derivation (ine) as shown in Figure 17.7.

The first member of the group was chlortetracycline (aureomycin), discovered in late 1940s by Dr. Benjamin Duggar from *Streptomyces aureofaciens*. Oxytetracycline (terramycin) was discovered shortly afterwards by AC Finlay from a similar soil bacterium called Streptomyces rimosus. Robert Burns Woodward revealed the structure of oxytetracycline enabling Lloyd H. Conover to successfully produce tetracycline as a synthetic product. Many chemically altered antibiotics developed in this group.

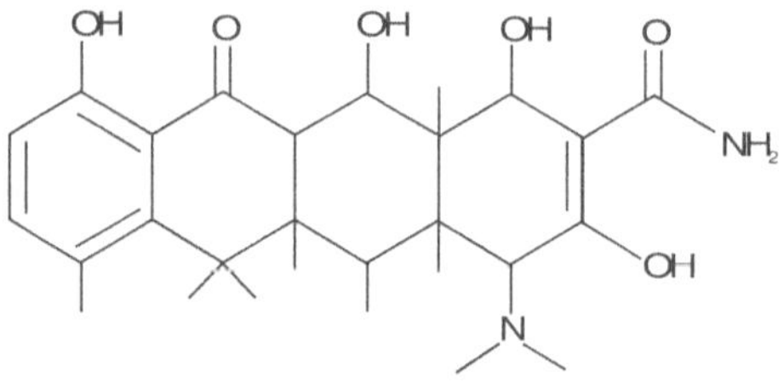

Figure 17.7   Basic tetracycline structure

Some naturally-occurring tetracycline groups are tetracycline, chlortetracycline, oxytetracycline, demeclocycline, etc.

Some semi-synthetic tetracycline groups are doxycycline, lymecycline, meclocycline, methacycline, minocycline, rolitetracycline, etc.

PRODUCTION OF ANTIBIOTICS

All antibiotics are produced in mechanically agitated and aerated vessels. The vessel must be able to run aseptically for prolonged time (maintaining positive pressure within the vessel).

A traditional fermenter (Figure 17.8) for production of antibiotics requires the following.

- ➲ Stainless steel construction
- ➲ Vessel height to tank diameter 2–4 : 1
- ➲ One Rushton impeller

⊃ Air sparged into the vessel below bottom impeller between 0.3 and 1.5 litre/minute

⊃ Power consumption in reactor in the order of 1–4 Watt/hour

The media formulation is carefully done to reduce the log phase of the organism since the antibiotics are secondary metabolites. Besides, the over production and accumulation of the antibiotic may cause feed back inhibition of production. Usually the media used for production are complex in nature. Many organisms produce their indigenous compounds on chemically defined media. However, complex media are cheaper to formulate and have higher yield and productivity.

The media must supply all the basic constituents for growth or product formation, namely: carbon source, nitrogen, phosphates and trace elements.

In antibiotic fermentation, the use of specific precursor molecules in media formulation is significant. For example, the production of penicillin G is greatly improved through the introduction of phenyl-acetic acid as a precursor into the fermentation medium. Nitrogen source, generally cheap supplements containing corn steep liquor is used for pencillin production.

Precursors   (phenylalanine and phenethylamine, soya flour and fish meal). These materials often supply trace elements, which aid production. Technical grade glucose or starch may be used as a carbohydrate source. Many antibiotics are not produced in the presence of excess carbon sources, especially glucose. In order to overcome catabolite repression, the addition of carbon source to the culture must be carefully controlled. The presence of excess nitrogen compounds or phosphates in fermentation may decrease antibiotic production severely.

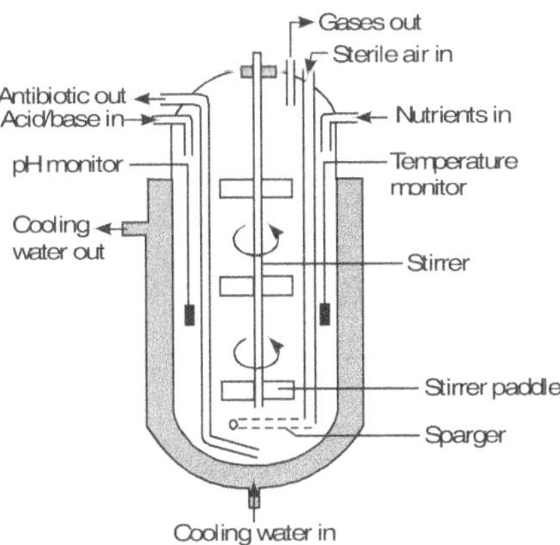

Figure 17.8   Fermenter used for antibiotic production

Antibiotics should be produced in sterile environment. The yield may be affected by the presence of antibiotic-resistant microorganisms or phages.

Extraction and recovery of antibiotics also requires concern. Antibiotics are usually present at very low concentrations (> 1g/l) and hence its recovery in pure form is a challenge.

PRODUCTION OF PENICILLIN

In 1929, Alexander Fleming noted that a fungal colony had grown as a contaminant on an agar plate streaked with the bacterium *Staphylococcus aureus*, and that the bacterial colonies around the fungus were transparent because their cells were lysing. Fleming recognized that a fungal metabolite might be used to control bacteria. The substance was named penicillin after *Penicillium notatum*. Fleming found that penicillin was effective against many gram-positive bacteria in laboratory conditions, and he even used crude preparations of this substance to control eye infections. However, he could not purify this compound because of its instability. However, during the Second World War (1939–1945), two other British scientists, Florey and Chain, working in the USA, managed to produce the antibiotic on an industrial scale. All three scientists shared the Nobel Prize for this work. Penicillin rapidly became the "wonder drug" to save millions of lives. It is still a "frontline" antibiotic for common bacterial infections although penicillin-resistance has developed in several pathogenic bacteria.

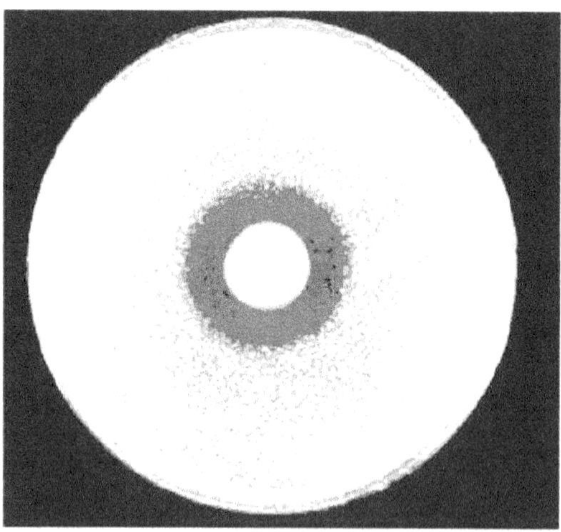

Figure 17.9   *Penicillium chrysogenum* shows antagonistic activity against streptococci

The two natural penicillins obtained from the culture filtrates of *Penicillium notatum* or *P. chrysogenum* are **penicillin G** and the more acid-resistant **penicillin V**. Penicillin is not made from *Penicillium notatum* any more; *Penicillium chrysogenum* is used instead because it has a higher yield. Figure 17.9 shows the antagonistic activity of *Penicillium chrysogenum* against streptococci.

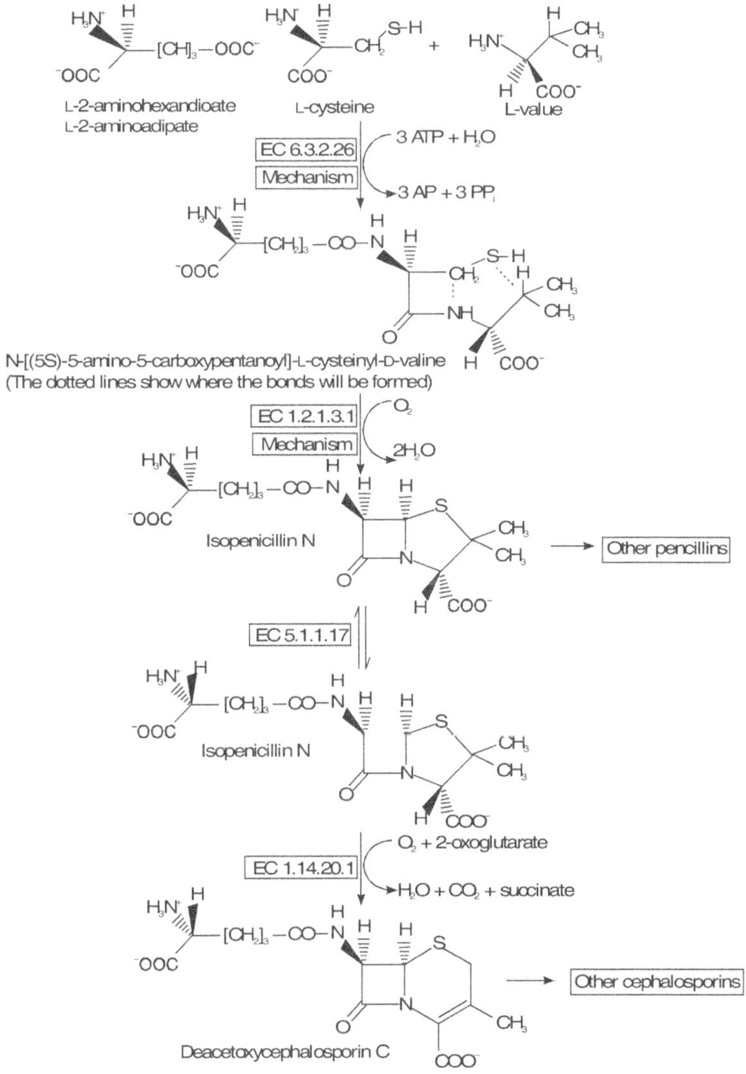

Figure 17.10 Biosynthesis of penicillin and cephalosporins

Table 17.1 Clinically important antibiotics

Antibiotics	Producer organism	Activity	Site or mode of action
Penicillin	*Penicillium chrysogenum*	Gram-positive bacteria	Wall synthesis
Cephalosporin	*Cephalosporium acremonium*	Broad spectrum	Wall synthesis
Griseofulvin	*Penicillium griseofulvum*	Dermatophytic fungi	Microtubules
Bacitracin	*Bacillus subtilis*	Gram-positive bacteria	Wall synthesis
Polymyxin B	*Bacillus polymyxa*	Gram-negative bacteria	Cell membrane
Amphotericin B	*Streptomyces nodosus*	Fungi	Cell membrane
Erythromycin	*Streptomyces erythreus*	Gram-positive bacteria	Protein synthesis
Neomycin	*Streptomyces fradiae*	Broad spectrum	Protein synthesis
Streptomycin	*Streptomyces griseus*	Gram-negative bacteria	Protein synthesis
Tetracycline	*Streptomyces rimosus*	Broad spectrum	Protein synthesis
Vancomycin	*Streptomyces orientalis*	Gram-positive bacteria	Protein synthesis
Gentamicin	*Micromonospora purpurea*	Broad spectrum	Protein synthesis
Rifamycin	*Streptomyces mediterranei*	Tuberculosis	Protein synthesis

Preparation of Inoculum

The inoculum preparation for *Penicillium chrysogenum* involves bringing the culture to fast-growing stage before it is inoculated into the fermentation broth. The initial inoculum preparation is done in solid media and later in stirred tank fermenter. The source of culture from natural environment is cultivated as lawn culture in agar plates. After reaching sufficient growth, the culture is inoculated in sporulation broth in shake flasks. Once the sufficient number of spores is formed, the culture from the sporulation broth is used as an inoculum for fermentation. The dry spores can be inoculated, wet pure cultures can be inoculated or the mycelial mass that are germinated from the spores can be inoculated.

Table 17.2   Composition of sporulation broth/agar (g/L)

Composition	Quantity (g/L)
Glycerol	7.5
Cane molasses	7.5
Corn steep liquor	2.5
Magnesium sulphate	0.05
KH_2PO_4	0.06
Peptone	5.0
Sodium chloride	4.0
Fe tartrate	0.005
Copper sulphate	0.004
Agar	2.0
Distilled water	1.0 L

The selection of raw material is crucial to the production of penicillin. The raw material should enhance mass mycelium and accumulate penicillium. It should be accessible to extraction and purification of penicillin. Usually, lactose is a satisfactory carbon source for the production of penicillin. Glucose and sucrose can be additional carbon sources. Nitrogen salts like sulphate, nitrates and ammonia are good sources of nitrogen. However, during production, the supply of ammonia gas can satisfy additional nitrogen source. Trace elements help grow the mould better. Corn steep liquor is the main ingredient, which contains sufficient amounts of magnesium and phosphates, so there is no need of additional phosphate and magnesium salts.

Penicillin is usually produced in large cylindrical vats, constructed of stainless steel, containing a liquid medium in which *Penicillium chrysogenum* is grown. If given a lot of nutrients, *Penicillium chrysogenum* will produce only small amounts of penicillin. This type of mould produces penicillin only if it competes with bacteria for food. If there is little food available, the genetics of the mould will allow for increased synthesis of penicillin. Usually the fermenters are operated in batch process.

After achieving a certain rate of fungal growth, followed by gradual production of the antibiotic, the contents are removed and processed to extract the antibiotics. Then the fermenter is cleaned, sterilized, and the process is repeated.

Penicillin Extraction

After 6 or 8 days of batch culture, the liquid medium is pumped out, filtered and concentrated. Rotatory vacuum filtration is often employed to separate the mycelium. The basic antibiotic (benzyl penicillin) is precipitated as crystals when potassium compounds are added. Another method of extraction is counter-current solvent extract method. In this method, the pH of the fermented broth is reduced to 2.0 by adding phosphoric or sulphuric acid. These favour to extract penicillin using solvents like amyl acetate, butyl acetate by increasing the partition coefficients. The solvent extraction against acidified fermented broth helps to purify the penicillin using other solvents.

The pencillin may then be modified by the action of other microorganisms or chemicals, before mixing with inert substances and pressed into tablets or converted into syrup or injectables. Although the molecular structure of penicillin is known, synthesizing it by chemical methods is not economical. The production process still relies on fungal fermentation based on biological principles, although modern strains are much more productive than the early strains. This has been achieved through screening programmes involving isolates from different sources, and treatment to encourage mutations.

PRODUCTION OF STREPTOMYCIN

Streptomycin is an amino glycoside antibiotic widely produced from *Streptomyces griseus*. Despite the great interest that streptomycin has evoked in recent years, very little is known about the biochemical production by *Streptomyces griseus*. It is believed that 28 enzymes are involved in the production of streptomycin from glucose through various metabolic activities. Glucose is converted into glucose-6-phosphate during glycolysis. Glucose-6-phosphate is further metabolized to form the precursor streptomycin-6-phosphate, which is subsequently converted into streptomycin.

The initial cultural conditions and inoculum preparations of *Streptomyces griseus* follow the routine. In the production medium, a number of carbon sources are used, including glucose, fructose, xylose, mannitol, maltose, starch and other sugars like arabinose, rhamnose, mannose, sorbitol, inositol and raffinose. Comparatively, polysaccharides bring lower yield than simple sugars. Glucose is better for the production of streptomycin. Peptones, meat extract, yeast extract, soy meal, ammonia and nitrate salts, and glycine are used as nitrogen sources. Proline gives better yield by slowing down the growth of *Streptomyces*. It is important that growth-stimulating compounds like phenyl acetic acid and alpha naphthalene acetic acid are added.

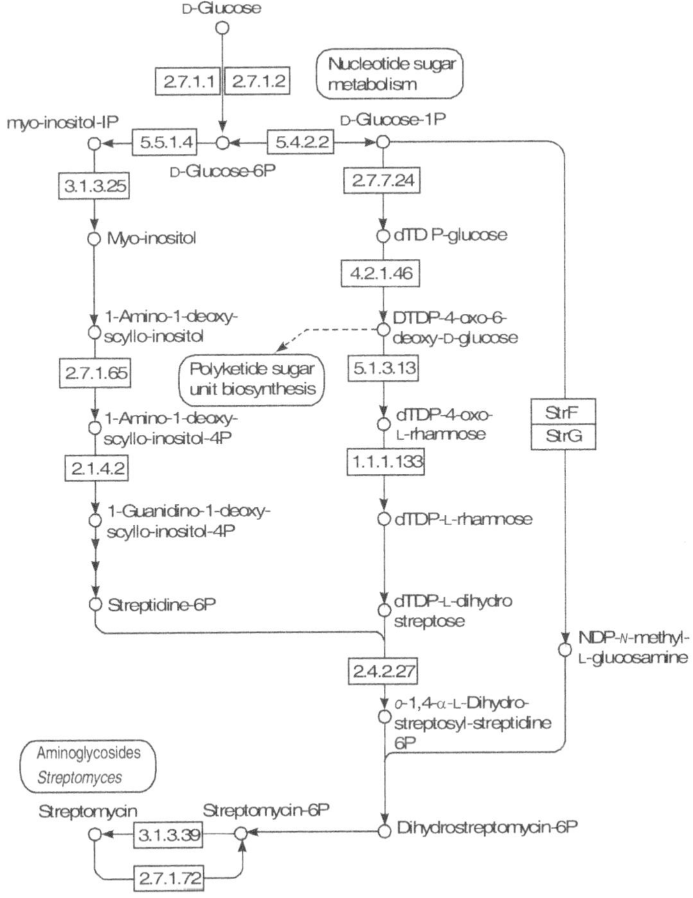

Figure 17.11   Streptomycin biosynthesis

In the fermenter, the production is carried out with optimum oxygen supply (0.5–1.0 vvm), temperature (28–30°C) and pH. Fermentation completes in four to seven days in three phases. The first phase is the growth phase in which mycelial growth occurs, which requires more oxygen and nutrients; in this phase, the pH is raised due to production of ammonia. In the second phase, the production of streptomycin occurs; in this phase, a gradual decrease of pH occurs due to reuse of the ammonia produced. This phase needs aeration. In the third phase, the mycelium undergoes autolyses because the medium is exhausted.

After cell separation, streptomycin is recovered by charcoal adsorption method. Further elution is done with aliphatic carboxylic acid, cyclo hexanol or phosphoric acid. The eluted product is further concentrated at 60°C under vacuum. Purification is done by ion exchange chromatography using aluminium oxide ion exchangers. Counter-current extraction with butanol is employed for extraction and purification of streptomycin.

PRODUCTION OF TETRACYCLINE

Tetracycline (TC) is a broad-spectrum antibiotic. It may be synthesized into some excellent semi-synthetic derivatives by modifying the chemical structure. Tetracycline is produced mainly from *Streptomyces aureofaciens* and *S. rimosus.*

Chlortetracyline and oxytetracycline are the major compounds produced from *Streptomyces*, where tetracycline is a minor product. In the case of *Streptomyces aureofaciens*, blocking chlorination will yield better production of tetracycline. In biosynthesis, glucose is converted into acetyl-CoA, which is further transformed into malonyl-CoA mediated with malonyl-CoA carboxylase. From malonyl-CoA, tetracycline is formed. There is a correlation between carbohydrate metabolism and tetracycline production. The high-yielding glycolysis strains will poorly yield tetracycline. The addition of benzyl thiocyanate can increase the production of tetracycline.

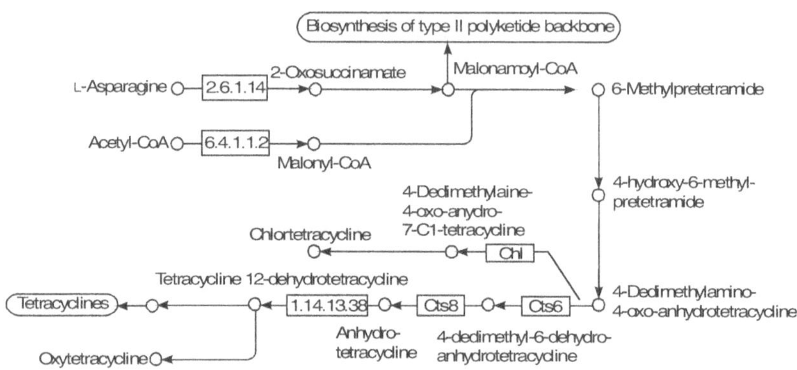

Figure 17.12   Tetracycline biosynthesis

In the process of inoculum preparation, the spores of *Streptomyces* taken from stock culture are inoculated in agar plates. The plates are incubated until fresh spores form on them. The second crop spores are inoculated in shake flask containing suitable inoculum broth. The inoculated broth is incubated until it forms fresh mycelium. The culture from the shake flask is used as the inoculum for production.

The immobilized cell inoculum can be used for the production of tetracycline. The inocula are incubated for 24 hours at 28°C in an orbital bench top shaker at 150 rpm within Erlenmeyer flask containing the growth medium. K-carrageenan gel beads are produced by mixing the inocula (50 ml containing 0.21 g of mycelia dry weight) with 30 ml solution of carrageenan, giving a final concentration of 1, 1.5, and 2% carrageenan, respectively. Each mixture is added to a sterile solution of 0.3, 0.7, and 1.1 M KCl. The beads remain in KCl solution for 20 minutes and then used for shake flask culture. The average gel bead diameter is 0.3 cm.

Table 17.3   Production medium (g/L)

Composition	Quantity
Corn steep liquor	20
Sucrose	30
$MgSO_4$	0.5
Calcium carbonate	0.33
Citric acid	2.55
Ammonium sulphate	3
Trace elements (solution)	1 ml

The culture medium and conditioning for production of tetracycline is the same as for penicillin and streptomycin. Molasses, starch and other simple sugars may be carbon sources. The nitrogen sources include corn steep liquor, groundnut flour, ammonia and urea. Usually, two kinds of tetracyclines are produced. They are chlortetracycline and tetracycline. The reduction of chloride in the medium leads to production of tetracycline than chlortetracycline. Since corn steep liquor and other nitrogenous materials contain chloride, they must be removed before production of tetracycline.

The duration of fermentation is usually not beyond 6 days. The temperature is maintained between 27–28°C with optimum pH.

Tetracycline is extracted from the cell free fermentation broth by charcoal adsorption method. Further elution is done with solvents like ethanol and methanol. The solvent extraction method using amyl alcohol is also used in the extraction of tetracycline.

PRODUCTION OF GRISEOFULVIN

Griseofulvin is a systemic antifungal antibiotic with low toxicity. This antibiotic is employed in treating a variety of fungal diseases including diseases of skin and hair. This antibiotic is also effective against fungal diseases of plants. Griseofulvin is a fungistatic compound active against dermatophytic fungi having a chitinous cell wall but has no effect on other fungi, yeast, actinomycetes, protozoa and bacteria. Griseofulvin was first isolated from *Penicillium griseofulvum*. It is now extensively produced from *P. patulum*.

The biosynthesis of griseofulvin involves the action of multi-enzyme complex. The antibiotic is produced from acetate, malonate, methyl group derivatives and methionine. The intermediate product of biosynthesis is the formation of polyketide, which is turned into griseofulvin by cyclization, chlorination and methylation. The main building block of griseofulvin is acetate, which contains neither sulphur nor nitrogen. So, the supply of chloride is essential for the production of griseofulvin, otherwise, dechlorgriseofulvin will form. Generally, during fermentation, chloride is supplied as potassium chloride.

The seed culture is prepared in shake flask. The composition of the medium is sucrose 30.0, yeast extract 5.0, K_2HPO_4 1.0, $NaNO_3$ 3.0, KCl 0.5, $MgSO_4.7H_2O$ 0.5, and $FeSO_4.7H_2O$ 0.01. The initial pH of the media is adjusted to 6.5 before sterilization. The seed culture medium is inoculated with 1 ml of spore's suspension (in sterile distilled water) containing 18×10^6 spores from a 9-day slant. The culture is incubated for 4 days on a rotary shaker at 160 rpm and 30°C.

Cultivation Medium and Culture Conditions

The production medium contains sucrose, starch or lactose as carbon sources. Supplements K_2HPO_4, $NaNO_3$ and $FeSO_4$ are added and the pH is adjusted between 5.5 and 7.5. The medium constitutes sucrose, $NaNO_3$, K_2HPO_4 and $FeSO_4.7H_2O$. Chemical parameters and physical parameters are varied according to experiment plan. The composition of the remainder of the medium consists of (in g/l) yeast extract 5.0, KCl 0.5, and $MgSO_4.7H_2O$ 0.5. The addition of potassium chloride is significant. For higher yield, glucose feed is carried out during fermentation.

Usually, aerobic submerged batch fermentation is conducted for the production of griseofulvin. The bioreactor is fitted with necessary controllers. The pH of the medium is adjusted with 2 M NaOH and/or 2 M HCl. Fermentation is carried out for 7 to 9 days with agitation at 150–250 rpm and aeration 1 vvm.

The griseofulvin is intracellular product and remain in the cell wall of the organism. The cells are removed from the broth using routing fungal separation methods and are extracted with acetone. The product recovered by acetone extraction and evaporation is a crude product. The crude product is washed several times with organic solvents to remove the acetone soluble impurities. The product is dissolved in acetone and re-dissolved in water for the precipitation of griseofulvin.

PRODUCTION OF BACITRACIN

Bacitracin is a peptide antibiotic produced from *Bacillus licheniformis*. This antibiotic inhibits bacterial cell wall formation and acts against gram-positive and gram-negative bacteria. It is an economically important antibiotic and often used as animal feed. *Bacillus licheniformis* produces bacitracin A, A′, B, C, D, E, F, F_1, F_2, F_3 and G, among which bacitracin A is the main component. In biosynthesis, the bacitracin synthetase is doing important role.

Bacitracin was initially produced from surface cultures. Presently, it is produced by aerated submerged culture process. Usually, isolation and cultivation is carried out using tryptone or peptone broth in shake flask. The inoculated medium is incubated at 37°C for 6 hours. The growth culture is used to prepare the seed for fermentation. A typical seed medium contains soy bean meal 4%, calcium carbonate 0.5% and starch 0.5%. The inoculum is prepared in shake flasks with aeration and constant agitation.

Mostly, repeated batch aerated type of fermentation is employed for the production of bacitracin. Recently, continuous fermentation is carried out using immobilized cells. Immobilization of *Bacillus licheniformis* is done by routine procedure. The cells are collected by centrifugation (10,000 × g for 30 min.) and washed twice in saline and re-suspended in saline (bacterial concentration was 200 mg wet weight per ml). To this suspension, 25% acrylamide monomer solution (dissolved in saline, 95% acrylamide, and 5% BIS) and saline is added. The system is blanketed with nitrogen, and polymerization is initiated with 5% ammonium persulphate (APS) and 1.0 ml of 5% N, N, N′, N′-tetramethylethylenediamine.

At the end of production, bacitracin is recovered by extraction with n-butanol, and bacitracin rich n-butanol is further re-extracted by buffers. The aqueous phase contains concentrated bacitracin. For purification, ion exchange resins are used. The extraction for the purpose of animal feed is quite simple, involving evaporation of whole fermentation broth, drying and blending with other animal feed.

REVIEW QUESTIONS

1. What is antibiotic?

2. What are beta lactam antibiotics?

3. Explain about types of antibiotics.

4. Give an account on general production of antibiotics.

5. Describe in detail about penicillin production.

6. Give a brief account on streptomycin production.

7. What are tetracyclines? How is it produced?

8. Describe the production of griseofulvin and bacitracin in detail.

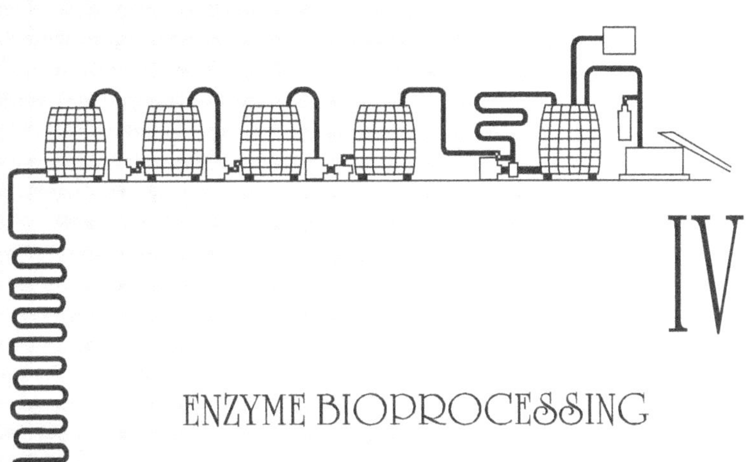

IV

ENZYME BIOPROCESSING

INDUSTRIAL ENZYMES

Most reactions in living organisms are catalysed by protein molecules called enzymes. Enzymes can rightly be called catalytic machinery of living systems. Enzymes are responsible for the biocatalytic fermentation of sugar to ethanol by yeasts, a reaction that forms the bases of beer and wine manufacturing. Enzymes oxidize ethanol to acetic acid. This reaction has been used in vinegar production since thousands of years. Similar microbial enzyme reactions of acid forming bacteria and yeasts are responsible for aroma forming activities in bread making and in preserving activities in sauerkraut preparation.

The fermentative activity of microorganisms was discovered only in 18th century and finally proved by the French scientist Louis Pasteur. The term enzyme comes from Latin words, which literally mean "in yeast". Enzymes are closely associated with yeast activity. The study of enzymes is a fairly recent activity. Scientists who found out that an alcohol precipitate of malt extract contained a thermolabile substance, which converted starch into sugar, made the first clear recognition of enzymes in 1833. They called the substance diastase. We now know that it was an enzyme called amylase. Sumner finally proved the protein nature of enzymes in 1926 when he was able to crystallize urease enzyme from jack bean.

The first application of cell free enzymes was probably the use of rennin isolated from calf or lamb stomach in cheese making. Rennin is an aspartic protease, which coagulates milk protein, and has been used for hundreds of years by cheese makers. Röhm in Germany prepared the first commercial enzyme in 1914. This trypsin enzyme isolated from animals degraded proteins and was used as a detergent. It proved to be so powerful as compared to traditional washing

18

powders. The real breakthrough of enzymes occurred with the introduction of microbial proteases into washing powders. The first commercial bacterial Bacillus protease was marketed in 1959, which became big business when Novozymes in Denmark started to manufacture it and major detergent manufactures started to use it around 1965.

In addition to cheese manufacturing, enzymes were used already in 1930 in fruit juice manufacturing. These enzymes, called pectinases, clarify the juice. They contain numerous different enzyme activities. The major usage of microbial enzymes in food industry started in 1960s in starch industry. The traditional acid hydrolysis of starch was completely replaced by alpha-amylases and glucoamylases, which could convert starch with over 95% yield to glucose. Starch industry became the second largest user of enzymes after detergent industry.

Presently, the enzyme companies sell enzymes for a wide variety of applications. Detergents (37%), textiles (12%), starch (11%), baking (8%) and animal feed (6%) are the main industries, which use about 75% of industrially produced enzymes. Enzymes are also indirectly used in biocatalytic processes involving living or dead and permeabilized microorganisms. The use of microorganisms as biocatalysts in chemical production is, however, an interesting and growing field. The techniques of genetic, protein and pathway engineering are making chemical production by living cells an interesting green alternative to replace traditional chemical processes.

ENZYME CLASSIFICATION

Presently, more than 2000 different enzyme activities have been isolated and characterized. The sequence information of a growing number of organisms opens the possibility to characterize all the enzymes of an organism on a genomic level. The smallest known organism, *Mycoplasma genitalium*, contains 470 genes of which 145 are related to gene replication and transcription. Baker's yeast has 7000 genes, coding for about 3000 enzymes. Thousands of different variants of natural enzymes are known. The number of reported 3-dimensional enzyme structures is rapidly increasing. In the year 2000, the structure of about 1300 different proteins were known. The enzymes are classified into six major categories based on the nature of the chemical reaction they catalyse:

1. Oxidoreductases catalyse oxidation or reduction of their substrates

2. Transferases catalyse group transfer

3. Hydrolases catalyse bond breakage with the addition of water

4. Lyases remove groups from their substrates

5. Isomerases catalyse intramolecular rearrangements

6. Ligases catalyse the joining of two molecules at the expense of chemical energy

Only a limited number of known enzymes are commercially available, and even a smaller number is used in large quantities. More than 75% of industrial enzymes are hydrolases. Protein-degrading enzymes constitute about 40% of all enzyme sales. Proteinases have found new applications, but their use in detergents is the major market. More than fifty commercial industrial enzymes are available and their number increases steadily. Table 18.1 provides the different sources of enzymes and Table 18.2 provides the enzymes used for industrial purposes.

Table 18.1 Different sources of enzymes and their applications

Enzyme	Source	Mode of production	Industrial use
Animal enzymes			
Catalase	Liver	Intracellular	Food
Chymotrypsin	Pancreas	Extracellular	Leather
Lipase	Pancreas	Extracellular	Food
Rennet	Abomasum	Extracellular	Cheese
Trypsin	Pancreas	Extracellular	Leather
Plant enzymes			
Actinidin	Kiwi fruit	Extracellular	Food
α-Amylase	Malted barley	Extracellular	Brewing
β-Amylase	Malted barley	Extracellular	Brewing
Bromelain	Pineapple latex	Extracellular	Brewing
β-Glucanase	Malted barley	Extracellular	Brewing
Ficin	Fig latex	Extracellular	Food
Lipoxygenase	Soybeans	Intracellular	Food
Papain	Pawpaw latex	Extracellular	Meat

(Contd.)

Table 18.1   (Continued)

Enzyme	Source	Mode of production	Industrial use
Bacterial enzymes			
α-Amylase	*Bacillus*	Extracellular	Starch
β-Amylase	*Bacillus*	Extracellular	Starch
Asparaginase	*Escherichia coli*	Intracellular	Health
Glucose isomerase	*Bacillus*	Intracellular	Fructose syrup
Penicillin amidase	*Bacillus*	Extracellular	Pharmaceutical
Protease	*Bacillus*	Extracellular	Detergent
Pullulanase	*Klebsiella*	Extracellular	Starch
Fungal enzymes			
Amylase	*Aspergillus*	Extracellular	Baking
Aminoacylase	*Aspergillus*	Intracellular	Pharmaceutical
Glucoamylase	*Aspergillus*	Extracellular	Starch
Catalase	*Aspergillus*	Intracellular	Food
Cellulase	*Trichoderma*	Extracellular	Waste
Dextranase	*Penicillium*	Extracellular	Food
Glucose oxidase	*Aspergillus*	Intracellular	Food
Lactase	*Aspergillus*	Extracellular	Dairy
Lipase	*Rhizopus*	Extracellular	Food
Rennet	*Mucor miehei*	Extracellular	Cheese
Pectinase	*Aspergillus*	Extracellular	Drinks
Pectin lyase	*Aspergillus*	Extracellular	Drinks
Protease	*Aspergillus*	Extracellular	Baking
Raffinase	*Mortierella*	Intracellular	Food

(Contd.)

Table 18.1 (Continued)

Enzyme	Source	Mode of production	Industrial use
Yeast enzymes			
Invertase	*Saccharomyces*	Intracellular/ extracellular	Confectionery
Lactase	*Kluyveromyces*	Intracellular/ extracellular	Dairy
Lipase	*Candida*	Extracellular	Food
Raffinase	*Saccharomyces*	Intracellular	Food

Table 18.2 Various industrial enzymes and their applications

Industry	Enzyme class	Applications
Detergents	Proteinase Amylase Lipase Cellulase Mannanase	Stain removal and colour clarification
Starch and fuel	Amylase Amyloglucosidase Pullulanase	Starch liquefaction and saccharification
	Xylanase	Viscosity reduction (fuel)
	Protease Lipase Lactase Pectinase	Milk clotting, flavouring, cheese flavour, lactose removal, fruit based products
	Amylase Xylanase Lipase Phospholipase Glucose oxidase Protease	Bread softness, dough conditioning, dough stability, dough strengthening of biscuits and cookies
	Xylanase	Phytate digestibility

(Contd.)

Table 18.2   (Continued)

Industry	Enzyme class	Applications
Beverage	Pectinase Amylase β-Glucanase Laccase	Depectinization, juice treatment, mashing, clarification, flavouring
Textile	Cellulase Amylase Catalase Laccase Peroxidase	Denim finishing, de-sizing, bleach termination, bleaching, excess dye removal
Pulp and paper	Lipase Protease Amylase Xylanase Cellulase	Pitch control, bio-film removal, starch coating, bleach boosting, de-inking
Leather	Protease Lipase	Dehairing, bating, de-pickling
Personal care	Amyloglucosidase Glucose oxidase Glucose peroxidase	Antimicrobial agents, bleaching

ENZYME PRODUCTION

Some enzymes are still extracted from animal or plant tissues. Commercial plant-derived enzymes include proteolytic enzymes, papain, bromelain and ficin and other special enzymes like lipoxygenase from soybeans. Animal-derived enzymes include proteinases like pepsin and rennin. Most of the enzymes are, however, produced with microorganisms in submerged cultures in fermenters. The enzyme production process can be divided into the following phases:

1. Selection of an enzyme

2. Selection of a production strain

3. Construction of an overproducing strain by genetic engineering

4. Optimization of culture medium and production conditions

5. Optimization of recovery process

6. Formulation of a stable enzyme product

The criteria used in the selection of an industrial enzyme include specificity, reaction rate, pH and temperature optima and stability, effect of inhibitors, and affinity to substrates. Enzymes used in paper industry should not contain cellulose-degrading activity because this activity would damage the cellulose fibres. Enzymes used in animal feed industry must tolerate and survive the hot extrusion process used in animal feed manufacturing. The same enzymes must have maximal activity at the body temperature of the animal. Enzymes used in industrial applications must usually be tolerant against various heavy metals and have no need for cofactors. They should be maximally active already in the presence of low substrate concentration so that the desired reaction proceeds to completion in a realistic time frame.

PRODUCTION OF MICROBIAL STRAINS

In choosing the production strain, several aspects have to be considered. Ideally, the enzyme is secreted from the cell. This makes the recovery and purification process much simpler as compared to the production of intracellular enzymes, which must be purified from thousands of different cell proteins and other components. Secondly, the production host should be GRAS, which means it is generally regarded as safe. This is especially important when the enzyme produced is used in food processes. Thirdly, the organism should be able to produce high amount of the desired enzyme in a reasonable time frame. Industrial strains typically produce over 50 mg/l of extracellular enzyme proteins. Most of the industrial enzymes are produced by a relatively few microbial hosts like *Aspergillus* and *Trichoderma* fungi, *Streptomyces* fungi imperfecti, and *Bacillus*. Yeasts are not good producers of extracellular enzymes and are rarely used for this purpose. Most of the industrially used microorganisms have been genetically modified to overproduce the desired activity and not to produce undesirable side activities.

General Method for Enzyme Production by Microbial Fermentation

Once the organism has been genetically engineered to overproduce the desired products, a production process has to be developed. The optimization of a fermentation process includes media composition, cultivation type and process conditions. This task often involves as much effort as the intracellular engineering of the cell. The bioprocess engineer may ask questions like: is the organism in question safe or are extra precautions needed, what kind of nutrients the organism needs and what is their optimal/economical concentration, how the nutrients should be sterilized, what kind of a reactor is needed (mass transfer, aeration, cooling, foam control, sampling), what needs to be measured and how is the process controlled, how is the organism cultivated (batch, fed-batch or continuous cultivation), what are the optimal growth conditions, what is the specific growth and product formation rate, what is the yield and volumetric productivity, how to maximize cell concentration in the reactor, is the product secreted out from the cells, how to degrade the cell if the product is intracellular, does any raw material or product inhibit the organism, and finally, how to recover, purify and preserve the product.

Large-volume industrial enzymes are produced in 50–500 m^3 fermenters. The extracellular enzymes are often recovered after cell removal (by vacuum drum filtration, separators or microfiltration) by ultrafiltration. If needed, the purification is carried out by ion exchange or gel filtration. The final product is either a concentrated liquid with necessary preservatives like salts or polyols, or alternatively granulated to a non-dusty dry product. Enzymes are proteins, which can cause allergic reactions. Therefore, protective measures are necessary in their production and application.

PRODUCTION OF AMYLASE

Amylases are enzymes that hydrolyse starch. Amylases are produced by a variety of living organisms, ranging from bacteria to plants, humans and animals. Bacteria and fungi secrete amylases from the outside of their cells to carry out extracellular digestion. When they have broken down the insoluble starch, the soluble end products such as glucose and maltose are absorbed into their cells. Starch, a glucose polymer, is one of the most widely available plant polysaccharides. The hydrolysis of starch with amylase results first in the production of short-chain polymers called dextrin, then the disaccharide maltose, and finally glucose.

Types of Amylase

Amylases are classified into three types based on how they break down starch molecules.

α-amylase [α-1,4-glucanohydrolases] They are also called endoamylase because they hydrolyse glucosidic linkage in the middle portion of starch molecules. *α*-amylases are extracellular enzymes, which hydrolyse

α-1, 4-glycosidic bonds at random, thereby producing varied sized chains of glucose. α-1, 6-glycosidic bonds do not inhibit their action although such bonds are not split. Many bacteria and fungi form α-amylases. They are classified according to their starch liquefying and saccharogenic effect, PH, temperature range and stability. Most α-amylases have molecular weight of about 50,000. Nowadays, α-amylases are purified in crystallization form.

β-*amylases   [α-1, 4-glucan-malto-hydrolases]* They are also called exoamylase. β-amylases break the glucose–glucose bonds down by removing two glucose units at a time, thereby producing maltose. It is usually of plant origin but some microbial producers also exist. Bacterial β-amylases have greater heat resistance (>70°C) than plant β-amylases.

Amyloglucosidase [α-1,4-glucan-glucohydrolases]   They have another name glucoamylases. Glucoamylases break successive bonds from the non-reducing end of the straight chain producing glucose. Maltose can be broken down only slowly, while 1,6-bonds in the branched polysaccharides are hardly attacked. Thus, glucose, maltose and dextrins are the end products of glucoamylase action.

SOURCES OF AMYLASE

Many microbial amylases usually contain a mixture of these amylases. Amylases are found in bacteria, fungi, plants, animals and humans. Bacteria like *Bacillus subtilis, B. cereus, B. licheniformis, B. amyloliquefaciens* and *B. megaterium* and fungi like *Aspergillus niger, Penicillium, Cephalosporium, Neurospora* and *Rhizopus* are major amylase producing microorganisms.

In animals, it is present abundantly in pancreas, salivary glands, and some quantities are also found in muscle, liver, urine and other tissues. In addition, amylase appears to be associated with the mitochondria in some cells (amoeba).

APPLICATIONS OF AMYLASE

Amylases have various commercial applications: industrial, laboratory, clinical and medical.

Industrial applications   Amylase has many industrial applications. These will be helpful in better development of the human race.

Starch industry   In the industrial production of glucose from starch, the bacterial amylase and fungal amyloglucosidase are mainly used. These enzymes completely convert the starch into glucose by breaking both the type of glucosidic bonds found in the starch. The conversion of insoluble starch granules in aqueous solution followed by hydrolysis of amylase is called starch liquefaction.

Alcohol industry In alcohol industry, especially in beer brewing industries, the starch present in the barley malt is converted into simple sugar followed by conversion of sugar into alcohol by yeast. The previous process is achieved by application of amylase enzyme. In general, conversion of polysaccharide into simple sugars is called saccharification.

Baking industry In the bread making process, the preliminary leavening process involves mixing of wheat flour with yeast and water. The enzymes of α-amylase and β-amylase naturally present in the wheat breaks the 1-4 glycosidic bond present in the starch.

Textile industries The amylase is used for continuous resizing of cloths and fibres. In textile weaving, starch past is applied for warping. This gives strength to the textile at weaving and also prevent the loss of strings. After weaving, the starch is removed by application of amylase enzyme.

Sugar industries The amylase enzyme is mainly used to produce the various sugars from starch. In food industry favourable sugar products like maltotetrose syrup, high molecular weight dextrins and also mixture which contains isomaltose, panose, isomaltotrioses and glucose are prepared using amylase enzymes.

Laundry and detergents The amylase enzyme is mixed with laundry detergents to increase the cleaning capacity of the detergent. The addition of amylase with detergent is mainly used to remove the starch based stains.

Waste treatment Starch is also present in the waste produced from food processing plants. Starch waste causes pollution problems. This waste is treated by growing useful amylase producing microorganisms in the waste water.

Laboratory, clinical and medical applications So many human diseases are diagnosed by measuring blood plasma amylase level through plasma amylase assay. The digestive enzyme α-amylase is produced in the pancreas, and this is found in increased amounts in plasma in certain diseases of the pancreas, particularly acute pancreatitis. α-amylase is the simplest to assay, since it utilizes starch as substrate and produces reducing sugars as products. An unusual feature of pancreatic α-amylase is that it is of the few proteins small enough to pass through the glomerulus of the kidney and be excreted in the urine. In a rare condition called macro-amylasaemia, the plasma α-amylase forms complexes with other proteins and can no longer be excreted in this way, which causes the plasma α-amylase activity to rise. However, macro-amylasaemia could not be confused with acute pancreatitis because of the different clinical symptoms.

STRAINS AND SCREENING FOR AMYLASE PRODUCTION

Microorganisms that are capable of spoiling starchy materials have the ability to produce amylase. The contaminated and spoiled starchy food materials and other decomposed starch materials are the best sources to isolate the amylase producing strains. The screening for the amylase production is generally carried out with starch agar plates (Figure 18.1). The pure cultures are streaked in sterile starch agar plates, which are incubated at necessary temperature. Generally, 1% iodine solution is over layered on the agar plates to observe the starch utility zone around the colony.

Composition of screening medium (g/100 ml distilled water)

Components	Quantity
Starch	0.5
Peptone	0.5
NaCl	0.5
Yeast extract	0.3
Agar	2.0

Figure 18.1   Amylase activity in starch agar plates

PRODUCTION PROCESS

The inoculum medium contains necessary nutrients, and starch acts as the inducer. For solid state fermentation, spore suspension is prepared in sterile distilled water and used as inoculum. For bacterial strains, an overnight culture from the inoculum medium is used for production. The inoculum preparation is carried out in shake flasks.

The enzyme production is generally carried out in fermenters. For laboratory scale, the shake flask fermentation is the preferable method.

Production Media

Lactose	4 g
Yeast extract	2 g
KH_2PO_4	0.005 g
$MnCl_2.4H_2O$	0.00015 g
$MgSO_4.7H_2O$	0.025 g
$CaCl_2.4H_2O$	0.005 g
$FeSO_4.7H_2O$	0.001 g
Distilled water	100 ml

Cultural Conditions

In fed-batch cultivation system, the temperature is maintained at 37°C for mesophilic bacteria and the pH of the medium maintained at 7.1 using a pH controller with 1M HCl and 10% ammonia water. Dissolved oxygen is measured using a DO controller. DO is usually observed to decrease with cell growth. When the DO falls, it is controlled by increasing the agitator rate or air flow rate. If the DO increases, it indicates nutrition depletion, and 20 ml of additional medium is provided.

Immobilization is carried out under sterile conditions. Twenty milligrams of wet cells is suspended in 12.5 ml of 2% (w/v) sodium alginate solution (0.16% (w/v) wet weight in gel). The mixture obtained is extruded drop-wise through a syringe into a 25 ml of 3.5% (w/v) calcium chloride solution. Alginate drops are solidified upon contact with $CaCl_2$, forming capsules and thus entrapping the bacterial cells. The capsules are allowed to harden for 30 minutes and then washed with sterile saline solution (0.9% (w/v) NaCl) to remove excess Ca^{2+} and cells.

For fungal fermentation, the initial pH is adjusted to 5 or 6 and sterilized at 121°C for 15 minutes. An inoculum of 5×10^6 spores per hundred ml is used. Fermentation is carried out with 150 rpm or 200 rpm at 30°C for 72 hours.

Analysis for Amylase Activity

After cultivation, the cells are separated either using centrifugation or filtration. The qualitative plate assay is performed using agar plates amended with starch. The agar plates are prepared with 2% of starch and 1.5% of agar. After solidification of the agar, a well of around 10 mm diameter is made aseptically with the help of cork borer. The well is filled with the culture filtrate and incubated at 37°C overnight. 1 % of iodine solution is over layered on the agar to visualize the zone formation due to the enzyme activity.

The quantitative assay for amylase activity is determined by the DNS (3,5-dinitro salicylic acid) method using starch as a substrate. The amount of reducing sugar concentration released during amylase and starch reaction is measured to calculate the enzyme activity. The reaction mixture containing 1% starch in sodium phosphate buffer, 1 ml of sodium chloride and enzyme solution is incubated at 37°C for 15 minutes to perform enzymatic reaction. After incubation, 1 ml of DNS (1%) is added as a colour developer and to stop the enzyme–substrate reaction. The content boils to develop colour reaction. The absorbency of the solution is made at 540 nm. The released sugar concentration is determined by comparing the absorbency value with the standard chart prepared using reducing sugar.

PRODUCTION OF PROTEASE

In the more recent history of proteolytic enzymes, attention has been focused on proteases that play regulatory roles in a great variety of physiological processes. These range from the processing and molecular assembly of nascent polypeptide chains to the development, fertilization and many other proteolytic processes important for cellular functions. The enzymes that hydrolyse peptide bonds are commonly called proteases. They include the well known families of serine, cysteine, aspartic acid and metallo- exo- and endopeptidases. When we compare the present investigated thousands of proteases with earlier investigations, only least number of proteases have been recognized in earlier days by their digestive action and regulatory of physiological process.

Proteolytic enzymes find wide applications in various pharmaceutical and industrial applications. Although proteolytic enzymes can be obtained from animal and higher plants, the microorganisms are preferred in industrial applications due to technical and economical advantage. Different species of bacteria produce acid, neutral and alkaline proteases. The production of extracellular proteases is governed, at least in part, by the concentration of available individual nutrients. Since microorganisms can be made to propagate rapidly and profusely, they are an ideal source for enzymes.

Proteases represent one of the most important groups of industrial enzymes and account for at least a quarter of the total global enzyme production. They have been classified into two types: (1) peptidases and (2) proteases.

Peptidases

Peptidases hydrolyse peptide bonds within the protein chain, previously called endopeptidases, while proteases hydrolyse large polypeptides into smaller molecules.

Endopeptidases   These cleave peptide bonds at points within the protein and remove amino acids sequentially from either N or C terminus, respectively. The term proteinase is used as a synonym for endopeptidase, and four mechanistic classes of proteinases are recognized by the IUBMB (International Union of Biochemistry and Molecular Biology, 1984).

Exopeptidases   The exopeptidases act only near the ends of polypeptide chains at the N or C terminus. Those acting at a free N terminus liberate a single amino acid residue (amino peptidases), a peptide (dipeptidyl peptidases) or a tripeptide (tripeptidyl peptidases). The exopeptidases acting at a free C terminus liberate a single amino acid (carboxy peptidases) or a dipeptide (peptidyl dipeptidases). Some exopeptidases are specific for dipeptidases or remove terminal residues that are substituted, cyclized or linked by isopeptide bonds. Isopeptide bonds are peptide linkages other than those of a carboxyl or amino acid groups, and this group of enzymes is redenominated omega.

SOURCES OF PROTEASES

Proteases are found in all forms of microbes, plants and animals.

Proteases from Microbes

Proteinases are found in several microorganisms such as viruses, protozoa, bacteria, yeast and fungi. The inability of the plant and animal proteases to meet market demands has led to an increased interest in microbial proteases.

Microorganisms represent an excellent source of enzymes owing to their broad biochemical diversity and their susceptibility to genetic manipulation. Proteins are degraded by microorganisms, and they utilize the initiated by proteinases (endopeptidases) secreted by microorganisms followed by further hydrolysis by peptidases (exopeptidases) at the extra or intracellular site. Several proteinases are produced by the same strain under various cultural conditions.

Candida albicans and *C. tropicalis* are the medically important opportunistic pathogens causing infections in immunocompromised patients. Their extracellular enzyme, an aspartic proteinase, is considered to be a major virulence factor. Most commercial serine proteases, mainly neutral and alkaline, are produced by organisms belonging to the genus *Bacillus*. Some of the gram-negative bacteria producing proteases identified were *Pseudomonas aeruginosa, Vibrio alginolyticus and Xanthomonas maltophila*. Some rare microorganisms produce alkaline proteases. *Kurthia spiroforme* was reported to produce protease. Halophiles were described to produce alkaline proteases, especially the *Halobacterium* species. Similar enzymes are also produced by other bacteria such as *Thermus caldophilus*, *Desulfurococcus mucosus, Streptomyces, Aeromonas* and *Escherichia* genera. Fungi produce several serine proteinases. Among these enzymes, many are produced by various strains of *Aspergillus oryzae*, etc. Cysteine proteinases are not so widely distributed as was seen with serine and aspartic proteinases.

The metalloproteases are found in members of family Trypanosomatidae and several genus such as *Crithidia, Phytomonas, Eretomonas, Trypanosoma, Leishmania, Leptomonas* and *Endotrypanum* are involved with the nutrition, life cycle and morphological differentiation of these parasites. *Trichomonas vaginalis*, a flagellated protozoan responsible for trichomonosis, one of the most common sexually transmitted diseases, has numerous cysteine and some metalloproteinases. The cysteine enzymes are involved in the damage to the host by the parasite.

Microbial proteases account to approximately 40% of the worldwide enzymes sales. In addition, proteases from microbial sources are preferred to the enzymes from plant and animal sources since they possess almost all characteristics desired for biotechnological applications.

Proteases from Plants

The proteolytic enzyme papain is obtained from the leaves and unripe fruit of the *Carica papaya*. Papain has the property to transform albuminoids into peptones in acid, alkaline or neutral medium which makes it superior to pepsin. Another plant-based proteolytic enzyme, bromalain, comes from the stems of pineapple.

Proteases from Animals

The most familiar proteases of animal origin are pancreatic trypsin, chymotrypsin, pepsin and rennin. These are prepared in pure form in bulk quantities. However, their production depends on the availability of livestock for slaughter. Rennet (mainly chymosin) obtained from the fourth stomach (abomasums) of unweaned calves has been used in the production of cheese. Digestive enzymes such as trypsin, chymotrypsin, etc., from animals are proteases.

APPLICATIONS OF PROTEASE

Proteolytic enzymes account for nearly 60% of the industrial market in the world. They find application in a number of biotechnological processes, *viz.* food processing, pharmaceuticals, leather industry, silk, bakery, soy processing, meat tendering, and brewery. However, its application in the production of peptide synthesis in organic media is limited by the presence of organic solvents.

Protease in detergent industry   Microbial alkaline protease dominates the commercial applications with a significant share of the market captured by subtilisin or alkaline protease from *Bacillus* sp. for laundry detergent applications. Alkaline proteases added to laundry detergents enable the release of pertinacious material from stains. The use of enzyme is mainly due to shorter period of agitation and lower wash temperature, often after a preliminary period of soaking. The interest in using alkaline enzymes in automatic dishwashing detergents has also increased recently.

Protease in leather and wool industries   The leather industry consumes a significant proportion of world's enzymes production. Alkaline proteases are used extensively in the removal of hair hides. The bating following the dehairing process involves the degradation of elastin and keratin, the removal of hair residues, and the deswelling of collagens, which produces good, soft leather useful in making leather cloths and goods. In addition, studies carried out by different researchers have demonstrated the successful uses of alkaline protease in leather tanning from *Aspergillus flavus*, *Streptomyces* sp. and *B. amyloliquefaciens*.

The applications of protease primarily are found in wool industry. Wool fibres are covered in overlapping scales pointing towards the fibre tip. A successful method involved the partial hydrolysis of scale tips with the protease papain. This method was abandoned few years ago, primarily for economic reasons.

Protease in silver recovery   Alkaline proteases find potential application in the bioprocessing of used X-ray films for silver recovery. The enzymatic hydrolysis of the gelatin layers on the X-ray film enables not only silver but also the polyester film base to be recycled. The alkaline protease from *Bacillus* sp. decomposes the gelatinous coating on the used X-ray films from which silver can be recovered.

Protease in food industry   Proteases play a prominent role in meat tenderization, especially beef. An alkaline elastase and thermophilic alkaline protease have proved to be successful and promising meat tenderizing enzymes, as they possess the ability to hydrolyse connective tissue proteins as well as muscle fibre proteins. A patented method used a specific combination of neutral and alkaline proteases for hydrolysing raw meat. The reason for this may be that the preferential specificity was favourable when metalloprotease and serine protease were used simultaneously. The current trend in similar research shows yet another alkaline protease from *B. amyloliquefaciens* resulted in the production of a methionine rich protein hydrolysate from chickpea protein.

Medical applications   Proteases regulate various metabolic processes such as blood coagulation, fibrinolysis, complement activation, phagocytosis, and blood pressure control. Collagenases with alkaline protease activity are increasingly used therapeutically in the preparation of slow-release dosage forms. Elastosterase, a preparation with high elastolytic activity from *B. subtilis* 316M, was immobilized on a bandage for the therapeutic application in the treatment of burns and purulent wounds, carbuncles and deep abscess. Alkaline proteases having fibrinolytic activity has been used as a thrombolytic agent.

STRAINS AND SCREENING FOR PROTEOLYTIC ACTIVITY

Strains that are present in the spoiled protein stuffs and related environment are the right strains to produce the protease enzyme. The strains isolated from the suitable environment using routine microbiological methods are screened for the production of protease using casein agar (Figure 18.2) or gelatine agar medium. The pure cultures are inoculated in sterile milk agar or gelatine agar plates and incubated at suitable temperature for 24 hours.

Around 5% of mercuric chloride solution is over layered on the medium to visualize the protein utilization zone around the positive organisms.

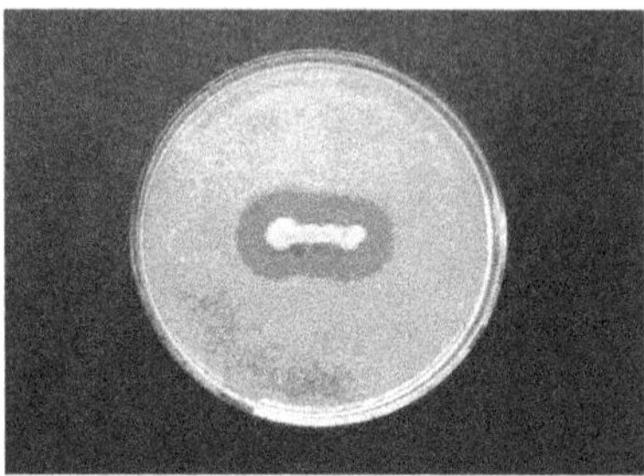

Figure 18.2   Proteolytic activity in casein agar plate

Culturing Condition

The inoculum for further production of enzyme is made using broth containing any one of the protein source such as gelatine, casein and peptone. For bacteria, the pure culture is inoculated into sterile inoculum broth and incubated at 37°C in a rotary shaker overnight. The fresh overnight culture is used as an inoculum in the production of enzyme. For fungi, spore suspension is prepared and used as inoculum.

The enzyme production is carried out using a production medium that contains a protein inducer or substrate. The temperature and agitation of the medium is maintained based on the nature of the strain.

For fungal solid state fermentation (SSF), various agro-industrial residues such as wheat bran, rice husk, rice bran, spent brewing grain, coconut oil cake, palm kernel cake, sesame oil cake, jackfruit seed powder and olive oil cake are used as potential substrates. SSF is carried out by taking the substrate in Erlenmeyer flask of trays, moistening it with salt solution containing 0.1% dipotassium hydrogen phosphate, 0.5% magnesium sulphate, 0.5% sodium chloride and 0.004% ferrous sulphate, and distilled water may be needed to adjust the required level of substrate moisture. All the flasks are autoclaved at 121°C for 20 minutes, and after cooling are inoculated with 1 ml of spore suspension and incubated at room temperature for the desired period.

In immobilized systems, entrapment of cells in alginate is one of the simplest, cheapest, non-toxic, and the most frequently used method of immobilization in protease production. Sodium alginate and calcium chloride are used to prepare the alginate beads containing the whole cells. Sodium alginate solution (3%) is prepared by dissolving sodium alginate in 100 ml hot water. The contents are stirred vigorously for 10 minutes to obtain thick uniform slurry without any undissolved lumps and sterilized by autoclaving. Both alginate slurry and cell suspension are mixed and stirred for 10 minutes to obtain a uniform mixture. The slurry is taken into sterile syringes and added drop-wise into 0.2M $CaCl_2$ solution from 5 cm height and kept for curing at 4°C for 1 hour. The cured beads are washed with sterile water 3–4 times. When the beads are not used, they are preserved in normal saline solution in a refrigerator. All these operations are carried out aseptically in a laminar flow unit.

Analysis for Protease Activity

The plate assay is performed using agar plates (Figure 18.3) amended with either casein or gelatin. The agar plates are amended 2% of substrate with 1.5% of agar. After solidification of the agar, a well of 10 mm diameter is cut out aseptically with the help of cork borer. The well is filled with the culture filtrate and incubated at 37°C for overnight. Observation is made to see the hydrolytic zone around the well. For better appearance of the zone, 5% mercuric chloride solution is added over the agar plates. The negative control is maintained by adding sterile water in the separate well.

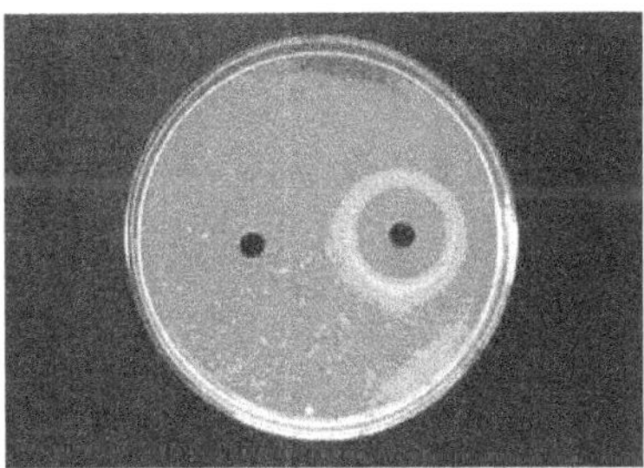

Figure 18.3 Plate assay for proteolytic activity

For chemical assay, the protease activity is determined using casein substrate. The reaction mixture containing the substrate (casein), enzyme solution and citrate phosphate buffer is incubated at 40°C for 60 minutes for enzymatic reaction. After incubation, the enzymatic reaction is terminated by the addition of trichloroacetic acid solution, and the formed precipitate is removed by centrifugation at 4°C. The supernatant is collected, and folin-phenol reagent is added to develop the colouration. The absorbance is read at 750 nm in UV-VIS spectrophotometer and the data are compared with tyrosine standard.

Protease Inhibitors

Extracts from cell disruption or microbial cultures may contain proteases. In the former case, proteases may be released from subcellular compartments, while in microbial culture filtrates, the proteases may have been secreted during the fermentation process. These proteases will need to be removed or inactivated to prevent the target protein from being degraded. This can be achieved through the inhibition of proteases and the reduction of storage temperature. Table 18.3 lists some of the inhibitors.

Table 18.3 List of inhibitors

Inhibitor	Enzymes inhibited
Phenylmethanesulphonyl fluoride (PMSF)	Serine proteases (e.g. chymotrypsin, trypsin, thrombinase) and thiol proteases (e.g. papain)
Ethylenediamine tetra acetic acid (EDTA)	Metalloproteases
Pepstatin A	Acid proteases such as pepsin, renin, cathepsin D and cymosin
Leupeptin	Serine and thiol proteases such as papain, plasmin and cathepsin B
Aprotinin	Serine proteases such as plasmin, kallikrein, trypsin and chymotrypsin

PRODUCTION OF LIPASES

Lipase [triacyl glycerol ester hydrolase, EC (3.1.3.3)] is a water-soluble enzyme that catalyses the hydrolysis of fats to produce monoglycosides, diglyceride free fatty acids and glycerol. Lipases possess the characteristic properties like substrate specificity, stereo specificity and the ability to catalyse heterogeneous reactions at the interphase of water-soluble and water-insoluble systems. Many lipases are active in organic solvents where they catalyse a number of useful reactions including esterification, *trans*-esterification and hydrolysis (Figure 18.4).

Lipases are produced by plants, animals, bacteria and moulds. Plant enzymes are not used commercially, while animal, bacterial and mould enzymes are used extensively. Most important animal sources are cattle, sheep and pigs; pancreatic lipases and dog pancreatic lipase are most widely used. Microbial lipases are from *Candida* and moulds *Aspergillus*.

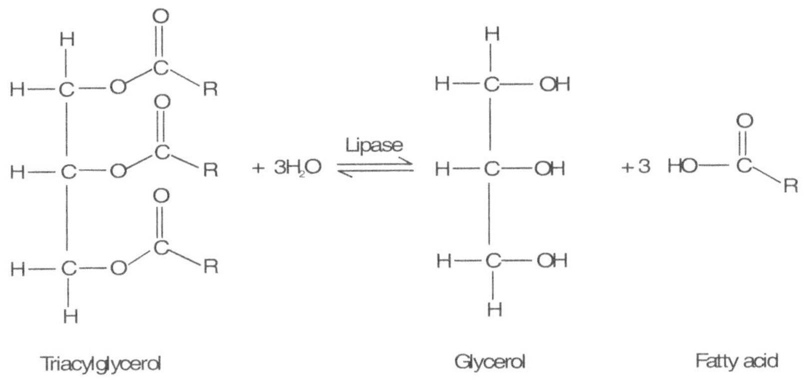

Figure 18.4 Hydrolysis of fat by lipase

Most lipases have an alkaline pH between 8 and 9. A few microbial lipases prefer more hydrogen ion having a pH of 5–6. The optimum temperature is 30–40°C; however, it should be recognized that most enzymes stop catalysis at –10°C because liquid substrates become solid ice. Lipases act on solid emulsion; so the fats degraded become rancid in the freezer. As indicated, lipases act on insoluble substrates, and they have next to no activity against soluble substrates; so the insoluble substrate must activate the enzyme.

SOURCES FOR LIPASES

Bacterial Lipases

A relatively smaller number of bacterial lipases have been well studied as compared to plant and fungal lipases. Bacterial lipases are glycoproteins, but some extracellular bacterial lipases are lipoproteins. Enzyme production in most of the bacteria is affected by certain polysaccharides. Most of the bacterial lipases reported so far are constitutive and are non-specific in their substrate specificity, and a few bacterial lipases are thermostable. Among bacteria, *Achromobacter* sp., *Alcaligenes* sp., *Arthrobacter* sp., *Pseudomonas* sp., *Staphylococcus* sp. and *Chromobacterium* sp. have been exploited for the production of lipases.

Fungal Lipases

Fungal lipases have been studied since 1950s. These lipases are being exploited due to their low cost of extraction, thermal and pH stability, substrate specificity, and activity in organic solvents. The chief producers of commercial lipases are *Aspergillus niger, Candida cylindracea, Humicola lanuginosa, Mucor miehei, Rhizopus arrhizus, R. delemar, R. japonicus, R. niveus* and *R. oryzae.*

Taxonomic Distribution of Fungal Lipases

Fungal lipases degrade lipids from palm oil. Among Mucorales, the lipolytic enzymes of the moulds *Mucor hiemalis, M. miehei, M. lipolyticus, M. pusillus, Rhizopus japonicus, R. arrhizus, R. delemar. R. nigricans, R. nodosus, R. microsporus* and *R. chinensis* have been studied in great detail. The thermophilic *M. pusillus* is well known as a producer of thermostable extracellular lipase. From a lipase-producing strain of *M. miehei*, two isoenzymes with slightly different isoelectric points but a high degree of antigenic identity could be isolated. A lipase of *M. miehei*, immobilized on a resin (Lipozyme™), has been commercialized by Novo Industries. Due to the 1,3-(regio)-specificity of *Rhizopus*, lipases that are especially suited for the conversion of triglycerides to their corresponding monoglycerides, and interesterification reactions of fats and oils, have food and pharmaceutical applications. *R. japonicus* lipase has been used to produce hard butter suitable for chocolate manufacture by interesterification of palm oil with methyl stearate. The lipases (40 to 45 kDa) of various *Rhizopus* species show maximum activity towards medium-chain fatty acids (C8-C10). In case of *R. delemar*, extracellular and intracellular lipase isoenzymes have been isolated. Lipase producers within the order Entomophthorales include *Entomophthora apiculata, E. coronata, E. thaxteriana, E. virulenta, Basidiobolus* spp. and *Conidiobolus* spp. The genera *Pichia, Hansenula* and *Saccharomyces* are also reported to produce lipase. Two kinds of cell-bound lipases were purified from *Saccharomyces lipolytica*. Lipases were reported from *Candida curvata, C. tropicalis, C. valida* and *C. pellioculosa*, and are non-specific towards different ester bonds in triglycerides, with the exception of *C. deformans*. The imperfect fungus *Geotrichum candidum* is responsible for acid formation in dairy products by lipolysing fat. The *G. candidum* lipase features specificity towards fatty acids with a *cis* double bond at C9, hence is applied for the structural analysis of triglycerides. The intracellular and extracellular lipases of *Aspergillus niger* are 1,3-(regio)-specific. *A. oryzae* was reported to be an efficient host for the heterologous expression of the lipase from *Rhizopus miehei* and *Humicola lanuginosa*. The lipase of *Penicillium roqueforti* is responsible for the flavour of Blue cheese. Lipolytic activity has also been detected in *P. camemberti*, the white surface mould of Brie and Camembert cheese. Lipases with specificity for butyric acid have been isolated from strains of the Penicillium species such as *P. cyclopium*, *P. verrucosum* var. *cyclopium* and *P. crustosum*. The *P. cyclopium* lipase has a much higher activity towards di- and monoglycerides than triglycerides. The lipase of *H. lanuginosa* DSM 3819 is suitable as a detergent additive because of its thermostability, high activity at alkaline pH, and stability towards anionic surfactants. *H. lanuginosa* lipases show a high degree of hydrolytic activity with coconut oil and oils having a high content of lauric acid. The two lipases differ in their positional specificity.

Industrial Applications of Lipase Enzyme

Dairy industry   Lipases are extensively used in the dairy industry for the hydrolysis of milk fat. Current applications include the flavour enhancement of cheeses, the acceleration of cheese ripening, the manufacturing of cheese like products, and the lipolysis of butterfat and cream. The free fatty acids generated by the action of lipases on milk fat endow many diary products, particularly soft cheeses with their specific flavour characteristics.

Detergents   Lipase can be used in the laundry detergents and automatic dish washing machines, which will work under alkaline conditions as fat stain removers.

Pharmaceutical industry   Lipase are a conventional tool in organic chemistry to synthesize the optically pure chemical compounds. Lipase is useful in the preparation of racemic chiral synthon compounds which are main skeleton structure of synthetic pharmaceutical products and agrochemicals. Besides, lipases are used as a industrial catalyst for the racemic alcohols and optically active intermediates that are useful in pharmaceutical industries.

Lipases in pulp and paper industry   The pitch is the hydrophobic component of wood mainly triglycerides and waxes that causes severe problems in pulp and paper manufacture. Lipases are used to remove the pitch from the pulp produced for paper making.

Lipases in food industry   Fats and oils are important constituents of foods. The nutritional value and the physical properties of a triglyceride are greatly influenced by factors such as the position of the fatty acid in the glycerol, the chain length of the fatty acid and its degree of unsaturation. Lipases are used to modify the properties of lipids by altering the location of fatty acid chains in the glyceride and replacing one or more of the fatty acids with new ones.

Cosmetic industry   Lipase has been used as a biocatalyst in solvent free esterification to produce compounds like isopropyl myristate, isopropyl palmitate and 2-ethylhexylpalmitate, and used as an emollient in personal care products such as skin and sun-tan creams, bath oils, etc.

Medical applications   Several medical applications of lipases are under consideration. The inhibition of the human enzyme as a method of reducing fatty acid absorption is being investigated for treatment for obesity. Lipases are used to acylate and deacylate the hydroxyl group of castanospermine, a promising drug for the treatment of AIDS.

ISOLATION AND SCREENING OF LIPASE PRODUCING MICROORGANISMS

Lipase-producing microorganisms have been found in diverse habitats such as industrial wastes, vegetable oil processing factories, dairies, soil contaminated with oil, oilseeds, decaying food, compost heaps, coal pits, and hot springs. Lipase-producing microorganisms include bacteria, fungi, yeasts and actinomycetes. A simple and reliable method for detecting lipase is the use of the surfactant Tween 80 in a solid medium. The formation of opaque zones around the colonies is an indication of lipase production by the organisms (Figure 18.5). Modifications of this assay use various Tween surfactants in combination with Nile blue or neet's foot oil and Cu^{2+} salts. Also, screening of lipase producers on agar plates is frequently done by using tributyrin as a substrate, and clear zones around the colonies indicate the production of lipase. The other screening system for the production of lipase is preparing the plates with chromogenic substances. The Rhodamine B agar is widely used for the screening of lipase production. The Rhodamine B is used as indicator of lipase production which can react with esters and fluorese under UV light.

Composition of Tween agar (g/l)

Components	Quantity
Peptone	10
NaCl	5
$CaCl_2$	0.01
Agar	20
Tween	10 ml

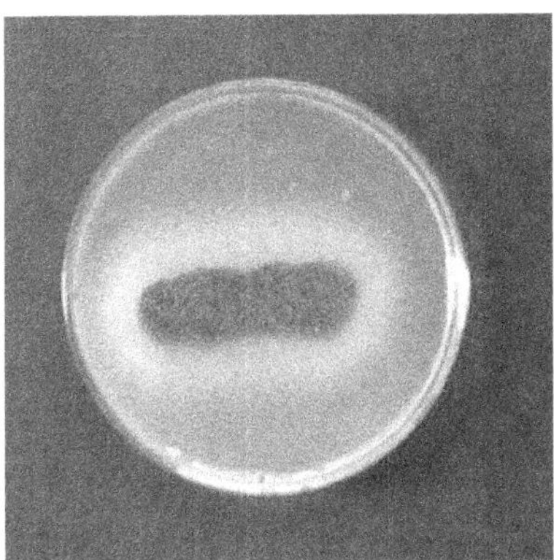

Figure 18.5 Lipase activity in Tween agar plate by *Aspergillus*

Production and Media Development for Lipase

Microbial lipases are produced mostly by submerged culture, but solid state fermentation methods can also be used. Immobilized cell culture has been used in a few cases. Many studies have been undertaken to define the optimal culture and nutritional requirements for lipase production by submerged culture. Lipase production is influenced by the type and concentration of carbon and nitrogen sources, the culture pH, the growth temperature, and the dissolved oxygen concentration. Lipidic carbon sources seem to be generally essential for obtaining a high lipase yield; however, a few authors have produced good yields in the absence of fats and oils. For example, it has been observed that the production of lipase enzyme from *Bacillus* sp. in 1% olive oil in the culture medium was normal. But the enzyme activity is observed in the prolonged culture even in the absence of olive oil. Little enzyme activity was observed in the absence of olive oil even after prolonged cultivation. Fructose and palm oil were reported to be the best carbohydrate and lipid sources, respectively, for the production of an extracellular lipase by *Rhodotorula glutinis.*

For nitrogen source, the peptone that is present in the culture medium gives considerable results in the extracellular lipase production. Nitrogen sources such as corn steep liquor and soybean meal stimulated lipase production but to a lesser extent than peptone. Urea and ammonium sulphate inhibited lipase synthesis.

Lipase production was higher when metal ions such as magnesium, iron, and calcium were added to the production medium. Besides, the production of an extracellular lipase is also enhanced when the medium was supplemented with Mg^{2+}, Ca^{2+}, Cu^{2+} and CO^{2+}.

Composition of production medium (g/l)

Components	Quantity
NH$_4$Cl	35
Glycerol	10 ml
K$_2$HPO$_4$	3
KH$_2$PO$_4$	1
MgSO$_4$	0.1
Glucose	2
MgCl$_2$	0.6 mM

Culturing Condition

Bacterial inoculums are prepared by growing the cells at 37°C for 10 hours in 500 ml shake flasks containing 100 ml of LB medium. Lipase production is performed in a tank fermenter. The fermentation condition is a three-step procedure: batch, fed-batch and repeated fed-batch. The conditions used in the batch phase are agitator speed 400 rpm, aeration rate 1 vvm, temperature 30°C, and optimum pH control. Cell concentration is measured by turbidimetry (600 nm) and correlated with dry cell weight.

In solid state fermentation of fungi, the cell suspension is prepared as follows. The organism is subcultured on potato dextrose agar medium and incubated for 72 hours at 28°C. The spore crop of each slant is scrapped into 5 ml sterile water and shaken well with sterile glass beads on a rotary shaker for 30 minutes to break the spore chains and to make a uniform suspension. This suspension is filtered through sterile cotton to remove the hyphal filaments. This spore suspension (5×10^6 spores per ml) is used as inoculum.

For the production of lipase in SSF, wheat bran moistened with mineral salt medium and oil is sterilized at 121°C for 30 minutes. After cooling, the substrates are inoculated with spore suspension containing 10^6 spores from 7-day-old culture. The medium is incubated at 30°C for appropriate time.

Preparation of Immobilized Cells

Sodium alginate entrapment method   Spores are immobilized in calcium alginate by the traditional external gelation method. About 20 ml of sodium alginate (3% w/v) and 5 ml of spore suspension (5×10^6 spores per ml) are mixed well, and this slurry is added drop-wise to 0.2 M $CaCl_2$ solution at room temperature. The beads formed are then cured in a refrigerator at 4°C for 1 hour. The beads are washed two to three times with sterile distilled water and used for the production of lipase.

κ-Carrageenan entrapment method   Carrageenan (4% w/v) maintained at 40–50°C is mixed with 2 ml of spore suspension (5×10^6 spores per ml) and poured into sterile 4-inch diameter petri plates and allowed to solidify. It is then cut into equal small blocks (approximately $5 \times 5 \times 4\,\text{mm}^3$). The carrageenan blocks are cured in 2% KCl for 1 hour in the refrigerator. These blocks are washed thoroughly two to three times with sterile distilled water and used for the production of lipase.

Assay for Lipase Activity

The qualitative plate is performed using Tween agar plates prepared by excluding peptone because for enzyme activity, the peptone is not needed. Besides, the presence of peptone in the plate assay medium may lead to the contamination of the plate assay medium. A well of around 10 mm diameter is cut out aseptically with the help of cork borer. The well is filled with the culture filtrate and incubated at 35°C overnight. The observation is made to see the precipitation band around the well, and the formation of precipitation band was considered as positive result.

The quantitative assay is done by photometric method using olive oil as a substrate and cupric acetate as colour developer. The reaction mixture is prepared by mixing enzyme solution and water emulsified substrate. The mixture is allowed to react to form the product. The reaction is stopped by adding 1 ml of 6N HCl, and iso-octane is added to release the free fatty acid, which was produced during reaction. Further, cupric acetate is added to develop the colour with fatty acid. The upper aqueous layer is removed and the absorbency measured at 715 nm. The amount of free fatty acid formed is calculated by comparing the absorbency with standard curve prepared using oleic acid.

PRODUCTION OF CELLULASE

Cellulase is an enzyme that catalyses the breakdown of cellulose. Cellulose, the largest renewable carbon source available (approximately 150 billion tons of organic material is photosynthesized annually), is frequently found in close association with other compounds, such as hemicellulose, lignin and other polysaccharides, which makes its bioconversion more difficult. Cellulose is an unbranched glucose polymer, composed of anhydro-D-glucose units linked by 1,4-β-D-glucoside bonds, which can be hydrolysed by cellulolytic enzymes produced by both bacteria and fungi. Cellulolytic bacteria include aerobic species such as *Pseudomonas* and *Actinomycetes*, facultative anaerobes such as *Bacillus* and *Cellulomonas*, and strict anaerobes such as *Clostridium*. The commercial possibility of using cellulase preparations to produce glucose, alcohol and protein from cellulose is under intensive study.

A cellulosic enzyme system consists of three major components: endo-β-glucanase (EC 3.2.1.4), exo-β-glucanase (EC 3.2.1.91) and β-glucosidase (EC 3.2.1.21).

The mode of action of each of these being:

1. Endo-β-glucanase, 1,4-β-D-glucan glucanohydrolase, CMCase, Cx: "random" scission of cellulose chains yielding glucose and cello-oligosaccharides.

2. Exo-β-glucanase, 1,4-β-D-glucan cellobiohydrolase, Avicelase, C1: exoattack on the non-reducing end of cellulase with cellobiose as the primary structure.

3. β-glucosidase, cellobiase: hydrolysis of cellobiose to glucose.

The exo-β-glucanase causes a disruption in cellulose hydrogen bonding, followed by hydrolysis of the accessible cellulose with endo-β-glucanase. The endo-β-glucanase acts randomly on the cellulose chain, while exo-β-glucanase acts on exposed chain ends by splitting off cellobiose or glucose. Cellobiose is subsequently hydrolysed by β-glucosidase to glucose. This indicates that three, rather than two enzymes, are essential for the decomposition of cellulosic biomass.

Bacteria and fungi have evolved complex enzymatic systems enabling their growth on plant material rich in cellulose, but these organisms typically require weeks, months, or even years to decompose a fallen log or a tilled corn stalk. For chemical or fuel production from these same materials, industry requires affordable chemical or enzymatic systems that can do the job in hours or in days.

Most cellulases studied have similar pH optima, solubility and amino acid composition. Thermal stability and exact substrate specificity may vary. However, it should be remembered that cellulase preparations generally contain other enzymatic activities besides cellulase, and these may also affect the properties of the preparations. Cellulase preparations are effective between pH 3 and 7. The optimum pH generally lies between

4 and 5 and the temperature 30–50°C. Cellulase is inhibited by its reaction products such as glucose; cellobiose inhibits cellulases completely, whereas Mn, Ag, Cu and Zn ions are only slightly inhibitory. The activity of cellulase preparations has been found to be completely destroyed after 10–15 minutes at 80°C. Solutions of cellulase at pH 5–7 are stable for 24 hours at 4°C and lyophilized preparations are stable for several months without significant loss of activity.

APPLICATIONS OF CELLULASE

Waste treatment   The production of ethanol and biogases from agricultural wastes like bagasses, straw, wood, etc., is a common waste recycling process. In this process, the conversion of polysaccharides into simple sugar will be technically called saccharification. The primary saccharification will be either done by adding a complex of enzymes including cellulase, or it will be achieved by growing cellulase producing organisms.

Food industry   The production of beer from cellulosic material is carried out *via* two steps. The primary step is conversion of complex sugar into simple sugar, and the secondary step is conversion of simple sugar into beer. The latter process is achieved by yeast, whereas the primary step is processed using enzymes such as cellulase, etc. Cellulase enzyme plays an important role in the preparation of malt adjunct. Besides, fruit and vegetable processing and extraction of venilla is carried out using cellulase. The complex of enzymes such as pectinase, cellulases, arabinase, galactase, mannase, alpha-galactosidase, protease and phytase are added as food additives to enhance the feed nutritional value and reduce the viscosity by hydrolysis of non-starch polysaccharides and proteins.

Textile industry   Biostoning of denim, biopolishing, depilling, bioscouring of cotton, and brightness of cloths can be achieved by cellulase treatment.

Pulp and paper industry   Pulp biobleaching, paper de-inking, etc., are mainly carried out using cellulase.

Detergent industry   Cellulase, lipase, amylase, etc., are added along with detergents for biodegradation of impurities, refreshment of textile materials, and depilling of cloths.

Strain development   The protoplasmic fusion is an important technique used widely for strain improvement of plants and fungi. The mixture of enzymes like cellulase, chitinase, pectinase, etc., is used to dissolve the cell wall of the organisms.

Strain Isolation and Screening

Mostly, the cellulose producing organisms are available in cellulose rich humus. Besides, the paper and pulp industry waste dumping is rich in cellulose producing microorganisms as native flora. The pure cultures are inoculated in sterile carboxymethyl cellulose (CMC) agar plates and incubated. The positive organisms form the cellulolytic zone around the colonies. The hydrolytic zone can be visualized by staining the agar plates with Congo red stain. Around 0.1% of Congo red solution is over layered on the medium and kept for 15 minutes. Destaining is done using 1M NaCl to make the zone clearly visible.

Composition of CMC agar (g/100 ml)

Components	Quantity
Carboxymethyl cellulose	0.5
$NaNO_3$	0.1
K_2HPO_4;	0.1
KCl	0.1
$MgSO_4$	0.05
Yeast extract	0.05
Glucose	0.1

Cultural Conditions

Enzyme production is carried out using the production medium comprising glucose as a carbon source and amended with peptone as a proteinacious substrate with pH 7 for bacteria. SSF is carried out in dry carbon source mixed with mineral solution. Carbon sources like wheat straw, rice straw, corn cobs, wheat bran, oat bran, Arundo donax and Populus tremuloides are used as raw substrates for cellulose production. The substrates are supplemented with the mineral solution prior to heat sterilization.

Analysis for Cellulase

The plate assay is performed using agar plates amended with hydroxyl methyl cellulose. The agar plates are prepared by mixing 1% carboxymethyl cellulose with 1.7% agar. After the solidification of agar, a well is made out aseptically with the help of cork borer. The well is filled with the culture filtrate and incubated at 37°C overnight. Observation is made to see the hydrolytic zone around the well. For better appearance of the zone, 0.1% of Congo red solution is over layered on the agar plates and kept for 15 minutes. Destaining is done using 1M NaCl to make the zone clearly visible.

The qualitative assay for cellulase activity is determined by estimating the reducing sugar produced during enzymatic reaction by dinitrosalicylic acid method.

Crude culture filtrate is used as an enzyme sample. Around 1 ml of enzyme solution is mixed with 1% of carboxy methyl cellulose (CMC) and incubated at 30°C for 15 minutes to perform the enzyme substrate reaction. The contents are cooled and dinitrosalicylic acid is added and heated at 90°C for 5–15 minutes till slight reddish brown is developed. The contents are cooled and potassium sodium tartrate is added. The absorbency of the contents is measured at 575 nm against the reaction mixture prepared using distilled water as blank.

PRODUCTION OF PHOSPHATASE

Phosphatases are enzymes that hydrolyse the complex organic phosphates into inorganic phosphates (Pi), by the process called dephosphorylation, which will be further used by the cell to construct the nucleic acid, phospholipids, ATP energy molecule, etc.

$$\text{Organic phosphate} \xrightarrow{\text{Phosphatase}} \text{Inorganic phosphate (Pi)} + \text{Organic residues}$$

Phosphatases are classified into three categories based on their reaction pH.

Acid phosphatase   Acid phosphatase is an enzyme that works under acid conditions. The main sources of acid phosphatases are the liver, spleen, bone marrow and prostate gland in animals. Besides, most of the phosphate solubilizing microbes such as *Bacillus* spp., *E.coli*, *Aspergillus* spp., etc. are capable of producing acid phosphatase. In animals, it converts an orthophosphoric monoester and water to alcohol and orthophosphate.

Neutral phosphatase   It is a cationic phosphatase enzyme that will show the peak enzymatic activity in neutral pH. Naturally, it is distributed in animals and the membrane surface of bacteria such as staphylococci, etc. It also has affinity towards attachment of immunoglobulins.

Alkaline phosphatase   Alkaline phosphatase is a non-specific metallo enzyme that hydrolyses many types of phosphate esters at an alkaline pH in the presence of zinc and magnesium ions. It is present in multiple molecular isoenzymes. In animals, there are four major isoenzymes of alkaline phosphatase, namely intestinal (IAP), corticosteroid (CAP), bone (BAP) and liver (LAP). Isoenzymes differ in catalytic sites and activity, immunogenicity, amino acid sequence and electrophoretic mobility.

SOURCES OF PHOSPHATASE

Animals

The enzyme is associated with microsomal membranes and is present in many tissues. Animal phosphatase is anchored to cell membranes by glycophosphatidylinositol (GPI) proteins. Cleavage of these proteins by bile acids, phospholipase D and proteases releases AP from membranes resulting in increased AP levels in serum/plasma.

Liver   The enzyme is produced by hepatocytes epithelium of the biliary tract. These cells are the source of LAP isoenzyme.

Bone   The isoenzyme is produced by osteoblasts and increases in serum in association with osteoblastic activity.

Intestinal, renal, placental tissues   These are not usually important sources of increased serum AP activity, though some placental isoenzyme is present in serum of normal pregnant queens and mares.

Shrimp alkaline phosphatase (SAP)   This phosphatase enzyme is extracted from a cold water shrimp *Pandalus borealis.*

Plant Sources

Phosphatase is widely found in plants having intracellular and extracellular activities. Phosphatase is important for phosphorous scavenging and remobilization in plants. The organic phosphate that is present in the soil is not directly taken by the plants. The extracellular phosphatase enzyme produced by the plants facilitate to convert the organic phosphate to inorganic phosphate form and mobilized inside the plant. Besides, it is providing adaptation during abiotic stress and regulating growth hormone during germination by maintaining the phosphate concentration level.

Microbial Sources

In microorganisms, alkaline phosphatase is located in the periplasmic space external to the cell membrane, which will be released during scarcity of phosphate in the cell and also presence of phosphate in the medium. Microbial alkaline phosphatase is comparatively resistant to inactivation, denaturation and degradation, and also has a higher rate of activity.

Mostly, bacterial strains like *E.coli.*, *Bacillus* spp., *Pseudomonas* spp., *Xanthomonas* spp., *Enterobacter* spp., *Acidovorans* spp., *Azotobacter* spp., *Arthrobacter* spp., *Acetobacter* spp., *Vibrio* spp., etc. and fungi like *Aspergillus* spp., *Cladosporium* sp., *Penicillium* spp., *Mucor* spp., *Fusarium* spp., *Curvularia* spp., etc., are producing extracellular alkaline phosphatase. Among this microbial population, some of the species are having symbiotic association with compatible plants and helping phosphate mobilization. Some bacterial populations are present in the rhizosphere soil of the plant and make the availability of inorganic phosphate to the plants by solubilizing the complex phosphate. Some of the fungi are making mycorrhizal association with plants, and thus increasing phosphate mobilization to the plants.

Microbial production of alkaline phosphatase enzymes has several advantages over the animal and plant enzymes productions.

- Most of the plants and animals are not producing extracellular alkaline phosphatase. In animals, the extraction of alkaline phosphatase from various organs or from other parts of the body is difficult. Sometimes they need to sacrifice their life.

- Economically, microbial production is feasible because large-scale production can be carried out within limited space and time.

- Microorganisms are capable of producing a wide variety of enzymes.

- They can grow in a wide range of environmental conditions.

- They show genetic flexibility.

- They have short generation time.

APPLICATIONS OF ALKALINE PHOSPHATASE

Medical applications   Certain recombinant human alkaline phosphatase isoenzymes are likely to have superior pharmacodynamic and pharmacokinetic properties. These are used for the treatment of inflammatory bowel disease (ulcerative colitis) and sepsis.

Microbiological applications   The detection of alkaline phosphatase is used as a tool to assess the quality of pasteurized milk. The milk contains heat-sensitive phosphatase enzymes that can be deactivated by heating. Pasteurized milk will not have phosphatase enzyme.

Immunological and diagnostic applications   Alkaline phosphatase is widely used as a conjugate for enzyme-linked immunosorbent assay (ELISA), enzyme immuno assay (EIA), western blotting, cell and tissue staining for microscopy, etc. The use of alkaline phosphate as a conjugate is having several advantages:

- ➲ It gives lower non-specific backgrounds than other enzyme conjugates.

- ➲ It catalyses enzymatic reactions, which generate either soluble or insoluble coloured (or fluorescent) products.

- ➲ Most alkaline substrates and products are non-hazardous.

Genetic engineering   Alkaline phosphatase has become a useful tool in molecular biology laboratories because the DNA normally possesses phosphate groups on the 5′-end. Removing these phosphates prevents the DNA from ligating (the 5′-end attaching to the 3′-end), thereby stabilizing the DNA until the next step of the process for which it is being prepared; also, removal of the phosphate groups allows radio labelling (replacement by radioactive phosphate groups) in order to measure the presence of labelled DNA through further steps in the process or experiment.

Agriculture   Even though the enzymes are not used directly in agriculture, the phosphate solubilizing microorganisms are used as biofertilizers and bioinoculants for several crops.

Environment   The phosphate solubilizing microorganisms are used in bioleaching and biomining processes to recover minerals. In the process of biomineralization, the sulphate reducing bacteria like *Thiobacillus* sp. and the phosphatase producing bacteria like *Citrobacter* sp. are coupled for mineralization. These organisms cumulatively recover phosphate and sulphur along with precipitation of cadmium.

ISOLATION AND SCREENING METHODS FOR PHOSPHATASE PRODUCTION

Phosphatase-producing organisms are widely present in the environment, namely in phosphate rich soils and water sources. The strains that are isolated from the environment can be screened for their phosphatase producing activity using the screening medium like hydroxyapatite agar (soil extract agar) and Pikovskaya agar. The medium contains calcium phosphate complex (hydroxyapatite), which can be hydrolysed by producing phosphatase enzymes. Thus, the phosphatase producing organism will form a clear zone around the colony by hydrolysing the calcium phosphate complex (Figure 18.6).

Composition of hydroxyapatite medium

Components	Quantity
Soil extract	100 ml
Glucose	10%
$CaCl_2$ (10%)	5 ml
KH_2PO_4 (10%)	5 ml
Agar	1.7%
pH	7.0

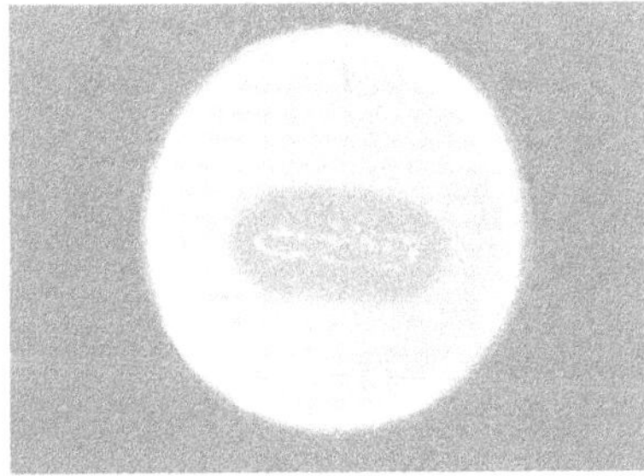

Figure 18.6 Phosphatase activity in soil extract agar

CULTURAL CONDITIONS AND PRODUCTION

The bacterial inoculum for the production of phosphatase enzyme is prepared using Luria Bertani (LB) broth. The production is carried out with the fermentation medium, which contains inorganic phosphate as inducer. The cultural condition for the production of phosphatase varies from strain to strain. The concentration of phosphate is the limiting factor of production. Generally, the production is carried out using both high phosphate and low phosphate medium. The concentration of phosphate optimization is an essential factor in the production of phosphatase enzyme. The increase of phosphate concentration level in the culture medium may retard the production of phosphatase.

Composition of high phosphate medium (g/l)

Contents	Quantity
Peptone	5
Glucose	2
Tris	6
NaCl	3
$MgCl_2$	0.05
$CaCl_2$	0.02
$FeSO_4$	0.001
Na_2HPO_4	1.42
pH	6.8

Note that the low phosphate medium can be prepared as per the composition of high phosphate medium excluding Na_2HPO_4.

PHOSPHATASE ENZYME ASSAY

Plate Assay

The plate assay is performed using hydroxyapatite medium that was prepared with 1.5% of agar in distilled water excluding glucose. After solidification of the agar, the enzyme solution is placed in a well formed on agar plates, and the plates are incubated at appropriate temperature for overnight in a humid chamber. Observation is made to see the phosphate solubilizing zone around the well.

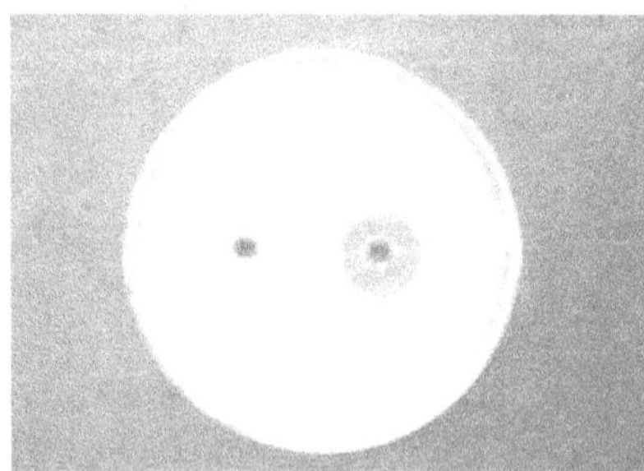

Figure 18.7   Plate assay for phosphatase activity

Chemical Assay

The reaction mixture is prepared by mixing the culture filtrate (enzyme), sodium carbonate–bicarbonate buffer, substrate (disodium phenyl phosphate) and magnesium chloride (0.2M). The mixture is incubated at 37°C for one hour for enzymatic reaction. The reaction is stopped by adding trichloro acetic acid (TCA) and centrifuged with supernatants ammonium molybdate and amino naphthal sulphonic acid (ANSA). The absorbance is measured at 650 nm with UV-VIS spectrophotometer using distilled water reaction mixture as a blank. The quantity of enzyme required to form a microgram of product is considered as one unit.

PRODUCTION OF PECTINASE

Pectinolytic enzymes or pectinases are a heterogeneous group of related enzymes that hydrolyse the pectic substances, present mostly in plants. Pectinolytic enzymes are widely distributed in higher plants and microorganisms. They are of prime importance to plants as they help in cell wall extension and softening of some tissues during maturation and storage. They also aid in maintaining ecological balance by decomposing and recycling waste plant materials. Plant pathogenicity and spoilage of fruits and vegetables by rotting are some other major manifestations of pectinolytic enzymes.

Pectinolytic enzymes may be divided in to three broader groups as follows:

1. Protopectinases degrade the insoluble protopectin and give rise to highly polymerized soluble pectin.

2. Esterases catalyse the de-esterification of pectin by the removal of methoxy esters.

3. Depolymerases catalyse the hydrolytic cleavage of α-(1-4)-glycosidic bonds in the D-galacturonic acid group of the pectic substances.

APPLICATIONS OF MICROBIAL PECTINASES

Biotechnological applications   Over the years, pectinases have been used in several conventional industrial processes, such as textile, plant fibre processing, tea, coffee, oil extraction, and treatment of industrial wastewater, containing pectinacious material, etc. They have also been reported to work on purification of viruses and in the making of paper. They are yet to be commercialized.

Fruit juice extraction   The largest industrial application of pectinases is in fruit juice extraction and clarification. Pectins contribute to fruit juice viscosity and turbidity. A mixture of pectinases and amylases is used to clarify fruit juices. It decreases filtration time up to 50%. Treatment of fruit pulps with pectinases also showed an increase in fruit juice volume from banana, grapes and apples. Pectinases in combination with other enzymes, *viz.* cellulases, arabinases and xylanases, have been shown to increase the pressing efficiency of the fruits for juice extraction. Vacuum infusion of pectinases has a commercial application to soften the peel of citrus fruits for removal. The infusion of free stone peaches with pectin methylesterase and calcium would result in four times firmer fruits. This may be applied to pickle processing where excessive softening may occur during fermentation and storage.

Textile processing and bioscouring of cotton fibres   Pectinases have been used in conjunction with amylases, lipases, cellulases and hemicellulases to remove sizing agents from cotton in a safe and eco-friendly manner, replacing the toxic caustic soda used earlier. Bioscouring is a novel process of removal of non-cellulosic impurities from the fibre with specific enzymes. Pectinases have been used for this purpose without any negative side effect on cellulose degradation.

Degumming of plant bast fibres   Bast fibres are the soft fibres formed in groups outside the xylem, phloem or pericycle, e.g. Ramie and sunn hemp. These fibres contain gum, which must be removed before its use in textile making. The chemical degumming treatment is polluting, toxic and non-biodegradable. Biotechnological degumming using pectinases in combination with xylanases presents an eco-friendly and economic alternative to the above problem.

Retting of plant fibres   Pectinases have been used in retting flax to separate the fibres and eliminate pectins.

Wastewater treatment   Vegetable processing industries release wastewaters that largely contain pectin. Pretreatment of these wastewaters with pectinolytic enzymes facilitates the removal of pectinaceous material and renders it suitable for decomposition by activated sludge treatment.

Coffee and tea fermentation   Pectinase treatment accelerates tea fermentation and also destroys the foam forming property of instant tea powders by destroying pectins. They are also used in coffee fermentation to remove mucilaginous coat from coffee bean.

Paper and pulp industry   In papermaking, pectinases can depolymerize pectins and subsequently lower the cationic demand of pectin solutions and the filtrate from peroxide bleaching.

Animal feed   Pectinases are used in the production of animal feeds. They reduce the feed viscosity, which increases the absorption of nutrients. They liberate nutrients either by hydrolysis of non-biodegradable fibres or by liberating nutrients blocked by these fibres, and subsequently reduce the amount of faeces.

Purification of plant viruses   In cases where the virus particle is restricted to the phloem, alkaline pectinases and cellulases can be used to liberate the virus from the tissues to give very pure preparations of the virus.

Oil extraction   Citrus oils such as lemon oil can be extracted with pectinases. They destroy the emulsifying properties of pectin, which interfere with the collection of oils from citrus peel extracts.

Improvement of chromaticity and stability of red wines   Pectinolytic enzymes added to macerated fruits before the addition of wine yeast in producing red wine resulted in improved visual characteristics (colour and turbidity) as compared to the untreated wines. Enzymatically-treated red wines presented chromatic characteristics, which are considered better than the control wines. These wines also showed greater stability as compared to the control.

STRAIN ISOLATION AND SCREENING FOR PECTINASE

Pectinase-producing organisms are widely present in dead trees and humus environments. The pectinolyic activity is screened using agar medium, which contains pectin as a substrate. The forming of clear pectinolytic zone is the indication of pectinase production.

The basal medium used in the fermentor contains (g/l): $(NH_4)_2SO_4$, 1.4; $MgSO_4$, 0.3; KH_2PO_4, 2; $CaCl_2$, 0.3; $NaNO_3$, 5; Tween 80, 1 ml; and 1 ml of trace element solution (g/l): $CoCl_2$, 2; $MnSO_4 \cdot H_2O$, 1.6; $ZnSO_4 \cdot H_2O$, 1.4; $FeSO_4 \cdot 7H_2O$, 0.5. The carbon sources like gruel, milled orange peel and gruel hydrolysate (GH, a solid by-product obtained after the hydrolysis of gruel using amylases) are mainly used as substrates in industrial fermentation. Gruel is principally composed of starch (60% of dry matter), polysaccharides (15%) and proteins (13%). The liquid medium is sterilized for 20 minutes at 121°C for enzyme production.

Assay for Pectinase

The plate assay for pectinolytic activity is carried out by well method in agar plate amended with pectin as a substrate. The quantitative assay is carried out by calculating the reducing sugar liberated by the enzyme during reaction with substrate. The liberated reducing sugar is estimated using DNS method. The polygalacturonic acid is used as standard.

GENERAL PURIFICATION METHODS OF PROTEINS

Most of the products of biotechnology are proteins, and these proteins must be prepared in large volumes in purified form. In general, the contaminants present, should be removed unless they are proved to be harmless. Most proteins are purified to where they compose 99.99% or more of the material in the sample. A protein must be purified from other proteins, but you cannot ignore the nucleic acids, carbohydrates or lipids in the sample.

In addition to purity, the protein must retain its biological activity. The process must produce the same amount and quality of protein every time. This would force the use of very robust purification processes, which may not be the very best methods used to purify the protein. A method that works exquisitely in a research laboratory may fail miserably on the production floor. For purifying proteins, we make use of their inherent similarities and differences. Protein similarity is used to purify proteins from other non-protein contaminants. Similarly, protein differences are used to purify one protein from another. Proteins vary from each other in size, shape, charge, hydrophobicity, solubility and biological activity.

Immuno-affinity Chromatography

Affinity chromatography relies on the protein binding specifically to an immobilized ligand, while the rest of the proteins pass through the column. The ligand of choice is monoclonal antibodies. Monoclonal antibodies are expensive. They have to be purified too. They normally bind with such tenacity that they are hard to make turn loose. So, you have to use harsh conditions to inactivate your protein or destroy a part of the monoclonal antibodies. Other proteins in your preparation may inactivate the antibodies (proteases) or bind non-specifically to them. Some of the monoclonal antibodies always leach off the column during purification, and then this must be removed from the protein. Affinity columns are used but usually only later in the process when the volume has been reduced and the majority of contaminants have already been removed. Generally, it will only be used as a last resort on an expensive product.

Affinity Chromatography

Some types of affinity chromatography can be very useful; however, they are used later in the purification process after much of the garbage has been removed. If the protein binds a specific carbohydrate or requires a specific cofactor, you may be able to get it to bind to a column on which that carbohydrate or cofactor is immobilized. The protein can then be eluted with a high concentration of the carbohydrate or cofactor. Mimics for binding sites can sometimes be used as affinity matrices; imagination is the only limit for some proteins. Antibodies can be purified by the use of bacterial proteins that bind to antibodies protein A and protein G.

Protein Precipitation

Proteins are usually soluble in water solutions because they have hydrophilic amino acids on their surfaces that attract water molecules and interact with them. This solubility is a function of the ionic strength and pH of the solution. Proteins have isoelectric points at which the charges of their amino acid side groups balance each other. If the ionic strength of a solution is either very high or very low, proteins will tend to precipitate at their isoelectric point. The solubility is also a function of ionic strength, and as you increase the ionic strength by adding salt, proteins will precipitate. Ammonium sulphate is the most common salt used for this purpose because it is unusually soluble in cold buffers. Ammonium sulphate fractionation is commonly used in research laboratories as a first step in protein purification because it provides some crude purification of proteins aside from non-proteins and also separates some proteins. The protein slurry formed during ammonium sulphate precipitation is usually very stable. So the prolongation of precipitation process will not affect the process.

In industrial uses, ammonium sulphate has some problems. Ammonia is very reactive with stainless steel, which is the primary building material for purification facilities. Other salts, such as sodium sulphate, are sometimes used to get over this problem but none are as good as ammonium sulphate. In addition to salting out, proteins can also be forced out of solution with polymers; the most frequently used polymer is polyethylene glycol (PEG). PEG is inert and, like ammonium sulphate, it tends to stabilize proteins. Increasing concentrations are added to cold protein solutions with gentle stirring. The precipitated proteins are removed by centrifugation or filtration. The slurry can be stored in the cold for long periods without detrimental effects on most proteins.

Protein precipitation is not a great method of purification; it typically will only give a few fold of purification where we may need 1000-fold or more by the time we are finished. On the good side, it does remove the protein from the growth medium or busted up cells where vicious proteases and other detrimental garbage may lurk.

Buffer Exchanges and Dialysis

The dialysis is a method that is employed to remove the salt from the protein-salt precipitate. This is the next step of ammonium sulphate precipitation of protein. After doing ammonium sulphate precipitation, the removal of protein binding ammonium sulphate is done by dialysis using cellophane dialysis membrane. This cellophane tube has small pore which will allow smaller salt molecules and not larger protein molecules to pass thorough it. Usually, set up is prepared by packing the ammonium sulphate precipitated protein in dialysis tube and is immersed in a suitable buffer bath. The salts from the precipitation will diffuse through the membrane and enter into the buffer. Here the periodical buffer exchange is essential because, the continuous diffusion of salt through membrane will lead into equilibrium state of salt concentration between inner side of the membrane and outer buffer. These may reduce the diffusion rate. So the periodical buffer change is giving continuous and complete salt removal from the protein that is present in the membrane.

Now, selectively permeable membranes are sandwiched between plates through which the protein solution is pumped. Buffer is pumped on one side of the membrane and the salt-laden protein solution on the other. By doing this under pressure, rapid equilibration occurs and the solution can be exchanged with another buffer in a matter of a few hours. This same type of membrane can be used to concentrate the protein by putting more pressure on the protein solution side and forcing water and salt through the membrane.

Every step in protein purification results in a loss of product, and buffer equilibration steps are often big losers. During buffer equilibration, protein bound to every surface is inactivated easily by shear forces, foaming or rapid changes in ionic strength results in product loss.

Ionic Exchange Chromatography

This is the most useful of all protein purification and concentration methods. By choosing different buffers, the same protein can adsorb to both anion exchangers (that bind negatively charged molecules) and cation exchangers (that bind positively charged molecules). Remember that proteins are composed of amino acids, and amino acids have different overall charges at different pH values. Most protein purification is done on anion exchange columns because most proteins are negatively charged at physiological pH values (pH 6–8); proteins can become inactivated at extreme pH, so they are avoided if possible. The matrix material for the column is formed from beads of some inactive material, often a carbohydrate such as cellulose or dextrans. This matrix is in the form of tiny beads, that can be slurried in buffer and poured into a vertical column formed by a glass tube. A screen at the bottom keeps the beads from flowing out with the buffer and the column is packed in this manner. When the protein solution is pumped onto the top of the column, the beads adsorb the proteins as they flow past. The proteins are then eluted from the column according to the tenacity with which they bind. Those that are the most highly charged at that pH will bind the tightest. To release the proteins in the order of binding tenacity, we can either increase the salt concentration or change the pH. Both methods are used in industry, but raising the salt concentration is by far the most common because it is easiest to control.

In research laboratories, we almost always elute an ion exchange column with a gradient of salt concentrations. Pumps add increasing amounts of salt to the buffer as it goes onto the column so that there is a continuous steady increase in the ionic concentration going through the column. The proteins then elute or come off the column matrix when the ionic strength of the buffer neutralizes their charge. The least charged molecules come off first, and the most highly charged come off last. In industrial production, this type of gradient is too difficult to control precisely, so the most common way to elute columns is by step elution. The column is thoroughly rinsed with buffers of increasing ionic strength until the protein you seek comes off; this exact sequence is repeated each time with the same amount of buffer to give reproducible yields and purification of the protein. Even with gradient elution a single anion exchange column is not going to purify protein; additional procedures are necessary.

Hydrophobic Interaction Chromatography

Proteins are made from both hydrophilic and hydrophobic amino acids. Most of the hydrophobic amino acids are located in the core of the protein structure away from the surface water molecules, but there are some hydrophobic residues on the surface of almost all proteins. It is these hydrophobic amino acids that are used in this form of column chromatography to separate proteins according to their relative hydrophobicity. The column beads are coated with hydrophobic fatty acid chains. The hydrophobic amino acid side groups are not normally exposed on proteins since hydrophilic amino acids attract so many water molecules that the entire protein is surrounded by water molecules, except in conditions of high salt concentration. In high salt, the hydrophobic areas are exposed and bind to the matrix. The column is eluted with decreasing concentrations of salt in the buffer. Proteins usually elute only when the salt concentration is very low.

The reduction of salt from the protein using this column is useful for further purification of protein by ion exchange column chromatography. The hydrophobic column does not require a buffer change before the protein is loaded; usually salt is added to the sample. The protein is eluted off in low ionic strength, so it is ready for the next purification step without a further buffer exchange. This saves time and protein loss.

Size-exclusion Chromatography

Proteins occur in many different sizes and the separation of the molecule is based on the size but it is not a very exacting process and only very gross separations are practical on a commercial scale. Beads are again used, but in this case the beads are made from a porous matrix into which the proteins can diffuse. The size of the pores in the matrix determines the rate at which proteins of various sizes diffuse into the beads, and some proteins are completely excluded. When the column is eluted by the buffer, the separation of the protein is based on the molecular weight. The higher molecular weight proteins are eluted first through this process. By choosing the correct pore sizes for your protein, you can achieve separation of your protein from some problem contaminants later in the purification scheme. This procedure is never used early because it requires that the sample be concentrated and applied as only a few per cent of the column volume. It requires long narrow columns to achieve separation efficiency, and this can pose problems in an industrial process.

Electrophoresis

This process is routinely used to monitor the complexity of the protein mixture by separating the proteins on acrylamide gels. These thin gels allow exquisite separations that unfortunately cannot be scaled up to production levels without major losses in efficiency. You should remember that electrophoresis separates on the same basis as ion exchange chromatography. The first proteins to elute from anion exchange gels will be the slowest moving on a normal acrylamide electrophoresis gel. When the gel is increased in thickness, the heat buildup causes disturbances in the flow and only small sample volumes can be applied to electrophoresis gels; these have so far made electrophoresis impractical for commercial separations. You can use electrophoresis to estimate where a protein will elute on a salt gradient from an anion exchange column; you can use it to determine the number of proteins in a mixture, and you can use it to separate proteins on the basis of size by the use of SDS gels. SDS is sodium dodecylsulphate, a detergent. SDS coats a protein, and when it is heated and denatured, the polypeptide becomes totally coated with SDS such that it has no native charge. The proteins all move at the same rate except that the gel is made of a small enough pore size that it restricts the flow of these now linear molecules, and they separate totally based on size and not on charge. SDS electrophoresis is extremely common in biochemistry laboratories and is used to monitor proteins during purification.

Assay and Specific Activity

The single most important idea that is central to protein purification is specific activity. You have to continually know how much of protein is present and how much contaminating material is present. The ratio is your specific activity. To be able to determine how much of a protein is present in a bunch of other proteins, you have to use a specific assay. A protein assay can tell you how much protein is present unless it is impure. You need to use some attribute of your protein to devise a specific assay. In the case of an enzyme that is usually easy, the protein quantitates that and you have an assay. For example, β-galactosidase breaks apart lactose into glucose and galactose; you can monitor the reaction for free glucose or you can use a specific compound X-gal that produces a colour when it is broken down into parts. The amount of colour generated in a certain period of time becomes a unit of activity. If you define that unit very well (0.1 optical density increase in a 1 cm cuvette at 650 nm at 37°C and pH 6.2), you can use it to estimate the amount of your enzyme present in the sample. The protein assay will give the amount of protein present in the sample and the activity of the protein. Generally the presence of one mg/ml of protein which shows one unit of activity is considered as pure. Assays can be complex or very simple, but they need to be relatively rapid and reproducible in order to be useful. An assay that requires a week to perform will slow down your purification process. Remember that all assays are only estimates of the amount present; protein assays are always inaccurate, but if they are reproducible and precise, they will work fine.

REVIEW QUESTIONS

1. What is enzyme? Give a brief account on classification of the enzymes.

2. Explain about enzymes of amylase and their applications.

3. Give an account on amylase production and purification.

4. Mention the types of proteases and their applications.

5. Describe in detail about protease enzyme production and their purification methods.

6. What are the commercial uses of lipase enzymes? Add notes on their sources and production of lipase enzymes.

7. Describe in detail the production of cellulase and their industrial applications.

8. Give a brief account on phosphatase enzyme production, purification and applications.

9. Write about the commercial uses of pectinase.

10. Give an account on production of pectinase.

11. Write in detail about general enzyme purification methods.

12. Describe various methods of chromatography techniques employed for protein purification.

IMMOBILIZATION

Immobilization may be defined as the restriction of enzyme or cell mobility in a fixed medium where it allows exchanging with other phases. However, enzymes are soluble in aqueous media and it is difficult and expensive to recover them from reactor effluents at the end of the catalytic process. Moreover, the use of enzymes in industrial applications has been limited by several factors, mainly the high cost, their instability and scarcity. Over the last few decades, intense research in the area of enzyme technology has provided many approaches that facilitate practical applications of enzymes. Among them, the technological developments in the field of immobilized biocatalysts offer the possibility of a wider and more economical exploitation of biocatalysts in industry, waste treatment, medicine and in the development of bioprocess monitoring devices like biosensor.

19

ADVANTAGES OF ENZYME IMMOBILIZATION

1. Immobilized cells or enzymes can be reused for several processes because they have high stability.

2. They have less chance to get denatured.

3. The cost of usage will be less and wastage is least.

4. The rate of reaction will be high due to high enzyme substrate ratio and less reaction time.

Criteria to Select the Carrier or Support

1. The binding capacity, porosity, functional group and hydrophobicity of support materials should be compatible for the enzyme.

2. Stability and retention of enzyme should be high for repeat usage.

3. It should not suppress and inactivate enzyme activity by blocking the active site of the enzymes.

4. The support material should not react with reactants.

5. The support material should be recyclable and comparitively less cost.

METHODS OF IMMOBILIZATION

There is a variety of immobilization methods. The type of immobilization for the enzymes or cultures is based on the nature of enzyme and the nature of reaction. Principally, there are three types of immobilization methods namely (i) carrier binding, (ii) cross linking and (iii) entrapment.

Carrier Binding or Surface Immobilization

The enzyme can be fixed externally on the surface of the support material by means of chemical bonds by the process called surface immobilization. The carrier-binding method is the oldest immobilization technique for enzymes. In this method, the amount of enzyme bound to the carrier and the activity after immobilization depend on the nature of the carrier. Figure 19.1 shows how the enzyme is bound to the carrier.

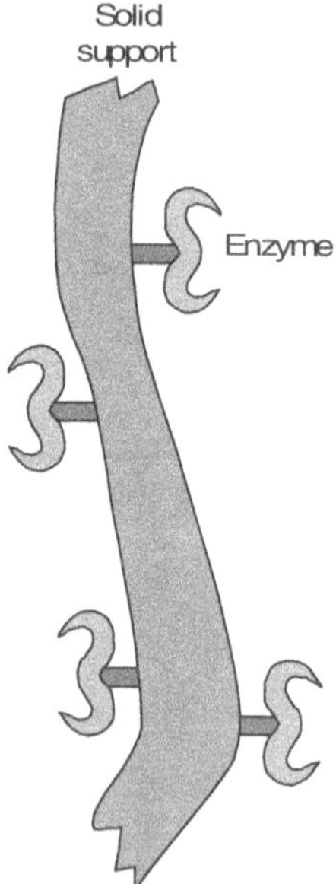

Figure 19.1   Immobilized enzyme in a carrier

The selection of the carrier depends on the nature of the enzyme as well as the:

- ➲ particle size
- ➲ surface area
- ➲ molar ratio of hydrophilic to hydrophobic groups
- ➲ chemical composition

In general, an increase in the ratio of hydrophilic groups and in the concentration of bound enzymes results in a higher activity of immobilized enzymes. Some of the most commonly used carriers for enzyme immobilization are polysaccharide derivatives such as cellulose, dextran, agarose and polyacrylamide gel.

According to the binding mode of the enzyme, the carrier-binding method can be further subclassified into:

- ➲ Physical adsorption
- ➲ Ionic binding
- ➲ Covalent binding

Physical adsorption Adsorption is attachment of enzyme over the surface of the support material by means of weak physical forces like hydrogen bonds, ionic bonds and van der Waals forces. The active sites of immobilized enzymes are usually unaffected and full activity is retained. However, desorption of the enzyme is a common problem over this method due to weak bond. Mineral support (e.g. aluminum oxide, clay, ceramics and diatomaceous earth), organic support (e.g. modified sepharose, carboxy methyl cellulose, DEAE cellulose and active carbon) and ion exchange resins (e.g. amberlite) are used as support materials for adsorption immobilization. Adsorption of enzymes can be stabilized and enhanced by glutaraldehyde but treatment with glutaraldehyde sometimes denature the proteins.

There are four methods to immobilize by adsorbent method:

i. *Static process* The support material is simply allowed to contact the enzyme solution without stirring.

ii. *Dynamic batch process or shaking bath loading* The support material will be placed inside the enzyme solution and constant agitation will be given.

iii. *Reactor loading process*   It is a dynamic process in the reactor.

iv. *Electro-deposition process*   The enzyme solution is placed in an enzyme bath and the supporting material is immersed inside the enzyme solution by connecting with electric terminus. When electricity is passed, the proteins migrate towards the support material and attach on the surface.

Of the four techniques, the most frequently used in lab is bath loading. For commercial purposes, the preferred method is reactor loading.

A major advantage of adsorption in immobilizing enzymes is that usually no reagents and only a minimum of activation steps are required. Adsorption tends to be less disruptive to the enzymatic protein than chemical means of attachment because binding is mainly by hydrogen bonds, multiple salt linkages and Van der Waal's forces. In this respect, the method bears the greatest similarity to the situation found in natural biological membranes and has been used to model such method.

Because of weak bonds involved, desorption of protein resulting from changes in temperature, pH, ionic strength or even the presence of substrate is often observed. Another disadvantage is non-specific, further adsorption of other proteins or other substances as immobilized enzyme are used. This may alter the properties of immobilized enzyme or, if the substance adsorbed is a substrate for the enzyme, the rate will probably decrease depending on the surface mobility of enzyme and substrate.

Adsorption of enzyme may be necessary to facilitate the covalent reactions. Stabilization of enzymes temporarily adsorbed onto a matrix has been achieved by cross-linking the protein in a chemical reaction subsequent to its physical adsorption.

Ionic binding   The ionic binding method relies on the ionic binding of enzyme protein to water-insoluble carriers containing ion-exchange residues. Polysaccharides and synthetic polymers having ion-exchange centres are generally used as carriers. The binding of an enzyme to the carrier is easily carried out, and the conditions are much milder than those needed for the covalent binding method. Hence, the ionic binding method causes little changes in the conformation and active site of the enzyme. Therefore, this method yields immobilized enzymes with high activity in most cases.

Leakage of enzymes from the carrier may occur in substrate solutions of high ionic strength or upon variation of pH. This is because the binding forces between enzyme proteins and carriers are weaker than those in covalent binding. The main difference between ionic binding and physical adsorption is that the enzyme to carrier linkages are much stronger for ionic binding although weaker than in covalent binding.

Covalent binding The enzyme will be bound with the surface of water-insoluble support materials by covalent bond formation. The enzyme molecules form the bond with support materials through certain functional groups like amino group, hydroxyl group, carboxyl group, sulphydryl groups, imidazole group, phenolic group, thiol group, threonine group and indole group. This method can be further classified into diazo, peptide and alkylation methods according to the mode of linkage. The conditions for immobilization by covalent binding are much more complicated and less mild than in the cases of physical adsorption and ionic binding. Therefore, covalent binding may alter the conformational structure and active centre of the enzyme, resulting in major loss of activity and/or changes of the substrate. However, the binding force between enzyme and carrier is so strong that no leakage of enzymes occurs, even in the presence of substrate or solution of high ionic strength. The drawback with this method is, some enzyme molecules share the same functional group as active site, and when active sites are interacting with support material, the enzymatic activity will be blocked.

Covalent attachment to a support matrix must involve only functional groups of the enzyme that are not essential for catalytic action. Higher activities result from prevention of inactivation reactions with amino acid residues of the active sites. A number of protective methods have been devised. They are

- covalent attachment of the enzyme in the presence of a competitive inhibitor or substrate

- a reversible, covalently linked enzyme–inhibitor complex

- a chemically modified soluble enzyme whose covalent linkage to the matrix is achieved by newly incorporated residues

- a zymogen precursor

Hence, covalent binding can be brought about by the following:

Diazotization	SUPPORT—N=N—ENZYME
Amide bond formation	SUPPORT—CO—NH—ENZYME
Alkylation and arylation	SUPPORT—CH_2—NH—ENZYME
	SUPPORT—CH_2—S—ENZYME
Schiff's base formation	SUPPORT—CH=N—ENZYME

Amidation reaction SUPPORT—CNH—NH—ENZYME

Thiol–disulphide interchange SUPPORT—S—S—ENZYME

Carrier binding with SUPPORT—$O(CH_2)_2$ $N{=}CH(CH_2)_3$
bifunctional reagents CH=N—ENZYME

The active site of the enzyme must not be hindered. There must be ample space between the enzyme and the backbone. It is possible in some cases to increase the number of reactive residues of an enzyme in order to increase the yield of immobilized enzyme. This provides alternative reaction sites to those essential for enzymatic activity. As with cross-linking, covalent bonding should provide stable, immobilized enzyme derivatives that do not leach enzyme into the surrounding solution. A wide variety of binding reactions and insoluble carriers (with functional groups capable of covalent coupling or being activated to give such groups) makes this a generally applicable method of immobilization. This is true even if very little is known about the protein structure or active site of the enzyme to be coupled. Table 19.1 provides the microbial cells covalently linked to various supports.

Table 19.1 Microbial cells covalently linked to various supports

Organisms	Support	Product
Acetobacter	Metal hydroxides	Acetic acid
Aspergillus niger	Glycidyl methacrylate	Gluconic acid
Micrococcus luteus	CM-cellulose	Urocanic acid
Saccharomyces cerevisiae	Aminopropyl silica	Ethanol
Saccharomyces cerevisiae	Hydroxyalkyl methacrylate	Killer toxin
Saccharomyces cerevisiae	Hydroxyalkylmethacrylate	β-Galactosidase
Zygosaccharomyces lactis	Cellulose	Ethanol

Cross-linking

Immobilization of enzymes has been achieved by intermolecular cross-linking (Figure 19.2) of the protein either to other protein molecules or functional groups on an insoluble support matrix. Cross-linking an enzyme to itself is both expensive and insufficient as some of the protein material will inevitably be acting mainly as a support. This will result in relatively low enzymatic activity. Generally, cross-linking is best used in conjunction with one of the other methods. It is used mostly as a means of stabilizing adsorbed enzymes and also for preventing leakage from polyacrylamide gels.

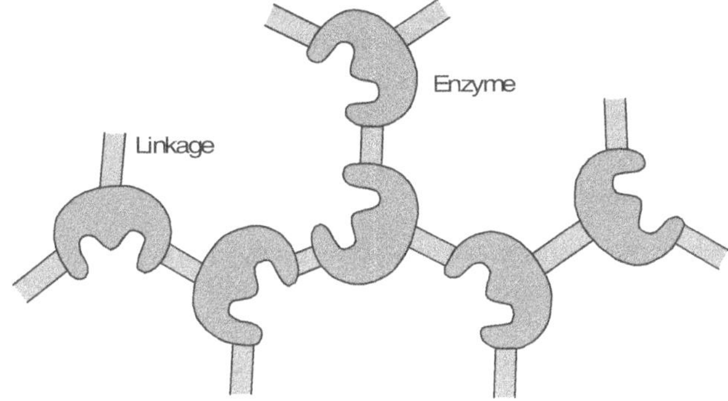

Figure 19.2   Enzyme cross-linking

Since the enzyme is covalently linked to the support matrix, very little desorption is possible with this method. It has been reported that carbamyl phosphokinase cross-linked to alkyl amine glass with glutaraldehyde lost only 16% of its activity after continuous use in a column at room temperature for 14 days.

The most common reagent used for cross-linking is glutaraldehyde. Cross-linking reactions are carried out under relatively severe conditions. These harsh conditions can change the conformation of active centre of the enzyme, and so may lead to significant loss of activity.

Entrapment

The entrapment method (Figure 19.3) is based on the localization of an enzyme within the lattice of a polymer matrix or membrane. It is done in such a way as to retain protein while allowing penetration of substrate. It may be classified into lattice and microcapsule types.

Figure 19.3   Entrapment (a) in matrix (b) in droplets

Entrapment differs from covalent binding and cross-linking in that the enzyme itself does not bind to the gel matrix or membrane. This results in wide applicability. The conditions used in chemical polymerization reaction are relatively severe and result in the loss of enzyme activity. Therefore, careful selection of the most suitable conditions for immobilization of various enzymes is required.

Lattice-type entrapment involves entrapping enzymes within the interstitial spaces of a cross-linked water-insoluble polymer. Some synthetic polymers, e.g. polyacrylamide and polyvinyl alcohol) and natural polymers (e.g. starch) have been used to immobilize enzymes using this technique.

Microcapsule-type entrapping involves enclosing the enzymes within semi-permeable polymer membranes. The preparation of enzyme microcapsules requires extremely well-controlled conditions. The procedures for microcapsulation of enzymes can be classified as:

Interfacial polymerization method In this procedure, enzymes are enclosed in semi-permeable membranes of polymers. An aqueous mixture of the enzyme and hydrophilic monomer are emulsified in a water-immiscible organic solvent. Then, the same hydrophilic monomer is added to the organic solvent by stirring. Polymerization of monomers occurs at the interface between the aqueous and organic solvent phases in the emulsion. The result is that the enzyme in the aqueous phase is enclosed in a membrane of polymer.

Liquid drying In this process, a polymer is dissolved in a water-immiscible organic solvent that has a boiling point lower than that of water. An aqueous solution of enzyme is dispersed in the organic phase to form a first emulsion of water-in-oil type. The first emulsion containing aqueous microdroplets is dispersed in an aqueous phase containing protective colloidal substances (such as gelatin) and surfactants, and a secondary emulsion is prepared. The organic solvent is removed by warming in vacuum. A polymer membrane is thus produced to give enzyme microcapsules.

Phase separation One purification method for polymers involves dissolving the polymer in an organic solvent and re-precipitating it. This is accomplished by adding another organic solvent that is miscible with the first but does not dissolve the polymer.

This form of an immobilized enzyme can be classified into four types: particles, membranes, tubes and filters.

Particles Most immobilized enzymes are in particle form for ease of handling and application. The preparation of enzyme immobilization is by microcapsule particles or droplets.

Membranes   Enzyme membranes can be prepared by attaching enzymes to membrane-type carriers or by moulding into membrane form. Moulding is done after the enzymes have been enclosed within semi-permeate membranes of polymer by entrapment.

Tubes   Enzyme tubes are produced using nylon and polyacrylamide carriers. The polymer tube is first treated in a series of chemical reactions and the enzyme is bound by diazo coupling to give a tube in final step.

Fibres   Enzymes that have been immobilized by entrapment in fibres are used to form enzyme fibres.

The solid supports used for enzyme immobilization can be inorganic or organic. Some organic supports include polysaccharides, proteins, carbon, polystyrenes, polyacrylates, maleic anhydride based copolymers, polypeptides, vinyl and allyl polymers, and polyamides. Table 19.2 provides the microbial cells encapsulated to various supports.

Table 19.2   Microbial cells encapsulated to various supports

Matrix	Species
Alginate	*Lactobacillus plantarum*
	Lactobacillus delbrueckii ssp.
	Bifidobacterium bifidum
	Bifidobacterium infantis
	Bifidobacterium longum
	Lactobacillus acidophilus
	Bifidobacterium lactis
κ-carrageenan	*Bifidobacterium bifidum*
	Bifidobacterium longum
Gellan/xanthan	*Bifidobacterium infantis*
Phthalate cellulose acetate	*Bifidobacterium pseudolongum*

IMMOBILIZATION OF CELLS

Biocatalysts may be immobilized using either the isolated enzymes or the whole cells. Immobilization of whole cells has been shown to be a better alternative to immobilization of isolated enzymes. It avoids the lengthy and expensive operations of enzyme purification and preserves the enzyme in its natural environment, thus protecting it from inactivation during immobilization and its subsequent use in continuous system. It may also provide a multipurpose catalyst, especially when the process requires the participation of a number of enzymes in sequence. The major limitations that may need to be addressed while using such cells are the diffusion of substrate and products through the cell wall and undesirable side reactions due to the presence of other enzymes. The cells may be immobilized either in a viable or a non-viable form. Immobilized non-viable cell preparations, which are normally obtained by permeabilizing the intact cells for the expression of intracellular activity, are useful for simple processes that require single-enzyme with no requirement for cofactor regeneration, like hydrolysis of sucrose or lactose. On the other hand, immobilized viable cells, which serve as "controlled catalytic biomass", have opened new avenues for continuous fermentation on heterogeneous catalysis by serving as self-proliferating biocatalysts.

The method of immobilization of the whole cells is similar to that of enzymes. Whole-cell immobilization has several advantages:

1. It provides high cell density and protection against shear damages.

2. It is reusable and easy to separate cells during downstream process.

3. Immobilization eliminates cell wash-out problem at high dilution rate.

4. It provides favourable microenvironmental conditions to the cells, such as gradient of pH and nutrient products, which result in high yield.

5. It maintains genetic stability to a certain extent.

The limitation in whole-cell immobilization is diffusion, which leads to the alteration of microenvironment including reduction of gas evaluation, growth and yield.

Active Immobilization

Active immobilization is by entrapment or binding of cells through matrix or porous polymers. Polymer beads should be porous enough to allow transport of nutrient substrates and product.

Polymers like agar, alginate, κ-carrageenan, polyacrylamide, gelation and collagen are widely used, and they are prepared by gelation, precipitation of polymer, ion exchange gelation, polycondensation and polymerization.

Passive Immobilization or Biofilm

Passive immobilization is forming biofilm over the support material. Growing multiple layers of the organism over the surface of support material is known as biofilm. The formation of natural biofilm is common in all hydrated environments. The formation of biofilm by single organism is uncommon, and it is generally by a consortium of organisms. Usually, initiation of biofilm formation is by some polymer producing organisms from the mixed culture, which facilitate the microenvironment for the growth of other organisms. The nutrient and product profile within the biofilm governs the cellular metabolism and physiology. Besides, the thickness of biofilm is another factor; thick biofilm will have high conventional rate due to high density of the cell mass whereas thin biofilm will have low rate of conversion.

IMMOBILIZED MULTI-ENZYMES AND ENZYME-CELL CO-IMMOBILIZATES

A common feature of metabolic pathways in intermediary metabolism is that the product of one enzyme in sequence is the substrate for the next and so forth. One of the advantages of such an arrangement is that a favourable, high local concentration of intermediates in the microenvironment of an enzyme system can be created. Such an approach perhaps will be useful in the development of enzymatic processes requiring multiple enzymes. The enzymes may be simultaneously bound to the same support or simultaneously immobilized through entrapment. Kinetic advantages of this system include reduction in the lag and significant enhancement in the final product.

Often a cell may not contain all the necessary enzyme compliments for carrying out a required process. Thus, *Saccharomyces cerevisiae*, a potential ethanol fermenter, cannot utilize lactose or starch since it lacks the enzymes β-galactosidase and amylase, respectively. There is considerable interest in genetically modifying such cells for the expression of deficient enzymes. The success will depend on plasmid stability and quite often on the ability of these cells to secrete these enzymes for their action on the raw materials present in growth medium. Recently, genetically engineered cells have been obtained, which can produce the deficient enzyme as well as transport it to the periplasmic space and naturally immobilize it. Another more practical approach is to immobilize the deficient enzyme directly on the cell wall. This approach, termed as enzyme-cell co-immobilizate, permits the tailoring of a whole cell for specific complex chemical conversion by combining the biochemical potential of both the cell and additional enzyme from another source. A few techniques developed include the binding of the glycoprotein glucose oxidase onto the microbial cell walls using Con A. Enzymes may also be immobilized on microbial cell surfaces coated with polyethylenamine through adsorption followed by cross-linking.

APPLICATION OF IMMOBILIZATION IN INDUSTRIAL BIOREACTORS

In an enzyme reactor, the highest specific enzyme activity is desirable. It is considered advantageous if the support aids in separation. One approach is to use a molecular sieve as the support and pulse the reactor bed with the alternating passage of substrate solution and water. The result is that bands of unused substrate and product progress down the column. If it happens so, then this technique will be useful otherwise it is better to immobilize the enzymes on a porous support.

For an industrial reactor, it is preferable to use supports that are non-biodegradable, e.g. glass, silica, celite, bentonite, alumina or titanium oxide. Even the linkages between enzyme and support can be non-biodegradable, as in the case of titanium. In some supports, the physical nature of the surface becomes a major problem. Thus, some supports that form excellent packed beds fail to do so when coated with enzyme. Particles that ideally self-suspend in a fluid bed may form aggregates during use, which will require more power to pump through substrate. Many problems were encountered with porous glass supports until someone realized that the glass itself could dissolve. This problem has been eliminated by the treatment of glass surface with zirconium.

Stirred Tank Reactors

A batch stirred tank reactor (Figure 19.4) is the simplest type of reactor. It is composed of a reactor and a mixer such as a stirrer, a turbine wing or a propeller.

The batch stirred tank reactor is useful for substrate solutions of high viscosity and for immobilized enzymes with relatively low activity. However, a problem may arise that an immobilized enzyme tends to decompose upon physical stirring. The batch system is generally suitable for the production of small amounts of chemicals.

A continuous stirred tank reactor (Figure 19.5) is more efficient than a batch stirred tank reactor but the equipment is slightly more complicated because periodical addition and removal of culture is made. During this process care should be taken to retain the immobilized enzyme at particular concentration level.

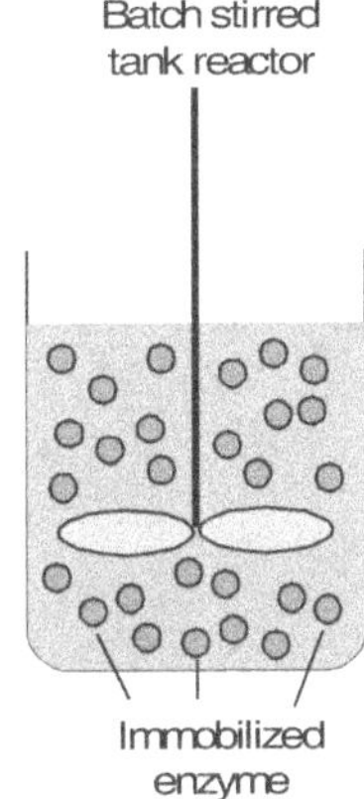

Figure 19.4   Batch reactor

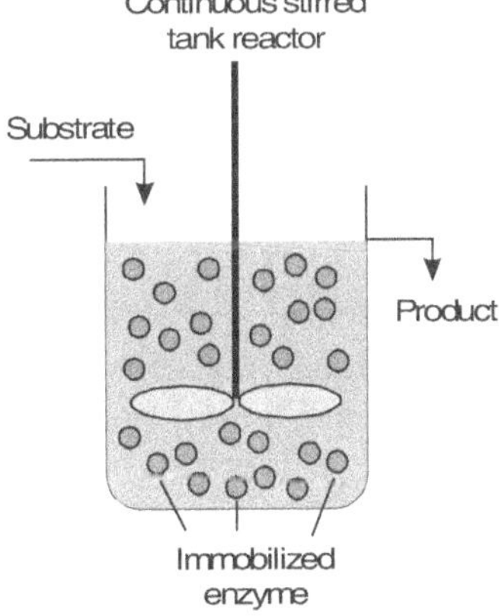

Figure 19.5   Continuous stir tank reactor

Combined CSTR/UF Reactor

A combined CSTR/UF reactor is a combination of a continuously stirred tank reactor and an ultra filtration unit. This type of reactor begins as a typical CSTR. However, the product passes through an ultra filtration unit where the enzyme is removed and recycled back into the reactor. An example of this type of rector is shown in Figure 19.6.

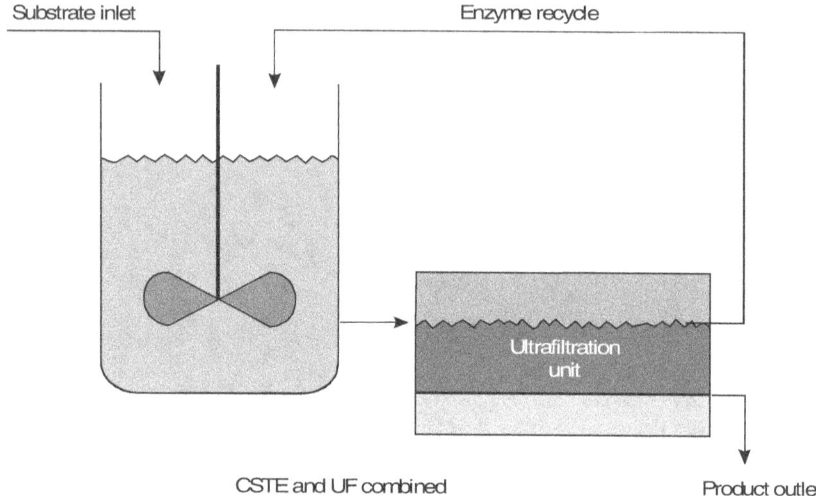

Figure 19.6   Combined CSTR/UF reactor

In a combined CSTR/UF reactor, the enzymes immobilized cannot leave the reactor because of the filtration unit. It allows continuous processing with free enzymes in the CSTR. The ultra filtration unit contains a membrane that provides a semi-permeable barrier, which allows products and un-reacted substrate, if any, to pass through while holding back the enzyme. There are other possibilities for similar reactors, such as combined reactor-separators. However, the combined CSTR/UF reactor has proven useful for several types of reactions where a typical immobilized enzyme would not be as effective. One example of this is the conversion of benzylpenicillin to 6-aminopenicillanic acid by penicillin amidase.

Packed Bed Reactor

Continuous packed bed reactors (Figure 19.7) are widely used for immobilized enzymes and microbial cells. In these systems, it is necessary to consider the pressure drop across the packed bed or column and the effect of the column dimensions on the reaction rate. There are three substrate flow possibilities in a packed bed:

1. Downward flow method

2. Upward flow method

3. Recycling method

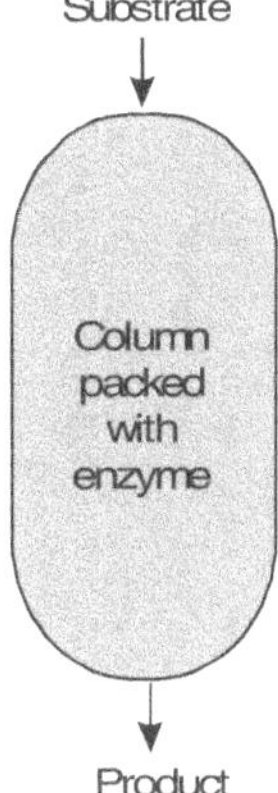

Figure 19.7   Packed bed reactor

The recycling method is advantageous when the linear velocity of the substrate solution affects the reaction flow rate. This is because the recycling method allows the substrate solution to be passed through the column at a desired velocity.

For industrial applications, upward flow is generally preferred over downward flow because upward flow does not compress the beds in enzyme columns as downward flow does. When gas is produced during an enzyme reaction, upward flow is preferred.

A continuous packed bed reactor has the following advantages over a batch packed bed reactor:

- ➲ easy, automatic control and operation

- ➲ reduction of labour costs

- ➲ stabilization of operating conditions

- ➲ easy quality control of products

Recycle Reactor

A recycle reactor is very important to chemical engineering because it allows those substrates, which could not be processed using other reactor types, to be processed. A recycle reactor is shown in Figure 19.8.

In a recycle reactor, a portion of the product stream is recycled and mixed with the inlet flow to the reactor. If the entire product stream is recycled back to the inlet stream, then it is called a total recycle reactor.

This can only be used in a batch process because, if the entire product stream is recycled back into the reactor in a continuous reactor, the volume of the reactor would increase to infinity.

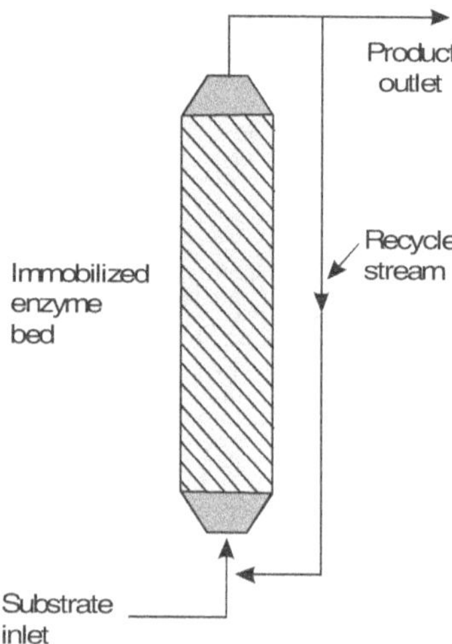

Figure 19.8   Recycle reactor

This type of reactor is used when a substrate cannot be completely processed on a single pass, such as an insoluble substrate. These reactors continue to move the same substrate through the reactor so that the effective contact time is high enough to allow the substrate to be processed. Recycle reactors also allow the reactor to operate at high fluid velocities. This is important because it minimizes the bulk mass transfer resistance to the transport of the substrate. It is important to remember that a recycle reactor is simply a reactor, such as a CSTR or fluidized-bed reactor, with a recycle stream.

Fluidized-bed Reactor

A fluidized-bed reactor is a combination of the packed-bed and stirred tank continuous flow reactors. It is very important to chemical engineering because of its excellent heat and mass transfer characteristics. The fluidized bed reactor is most suitable when a high viscosity substrate solution and a gaseous substrate or product are used in a continuous reaction system.

In a fluidized-bed reactor (Figure 19.9), the substrate is passed upward through the immobilized enzyme bed at a high velocity to lift the particles. However, the velocity must not be so high to sweep away enzymes from the reactor entirely. This causes an inadequate mixing than the piston flow model. In piston flow model, the substrate is passed through the peristaltic pump with high velocity which will lead to complete mixing of substrate with immobilized enzymes. This type of reactor is ideal for highly exothermic

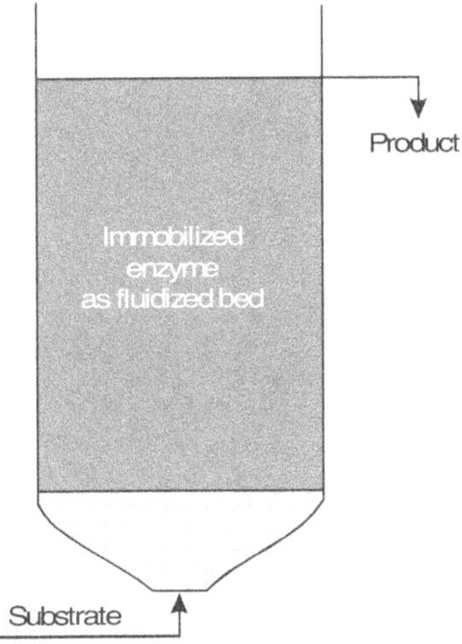

Figure 19.9 Fluidized bed reactor

reactions because it eliminates local hot-spots due to its mass and heat transfer characteristics. It is often applied in immobilized-enzyme catalysis where viscous, particulate substrates are handled. In this system, care must be taken to avoid destruction and decomposition of immobilized enzymes. The particle size of immobilized enzymes is an important factor in the formation of a smooth fluidized bed.

REVIEW QUESTIONS

1. What is immobilization? Give their advantages.

2. Give the criteria about selection of carriers for immobilization.

3. Write in detail about different types of immobilization.

4. Give the difference between carrier binding and cross-linking immobilization.

5. Give an account on entrapment.

6. Explain about immobilization of cells.

7. Explain in detail about application of immobilization on various types of bioreactors.

ENZYME KINETICS

20

INTRODUCTION TO ENERGETICS

All reversible reactions tend to proceed in the direction that decreases the energy of the participants. However, even in this direction, the reaction may not proceed very quickly because very few molecules overcome the energy potential barrier and make the transition toward lower energy state.

Figure 20.1 shows energy diagram for a reversible chemical reaction:

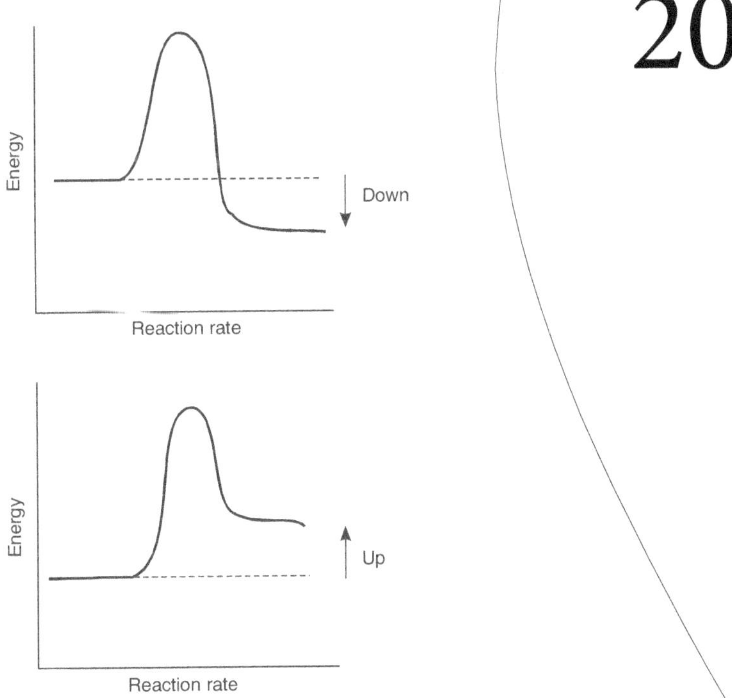

Figure 20.1   Energy diagram for a reversible chemical reaction

In a population of molecules, there is distribution of energies because individual molecules are subject to collisions with other molecules, atomic vibrations and external energy sources (such as light). Most molecules have energy near average energy, not enough to jump over the potential barrier. Only a small percentage of molecules have enough energy to make the transition over the barrier.

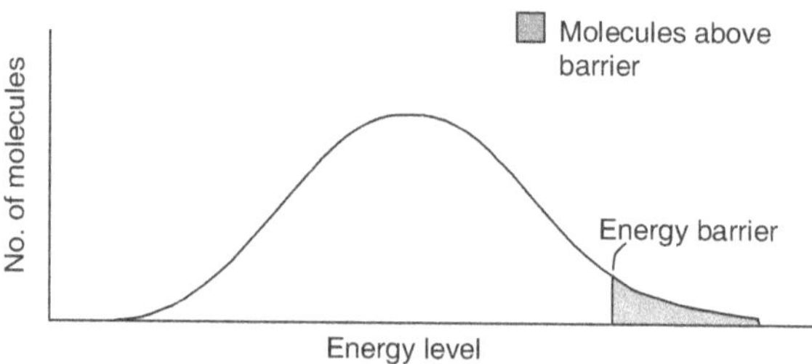

Figure 20.2   Energy distribution curve

Consider a reaction proceeding in the opposite direction, where the end products are at a higher energy level than the starting materials. For example, to convert a mixture of amino acids into a protein requires energy. It is called an endothermic reaction, which are not spontaneous, even with enzymes to catalyse it.

In general, a series of reactions involved in the synthesis of large molecules are endothermic. For example, the conversion of sugars to sugar-phosphates requires an input of energy. Without energy, even a simple reaction cannot proceed, as for example,

$$\text{Glucose} + \text{Pi} + \text{energy} \longrightarrow \text{Glucose 6-phosphate}$$

$$\text{Glucose} + \text{Pi} \longrightarrow\!\!\!\!\!\times \text{No product}$$

How is energy made available in cells? In what form is it stored and used are as follows.

CHEMICAL POTENTIAL

In a chemical reaction, the chemical potential (or potential difference) that drives it involves two things: first, the intrinsic properties of the substrate and product molecules; second, their concentrations. If a reaction $A + B = C + D$ is reversible, the direction in which it goes must depend on the properties and quantities of A, B, C and D. The more A and B added, the more C and D are formed, and vice versa. Also, if C or D is removed as fast as it is made, the reaction is driven to the right.

When a cell needs a reaction to go in an energetically unfavourable (endergonic) direction, it uses the mechanism called coupling: it converts the substrate to something with higher chemical potential by using another reaction that is energetically favourable, i.e., exergonic. The chemical potential of the modified substrate is higher than that of the product so that the reaction will now tend to proceed spontaneously in the desired direction.

For example, the conversion glucose + Pi $\rightarrow$ Glucose 6-P + H_2O requires energy. The reaction $ATP + H_2O \rightarrow ADP + H_3PO_4$ releases more energy. They are coupled (on an enzyme surface) into the overall reaction: Glucose + ATP $\rightarrow$ Glucose 6-P + ADP, which is still exergonic. Here, to carry out the reaction, the cell uses a phosphate group that is trapped into the ATP molecule, which acts a donor of phosphate to glucose.

Cells use similar mechanisms to make all sorts of energetically unfavourable reactions proceed. In protein synthesis, for example, the first step, aa + ATP $\rightarrow$ AMP-aa + PPi (inorganic pyrophosphate), is exergonic and converts the amino acid (aa) into a new form, which still retains some of the high chemical potential and serves as a donor of amino acid for making protein. Note that ATP was used in both the above examples of coupled reactions. In the first case, it donated one of its phosphate groups to glucose and thereby increased the reactivity of the glucose so that it could be used for other reactions. In the second case, ATP donated another part of its molecule to an amino acid.

ATP AND HIGH-ENERGY BONDS

ATP is the most important of a group of compounds that have traditionally been called high-energy compounds. More correctly, it is the prototype of a group of molecules with high group donor potential, that is, molecules that release much energy when they donate some parts of themselves to water or other acceptor molecules. The parts donated may be atoms or groups of atoms or, in some cases, just pairs of electrons.

High-energy Bonds

In ATP, there are at least two binding places that have a tendency to participate in the donation of parts of the molecule: AMP ~ P ~ P. Traditionally, but incorrectly, these are referred as high-energy bonds and represented by the symbol ~: AMP ~ P ~ P. The energy (chemical potential), the tendency to transfer chemical groups, is in the molecule as a whole, not in the bond. The same is true, for example, of NADH: it has a high electron donor potential and can therefore 'reduce', i.e., transfer electrons to a variety of substances.

The chemical machinery of the cell is driven by substances with high group donor potential. Cells use only two kinds of energy:

1. light energy—trapped and used by plants and some bacteria for photosynthesis

2. chemical energy—the energy held in the bonds of various chemicals

Cells do not use thermal or electrical energy because they do not have thermal or electrical converters. Thermal potential (temperature) affects the rate of chemical reactions but does not provide any energy. The electrical signals of nervous impulses use energy in the form of ATP to generate electric potentials in the membrane of nerve cells and fibres.

FREE ENERGY AND CHEMICAL EQUILIBRIUM

How can we measure energetic budget of a cell? The change (G in usable energy (chemical potential) in a reaction is described by the equation

$$E = T^* \Delta S + \Delta G$$

where,

E   = overall change in energy,

T   = absolute temperature,

ΔS = change in entropy (which measures changes in the order of the system), and

ΔG = change in free energy or chemical potential. The chemical potential (or Gibb's free energy) is the amount of energy that is available to do work.

The change in free energy for a reversible reaction, A + B = C + D, is given by

$$\Delta G = \Delta G_0 + RT \ln \frac{[C][D]}{[A][B]}$$

where,

R   = gas constant,

T   = absolute temperature and

ΔG_0 = is a constant characteristic of the substances involved.

Let the reaction be in equilibrium. Then, the ratio [C][D]/[A][B] equals Keq. By definition, at equilibrium, there are no changes in concentration and $\Delta G = 0$. Therefore,

$$\Delta G = \Delta G_0 + RT \ln Keq$$

$$\Delta G_0 = -RT \ln Keq$$

This relationship makes it possible to measure the change in free energy for a given chemical reaction if one can measure the equilibrium concentration of products and substrates. A chemical reaction always goes in the direction that tends to decrease free energy (the direction for which $\Delta G < 0$).

A reaction will not go in the opposite direction unless it is pushed or pulled. One can push or pull by varying the concentrations of reactants and/ or products. One way, as mentioned earlier, is to remove one product. For example, if the reversible reaction is A + B = C + D, another reaction may remove D as soon as it is made. Alternatively, another reaction can produce more and more A or B. Increased production of C or D would push the reaction backward.

It is very important to remember that the two factors that determine the change in free energy in a reaction are the intrinsic properties of the substances and the concentrations of the reactants and products, because it helps us to understand why certain reactions go. For example, in the conversion of glucose to lactic acid, which serves to produce ATP for muscle contraction and many other processes, there is a series of reactions, some of which are uphill (energy-requiring $\Delta G > 0$) and others are downhill (energy-releasing $\Delta G < 0$). The uphill reactions can proceed only when the products are continuously removed for use in later downhill reactions. If you have to go over a series of hills, the only thing that matters is that the initial level is higher than the ultimate level so that the process as a whole represents a descent. Water in a completely filled tube can go up and downhill provided the initial pressure is higher than the final pressure. The ΔG for the entire series of reactions is the algebraic sum of the ΔG's of the individual reactions.

The relation of the concentration of a substance to free energy becomes clearer when we consider active transport across a membrane, that is, the transfer of a substance from a region of lower concentration to one of higher concentration. Energy is needed to overcome the gradient of concentrations (since the substance would tend to flow down the gradient till the concentrations are equal):

$$\Delta G = RT \ln \frac{conc_{out}}{conc_{in}}$$

If $conc_{in}$ is higher than $conc_{out}$, the ΔG for transport inward is positive and energy is required. In living cells, this energy comes from ATP or other substances with high group donor potential, but how it is actually used for transport remains obscure. This energy is most likely used to distort or rotate molecules of specific transporter proteins.

Substances like ATP with high group donor potentials when ΔG_o for hydrolysis lower than –7 kcal/mole. The transfer of one phosphate or of two phosphates (P—P) or of the AMP group to water or to some other substance releases more than 7 kcal/mole. For example, in the hydrolysis of ATP in water,

$$ATP + H_2O \rightarrow ADP + H_3PO_4$$

About 8,000 calories are released as heat (measurable in a calorimeter) for every mole of ATP that disappears. Sometimes, ATP donates a phosphate to another substance to generate a compound that itself has a high group donor potential. For example, acetyl phosphate formed in the reaction

$$ATP + Acetic\ acid \rightarrow ADP + Acetyl\ phosphate$$

is itself capable of donating phosphate to other substances. In general, any phosphate group that is attached to an acid through an anhydride bond —(CO)—O—P is very reactive. An anhydride bond is formed by joining two acids face to face with removal of a molecule of water. It has a high tendency to be split by water to reform the two acids or to donate one of the acid groups. When ATP donates a phosphate to an alcohol (—C—OH) as in the already familiar reaction

$$Glucose + ATP \rightarrow Glucose\ 6\text{-phosphate} + ADP$$

the substance that is formed does not have a high group donor potential. For example, if glucose 6-P were to hydrolyse in water, it would release only about 3 kcal/mole; in fact, it has little tendency to do so since it has weak donor potential.

ENZYME MECHANISMS

Enzymes perform the chemical reactions in cells. Not all proteins are enzymes, but most enzymes are proteins (the exception is catalytic RNA). A catalyst is a molecule which increases the rate of a reaction but is not the substrate or product of that reaction. A substrate is a molecule upon which an enzyme acts to yield a product.

$$A \xrightarrow[\text{Enzyme}]{} B$$

The free energy of this reaction is not changed by the presence of an enzyme but, for a favoured reaction (where ΔG is negative), the enzyme can speed it up.

The free energy against the reaction progress is shown in Figure 20.3.

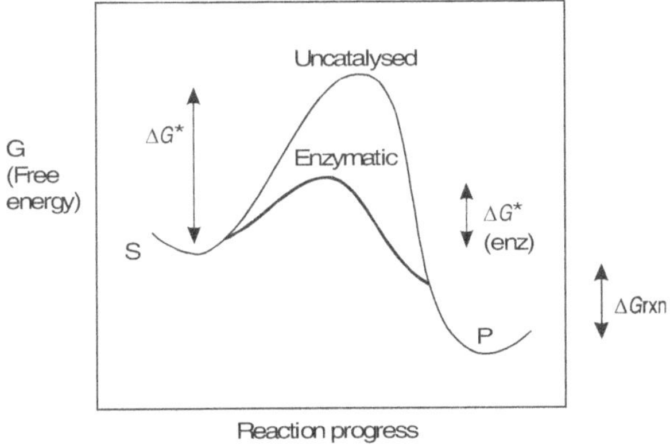

Figure 20.3   Free energy vs. reaction progress

ΔG^* is activation energy

ΔG is negative overall for forward reaction

ENZYME-CATALYSED REACTIONS

Enzymes have affinity for the substrate in a transition state. They get the substrate into the right conformation, which will lead to breakdown into products.

Alternatively, for a reaction $A + B \rightarrow C$, the enzyme may increase the local concentration of the two substrates A and B, driving the reaction forward.

The part of the enzyme that does the work is called the active site. The residues in this site are in the right 3-D conformation to accomplish the enzyme's work.

Nomenclature

Enzymes are named in a variety of ways. The general rules of enzyme nomenclature are listed below.

Based on substrate   Some enzymes are named after their specific substrate in which the enzyme will act. Generally the 'ase' is suffixed with the name of the substrate. For example, lactase, an enzyme that acts on lactose catalyses the following reaction.

$$\text{Lactose} \rightarrow \text{Glucose} + \text{Galactose}$$

If you are lactose-intolerant, you may use lactase in a powdered form to help digest foods. An enzyme with this function, produced by the bacterium *E. coli*, is called β-galactosidase.

Based on action   Some of the enzymes are named based on their catalytic reaction like addition or removal of particular group from the substrate and are called with the name behind the group which are removed or added along with substrate. For example, an enzyme that removes hydrogen (water) group from the substrate is called dehydrogenase. For example, alcohol dehydrogenase catalyses the following reaction.

$$\text{Alcohol} \xrightarrow{\text{Alcohol dehydrogenase}} \text{Acetaldehyde}$$

ASSAYING ENZYMES

To study an enzyme, an assay is necessary. The assay is a measurement of a chemical reaction, which might involve measuring the formation of the product. For example, β-galactosidase catalyses the following reaction:

$$\text{lactose} \xrightarrow{\text{β-galactosidase}} \text{glucose} + \text{galactose}$$

For this reaction, measuring the formation of glucose would constitute an assay. Because this is technically difficult, an easier way to follow the reaction is to use a substrate that gives a coloured reaction product. β-galactosidase can catalyse many reactions of the following general type:

$$y\text{-galactose} \xrightarrow{\text{β-gal}} y + \text{galactose}$$

In a "normal" reaction, y-galactose = lactose = glucose–galactose (y = glucose).

The two alternative substrates for β-galactosidase are:

1. ONPG = ONP-galactose (ONP = *o*-nitrophenol)

2. X-gal   = X-galactose (X = 4-chloro-3-bromo indole)

Both ONPG and X-gal are colourless, but when hydrolysed by β-galactosidase, they produce coloured products:

$$\text{ONPG} \xrightarrow{\;\beta\text{-gal}\;} \text{galactose} + \text{ONP}$$

(colourless)   (colourless)   (yellow)

$$\text{X-gal} \xrightarrow{\;\beta\text{-gal}\;} \text{galactose} + 4\text{-Cl-3-Br-indigo}$$

(colourless)   (colourless)   (deep blue)

The coloured products of these reactions are much easier to detect and can be used to assay (measure) enzyme activity more easily than the "normal" reaction catalysed by β-galactosidase.

Enzyme Catalysis

The enzyme converts S to P (substrate to product). Initially, [P] is small, so the majority reaction is S → P. Later, as [P] grows, the back reaction rate increases, until equilibrium is reached (enzymes catalyse both forward and reverse reactions). Figure 20.4 is an example of an enzyme catalysed reaction.

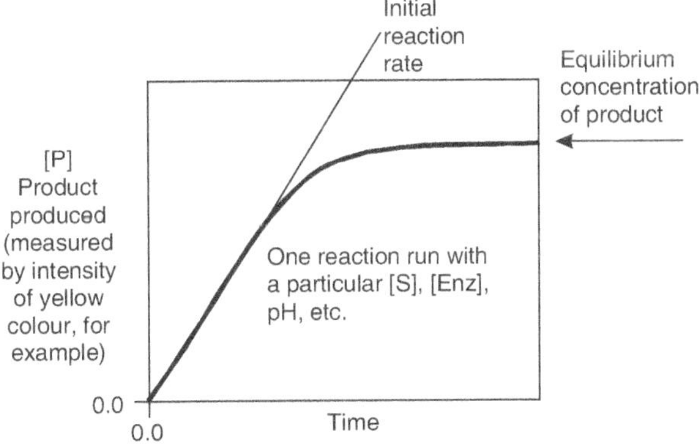

Figure 20.4   Enzyme-catalysed reaction

To measure the kinetic properties of a given enzyme, you must perform many experiments like the one above. Keep the enzyme concentration constant and measure the initial rate of product formation (before the reaction is anywhere near equilibrium) at several different initial substrate concentrations. Then, plot the initial velocity of the reaction: V_o as a function of [S].

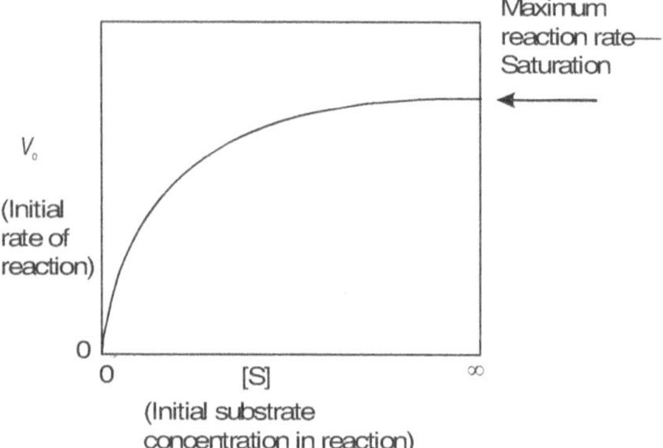

Figure 20.5   Relationship between rate of reaction and substrate concentration

Note that the graph (Figure 20.5) is not linear, as [S] gets large the enzyme is in constant level and all the enzymes are occupied by substrates. The rate of reaction depends on the amount of enzyme and not on the amount of substrate.

From the graph we can infer that

➲   Velocity is dependent on [S].

➲   The more the enzyme added, the faster the reaction goes.

Note   Figures 20.4 and 20.5 look very similar but mean quite different things. It is important to understand the difference between an individual reaction: [P] vs t (from which you get V_o) and a kinetic graph of many such reactions: V_o vs [S]. This kinetic behaviour can be modelled mathematically.

BASIC KINETICS

Consider the following reaction.

$$A \longrightarrow B$$

The equation for this first order reaction is:

$$\frac{d[A]}{dt} = k_1[A]$$

k_1 is negative because A is disappearing.

Consider the second reaction

$$A + B \rightarrow C$$

The rate equation for this second order reaction is:

$$\frac{d[A]}{dt} = k_1[A]^2$$

Consider the enzymatic reaction

$$E + S \underset{k_{-1}}{\overset{k_1}{\rightleftharpoons}} ES \underset{k_{-2}}{\overset{k_2}{\rightleftharpoons}} P + E$$

where,

E is the enzyme (β-galactosidase)

S is the substrate (lactose)

ES is the enzyme–substrate complex

P is the product (glucose or galactose)

First consider the initial velocity of the reaction. In this case, there will be a negligibly small amount of product present ([P] < 5% of [S] is considered negligible). Under these conditions, the back reaction is negligible, i.e., k–2[P] = 0 (approximately). The initial velocity is simply:

$$V_o = k_2[ES] \tag{1}$$

The problem with this equation is that the quantity [ES] cannot be measured. However, [S] (the initial concentration of substrate) is known, [P] (product produced) can be measured, and $[E]_{tot}$ (the amount of enzyme added to the reaction) is known.

Let us now use the rate equations and few other assumptions to derive an expression for [ES] (which we cannot measure) in terms of quantities which we can measure ([S], [P] and $[E]_{tot}$).

Michaelis–Menten Kinetics

For the purposes of our analysis, we shall assume "steady state kinetic conditions". That is, [S] and [P] are changing, but [ES] does not change (a constant flux of S through the enzyme). Mathematically, this can be written as:

$$\frac{d[ES]}{dt} = 0$$

Also (from conservation of matter):

$$[E]_{tot} = [E]_{free} + [ES] \tag{2}$$

(the total enzyme is either bound to substrate or free)

Dividing equation 1 by $[E]_{tot}$,

$$\frac{V_o}{[E]_{tot}} = \frac{k_2[ES]}{[E]_{tot}}$$

$$V_{formation} = k_1\,[E]_{free}[S] \text{ (second order rate equation)}$$
$$V_{breakdown} = k_2\,[ES] + k{-}1\,[ES]$$
$$= (k_2 + k{-}1)\,[ES] \text{ (Two first order rate equations)}$$
$$k_1\,[E]_{free}[S] = (k_2 + k{-}1)\,[ES] \text{ (Rates must be equal)}$$

Rearranging, solving for $[ES]$:

$$[ES] = \frac{k_1[E]_{free}[S]}{(k_2 + k_1)} \tag{3}$$

Let us define the Michaelis–Menten constant, K_m (m stands for Michaelis–Menten; these equations were formulated by Leonor Michaelis and Maud Menten in 1913):

$$K_m = \frac{(k_2 + k_1)}{k_1}$$

Substituting it into equation 3, we get:

$$[ES] = \frac{[E]_{free}[S]}{K_m} \tag{4}$$

From equation 2, $[E]_{free} = [E]_{tot} - [ES]$

and then substitute into equation 4, giving:

$$[ES] = \frac{([E]_{tot} - [ES])[S]}{K_m}$$

Then, solving for $[ES]$ gives:

$$[ES] = [E]_{tot}\left(\frac{\dfrac{[S]}{K_m}}{\left(1 + \dfrac{[S]}{K_m}\right)}\right) \text{ Then multiply top and bottom by } K_m \text{ to get:}$$

$$[ES] = [E]_{tot} \left(\frac{[S]}{[S] + K_m} \right)$$

Finally, substitute into equation 1 to get:

$$V_o = k_2 [E]_{tot} \left(\frac{[S]}{[S] + K_m} \right) \tag{5}$$

Three cases will illustrate how this equation is made,

Case 1:   [S] is very large, i.e., $[S] \gg K_m$. The enzyme is saturated with substrate.

In this case, $[S] + K_m = [S]$ (approximately), so equation 5 becomes:

$$V_o = k_2 [E]_{tot} \left(\frac{[S]}{[S]} \right)$$

or

$$V_o = k_2 [E]_{tot}$$

This is the rate at large substrate concentration, the maximal rate for [E], which is called V_{max}.

$$V_{max} = k_2 [E]_{tot}$$

Substituting into equation 5 gives the Michaelis–Menten equation:

$$V_o = \frac{V_{max} [S]}{K_m + [S]} \tag{6}$$

Remember, this equation is derived for V_o when very little product has formed and the back-reaction can be ignored.

Case 2:   [S] is small, i.e., $[S] \ll K_m$. The activity is in the linear range.

In this case, $[S] + K_m = K_m$ (approximately); so equation 5 becomes:

$$V_o = \frac{V_{max} [S]}{K_m}$$

or

$$V_o = [S]$$

So, at low [S], V_o is linearly proportional to [S].

Case 3: $[S] = K_m$

By definition of K_m,

$$V_o = \frac{V_{max}[S]}{[S]+[S]}$$

or

$$V_o = \frac{V_{max}}{2}$$

K_m is defined as the [S] that results in half-maximal rate.

Significance of K_m and V_{max}

$$K_n = \frac{k_2 + k_1}{k_1}$$

When k_2 is very small ($k_2 << k_1$),

$$E + S \leftrightarrow ES \xrightarrow{\text{slow, infrequent}} P$$

which is equal to the dissociation constant K_d for the equilibrium:

$$E + S \leftrightarrow ES$$

$$K_d = \frac{[E]_{tot}[S]}{[ES]}$$

K_d is a measure of how tightly the enzyme binds to the substrate. (Note the similarity to K_a for dissociation of an acid). Therefore, K_m is an approximate measure of the affinity of the substrate for the enzyme.

V_{max} and K_m are the two parameters that define the kinetic behaviour of an enzyme as a function of [S].

V_{max} is a measure of how fast the enzyme can go at full speed.

V_{max} is a rate of reaction. It will have units of: moles/hour, moles/minute, or moles/second, etc.

K_m is a measure of roughly how much substrate is required to get to full speed. If [S] >> K_m, then V_o will be close to V_{max}.

K_m is a concentration. It will have units of: moles/l or moles/ml, etc.

Measuring K_m and V_{max}   The quantities K_m and V_{max} are experimentally determined and are different for each enzyme. Once you have an assay for enzyme activity, you can determine these parameters.

You can estimate K_m and V_{max} from the graph of initial velocity versus [S] (Figure 20.6).

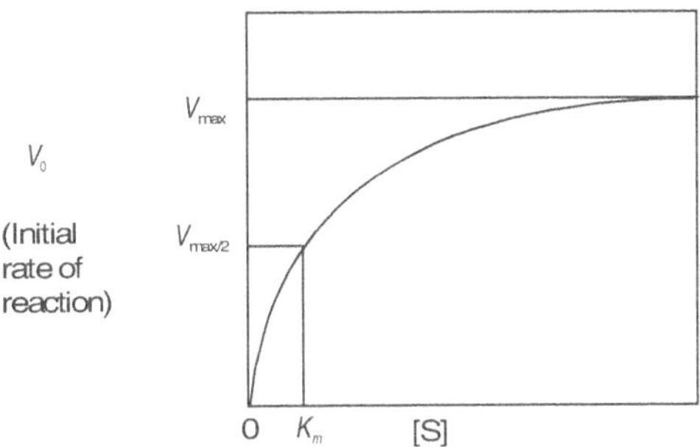

Figure 20.6 $V_{max} - K_m$ curve for initial velocity

1. Run a series of reactions with constant $[E_{tot}]$, varying [S], and measure V_o.

2. Draw a graph of V_o vs. [S].

3. Estimate V_{max} from asymptote.

4. Calculate $\dfrac{V_{max}}{2}$.

5. Read K_m from graph.

V_{max} is difficult to determine if data are graphed this way. Since the graph is hyperbolic, it is hard to extrapolate to infinite [S] and guess V_{max}.

For this reason, the Lineweaver–Burke plot is used.

You can rearrange equation 6 into linear form, $y = mx + b$, if you let

$$x = \left(\frac{1}{[S]}\right) \quad \text{and} \quad y = \left(\frac{1}{V_0}\right)$$

Then, equation 6 rearranges to:

$$\left(\frac{1}{V_o}\right) = \frac{K_m}{V_{max}}\left(\frac{1}{[S]}\right) + \frac{1}{V_{max}}$$

If the data are plotted in this way, it looks like:

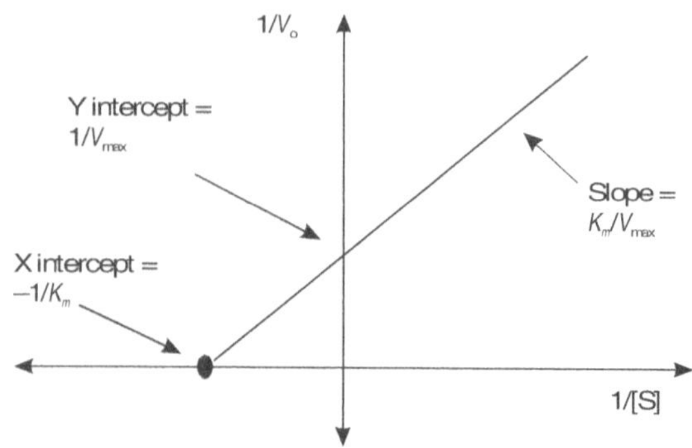

Figure 20.7   Lineweaver-Burke plot

From this type of graph (Figure 20.7), it is easy to estimate K_m and V_{max}.

COMPETITIVE AND NON-COMPETITIVE INHIBITION

There are two main types of reversible enzyme inhibition.

Competitive inhibition   The inhibitor binds to the active site and competes with the substrate for binding. In this case, at low [S], the inhibitor competes for the active site and effectively lowers the [S] at the active site. This lowers the rate of [S] and increases the apparent K_m. At very high [S], the level of S overcomes the inhibition by mass action and V_o approaches V_{max}. V_{max} is unchanged because all the enzyme molecules are active.

Non-competitive inhibition   The inhibitor binds to another site on the enzyme and inactivates the enzyme molecule. This effectively reduces the $[E]_{tot}$ available for catalysis. Since V_{max} is proportional to $[E]_{tot}$, V_{max} is reduced. Since the remaining active enzyme molecules are unaltered, K_m is unchanged.

Solving Enzyme Kinetics Problems

Problem 1   Two strains of bacterium sweetans, A and B, use sucrose (table sugar) as the sole carbon source. The first step in the process of sucrose utilization is the passage of sucrose through a sucrose transporter protein in the membrane. The characteristics of the two transport proteins are as follows (assuming $[E]_{tot}$ is the same in both):

Strain	A	B
K_m	1000 mm	10 mm
V_{max}	1000 mmol/min	100 mmol/min.

a) Assuming that the rate of sucrose uptake is the rate limiting step in growth, which strain will grow faster if the concentration of glucose is 10 mM? 100 mM? 1000 mM?

b) One strain is isolated from normal environmental soil and other strain is isolated from sugar industry waste soil. Which organism will have the fastest uptake of sucrose?

Solution   Using the Michaelis–Menten equation, you can calculate the initial velocity of sucrose uptake for each strain under the conditions given. Note that this is the initial rate; as sucrose was depleted from the medium, the rate would decrease. Assume that the number of cells is small so that the sucrose concentration does not change appreciably during growth, and that the transport protein is always running at V_o.

For these calculations, substrate = sucrose.

[Sucrose] nM	V_o, enzyme of A $K_m = 1000$ nM $V_{max} = 100$ nmol/min.	V_o, enzyme of B $K_m = 10$ nM $V_{max} = 100$ nmol/min.	Faster growing strain
10	9.9 mmol/min.	50 mmol/min.	B
100	91 mmol/min.	91 mmol/min.	neither
1000	500 mol/min.	99 mmol/min.	A

The sucrose level in the soil is very low and the strain isolated from the soil will have the low sucrose transport machinery for its adaptation, whereas the strain isolated from sugar industry waste will have fast sucrose uptake machinery adaptation since the sucrose level in the sugar waste soil is high. Under this condition, the isolate from sugar waste soil would be an advantage.

Problem 2   Given an enzyme with $K_m = 10$ mM and $V_{max} = 100$ mmol/min.

i)   If [S] = 100 mM, which will increase the velocity more: a 10-fold decrease in K_m or a 10-fold increase in V_{max}?

ii) If [S] = 10 mM, which will increase the velocity more: a 10-fold decrease in K_m or a 10-fold increase in V_{max}?

Solution   One can answer these questions qualitatively and quantitatively. Having a qualitative understanding of what K_m and V_{max} mean is more important than the Michaelis–Menten equation.

Qualitatively

a)   With [S] >> K_m, the enzyme is running close to V_{max}. Because of this, decreasing K_m would hardly increase V whereas increasing V_{max} 10-fold would increase V 10-fold.

b)   With [S] = K_m, the enzyme is running at 50% of V_{max}. Decreasing K_m 10-fold will raise V to near V_{max}, in fact doubling V. Increasing V_{max} 10-fold would also increase V 10-fold.

Quantitatively

Part	K_m	V_{max}	[S]	V_0	Result
i.	10 mM	100 mmol/min.	100 mM	91 mmol/min.	
i.	1 nM	100 mmol/min.	100 mM	99 mmol/min.	
i.	1 nM	100 mmol/min.	100 mM	910 mmol/min.	Greater effect
ii.	1 nM	100 mmol/min.	100 mM	50 mmol/min.	
ii.	1 nM	100 mmol/min.	100 mM	91 mmol/min.	
ii.	1 nM	100 mmol/min.	100 mM	500 mmol/min.	Greater effect

Problem 3   The enzyme carbonic anhydrase (CA) catalyses the following reaction:

$$CO_2 + H_2O \leftrightarrow H_2CO_3$$

Given two identical sealed tubes shown in the figure with CO_2 gas above buffer containing H_2O and CO_2—one with added carbonic anhydrase, one without. Allow the contents to come to equilibrium, then:

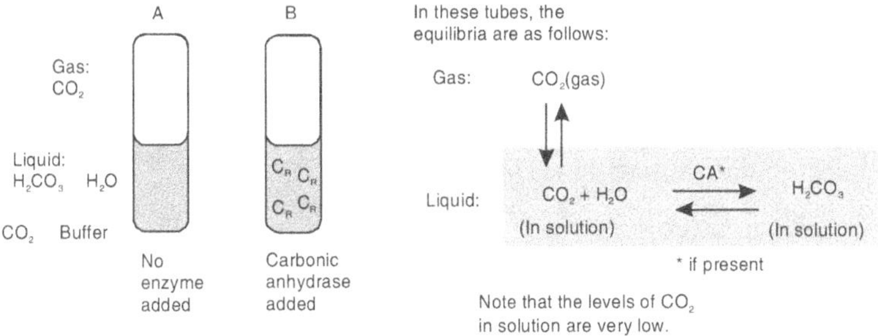

i. At equilibrium, which tube will have more CO_2 in the gas phase, A or B?

ii. Add CO_2 to the gas phase of both tubes. Which tube will absorb it faster into the liquid, A or B? Why?

iii. Remove CO_2 from the gas phase of both tubes. Which will replenish the CO_2 faster, A or B? Why?

iv. Take tube A, allow it to come to equilibrium, and inject a negligible volume of highly concentrated CA into the liquid layer. Will CO_2 gas be produced, absorbed, or will there be no change?

v. Take tube B, allow it to come to equilibrium, and inject a negligible volume of highly concentrated ATP. Will CO_2 gas be produced, absorbed, or will there be no change?

Solution

i. Since the tubes are at equilibrium and enzymes do not change the equilibrium concentrations of reactants and products (i.e., they do not change (G), both tubes will have identical $[CO_2]$ in the gas phase.

ii. The CA in tube B will accelerate the absorption of CO_2 by accelerating the conversion of CO_2 into H_2CO_3.

iii. The CA in tube B will accelerate the production of CO_2 by accelerating the conversion of H_2CO_3 into CO_2.

iv. This is same as (a) above; the CA cannot affect the equilibrium concentrations of reactants or products, but only alter the rate. So, the addition of CA to a tube at equilibrium will have no effect. However, addition of CA to a tube that was not at equilibrium could cause rapid absorption or production of CO_2. Normally, enzyme assays are carried out in reactions where the substrates have not had enough time to reach equilibrium, so that the effect of adding enzyme will be detectable.

v.   Nothing will happen. Since ATP is not a substrate of CA, the concentrations of reactants and products have not changed, so the equilibrium will not shift.

If ATP acts as a substrate for CA, then adding it to the reaction would change the amount of CO_2 in the gas phase.

Problem 4   The enzyme lysozyme is present in human nasal mucus and catalyses the breakdown of a component of the cell wall of certain bacteria, causing them to burst. Lysozyme cleaves a polymer of N-acetyl glucosamine (NAG)n into smaller fragments:

NAG-NAG-NAG-NAG-NAG-NAG-NAG-NAG-NAG-NAG-NAG-NAG

$$+ \, H_2O \xrightarrow{\text{lysozyme}}$$

NAG-NAG-NAG-NAG-NAG-NAG + NAG-NAG-NAG-NAG-NAG-NAG

i.   Given that the reaction proceeds as written, what can you conclude about the relative stability of (NAG)n?

ii. Given the following information, what can you conclude about the binding of lysozyme to its substrate?

Substrate	Relative rate of hydrolysis
(NAG)2	0
(NAG)3	1
(NAG)4	8
(NAG)5	4,000
(NAG)6	30,000
(NAG)8	30,000

iii. If you treated a solution of (NAG)100 with lysozyme, what would the majority of the product be?

Solution

i. Since lysozyme cannot affect ΔG of the reaction, (NAG)n must be unstable (i.e., the reaction where it breaks into monomers has a $\Delta G < 0$).

Then why do bacteria use an unstable molecule as a cell wall? Because, even though the (NAG)n will eventually come to equilibrium and decompose into monomers, the uncatalysed rate of hydrolysis is very low, so reaching equilibrium might take many decades. Lysozyme increases this rate enormously, and acts as a bacteriocidal agent.

ii. Even though the reaction catalysed by lysozyme appears to be:

$$NAG\text{-}NAG + H_2O \rightarrow NAG + NAG$$

the substrate binding site must have a pocket designed to hold 5 or 6 NAG units before the enzyme can get into active conformation. This is an example of substrate specificity.

iii. After a short time, it would mostly be (NAG)4, since any larger fragments will be rapidly cleaved. After a long time, it will be (NAG)2 and (NAG)3.

Feedback Inhibition

A biosynthetic pathway is made up of a series of enzymes that take some molecule and convert it into another molecule through a sequence of catalysed reactions. For example, Figure 20.8 is a generic biosynthetic pathway composed of four enzymes, labelled 1 through 4, that convert molecule A into molecule E through a series of intermediates (B, C and D):

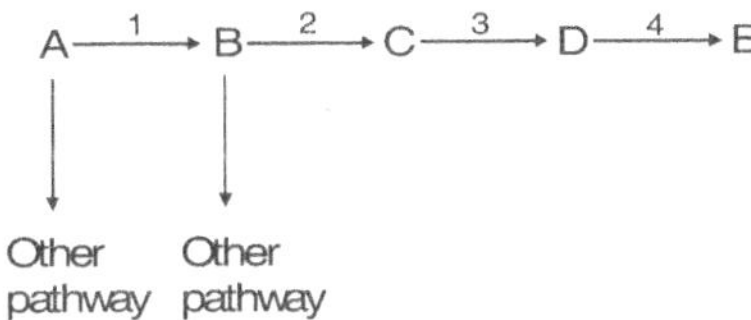

Figure 20.8   Generic biosynthetic pathway

Biosynthetic pathways are necessary to make major molecules in cells: nucleotides, amino acids, sugars and lipids. Catabolic pathways are organized in the same manner, but they break down molecules.

Feedback regulation is the mechanism by which biosynthetic and catabolic pathways regulate themselves. These pathways have evolved self-regulation such that, if too much of the end product is around (E in the pathway above), it, or a by-product, acts as an inhibitor of an earlier reaction. Generally, the end product acts as a non-competitive inhibitor of the first committed step in the pathway. For example, if both A and B are involved in other pathways as well as the one shown above, and there is an excess of E in the cell, E (or a by-product of E) will inhibit enzyme 2, preventing it from converting any more B to C. In this way, the cell keeps from synthesizing excessive amounts of E and keeps A and B available for use in other pathways.

REVIEW QUESTIONS

1. Give a brief account on free energetics. Add a note on free energy in chemical equilibrium.

2. Give a brief account on mechanism of enzyme action and enzyme nomenclature.

3. Derive Michaelis-Menten kinetics.

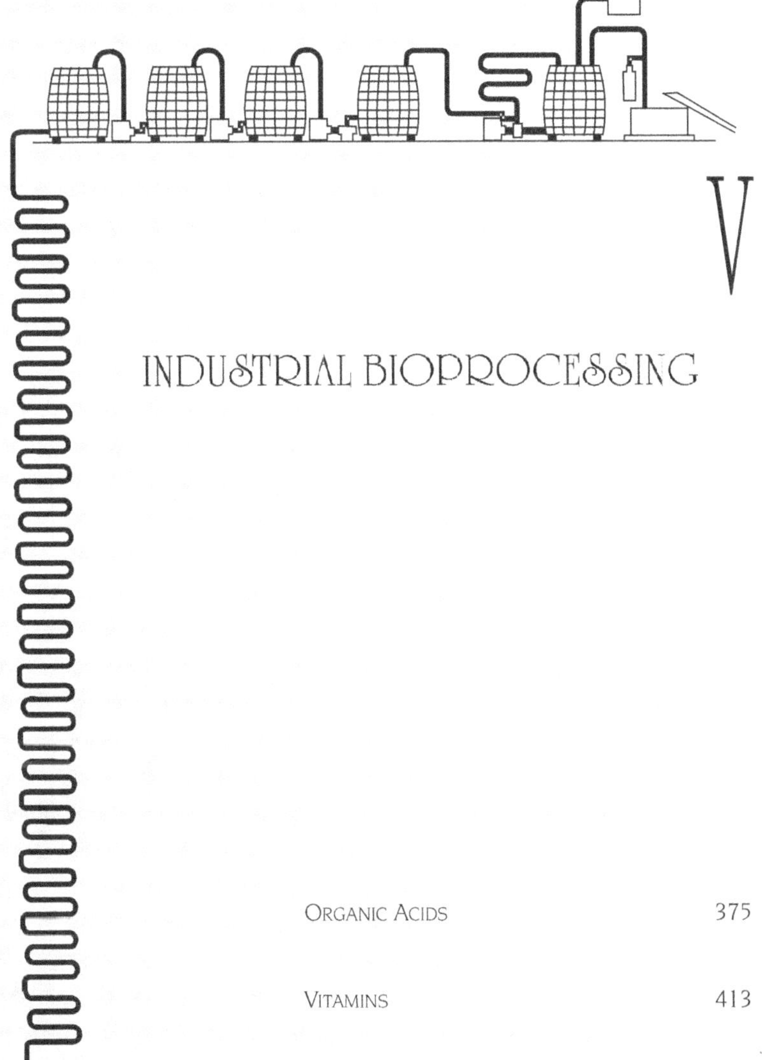

INDUSTRIAL BIOPROCESSING

ORGANIC ACIDS

INTRODUCTION

Many commercial production processes for organic acids are excellent. However, unlike penicillin, organic acids have had less visible impact on human wellbeing. Indeed, organic acid fermentations are often not identified as fungal bioprocesses, having been overshadowed by the successful deployment of β-lactam processes. Yet, in terms of productivity, fungal organic acid processes may be the best example of all. For example, commercial processes using *Aspergillus niger* in aerated stirred-tank-reactors can convert glucose to citric acid with greater than 80% efficiency and final concentrations in hundreds of grams per litre. Surprisingly, this phenomenal productivity has been the objective of relatively few research programmes. Perhaps, a greater understanding of this extraordinary capacity of filamentous fungi to produce organic acids in high concentrations will allow greater exploitation of these organisms via application of new knowledge in this era of genomics-based biotechnology. In this chapter, we will explore the biochemistry and modern genetic aspects of the current and potential commercial processes for making organic acids. The organisms involved, with a few exceptions, are filamentous fungi. Although yeasts including *Saccharomyces cerevisiae*, species of *Rhodotorula*, *Pichia* and *Hansenula*, are important organisms in fungal biotechnology, they have not been significant for commercial organic acid production, with one exception. The yeast, *Yarrowia lipolytica*, and related yeast species, may be in use commercially to produce citric acid (Lopez-Garcia, 2002). Furthermore, in the near future, engineered yeasts may provide new commercial processes to make lactic acid (Porro *et al.*, 2002). This chapter

21

contains a review of the commercial aspects of current and potential large-scale processes for fungal organic acid production and presents a detailed review of current knowledge of the biochemistry and genetic regulation of organic acid biosynthesis. The organic acids considered are limited to polyfunctional acids containing one or more carboxyl groups, hydroxyl groups, or both, that are closely tied to central metabolic pathways. A major objective of the chapter is to link the biochemistry of organic acid production to the available genomic data.

COMMERCIAL PROCESS

Although many organic acids are made from living cells, few are produced commercially. Citric, gluconic, itaconic and lactic acids are manufactured *via* large-scale bioprocesses. Oxalic, fumaric and malic acids can be made through fungal bioprocesses, but the market demand is small since competing chemical conversion routes are currently more economical. A few other organic acids have been explored for the development of novel processes. To date, the largest commercial quantities of fungal organic acids are citric acid and gluconic acid, both of which are prepared by fermentation of glucose or sucrose by *A. niger.* Another Aspergillus species, *A. terreus,* is used to make itaconic acid. A significant commercial source of lactic acid at the time of writing this book is a bioprocess employing the Zygomycetes such as *Rhizopus oryzae.*

These three species of fungi were initially chosen for process development because they exhibited the ability to produce large amounts of a particular organic acid. The reason why these fungi from the soil is producing large quantity of organic acids is due to their natural habitats. In soil, very often the availability of the sugar is not possible and the acid production also will be less. When the strains are placed in the synthetic medium with enough sugar concentration, the high uptake of sugar and high production of organic acids are formed. The acidification of environment by producing citric acid is a main advantage for the soil fungi to prevent the growth of the competitors.

The three fungal species produce a variety of organic acids, which may reflect different strategies to compete with other microorganisms. Many strains of *A. niger* can lower the pH of their environment by oxidizing glucose outside the cell wall, converting it to gluconic acid *via* the action of the enzyme glucose oxidase. The ability to catabolize gluconic acid is more unusual than the catabolism of glucose, and gluconic acid is also an effective chelator and acidulant. Other strains of *A. niger* produce citric acid intracellularly and export the acid, perhaps, as a chelator and acidifier that can also be reabsorbed for use as a carbon source. *A. terreus* acidifies the environment by producing itaconic acid. Itaconic acid is not a primary metabolite, so both the anabolism and the catabolism of this acid are relatively rare metabolic attributes. Once again, acidification of the environment with itaconic acid will inhibit the growth of many microorganisms. Subsequently, the relatively unusual nature of itaconic acid would permit *A. terreus* and only a few other species to catabolize the acid. It is interesting to note that the aspergilli and all the other filamentous fungi of the phylum Ascomycota fail to produce lactic acid. The ability to produce large amounts of lactic acid appears to be restricted to the phylum Zygomycota. Perhaps, fungi classified as Zygomycetes, including *R. oryzae*, have developed a different strategy for acidifying the environment by producing lactic acid to compete with fungi unable to metabolize lactic acid. These fungi often produce both ethanol and lactic acid, a combination that would discourage many competitors.

The four commercial organic acids (citric acid, lactic acid, gluconic acid and fumaric acid) produced by fungi are employed in high-volume, low-value applications. For example, they are used in industrial metal cleaning or other metal treatments and in the food and feed industry as flavour enhancers, acidifiers, stabilizers or preservatives. The commercial success of fungal bioprocesses is ultimately based on rapid and economical conversion of sugars to acid, but that alone does not explain the commercial situation for each of these acids. An understanding of the economic and business parameters that have contributed to the success of these four products may be useful in the development and commercialization of new organic acid products from filamentous fungi.

PRODUCTION OF CITRIC ACID

Citric acid (Figure 21.1) or 2-hydroxy-1, 2, 3-propanetricarboxylic acid (77-92-9) was first isolated from lemon juice. Although it occurs in rather high concentrations in citrus fruits, citric acid is ubiquitous in nature as an intermediate in the citric acid (Krebs) cycle (Figure 21.2) whereby carbohydrates are oxidized to carbon dioxide. The widespread presence of citric acid in the animal and plant kingdoms is an assurance of its non-toxic nature, and it has long been used as an acidulant in the manufacture of soft drinks, as an aid to the setting of jams and in other ways in the confectionery industry.

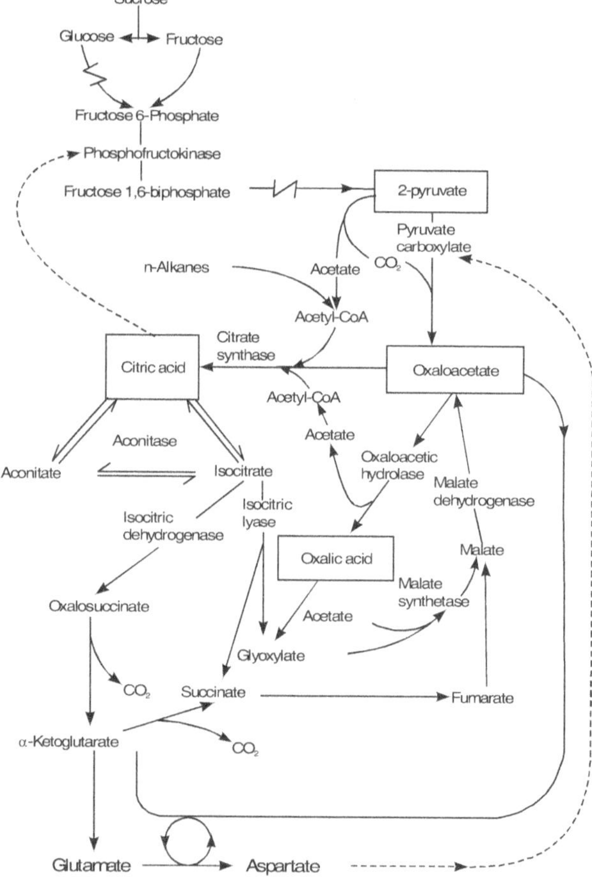

Figure 21.1   Citric acid

Figure 21.2   Biosynthetic pathway of citric acid (organic acids)

PROPERTIES OF CITRIC ACID

Citric acid is produced either in the anhydrous form or as monohydrates. The transition temperature between the two forms is 36.6°C. Thus, the anhydrous form is obtained by crystallization from hot aqueous solutions whereas the monohydrate is obtained by crystallization at temperature below 36.6°C. Both forms are utilized commercially. The application of citric acid worldwide is divided amongst the principal fields as follows: food, confectionery and beverages (75%), pharmaceutical (10%) and industrial (15%).

In food, confectionery and beverages, citric acid is the most versatile and widely used food acidulant. The use of citric acid as an acidulant depends in part on its strength as an acid. However, its pleasant taste and its property of enhancing existing flavour have ensured its dominant position in this market. Citric acid is able to make complex with heavy metals such as iron and copper. This property has led to its increasing use as a stabilizer of oils and fats where it greatly reduces oxidation catalysed by these metals. The ability to complex metals combined with its low degree of attack on special steels allows the use of solutions of citric acid in the cleaning of power station boilers and similar installations.

The sequestering action of citric acid is also used in the pharmaceutical industry, for example in the stabilization of ascorbic acid. Another pharmaceutical use is the effervescent effect it produces when combined with carbonates and bicarbonates, for example in antacid and soluble aspirin preparations. Citric acid is often used as an anion in pharmaceutical preparations.

Citric acid forms a wide range of metallic salts, many of which are commercial products. In terms of volume, trisodium and tripotassium citrates are the most important. Trisodium citrate is widely used as a blood preservative, where it prevents clotting by complexing calcium. It is also employed as an aid to emulsification in the manufacture of processed foodstuff, e.g. cheese. In regions where there are restrictions on phosphates in detergents, trisodium citrate replaces phosphates in specialty cleaners and heavy-duty liquids. Ferric ammonium citrate is used in the treatment of anaemia although other iron salts are increasingly preferred. Mixtures of citric acid and its salts have good buffering capacity and are extensively used for this purpose in the pharmaceutical, toiletry and food industries. In a process for the removal of sulphur dioxide from the fuel gases of power stations and metal smelters, a buffer solution principally containing H_2Cit is used as a scrubbing agent. A complex ion, $H_2CitHSO$, which is formed in the second stage reacts with H_2S to produce elemental sulphur. Sulphur is readily separated from the citrate solution, which may be recycled. The advantage of using citrate solution in this process is that certain side reactions yielding sulphate and thiosulphate, both giving rise to disposal problems, are substantially inhibited by citrate.

Citric acid esters of a wide range of alcohols are known. In particular, the triethyl, tributyl and acetyltributyl esters are employed as non-toxic plasticizers in plastic films used to protect foodstuff. Monostearyl citrate can be used instead of citric acid as an antioxidant in oils and fats.

PRODUCTION

Citric acid is mainly produced from *Aspergillus niger* and yeast either by surface fermentation using beet molasses or submerged fermentation using beet, cane molasses or glucose syrup. Submerged processes using sucrose as carbohydrate source are popular in areas where sugar is cheap. For yeast, submerged fermentation is carried out using beet molasses or glucose syrup.

The quality of both beet and cane molasses vary between seasons and from one refinery to another. In spite of many investigations, no clear reason for this has emerged. One probable reason may be that the composition of molasses is so complex. It is necessary, therefore, to make a selection of available molasses on the basis of performance. In order to obtain good yields of citric acid, particularly when *A. niger* is used, it is important to keep available levels of heavy metals, including iron and manganese, below critical levels. This is done in molasses media by additions of sodium or potassium ferrocyanide. Other inorganic nutrients are supplemented when necessary but most inorganic nutrients are already present in molasses. Where glucose syrup is employed as carbohydrate source, heavy metals are removed by ion exchange. Pretreatment of sucrose-based media by addition of ferrocyanide has also been proposed. In processes where pure glucose or sucrose is used as the substrate, additions of nitrogen, phosphate and other essential nutrients are made.

The yield of citric acid in the fermentation is expressed as kg citric acid monohydrate per 100 kg carbohydrate supplied. Yields in the range 70–90% on this basis have been reported. It should be mentioned that the theoretical yield of citric acid monohydrate from sucrose assuming no carbon is diverted to biomass, carbon dioxide or other by-products is 123% and that from anhydrous glucose is 117%. Thus, up to three quarters of the supplied carbon is converted to citric acid in a good fermentation.

Solid State Fermentation of *Aspergillus niger*

Surface fermentation using *A. niger* with beet molasses as raw material is extensively employed by major manufacturers. Although labour intensive, the power requirements are less than that in submerged fermentation. Beet molasses is diluted with water to a suitable sugar concentration, e.g. 150 kg m^{-3}, and the pH adjusted. An initial pH of 5 to 7 is usually employed because *A. niger* will not germinate at higher hydrogen ion concentrations. This effect is unknown when the media based on sucrose are being used and starting pH values as low as 2 are employed. The lack of germination in molasses at low pH is ascribed to the presence of acetic acid, which is a normal constituent of molasses. It appears that unionized acetic acid is what prevents germination, acetate being harmless in this respect, and hence the effect of pH.

Additional nutrients and alkali ferrocyanide are added and the whole content is boiled or sterilized. After cooling, the prepared medium is run down into a series of trays supported on racks in a ventilated chamber. The trays, which are usually made of very high purity aluminium, are filled to a depth of between 0.05 and 0.20 m. Spores of *A. niger* are obtained by growing a selected strain on a sporulation medium. The spores are collected and distributed over the surface of the medium in the trays. Sterile air is supplied to the fermentation chamber. The air performs the dual function of supplying oxygen and carrying away fermentation heat, and the rate of flow of air is regulated accordingly. A temperature in the region of 30°C is often employed. The mycelium forms a coherent on the surface of the liquid and becomes progressively more convoluted. The removal of heavy metals with ferrocyanide severely restricts sporulation. After a period of 7 to 15 days, the trays are emptied, the mycelium being at the same time separated from the fermented liquor. The liquors are pumped forward to the recovery section. Unwanted by-products of the process are gluconic and oxalic acids.

Submerged Fermentation

Cultivation of A. niger   The growth of *A. niger* requires, in addition to a source of carbon, the supplies of nitrogen, phosphate, potassium, magnesium and sulphur. In addition, small quantities of zinc, iron, copper and manganese are necessary, as well as molybdenum if nitrate is to be metabolized. The basal production medium for citric acid production is give in Table 21.1. When sufficient nutrients are available, the mould grows to its full extent, finally entering the sporulation phase. In order that citric acid accumulation can occur to the extent required by the commercial process, neither full growth nor sporulation must take place. This implies a limitation of nutrients. Researchers grew a mould of the *A. niger* group in continuous culture. They found a sharp maximum production of citric acid at 0.8 kg m^{-3} NH$_4$NO$_3$ and concluded that nitrogen limitation is an essential requirement for citric acid production. However, they demonstrated that cultures otherwise conditioned to citrate accumulation moved out of the growth phase when phosphate is exhausted even when nitrogen is not limiting. Probably either nitrogen or phosphate limitation is effective in inducing citric acid production depending on strain and other conditions. The positive effect of phosphate limitation would explain the successful use of beet molasses that contain relatively large amounts of metabolizable nitrogen.

Table 21.1   Basal production medium for citric acid production

Sugar (sucrose)	150–200.00 g
Sodium nitrate (NaNO$_3$)	4.00 g
Di-potassium phosphate (K$_2$HPO$_4$)	1.00 g
Magnesium sulphate (MgSO$_4$.7H$_2$O)	0.50 g
Potassium chloride (KCI)	0.50 g
Ferric chloride (FeCI$_3$)	0.02 g
Zinc sulphate (ZnSO$_4$)	0.10 g
Water to make up to 1 litre	

The importance of correct levels of trace elements like zinc, iron, copper and manganese to citric acid fermentation has long been appreciated. Optimum addition of iron and manganese was found necessary for best citric acid yields, and the optima varied from strain to strain. The medium containing purified glucose with 03–30 ppm zinc and 1.3 ppm iron is optimal for the better production of citric acid for *A. niger.* Addition of manganese at any level reduce the production. Researchers found that the pre-accumulation of manganese in the spores of *A.niger* when it is cultivated in the medium containing manganese that leads to reduction in production.

The importance of zinc, iron and manganese concentrations in the medium thus explains the necessity for the ion-exchange purification of glucose solutions or the treatment of molasses with ferrocyanide ions. Scientists have examined the effect of temperature and pH on the precipitation of iron, copper and manganese from molasses solutions. They found that addition of as little as 1 ppb manganese to ferrocyanide-treated beet molasses reduced the yield of citric acid by 10%. Hence, they concluded that ferrocyanide ion directly inhibited mould growth as well as removed unwanted heavy metals from molasses. Researchers found that a small excess of free ferrocyanide in the medium after removal of metals led to increased yields.

The medium preparation (suitable for molasses) shows an inline sterilization step, but sterilization of the medium in the fermenter is a possible alternative. Where inline sterilization is used, the fermenters are sterilized separately using steam at not less than 1 bar.

In the process flowchart shown in Figure 21.3, a vegetative inoculum stage is used where spores of *A. niger* are allowed to germinate in an inoculum medium before being transferred to the fermentation medium in a larger vessel. In some processes, the spores are introduced directly into the main fermentation but in other cases to the inoculum stage or the initial growth stage in the main fermentation. According to many reports, the morphology of the mycelium at this point is crucial, not only in relation to the shape of the hyphae but also in the aggregation of the growth of small spherical pellets. Thus, the hyphae should be abnormally short, stubby, forked and bulbous. This state of affair is brought about by the deficiency of manganese in the medium or obviously the related addition of ferrocyanide ion. The mycelial pellet should be small (0.2 to 0.5 mm) with hard, smooth surface. Factors leading to the production of such pellets are correct ferrocyanide level, ion concentration of less than 1 ppm, adjustment of pH, aeration and agitation, and concentration of manganese and the amount of spore inoculum. Whether the aggregation of the deformed hyphae into pellets is really necessary is doubtful, especially when stirred fermenters are used, but the pellet form does give rise to a broth that is more readily mixed.

When a separate inoculum stage is employed, a suspension of spores of *A. niger*, usually grown on a solid medium, is introduced into the sterilized medium in the inoculum fermenter. The medium is aerated and, in some processes, agitated and the mould is allowed to grow at a temperature of about 30°C for 18 to 30 hours and the growth rate is judged by the pH level reached or in other ways such as measuring biomass.

The fermentation medium is prepared and transferred to the main fermentor and the grown inoculum is incorporated at the rate of about 1 m³

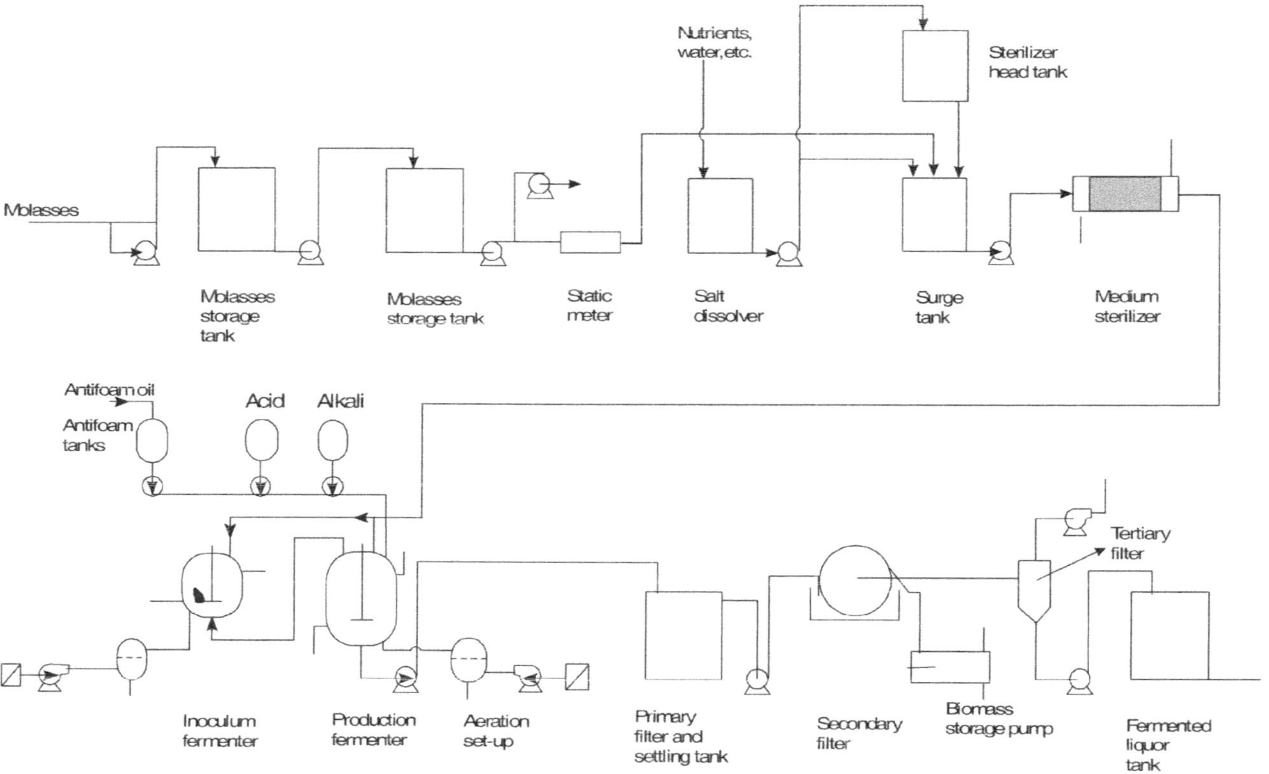

Figure 21.3   Process of citric acid production

inoculum to 10 m³ fermentation medium. When the inoculum or fermentation medium is based on molasses, the initial pH is normally in the range of 5 to 7. As mentioned above, most *A. niger* strains do not germinate or grow at lower pH values in this type of medium. On the other hand, lower initial pH values can be tolerated in glucose- or sucrose-based media and are often used with advantage. There is a lesser chance of infection by adventitious organisms. The fermentation is conducted at about 30°C.

Two types of fermenters are in common use, namely the stirred, aerated, baffled tank and the aerated tower fermenter, which has a much higher aspect ratio than the former and often contains an internal draught tube to promote circulation. Both types are constructed with high grade stainless steel and have provision for cooling. Both designs are sparged from the base with sterile air, although extra oxygen is sometimes used in the tower type. Sometimes, superatmospheric pressure is used in the fermenter to increase the oxygen solution rate. The effect of dissolved oxygen tension (DOT) on citric acid fermentation is studied by DO electrode provided in the fermenter. The kinetics of citric acid fermentation with *A. niger* have been reported. A growth phase is followed by the citric acid producing phase during which only a small amount of growth occurs.

Cultivation of yeast   The yeast Candida is commercially used to produce citric acid. Glucose and molasses are the suitable media for the production of citric acid. The fermentation is carried out at a neutral pH maintained by incorporating calcium carbonate in the medium. Around 65% from black strap molasses containing calcium carbonate is found to be suitable for the production of citric acid for *Candida oleophila*. However, a limitation of citric acid yield is the production of quantities of L-(+)-isocitric acid. Efforts should be made to select yeast strains without this defect. The subspecies of *Candida guilliermondii* produces only small amounts of isocitrate. Some mutants of yeasts give negligible quantities of isocitric acid wherein the mutants are selected for inability to grow on media containing halocitric acid or a precursor thereof. A mutant of *Debaryomyces claussenii* produces 73% yield from the medium containing glucose. The mutant species of *Candida zeylanoides* requires iron content in the medium for the higher production. This is the unique feature of the mutant than the wild strains of *C. zeylanoides*. It has been proposed that species of *Candida* may be cultivated in media containing glucose or blackstrap molasses to which 1.5 kg m⁻³ lead acetate is added, obtaining higher yields of citric acid. Addition of lead acetate, together with either n-hexadecyl citric acid or *trans*-aconitic acid, has also been proposed. By using strains of osmophilic yeasts, initial concentrations of sugar as high as 280 kg m⁻³ may be employed. In this particular process, no pretreatment of medium prepared from crude carbohydrates to remove metal ions is apparently necessary. Calcium carbonate, lead acetate and fluoroacetamide are also added.

A continuous fermentation process for the production of citric acid from blackstrap molasses using *Candida guilliermondii* has been described. The advantages of using yeast, rather than *A. niger*, are the possibility of using very high initial sugar concentrations together with a much faster fermentation. This combination gives a high productivity per run. It has been reported that the sugars containing various heavy metal content of the crude carbohydrate were used.

Fermenters of the tower type are suitable, but more effective cooling is required because of the rapid rate of heat output during shorter fermentation. An inoculum culture is prepared in a smaller fermenter by introducing cells of the selected yeast strain previously grown in slope culture. When the inoculum is sufficiently grown, it is transferred to the fermentation medium in the main fermenter. The fermentation is conducted at a temperature between 25 and 37°C, depending on the organism. The initial pH must not be too low, or else yeast growth will be impaired. This is presumably the reason for adding calcium carbonate to the medium. The pH can subsequently be allowed to fall. (In the continuous process, the pH is controlled at 3.5 by means of aqueous ammonia.) The broth is harvested when citric acid accumulation has become uneconomically low.

Product Recovery

The removal of the fermenting organism from the final broth leaves an aqueous solution of citric acid contaminated by various organics and inorganics depending on the initial carbon source.

The classical citric acid recovery process, which is particularly suitable with very impure liquors derived from molasses, is to heat the fermented liquor and add lime. The insoluble calcium citrate tetrahydrate is precipitated. The washed precipitate is treated in aqueous suspension with H_2SO_4 yielding an aqueous solution of citric acid and a precipitate of the by-product $CaSO_4$ (gypsum). This set of operations effectively removes most of the impurities derived from the substrate or generated in fermentation. As mentioned before, the conditions of concentration or crystallization steps can be varied to produce either the anhydrous acid or the monohydrate.

Solvent extraction is a possible alternative to the classical method but, because the available solvents tend to extract some impurities contained in molasses-derived liquors, it is easier to be applied to the products from glucose or alkane-based substrates. The advantage of the solvent extraction method is that it avoids the use of lime and H_2SO_4 and the concomitant problem of gypsum disposal. The use of butan-2-ol as an extractant has been proposed. But tributyl phosphate diluted with a minor amount of kerosene could be used. In this process, a better recovery could be realized by extracting citric acid into the solvent at a low temperature, subsequently stripping the solvent with hot water.

Another variant of the solvent extraction process is the ion pair extraction system in which the extractant consists of secondary or tertiary amines having in total at least 20 carbon atoms dissolved in a water immiscible solvent. Again the extraction of citric acid from the fermented liquor is done at a lower temperature (20°C) and the stripping state at a higher (80°C). In fermentation with n-alkanes and in particular where sodium hydroxide is used for pH control, monosodium or trisodium citrate can be directly crystallized from the clarified fermented liquor. In this case, citric acid may be produced from sodium citrate by electrodialysis.

LACTIC ACID

Lactic acid (2-hydroxypropanoic acid, 2-hydroxypropionic acid) is an organic hydroxy acid whose occurrence in nature is widespread. It is a three carbon organic acid: one terminal carbon atom is part of an acid or carboxyl group; the other terminal carbon atom is part of a methyl or hydrocarbon group; and a central carbon atom having an alcohol carbon group. Lactic acid exists in two optically active isomeric forms (Figure 21.4). It is soluble in water and water miscible organic solvents but insoluble in other organic solvents. It exhibits low volatility. The properties of lactic acid are given in Table 21.2.

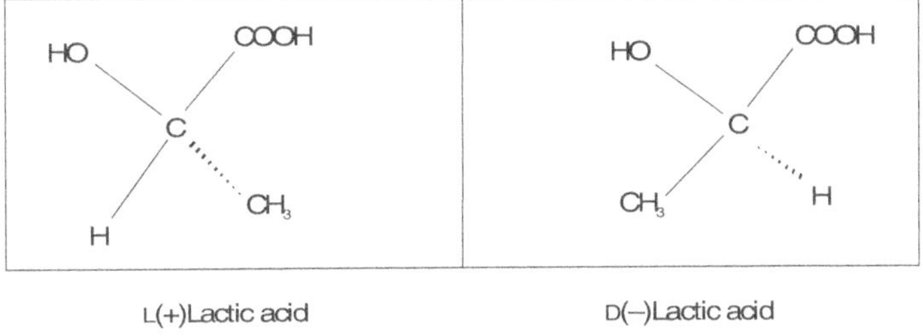

Figure 21.4   Structure of lactic acid

Table 21.2 Properties of lactic acid

Molecular weight	90.08
Melting point	16.8°C
Boling point	82°C at 0.5 mm Hg, 122°C at 14 mm Hg.
Dissociation constant, K_a at 25°C	1.37×10^{-4}
Heat of combustion, $D\ H_c$	1361 kJ/mole
Specific heat, C_p at 20°C	190 J/mole/°C

USES AND APPLICATIONS

Lactic acid is sold in technical, food and pharmaceutical grades, although most lactic acid meets the food and pharmaceutical requirements. The most common concentrations are 80 and 50% lactic acid. Lactic acid concentration above 90% is difficult to pump and handle. Higher quality grades have lower concentrations of contaminants such as sugar, metals, chloride, sulphate and ash. The fermentation grade acid generally contains some residual sugars and other impurities from carbohydrates and nitrogenous nutrients used in fermentation. Fermented lactic acid generally has yellow colour, which may be darkest for the concentrated technical grade and pale yellow for the food grade. The corrosiveness and liquid form of lactic acid may cause a handling problem. Thus, some manufacturers supply a powdered form of lactic acid on calcium lactate base. Aqueous solutions of lactic acid are sold in plastic-lined tank cans and drums and in plastic carboys.

Lactic acid finds applications as an intermediate in pharmaceutical manufacturing for adjusting the pH of preparations and in topical wart medications. Biodegradable plastic made of polylactic acid is used for sutures; they do not need to be removed surgically, and has been evaluated for use as a biodegradable implant for the repair of fractures and other injuries. The largest use of high-quality "heat stable" food or pharmaceutical grade lactic acid is the production of stearoyl-2-lactylates. Approximately $5.0–6.0 \times 10^6$ kg/yr of lactic acid is used for this purpose in the world. This acid is mostly produced synthetically, because residual sugar from the fermentation causes a caramelization colour, odour and flavour during the manufacture of stearoyl-2-lactylates. Calcium stearoyl-2-lactylate (CSL) is used mostly in baking. CSL acts as a "dough conditioner" by combining with the gluten in the dough, making it more tolerant to mixing and processing conditions as well as allowing a wider variation of bread ingredients. It also acts as a "crumb softener" by complexing the starch in the flour, which produces a softer texture when baked. Sodium stearoyl-2-lactylte (SSL) behaves similarly as CSL and acts as an emulsifier as well. Both CSL and SSL help extend the shelf life of baked products. Stearoyl-2-lactylates are also used as starch conditioners in other food products such as dehydrated potatoes, and as emulsifiers in cosmetics and food products. Stearoyl-2-lactic acid is hard to handle and the lactylate moiety depolymerizes easily; however, a small amount is used in food mixes. Lactylated fatty acid esters of mono- and diglycerides such as glycerol lactopalmitate and glycerol lactosterate are used as emulsifiers in cake mixes, bakery products, liquid shortenings and cosmetics. Stearoyl-2-lactylates and lactylated fatty esters compete with other emulsifiers such as ethoxylated mono- and diglycerides and succinylates monoglycerides. The principal competition in the US for stearoyl-2-lactylates in baking purposes is a combination of monoglycerides and ethoxylated monoglycerides. This combination is cheaper to use, but gives an inferior performance. Diacetyltartaric acid esters is also used in preference to stearoyl-2-lactylates.

Lactic acid is mainly used directly as a food ingredient. Perhaps, 50% of lactic acid production is used for this purpose. Both synthetic and fermentation derived lactic acid are used for this purpose. Lactic acid is also used as a food acidulant because it naturally occurs in many foodstuff, has a mild acid taste, and has no strong flavour or odour of its own. It is also used as a preservative, sometimes, in combination with other food acids such as propanoic and acetic acid. As a food acidulant, lactic acid experiences competition primarily from citric, acetic and phosphoric acid and to a lesser extent from malic, adipic, fumaric, propanoic, formic and tartaric acid. Lactic acid is generally more expensive than other food acids, but it is sometimes preferred because it adds less of its own flavour to the food. Lactic acid is used in brines for processing and packaging foods such as olives, pickles and sauerkraut. Many cheese products such as cheese spreads, cold pack cheese and processed cheese contain lactic acid. Creamy salad dressings in both liquid and powder form often contain lactic acid. Some powder mixes for preparation of dips, sour cream and cheese cake contain lactic acid. A few meat products such as salami contain lactic acid. Lactic acid's use in soft drinks has been largely replaced by citric acid, phosphoric acid and other food acids. Lactic acid is used directly in the production of some rye and sourdough breads. It is also used in a few bakery products as a preservative. The use of lactic acid in jams, jellies, pie filling and packaged pectin powders has been largely replaced by citric, malic and fumaric acid in the US, although it is still used to some extent in Europe. Lactic acid competes with phosphoric acid for use in adjusting the pH of water used in beer brewing. Lactic acid buffered with sodium lactate is used in the production of confectionery products. Some animal feeds contain lactic acid. The use of crude ammonium lactate from fermentation as a potential animal feed supplement has attracted attention in recent years. Lactic acid is sometimes used in the production of wine. The technical uses of lactic acid comprise a relatively small portion of the world's production. It is used in the manufacture of cellophane to control the pH in the film coating bath. It finds some application in plastics for the production of phenol-formaldehyde resins, and can be reacted with alcohols and acids to make polyesters that are useful as plasticizer. It is used in treating metal surface, manufacture of rubber products, electrostatic painting, textile and paper printing, "brightening" of silk and rayon, and textile dyeing. It was once widely used in the deliming of hides and in other leather processes; however, its use has been largely replaced by sulphuric and formic acids. It has potential for use in combination with other copolymers for the production of biodegradable plastics. Its use is likely to be limited to high-value, low-volume specialty plastics in medical applications and controlled release of pesticides. Lactic acid is also used to manufacture some herbicides, fungicides and pesticides. Stereo-specificity may have some advantages here as well as in some pharmaceutical products.

The calcium salts of lactic acid are produced in granular and powder form. Calcium lactate trihydrate is used in pharmaceuticals primarily as a dietary calcium source and also as a blood coagulant in the treatment of haemorrhage and to inhibit bleeding during dental operations. Calcium lactate pentahydrate is used primarily to firm potatoes and other vegetables and fruits for packaging. Calcium lactate is also used in baking powders. Sodium lactate is very hygroscopic and is typically marketed as a 60% solution. It is used primarily as an electrolyte replenisher in intravenous feeding formulations. It is also used in the production of some antibiotics and to buffer some pharmaceutical preparations. As a humectant, it is used in the production of cosmetics, personal care products, paper and tobacco. The world production of lactic acid salts is of the order of 106 kg/year. Most of this is sodium lactate, followed by calcium lactate. Potassium lactate and metal salts of lactic acid are produced in much smaller quantities.

The alkyl esters of lactic acid find uses in pharmaceuticals, foods and industry. The ethyl and butyl lactates are used as flavour ingredients and as solvents. Ethyl lactate is also used as a tablet lubricant, in cryogenic greases, in the manufacture of pesticides and herbicides, and in the manufacture of some anti-inflammatory drugs. Isopropyl lactate is used in pesticides and herbicides. Alkyl lactates are also used in brushing and stripping lacquers, in printing inks and in scented ink.

Microorganisms

Lactic acid bacteria are gram-positive, non-sporulating, low guanine, cytosine (GC), rods and cocci, which produce lactic acid as a major fermentation product. Members of these group lack porphyrins and cytochromes, do not carry out electron transport phosphorylation, and hence obtain energy only by substrate level phosphorylation. All lactic acid bacteria grow anaerobically. Unlike many anaerobes, however, most lactic acid bacteria are not sensitive to oxygen and can grow in its presence as well as in its absence. Thus, they are aerotolerant anaerobes. They usually have only limited biosynthetic ability, and their complex nutritional requirements include amino acids, vitamins, purines and pyrimidines. They are of two types: (1) Homolactic, which includes all members of genera *Pediococcus*, *Streptococcus* and *Lactococcus*, and (2) Heterolactic including all members of genera *Leuconostoc* and *Lactobacillus*.

The homofermentation produces a single fermentation product, i.e., lactic acid, whereas heterofermentation produces other products mainly ethanol, carbon dioxide as well as lactate. The differences observed in the fermentation patterns are determined by the presence or absence of the enzyme aldolase. Homofermenters lacking aldolase cannot breakdown fructose bisphosphate to triose phosphate. Instead, they oxidize glucose 6-phosphate, which is broken down to triose phosphate and acetylphosphate by means of enzyme phosphoketolase. In heterofermenters, triose phosphate is converted ultimately to lactic acid with the production of 1 mol of ATP, while acetylphosphate accepts electrons from NADH generated during the production of pentose phosphate and is thereby converted to ethanol without yielding ATP. Because of this, heterofementers produce only 1 mol of ATP from glucose instead of 2 mol produced by homofermenters. Also they produce carbon dioxide as end product from decarboxylate 6-phosphate. But the homofermenters produce little or no carbon dioxide. Therefore, one simple way of detecting a heterofermenter is to observe the production of carbon dioxide.

Homofermentation

Glucose + 2ATP — $\longrightarrow$ Fructose 1,6-bisphosphate + 2ADP

Fructose 1,6-bisphosphate $\xrightarrow{\text{Aldolase}}$ Glyceraldehyde 3-phosphate + Dihydroxyacetone phosphate

Glyceraldehyde 3-phosphate + NAD^+ + Pi $\longrightarrow$ 1,3-bisphosphoglycerate + NADH + H^+

1,3-bisphosphoglycerate + 2ADP — $\longrightarrow$ Pyruvate + 2ATP

Pyruvate + NADH + H^+ — $\longrightarrow$ Lactate + NAD^+

Heterofermentation

Glucose + ATP — $\longrightarrow$ Glucose 6-phosphate + ADP

Glucose 6-phosphate + NAD^+ $\longrightarrow$ 6-phosphogluconate + NADH + H^+

Ribulose 5-phosphate — $\longrightarrow$ Xylulose 5-phosphate

Xylulose 5-phosphate + Pi — $\longrightarrow$ Glyceraldehyde 3-phosphate + acetylphosphate + H_2O

Acetylphosphate + CoASH — $\longrightarrow$ Acetyl-CoA + Pi

Acetyl-CoA + NADH + H^+ $\longrightarrow$ Acetaldehyde + CoASH + NAD^+

Acetaldehyde + NADH + H^+ $\longrightarrow$ Ethanol + NAD^+

The selection of an organism depends primarily on the carbohydrate to be fermented. *L. bulgaricus, L. casei* and *S. lactis* are used to ferment lactose. Adapted strains of *L. delbrueckii and L. leichmannii* are typically used to ferment glucose. *L. pentosus* has been used to ferment sulphite waste liquor. Researchers have isolated a homofermentative strain called *L. amylophilus* that is capable of fermenting starch to L(+)-lactic acid with 90 wt% yields. Mixtures of strains as well as pure cultures have been used for commercial production of lactic acid.

Some fungi of the species *Rhizopus*, particularly *R. oryzae*, can be used to produce L(+)-lactic acid. This organism has less complex nutritional requirements than the lactic acid bacteria. Yields and fermentation rates of glucose are comparable with those lactic acid bacteria *Lactobacillus*. The *Rhizopus* sp. also can utilize starch feed stock like *Lactobacillus* and these has never lead in to fall of commercial success.

Raw Materials

A large number of carbohydrate materials have been used, tested and proposed for use in the fermentation of lactic acid. It is useful to compare feedstocks on the basis of the following desirable characteristics: 1) low cost, 2) low levels of contaminants, 3) fast fermentation rate, 4) high lactic acid yields, 5) little or no by-product formation, 6) ability to be fermented with little or no pretreatment, and 7) year-round availability. Crude feedstocks have historically been avoided because high levels of extraneous materials can cause separation problems in the recovery stage. The use of pentose sugars would result in the production of acetic acid, which will incur extra process equipment for separation.

Sucrose from cane and beet sugar, whey containing lactose, and maltose and dextrose from hydrolysed starch are presently used commercially. Refined sucrose, although expensive, is the most commonly used substrate, followed by dextrose. Not surprisingly, some lactic acid manufacturers are connected with beet or cane sugar business. Dextrose from corn starch was the most commonly used feedstock. Concentrated whey has been used without any pretreatment. Cellulosic materials such as corn cobs, corn stalks, cottonseed hulls, and straw and sulphite waste liquor have also been used. Acetic acid is produced as a by-product from cellulose derived sugars.

Fermentation Process

Batch fermentation is the method used industrially. Fermenters have been constructed with wood or stainless steel, and equipped with heat transfer coils for temperature control. Minimal agitation is provided by top or side mounted stirrers in order to keep the contents mixed. Fermenters are typically steam heated with water for boiling or chemically sterilized before filling with a pasteurized medium. Often, the fermenter lid is fitted in a loose condition. Contamination is not a large problem; the most serious contamination problems are due to the growth of butyric acid bacteria at the end of fermentation. Final product concentrations are less than 12–15% depending on other fermentation conditions to prevent precipitation of calcium lactate.

Fermentation conditions are different for each industrial producer but are typically in the range of 45–60°C with a pH of 5.0–6.5 for *L. delbrueckii*; 42°C and a pH of 6–7 for *L. bulgaricus*, and 30–50°C and a pH below 6 for *Rhizopus*. The inoculum size is always 5 –10% of the medium volume taken in the fermenter. The inoculum can be propagated in seed tanks or taken from a completed large-scale fermentation. The acid formed is neutralized by calcium hydroxide or calcium carbonate. The neutralizing agent can be added in excess as a slurry at the beginning of fermentation or added intermittently during fermentation on the basis of pH or acid titration measurements. The fermentation time is 1–2 days for a 5% sugar source such as whey and 2–6 days for a 15% sugar source such as glucose or sucrose. Reactor productivities are in the range of 1–3 kgm^{-3} h^{-1}. Under optimal laboratory conditions, fermentation takes one or two days. The yield of lactic acid after the fermentation stage would be 90–95 wt% based on the initial sugar or starch concentration (Figure 21.5). The residual sugar concentration is typically less than 0.1%. Cell mass yields can be as large as 30 wt% but generally are less than 15 wt% based on the initial sugar concentration. The yield of cell mass depends heavily on the amount of nitrogenous nutrients used. The fermentation rate depends primarily on the temperature, pH, concentration of nitrogenous nutrients, and lactic acid concentration. A controlled pH batch fermentation will proceed quickly at first. The minimum cell mass doubling time is about one hour, but this is not achieved under industrial operating conditions where the amounts of nitrogenous nutrients are suboptimal.

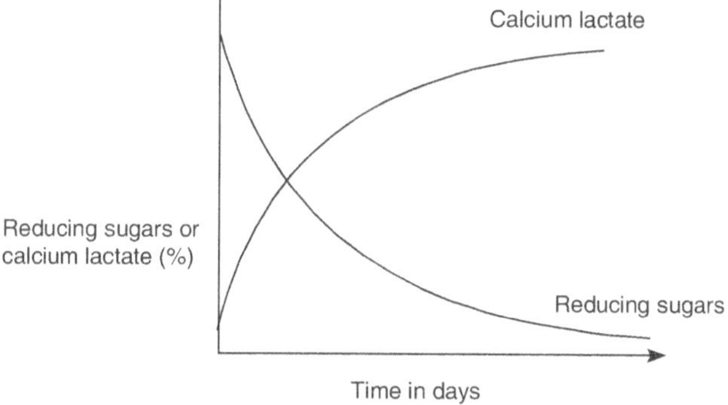

Figure 21.5   Reduction of reducing sugar during the production of lactic acid

Product Recovery Processes

We have discussed previously that lactic acid is sold in three major grades: technical, food and pharmaceutical. The grades are listed in the order of increasing purity, and more elaborate recovery processes are needed to produce higher quality material. In addition, heat stable lactic acid, which does not discolour significantly upon heating to about 200°C for a few hours, has a large demand. The recovery of lactate or lactate salts are cost effective. So, synthetically prepared lactic acid may be purified with less effort and less cost. So the synthetic production of lactic acid is preferred.

The materials of construction for fermentation and recovery equipment are limited by the corrosive nature of lactic acid, and contribute significantly to the products' final cost. Iron, copper, copper alloys, steel and chrome steel are unsatisfactory. Inconel and nickel are better but not recommended. Low iron alloys with large amounts of nickel and chromium corrode at an even slower rate. High molybdenum stainless steels such as 316 SS are satisfactory but still encounter problems, especially at improperly annealed welds and at gas or liquid interfaces where oxygen is present. Silver and tantalum are suitable, but too expensive for general use. In addition to equipment failure, corrosion increases the number of metal ions in the product, which must be removed for some end users. Wood, especially cypress and pitch pine, are satisfactory for dilute solutions but become dried out when exposed to concentrated solutions. Rubber is suitable for low temperature applications. Glass and ceramics are resistant, but their brittleness and poor heat transfer properties limit their usefulness. Some plastics are softened by warm concentrated lactic acid; however, heresite-lined, saran-lined, teflon-lined and polyester materials have been used. Plasticizer and other additives in plastic and rubber materials may be extracted or decomposed by lactic acid under some conditions. Advances in plastic, rubber ceramic, composite materials and metal alloys may provide some new useful materials with more attractive cost and acid resistance.

The first step in the recovery processes is to raise the fermentation liquor's temperature to 80–100°C and increase the pH to 10 or 11. This procedure kills the organisms, coagulates the proteins, solubilizes calcium lactate, and degrades some residual sugars. The liquid is then decanted or filtered. For some purposes, acidification of this liquor yields a usable product; however, for most applications, further processing with one of the following methods is required. It should also be noted that the use of cheap but impure raw materials must be weighed against higher purification costs.

Filtration, Carbon Treatment and Evaporation

One of the common methods for commercially producing lactic acid relies on the fermentation of relatively pure sugars with minimal amounts of nitrogenous nutrients. Thus, by using a pure feedstock, the recovery process is simplified. After the fermenter broth is filtered, activated vegetable carbon is used to bleach calcium lactate for the production of food grade acid. No carbon treatment is used for the technical grade. Next, the calcium lactate is evaporated to a 37% concentration at 70°C and 0.57 atm. The concentrated lactate is then acidified with 63% sulphuric acid, and the calcium sulphate precipitate is removed by a continuous filter and sent back to the first filter that treats the fermenter liquor. The filtered acid is then treated with activated carbon from the filter cakes of the first, third and fourth carbon treatments. The carbon from this step is discarded. The lactic acid is then evaporated from 8% to 52% or 82% in 316 stainless steel evaporators. Technical grade acid is then diluted to 50% or 80% and treated with sodium sulphide to remove heavy metals, if needed. Edible grade acid is diluted to 50 or 80%, bleached with activated carbon for a third time, and treated with sodium sulphide to remove heavy metals. It is then bleached a fourth time with carbon before packaging. Many manufacturers use fewer carbon treatments. Heavy metals could be removed by ion exchange, which may also remove some of the amino acids present. Heavy metals can also be removed by the stoichiometric addition of calcium or sodium ferrocyanide to form insoluble ferrocyanide salts of the heavy metals.

Calcium Lactate Crystallization

Lactic acid may be recovered from fermentations utilizing cruder raw materials such as whey or molasses. Several grades of calcium lactate and lactic acid can be produced from whey. The filtered liquor from the fermenters is treated with carbon first under slightly alkaline and then under slightly acidic conditions. The crude calcium lactate liquor is then evaporated under vacuum to a density of about 1.12 kg m^{-3}. Technical grade acid can be made from this liquor after evaporation, acidification and filtration of the precipitated calcium sulphate, and finally carbon treatment and heavy precipitation. To produce higher grades of acids, the liquor is cooled, crystallized and washed. The mother liquor and wash water are also cooled, crystallized and washed. The crystals are redissolved and recrystallized to get pure grades. In another case, the fermenter liquor is filtered and evaporated to 25% lactic acid. Calcium lactate is then crystallized and separated from the mother liquor. The mother liquor can be used for technical acid. The crystals are made into acid by a series of treatments.

These methods create a product that is low in unfermented carbohydrates but may contain some ash, which is mainly calcium sulphate. The crystals tend to form clusters and can be difficult to wash. The wash water and mother liquor contain high amounts of calcium lactate due to its high solubility, and must be recycled. Important costs in the purification process are related to the energy for water removal, losses in yield, and labour.

Liquid–Liquid Extraction

The extraction of lactic acid into an immiscible solvent phase has been researched by many investigations. Lactic acid can be purified in this way from fermentations using crude raw materials. In all such processes, the acid must first be extracted from the crude liquor by the solvent and then recovered from the solvent by methods such as back extraction into water or distillation of the solvent–lactic acid mixture.

The extraction solvent should have low water solubility, high distribution coefficient for lactic acid, and low distribution coefficient for impurities such as residual sugars. The distribution coefficient is defined as the concentration of lactic acid in the solvent phase divided by the concentration of lactic acid in the water phase.

The fermentation liquor is filtered and then acidified with sulphuric acid, and calcium sulphate is filtered off. Next, the crude lactic acid is decolorized with activated carbon, and then heavy metals; calcium and amino acids are removed by ion exchange. The acid is then evaporated under vacuum before it enters the countercurrent extraction columns. The acid is recovered from the solvent by countercurrent extraction into water. Next, the acid is given additional activated carbon and ion exchange treatments as needed. Lastly, the acid is evaporated to its final concentration. The solubility of isopropyl ether in water is low and the loss of solvent is tolerable. Lactic acid refined by liquid–liquid extraction is substantially free from ash, but contains other impurities from the raw materials and needs additional treatment by activated carbon, oxidation and other means.

Distillation of Lactate Esters

High quality lactic acid, substantially free from residual sugars and other impurities, can be prepared by the esterification of lactic acid with a low molecular weight alcohol. The counter current extraction is carried out using isopropyl ether as solvent. The lactic acid is produced by initial distillation of lactic ester and followed by hydrolysis of lactate ester. The hydrolysis result yield lactic acid and alcohol. The by-product alcohol is removed by the catalytic action of sulphuric acid. In this method the purity of extract is around 85%. Another method of extraction of lactic acid esters is using ammonium salts in the form of ammonium lactate. The ammonium lactate is a crude product and is derived by addition of ammonia or one of its salt directly to the fermentation broth. The production of lactic acid by the distillation of methyl lactate is another method of extraction of lactic acid which has wide commercial application. The product obtained is ash free and the level of other impurities is low. Some time, the corrosive stainless steel column may provide the iron contamination.

PRODUCTION OF GLUCONIC ACID

Gluconic acid (pentahydroxycaproic acid, Figure 21.6) may be produced from glucose through a simple dehydrogenation reaction catalysed by glucose oxidase. Oxidation of the aldehyde group on the C-1 of β-D-glucose to a carboxyl group results in the production of glucono-D-lactone ($C_6H_{10}O_6$) and hydrogen peroxide. Glucono-D-lactone is further hydrolysed to gluconic acid either spontaneously or by lactone hydrolysing enzyme, while hydrogen peroxide is decomposed to water and oxygen by peroxidase.

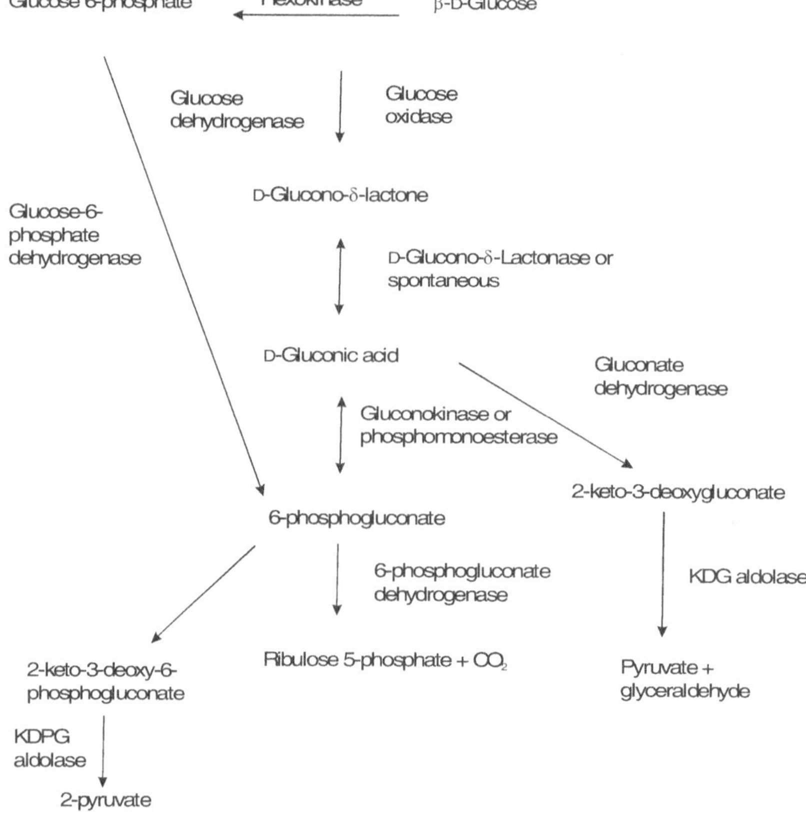

Figure 21.7   General gluconate pathways

The gluconate pathway is described in Figure 21.7. The conversion process could be purely chemical, but the most commonly used method is the fermentation process. The enzymatic process could also be conducted, where the conversion takes place in the absence of cells with glucose oxidase and catalase derived from *A. niger*. Nearly 100% of the glucose is converted to gluconic acid under appropriate conditions. The production of gluconic acid using enzymes has the potential advantage that no product purification steps are required if the enzyme is immobilized, e.g. the use of a polymer membrane adjacent to anion-exchange membrane of low-density polyethylene grafted with 4-vinylpyridine. However, this approach is not yet common in the industry. The properties of gluconic acid are given in Table 21.3.

Table 21.3   Properties of gluconic acid

Nature	Non-corrosive, mildly acidic, less irritating, non-odorus, non-toxic, easily biodegradable, non-volatile, organic acid
Relative molecular mass	196.16
Chemical formula	$C_6H_{12}O_7$
Synonym	2,3,4,5,6-pentahydroxyhexanoic acid
pK_a	3.7
Melting point (50% solution)	Lower than 12°C
Boiling point (50% solution)	Higher than 100°C
Density	1.24 g/mL
Appearance	Clear to brown
Solubility	Soluble in water
Sourness	Mild, soft, refreshing taste
Degree of sourness (sourness of citric acid is regarded as 100)	29–35

APPLICATIONS

Gluconic acid is a mild organic acid that finds vast applications in the food industry (Figure 21.4). It is a natural constituent in fruit juices and honey and is used in the pickling of foods. Its inner ester, glucono-D-lactone, imparts an initially sweet taste, which later becomes slightly acidic. It is used in meat and dairy products, particularly in baked goods, as a component of leavening agent for preleavened products. It is used as a flavouring agent (for example, in sherbets), and it is used in reducing fat absorption in doughnuts and cones. Foodstuffs that contain glucono-D-lactone include bean curd, yogurt, cottage cheese, bread, confectionery and meat.

Table 21.4 Applications of gluconic acid and its salts

Components	Applications
Gluconic acid	Prevention of milk stone in dairy industry, cleaning of aluminium cans
Glucono-D-lactone	Latent acid in baking powders for use in dry cakes and instantly leavened bread mixes; slow acting acidulant in meat processing such as sausages; coagulation of soybean protein in the manufacture of tofu. In dairy industry for cheese curd formation and for improvement of heat stability of milk.
Sodium salt of gluconic acid	Detergent in bottle washing, metallurgy (alkaline derusting), additive in cement as derusting agent, textile to prevent iron deposits, paper industry for processing paper pulp.
Calcium salt of gluconic acid	Calcium therapy; animal nutrition
Iron salt of gluconic acid	Treatment of anaemia; foliar feed formulations in horticulture

Generally speaking, gluconic acid and its salts are used in the formulation of food, pharmaceutical and hygienic products. They are also used in mineral supplements to prevent the deficiency of calcium, iron, etc., and as buffer salts. The sodium salt of gluconic acid has the outstanding property to chelate calcium and other di- and trivalent metal ions. It is used in the bottle washing preparations, where it helps to prevent scale formation and its removal from glass. It is well suited for removing calcareous deposits from metals and other surfaces, including milk or beer scale on galvanized iron or stainless steel. Its property of sequestering iron over a wide range of pH is exploited in the textile industry, where it prevents the deposition of iron, and for desizing polyester and polyamide fabrics. It is also used in metallurgy for alkaline derusting as well as in the washing of painted walls and removal of metal carbonate precipitates without causing corrosion. It also finds application as an additive to cement, controlling the setting time and increasing the strength and water resistance of cement. It helps in the manufacture of frost and crack-resistant concretes. It is also used in the household cleaning compounds (for example, mouthwash).

Calcium gluconate is used in pharmaceutical industry as a source of calcium for treating calcium deficiency by oral or intravenous administration. It also finds a place in animal nutrition. Iron gluconate and iron phosphogluconate are used in iron therapy. Zinc gluconate is used as an ingredient for treating common cold, wound healing and various diseases caused by zinc deficiencies such as delayed sexual maturation, mental lethargy, skin changes, and susceptibility to infections.

ORGANISMS

Gluconic acid is abundantly available in plants, fruits and other foodstuff such as rice, meat, dairy products, wine (up to 0.25%), honey (up to 1%) and vinegar. It may be produced by different kinds of microorganisms as well, which include bacteria such as *Pseudomonas ovalis, Acetobacter methanolicus, Zymomonas mobilis, Acetobacter diazotrophicus, Gluconobacter oxydans, Gluconobacter suboxydans, Azospirillum brasiliense,* fungi such as *Aspergillus niger, Penicillium funiculosum, P. variabile, P. amagasakiense* and various other species such as *Gliocladium, Scopulariopsis, Gonatobotrys, Endomycopsis,* and yeasts such as *Aureobasidium pullulans.* The ectomycorrhizal fungus, *Tricholoma robustum,* which is associated with the roots of *Pinus densiflora,* was found to synthesize gluconic acid.

PRODUCTION

There are different approaches to produce gluconic acid, namely chemical, electrochemical, biochemical and bioelectrochemical. There are several oxidizing agents available, yet the process appears to be costlier and less efficient as compared to the fermentation process. Although a simple one-step process, the chemical method is not favoured. Thus, fermentation has been one of the efficient and dominant techniques in manufacturing gluconic acid. Among various microbial fermentation processes, the method utilizing the fungus *A. niger* is most widely used. However, the process using *G. oxydans* has also gained importance. Irrespective of whether fungi or bacteria are used, the significance lies on the product, for example, sodium gluconate or calcium gluconate. As the reaction leads to an acidic product, it is required to be neutralized by the addition of neutralizing agents, otherwise the acidity inactivates glucose oxidase, resulting in the arrest of gluconic acid production. The conditions for the fermentation processes in the production of calcium gluconate and sodium gluconate differ in many aspects such as glucose concentration (initial and final) and pH control. In the process involving calcium gluconate production, the control of pH results from the addition of calcium carbonate slurry. Another important point to be noted is the solubility of calcium gluconate in water (4% at 30°C). At high glucose concentration, above 15%, supersaturation occurs, and if it exceeds, the calcium salt precipitates on the mycelia and inhibits oxygen transfer. The neutralizing agent should also be sterilized separately from the glucose solution to avoid Lobry de Bruyn-van Ekenstein reaction, which alters the conformation of glucose and results in the reduction of yield to about 30%. On the contrary, the process involved in sodium gluconate production is highly preferable as the glucose concentration of up to 350 g/L can be used without any such problems. The pH is controlled by the addition of NaOH solution. Sodium gluconate is readily soluble in water (39.6% at 30°C).

Production from *Aspergillus Niger*

A. niger produces all the enzymes required for the conversion of glucose into gluconic acid, which include glucose oxidase, catalase, lactonase and mutarotase. Although crystalline glucose monohydrate, which is in the alpha form, is converted spontaneously into beta form in the solution, *A. niger* produces the enzyme mutarotase, which serves to accelerate the reaction. During the process of glucose conversion (Figure 21.8), glucose oxidase present in *A. niger* undergoes self-reduction by the removal of two hydrogens. The reduced form of the enzyme is further oxidized by the molecular oxygen, which results in the formation of hydrogen peroxide, a by-product in the reaction. *A. niger* produces catalase, which acts on hydrogen peroxide releasing water and oxygen. Hydrolysis of glucono-D-lactone to gluconic acid is facilitated by lactonase. The reaction can be carried out spontaneously as the cleavage of lactone occurs rapidly at pH near neutral, which is brought about by the addition of calcium carbonate or sodium hydroxide. The removal of lactone from the medium is necessary as its accumulation may have a negative effect on the rate of glucose oxidation and the production of gluconic acid and its salts. Previous research points out that the enzyme gluconolactonase is also present in *A. niger*, which increases the rate of conversion of glucono-D-lactone to gluconic acid.

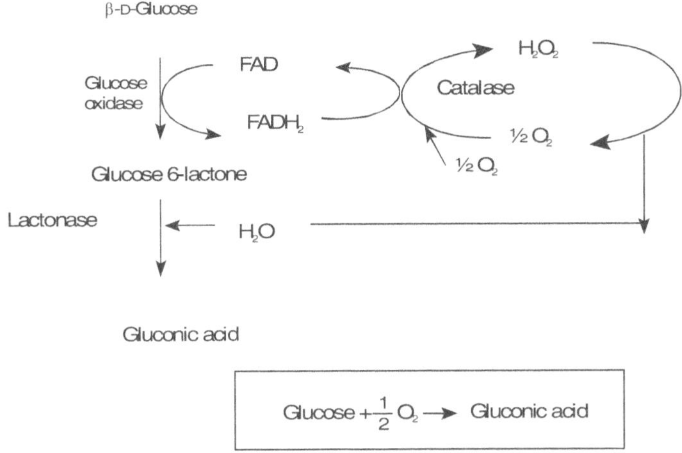

$$\text{Glucose} + \frac{1}{2}\,O_2 \longrightarrow \text{Gluconic acid}$$

Figure 21.8 Production of gluconic acid by *Aspergillus*

The production of gluconic acid is directly linked with the glucose oxidase activity. Depending on the application, the fermentation broths containing sodium gluconate or calcium gluconate are produced by the addition of solutions of sodium hydroxide or calcium carbonate, respectively, for neutralization. The general optimal conditions for gluconic acid production are as follows: glucose at concentrations between 110–250 g/l, nitrogen and phosphorus sources at a very low concentration (20 mM), pH value of medium around 4.5–6.5 and very high aeration rate by the application of elevated air pressure (4 bar). There are two key parameters that influence the production of gluconic acid. These are oxygen availability and pH of the culture medium. Oxygen is one of the key substrates in the oxidation of glucose since glucose oxidase uses molecular oxygen in the bioconversion of glucose. The concentration of oxygen gradient and the volumetric oxygen transfer coefficient are the critical factors that monitor the availability of oxygen in the medium. These two factors highly influence the rate of transfer of oxygen from gaseous to aqueous phase. The aeration rate and the speed of agitation are the two parameters that affect the availability of oxygen in the medium. Gluconic acid production is an extremely oxygen-consuming process with a high oxygen demand for the bioconversion reaction, which is strongly influenced by the dissolved oxygen concentration. Oxygen is generally supplied in the form of atmospheric air; however, in some studies, high-pressure pure oxygen has also been provided.

Substrates

Glucose is generally used as carbon source for microbial production of gluconic acid. However, hydrolysates of various raw materials, such as agro-industrial waste, have also been used as substrate. The high yield of gluconic acid would be obtained in media containing glucose or starch hydrolysate as the sole carbon source. Other substrates like hydrol (corn starch hydrolysate), whey and cane molasses are also used as a source of glucose. Cane molasses is generally subject to different pretreatments such as acid treatment, potassium ferrocyanide treatment and salt treatment. Potassium ferrocyanide treatment would give better results. Gluconic acid synthesis may be influenced by various metal ions such as copper, zinc, magnesium, calcium and iron. The production medium for *Aspergillus* is provided in Table 21.5.

Table 21.5 Production medium for *Aspergillus*

Production medium	(g/L)
Glucose	150
Maize steep liquor	3.7
KH_2PO_4	0.2
$MgSO_4 \cdot 7 H_2O$	0.17
Urea	0.10
Trace element solution	1ml

Production of Gluconic Acid by Bacteria

Acetic acid bacteria and *Pseudomonas savastanoi* were cultures initially observed to produce gluconic acid. Unlike fungi, the reaction in bacteria is carried out by glucose dehydrogenase that oxidizes glucose to gluconic acid, which is further oxidized to 2-ketogluconate by gluconic acid dehydrogenase (GADH). The final oxidation step to 2,5-diketogluconic acid (DKG) is mediated by 2-ketogluconate dehydrogenase (KGDH). All the three enzymes are localized in the membranes of the cells and are induced by high glucose concentrations. GADH is an extracellular protein and has PQQ (pyrroloquinoline quinine) as a coenzyme. Also, there is an intracellular enzyme, an $NADP^+$-dependent glucose dehydrogenase, which is less involved in the gluconic acid formation than the extracellular enzyme. Gluconic acid produced is exported to the cell and further catabolized *via* the reactions in pentose phosphate pathway. When glucose concentration in the medium is greater than 15 mM, pentose phosphate pathway is repressed and thus gluconic acid accumulation takes place.

Gluconobacter oxydans is an obligate aerobic bacterium that oxidizes glucose *via* two alternative pathways. The first pathway requires an initial phosphorylation followed by oxidation *via* the pentose phosphate pathway. The second is the direct glucose oxidation pathway, which results in the formation of gluconic acid and ketogluconic acid. *G. oxydans* converts D-glucose into 2,5-diketogluconic acid by the action of three membrane-bound $NADP^+$-independent dehydrogenases. The acid-tolerant acetic acid bacterium, *Acetobacter diazotrophicus*, would exhibit a high rate of gluconic acid formation. Glucose oxidation by this organism is less sensitive to low pH values than glucose oxidation by *G. oxydans*. Both phosphorylative and direct oxidative pathways of glucose metabolism appear to be operative. In addition to the pyridine nucleotide (strictly NAD^+-dependent glucose dehydrogenase), *A. diazotrophicus* contains a PQQ-dependent glucose dehydrogenase, which is primarily responsible for gluconic acid formation. Bacterial gluconic acid production has limited success at industrial scale, since the oxidation proceeds with the secondary reactions leading to oxogluconic acids. The ability of *Pseudomonas* and *Gluconobacter* spp. to produce gluconolactone and gluconic acid has been exploited and the process is used commercially in the production of lactone.

Acetobacter methanolicus is also used to catalyse the conversion of glucose into gluconic acid. The key advantage of using this facultatively methylotrophic microorganism as a catalyst is that the gluconic acid formed is a metabolic dead-end product, and unlike in other bacterial fermentation processes, this organism uses methanol, a cheap raw material, as a substrate. Further, in the process, glucose is not assimilated or consumed for growth, and so maximum theoretical yield coefficient is achieved consequently.

Yeast

Aureobasidium pullulans, a yeast-like form of the dimorphic fungi, is used for the production of gluconic acid. Various process parameters for continuous and discontinuous production of gluconic acid, such as pH, oxygen, temperature, medium composition, air saturation, etc., have been studied. The highest glucose conversion of 94% and product yield of 87.1% would be achieved at an optimum pH of 6.5. At pH 4.5, the product selectivity and yield are very poor, reaching 67.8% and 20.7%, respectively. The temperature range of 29 to 31°C is suitable for the production of gluconic acid by the yeast. An increase of temperature by 1°C (i.e., to 32°C) would dramatically reduce the steady state concentration of the biomass and the product.

Recovery of the Product

The recovery process depends on the method followed for broth neutralization and the nature of carbon sources used. Generally, the downstream process is similar to the fermentation processes using fungal and bacterial species. Gluconic acid, glucono-D-lactone, calcium gluconate and sodium gluconate are some important products and their extraction processes are briefly outlined below.

For the recovery of free gluconic acid from calcium gluconate, the broth is clarified, decolorized, concentrated and exposed to −10°C in the presence or absence of alcohol. Thus, the calcium salt of gluconic acid crystallizes, which is recovered and further purified. The gluconic acid can also be obtained from calcium gluconate precipitate that are derived from hyper saturation solution in cold condition. The releasing of gluconate from calcium gluconate is achieved by addition of sulphuric acid which remove the calcium as calcium sulphate.

For obtaining calcium gluconate as a product, calcium hydroxide or calcium carbonate is used as the neutralizing agent. They are added to the nutritive broth accompanied by heating and vigorous stirring. The broth is concentrated to a hot supersaturated solution of calcium gluconate, followed by cooling at 20°C and adding water miscible solvents, which crystallizes the compound. A treatment with activated carbon facilitates the crystallization process. Finally, they are centrifuged, washed several times and dried at 80°C.

Sodium gluconate, the principle material of gluconic acid is generally prepared by ion exchange. Sodium gluconate from the filtered fermented broth is concentrated to 45% (mass per volume) followed by the addition of sodium hydroxide solution raising the pH to 7.5 and drum drying. Carbon treatment of the hot solution before drying is practised for obtaining a refined product. Glucono-D-lactone recovery is a very simple process. Aqueous solutions of gluconic acid are an equilibrium mixture of glucono-δ-lactone, glucono-λ-lactone and gluconic acid. At temperature between 30–70°C, the crystal that is separated from the supersaturated solution is glucono-δ-lactone. At temperature below 30°C, gluconic acid results even above 70°C, and the resulting product would be glucono-λ-lactone.

PRODUCTION OF FUMARIC ACID

Fumaric acid is a naturally occurring four-carbon dicarboxylic acid that finds increasing use as a food acidulant and beverage ingredient. Because of its double bond and two carboxylic groups, fumaric acid has many potential industrial applications, ranging from the manufacture of synthetic resins and biodegradable polymers to the production of intermediates for chemical syntheses. Fumaric acid is derived exclusively from petroleum-based materials, since large-scale biological production of fumaric acid is too expensive to compete with the synthetic route. Fumaric acid esters are sometimes used to treat psoriasis, as it has been suggested that the condition is caused by an impairment of fumaric acid production in the skin. Fumaric acid is a food acidulant and it is non-toxic. It is used in beverages as preservatives, such as Welch's Grape Drink. It is also used as a substitute for tartaric acid and occasionally in place of citric acid, at a rate of 1.36 g of citric acid to every 0.91 g of fumaric acid for the same taste.

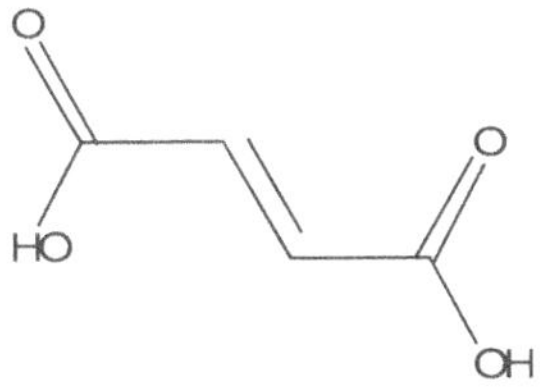

Figure 21.9   Structure of fumaric acid

ORGANISMS AND CULTURAL CONDITIONS

The *Rhizopus* spp. are the best identified fungal sources for fumarate production. Ehrlich (1911) first identified fungal fumaric acid production in a strain of *Rhizopus nigricans*. A later survey showed that the *Rhizopus, Mucor, Cunninghamella* and *Curcinella* spp. are found to produce fumaric acid. The nutritional and physical requirements of *R. oryzae* leading to maximum yields and minimal biomass accumulation are much like other fungal fermentations including high concentrations of organic acids, high carbohydrate concentrations, and high carbon to nitrogen ratios. A conversion of 60% to 70% of sugar to fumaric acid (w/w) could be achieved in vigorously agitated submerged cultures containing 10–12% glucose and C : N ratios ranging from 120 : 1 to 150 : 1. Standard minerals and calcium carbonate are added after 3–8 days, and the cultures are incubated at 33°C. Recently, pH and metal (magnesium, zinc, iron and manganese) concentrations were varied with the result that consistent pellet morphology (about 1 mm pellets) and relatively high fumarate output was obtained.

The carbonate in fumaric acid production media is required to neutralize and precipitate the fumaric acid. In addition, carbonate is necessary for the formation of oxaloacetate by pyruvate carboxylase. The metabolic model developed for an L-lactate synthesizing strain of *R. oryzae* has bearing also on fumarate synthesis. The results of these modelling studies indicate that increasing carbonate concentrations raise fumarate and malate yields at the expense of lactate yields. This is consistent with the requirement for high concentrations of carbonate in fumarate production strains. An increase in pyruvate carboxylase activity was observed to correlate with glucose utilization and fumarate production in *R. oryzae*. Pyruvate carboxylase is located in the cytosol of *R. oryzae* as one of two fumarate hydratase isozymes.

Many species of organisms produce small amounts of fumaric, malic and other organic acids as metabolic by-products during oxidative metabolism. In some instances, certain mycelial fungi are capable of producing significant quantities of fumaric acid from glucose and carbon dioxide. For maximum productivity of fumarate, it should be produced through reduction of tricarboxylic acid. By limiting the nitrogen source, the growth of *Rhizopus* can be minimized. During this non-growth stage, fumaric acid can be accumulated with a maximal yield of 2 mol/mol of glucose consumed or 1.29 g of fumarate per g of glucose consumed on a weight basis. In reality, however, the reductive process requires the supply of NADH from the tricarboxylic acid cycle. Therefore, the obtainable yield is about 1.45 mol of fumarate per mol of glucose (as much as 0.93 g/g of glucose consumed).

The production is mostly carried out in stirred tank fermenters equipped with a pH controller to produce fumaric acid from glucose with the addition of calcium carbonate. The spore suspension is inoculated into the production medium consisting of 100 g of glucose, 0.6 g of KH_2PO_4, 0.25 g of $MgSO_4 \cdot 7H_2O$, and 0.088 g of $ZnSO_4 \cdot 7H_2O$ in 1 litre of distilled water. Sterile $CaCO_3$ is added, whenever needed, to maintain the pH at 5. Growth is normally carried out at 30°C and 215 rpm. After cultivation, the mycelial pellets are harvested by filtration.

In typical fumaric acid fermentation with *Rhizopus oryzae*, the presence of a neutralizing agent is required not only to neutralize the acid but also to remove fumaric acid produced from the fermentation broth. The recovery of calcium fumarate and regeneration of free acid from fumarate are complicated, tedious, and expensive. One technique is the use of adsorbents such as polyvinyl pyridine (PVP) to remove acid while it is being produced. PVP resin is selective in that it adsorbs free acid but not the salt. There have been no reports regarding the biological production of fumaric acid with a simultaneous production-recovery system.

PRODUCTION OF ITACONIC ACID

The production of itaconic acid was first reported with *Aspergillus itaconicus* and later shown to be overproduced by *A. terreus*. Sugars such as glucose and sucrose are generally used in industrial production processes, and several biosynthetic pathways leading to itaconic acid accumulation have been postulated. While it is generally agreed that glucose is metabolized to pyruvate mainly *via* the Embden–Meyerhof–Parnas (EMP) pathway, the conversion of pyruvate to itaconate is still unidentified. Citric acid is generally considered to be a poor precursor of itaconic acid. Figure 21.10 shows the biosynthesis of itaconic acid.

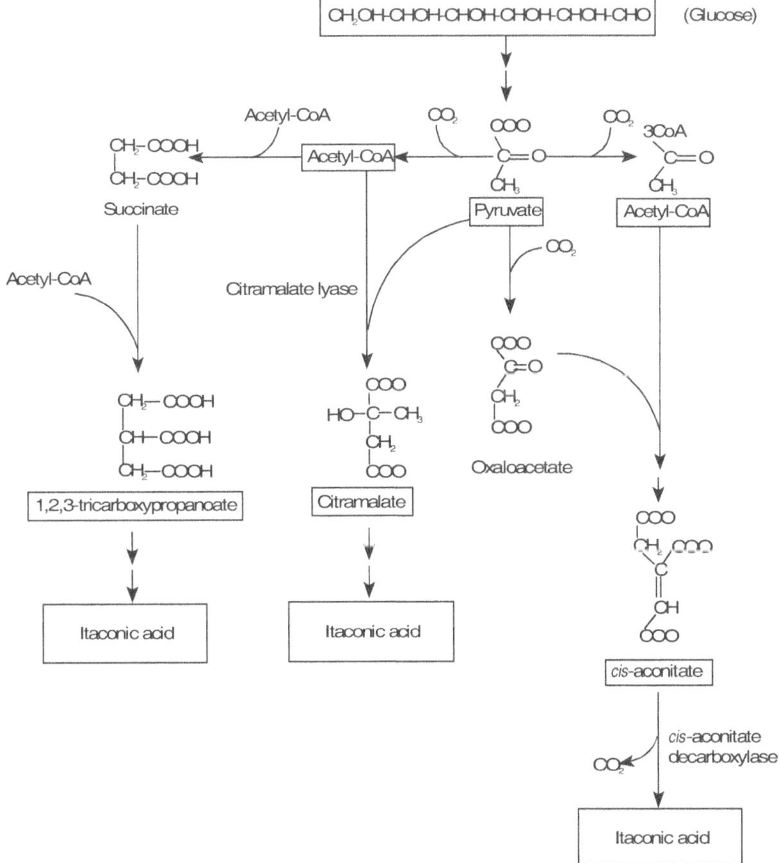

Figure 21.10 Biosynthesis of itaconic acid

A. terreus is employed for itaconic acid production in a process similar to that for citric acid. Both processes were invented about the same time (Kane 1945; Nubel and Ratajak, 1962; Batti and Schweiger, 1963), and both can be conducted in the same manufacturing facility. Although the processes are similar, as would be expected for a by-product of the citric acid cycle, there is a significant difference: the sensitivity of the organism to the acid necessitates neutralization to obtain yields above 80 g/l. The recovery of itaconic acid can be accomplished with the technology used for citric acid, and the final product can be prepared as a dry crystalline powder. In contrast to citric, gluconic and lactic acids, itaconic acid is used exclusively in non-food applications. Its primary application is in the polymer industry where it is employed as a co-monomer at a level of 1–5% for certain products. Itaconic acid is also an important ingredient in manufacturing synthetic fibres, coatings, adhesives, thickeners and binders. To date, very little research has been directed at the improvement of itaconic acid production. In contrast, there has been a larger research effort directed at lactic acid production to feed the market for biodegradable plastic.

Cultural Conditions

Simple sugars such as glucose and sucrose are the carbon sources for the production of itaconic acid. In large-scale production, cane molasses is used to satisfy the carbon sources. Usually 15–25% of molasses is used in the production medium. Beet molasses is used as the carbon source for spore germinating medium.

Ammonium salts like NH_4NO_3 and NH_42SO_3 are used as nitrogen sources. The mineral sources required are iron, copper, magnesium and zinc. At the initial stage of fermentation, the pH ranging from 1.8 to 2.0 is maintained. The pH is adjusted by adding strong acids like sulphuric acids, nitric acids, etc. Sulphuric acid is preferable. During production, the itaconic acid produced gradually reduces the pH of the medium. The neutralization of the medium pH is achieved by periodical addition of ammonia. The optimum temperature in fermentation is maintained at 30–35°C. Fermentation may take 3–5 days based on the organisms used. The recovery of itaconic acid includes steps such as acidification, clarification and other purification treatments followed by concentration and crystallization. Solvent extraction method is also employed in the recovery of itaconic acid.

OTHER ORGANIC ACIDS

SUCCINIC ACID

The emphasis on biological succinic acid production has been on bacterial fermentations especially with anaerobic bacteria. However, *Fusarium* spp., *Aspergillus* spp. and *Penicillium simplicissimum* are known to produce this acid. The L-malic acid producing *Aspergillus* spp. secretes succinic acid as a secondary product at lower concentrations; the highest titer cited was only 1.3%, representing 25% of the total organic acid production. The formation of succinate from glucose by P. *simplicissimum* was investigated under aerobic and anaerobic conditions. This fungus secretes low levels of succinate; the highest rate of succinate produced by the fungus is 0.063 g of succinic acid per gram of dry mycelial weight under the respiratory inhibitory condition. Nevertheless, the results were interesting for three reasons: succinic acid was the predominant acid produced under anaerobic conditions; pellet formation (diameter not reported) was shown to be important in obtaining maximum succinate production rates; and it raised the possibility of fumarate respiration as a biochemical mechanism for succinate production under anaerobic conditions.

There are three possible metabolic mechanisms in the production of succinate: the oxidative portion of the TCA cycle, the reductive portion of the TCA cycle, and the glyoxylate bypass. Metabolism by either the oxidative portion of the TCA cycle or the glyoxylate bypass pathway conserves only four of the six carbons from glucose in the four-carbon succinic acid product. On the other hand, the reductive portion of the TCA cycle produces two four-carbon acids for every glucose molecule metabolized *via* glycolysis operating in conjunction with pyruvate carboxylase. Thus, anaerobic metabolism is preferred for succinic acid producing microorganisms. The succinate production in P. *simplicissimum* may occur *via* fumarate respiration, but this has not been demonstrated. The physiology of succinate production by filamentous fungi is an emerging field. A screening strategy might identify more promising fungal strains that combine the features of high succinate rates and final titers, low side-product concentrations, and tolerance to low pH.

MALIC ACID

L-Malic acid production has been observed in *R. oryzae* and *Aspergillus* spp. Generally, L-malate accumulation in *R. oryzae* is minor as compared to L-lactate or fumarate. Clearly, the mechanism leading to malate production in *R. oryzae* is the same as the pathway leading to fumarate, abbreviated by one step. If the cytosolic isozyme of fumarate hydratase could be decreased, the transformation of a fumarate producing strain of *R. oryzae* into an L-malate producing strain would be possible. However, the ambiguity regarding the mechanism leading to increased cytosolic fumarate hydratase activity in *R. oryzae* renders the manipulation of the organism problematic. A variety of *Aspergillus* spp. has been found to produce L-malic acid. Bercovitz *et al.* (1990) tested 13 strains, representing nine species of Aspergillus, for L-malic acid production and found yields of 1–4% (w/v). A strain of *A. flavus* (ATCC 13697) was found to be the best producer of L-malic acid, confirming the earlier results of Abe *et al.* (1962). However, this strain had some of the lowest levels of cytosolic pyruvate carboxylase and NAD-malate dehydrogenase activities of the strains tested, suggesting that flux through this portion of the metabolic pathway is not rate-limiting. Through manipulation of standard fermentation parameters (agitation, aeration, glucose, nitrogen, phosphate and metals), an efficient process for production of L-malate was developed and yields up to 128 mole per cent (95 weight per cent) were reported (Battat *et al.*, 1991). Interestingly, the addition of 50 ppb of Mn^{2+} led to a precipitous decline in acid production, which is consistent with the effect of this metal on citric acid production by *A. niger*. Unfortunately, the use of *A. flavus* has undesirable implications in the production of food grade L-malate, such as the possibility of aflatoxin contamination. Another characteristic of this organism is the production of significant quantities of succinic acid and, to a lesser extent, fumaric acid. *Aspergillus sojae* (ATCC 4625, a soy sauce producing strain) appears to be the best candidate with regard to minimal contaminant, acid production and food safety. Perhaps, a broad survey of non-aflatoxigenic *Aspergillus* spp. would prove useful in identifying an organism producing a high yield of L-malic acid and low levels of by-product acids.

REVIEW QUESTIONS

1. What are organic acids? Explain in detail about their commercial productions.

2. Write in detail about citric acid production and their product recovery process.

3. What are the commercial uses of lactic acid? Add notes on production and purification of lactic acid.

4. Explain in detail about gluconic acid production.

5. Describe in detail about the production of fumaric acid and itaconic acid.

VITAMINS

Vitamins are nutrients required in very small amounts for essential metabolic reactions in the body. The term vitamin does not refer to the large number of other nutrients that promote health including dietary minerals and essential fatty acids or amino acids.

Vitamins are biomolecules that act as both catalysts and substrates in chemical reactions. When acting as catalysts, vitamins are bound to enzymes and hence are called cofactors. For example, vitamin K forms part of proteases involved in blood clotting. Vitamins also act as coenzymes that carry chemical groups between enzymes. For example, folic acid carries various forms of carbon groups (methyl, formyl or methylene) in the cell.

Vitamins are classified as either water-soluble (capable of dissolving easily in water) or fat-soluble (absorbed through the intestinal tract with the help of lipids). Each vitamin is typically used in multiple reactions. In humans, there are thirteen vitamins: four fat-soluble (vitamins A, D, E and K) and nine water-soluble (eight B vitamins and one vitamin C). Table 22.1 provides the different types of vitamins and their chemical names.

VITAMIN A

Retinol is a yellow, fat-soluble, antioxidant vitamin important for vision and bone growth. It belongs to the family of chemical compounds known as retinoids. Retinol is usually ingested in a precursor form; milk and eggs contain retinyl esters, whereas plants (carrots, spinach) contain provitamin A carotenoids. Hydrolysis of retinyl esters would result in retinol while provitamin A carotenoids may be cleaved to produce retinal. Retinal, also known as retinaldehyde, may be reversibly reduced to produce retinol, or irreversibly oxidized to produce retinoic acid. The best described active retinoid metabolites are 11-*cis*-retinal and all-*trans* and 9-*cis*-isomers of retinoic acid. The structure of vitamin A is shown in Figure 22.1.

Table 22.1 Types of vitamins and their chemical names

Vitamin	Chemical name	Solubility
Vitamin A	Retinoids (including retinol, retinal, retinoic acid, 3-dehydroretinol and its derivatives)	Fat
Vitamin B_1	Thiamine	Water
Vitamin B_2	Riboflavin	Water
Vitamin B_3	Niacin	Water
Vitamin B_5	Pantothenic acid	Water
Vitamin B_6	Pyridoxine	Water
Vitamin B_7	Biotin	Water
Vitamin B_9	Folic acid	Water
Vitamin B_{12}	Cyanocobalamine	Water
Vitamin C	Ascorbic acid	Water
Vitamin D_2–D_4	Lumisterol, ergocalciferol, cholecalciferol, dihydrotachysterol, 7-dehydrocholesterol	Fat
Vitamin E	Tocopherol, tocotrienol	Fat
Vitamin K	Naphthoquinone	Fat

Many different geometric isomers of retinol, retinal and retinoic acid are possible as a result of either *trans* or *cis* configuration of the four double bonds found in the polyene chain. The *cis* isomers are less stable and can readily convert to all-*trans* configuration. Nevertheless, some *cis* isomers are found naturally, which carry out essential functions. For example, the 11-*cis*-retinal isomer is the chromophore of rhodopsin, the vertebrate photoreceptor molecule. Rhodopsin comprises 11-*cis*-retinal covalently linked *via* a Schiff base to the opsin protein (either rod opsin, or blue, red or green cone opsins). The process of vision relies on the light-induced isomerization of the chromophore from 11-*cis* to all -*trans*, resulting in the change of conformation and activation of the photoreceptor molecule. One of the earliest signs of vitamin A deficiency is night-blindness followed by decreased visual acuity.

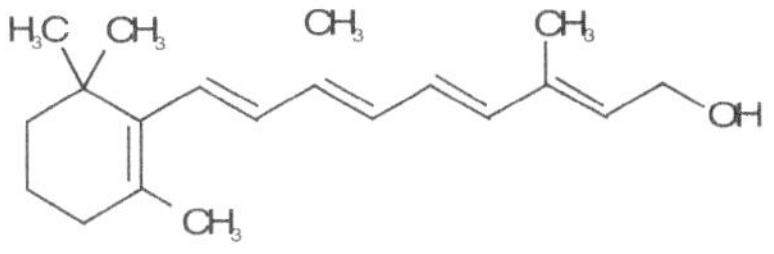

Figure 22.1   Vitamin A

B VITAMINS

Vitamin B comprises eight water-soluble vitamins, which are active in cell metabolism (Figure 22.2). Vitamin B was once considered a single vitamin, much like Vitamin C or D. Research has shown that a complex of chemically distinct vitamins often coexist in the same food. Hence the term vitamin B has gradually declined in use, being replaced by the generic term "B vitamins" or "vitamin B complex". B vitamins include the following:

- Vitamin B_1 (Thiamine)

- Vitamin B_2 (Riboflavin)

- Vitamin B_3 also vitamin P or vitamin PP (Niacin)

- Vitamin B_5 (Pantothenic acid)

- Vitamin B_6 (Pyridoxine and pyridoxamine)

- Vitamin B_7 also vitamin H (Biotin)

- Vitamin B_9 also vitamin M and vitamin Bc (Folic acid)

- Vitamin B_{12} (Cyanocobalamin)

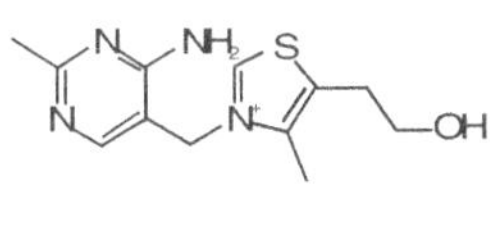

a. Thiamine

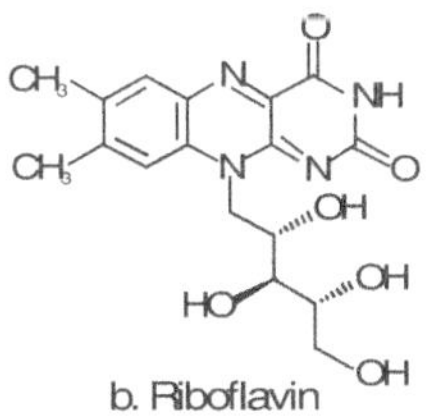

b. Riboflavin

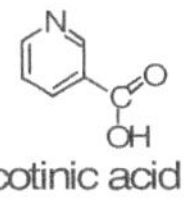

Nicotinic acid 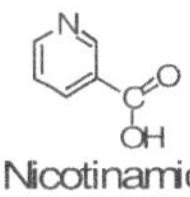   Nicotinamide

c. Niacin

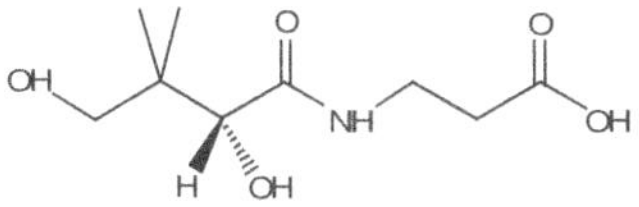

d. Pantothenic acid

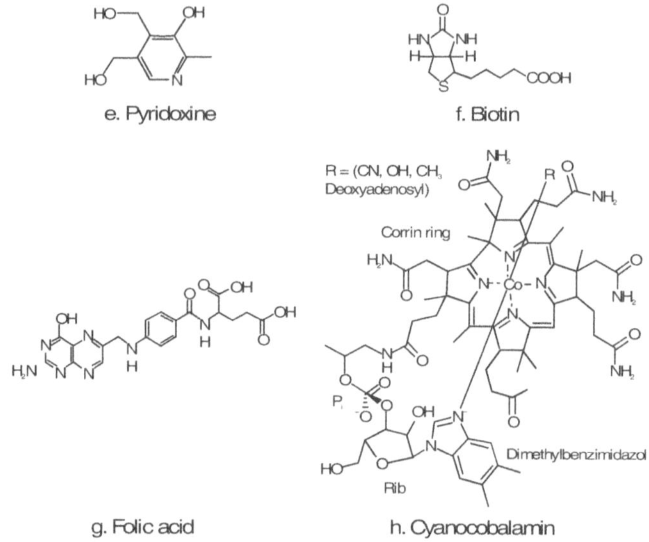

e. Pyridoxine

f. Biotin

R = (CN, OH, CH, Deoxyadenosyl)

Corrin ring

Dimethylbenzimidazol

Rib

g. Folic acid

h. Cyanocobalamin

Figure 22.2 Structure of B vitamins

VITAMIN C

Vitamin C is a water-soluble nutrient essential for life and maintaining optimal health. It is also known by the chemical name L-ascorbic acid.

Principally, vitamin C is a weak acid, called ascorbic acid, or a salt, called ascorbate. It is the L-enantiomer of ascorbic acid. The D-enantiomer shows no biological activity. However, both are mirror image forms of the same chemical molecular structure.

The active part of the substance is ascorbate ion, which can express itself as either an acid or a salt of ascorbate that is neutral or slightly basic. Commercial vitamin C is often a mix of ascorbic acid, sodium ascorbate and/or other ascorbates. Some supplements contain in part of the D-enantiomer. Figure 22.3 shows the structure of vitamin C.

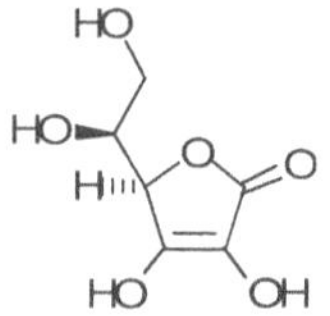

Figure 22.3 Structure of vitamin C

VITAMIN D

Vitamin D belongs to a group of fat-soluble prohormones and their metabolites and analogues. Vitamin D_2 (ergocalciferol) and D_3 (cholecalciferol) are the two major forms of vitamin D (Figure 22.4). Vitamin D_3 is produced when skin is exposed to sunlight, specifically ultraviolet B radiation. Milk and cereal grains are rich in vitamin D.

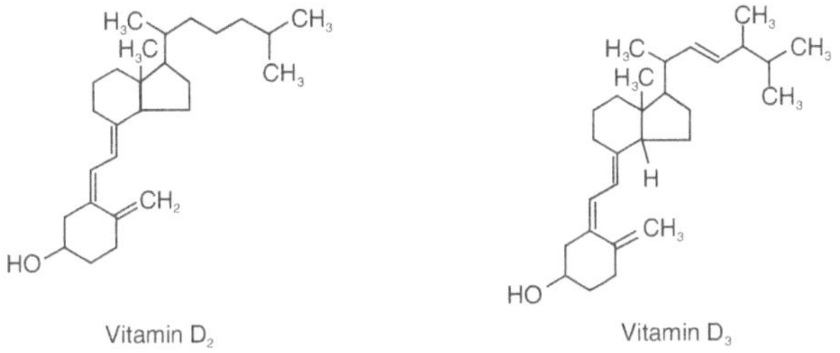

Figure 22.4   Structure of vitamin D

Ergocalciferol is derived from fungal and plant sources and is not produced in the human body. Cholecalciferol, on the other hand, is produced in the skin of animals when 7-dehydrocholesterol reacts with UVB of wavelengths 290–315 nm. These wavelengths are present in sunlight when the sun is more than 45 degrees above the horizon or when the UV index is greater than 3. Typically, 250 µg can be produced in the skin after one minimal erythemal dose of exposure, or until the skin begins to turn pink. However, longer exposure to UVB simply degrades the product as fast as it is generated.

The structural difference between vitamin D_2 and vitamin D_3 is in their side chains. The side chain of D_2 contains a double bond between carbons 22 and 23, and a methyl group on carbon 24. In most mammals, including humans, D_3 is more effective than D_2, increasing the levels of vitamin D hormone in circulation. In rats, D_2 is more effective than D_3. However, both vitamin D_2 and D_3 are used for nutritional supplementation. Pharmaceutical forms of vitamin D include calcitriol (1-alpha, 25-dihydroxycholecalciferol), doxercalciferol and calcipotriene.

VITAMIN E

Vitamin E or tocopherol (Figure 22.5) is a fat-soluble vitamin and an important antioxidant. Vitamin E is often used in skin creams and lotions because it encourages skin healing and reduces scarring after injuries like burns.

Natural vitamin E exists in eight different forms or isomers: four tocopherols and four tocotrienols. All isomers have a chromanol ring with a hydroxyl group, which can donate a hydrogen atom to reduce free radicals, and a hydrophobic side chain, which allows for penetration into biological membranes. There is an alpha, beta, gamma and delta form of both tocopherols and tocotrienols, determined by the number of methyl groups on the chromanol ring. Each form has its own biological activity in the body.

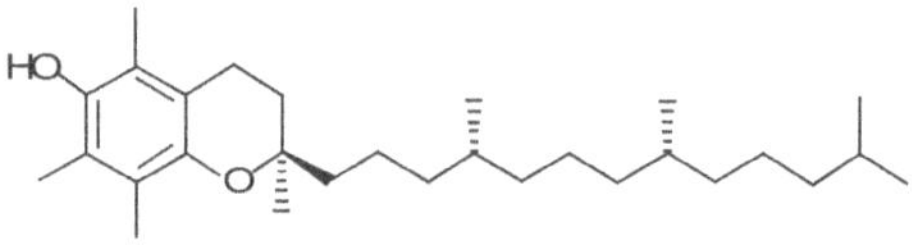

Figure 22.5   Structure of tocopherol

VITAMIN K

Vitamin K denotes a group of 2-methyl-naphthoquinone derivatives. They are human vitamins, lipophilic (soluble in lipids) and, therefore, hydrophobic (poorly soluble in water). They are needed for post-translational modification of certain proteins and for blood coagulation.

Vitamin K ('Koagulation' in German) includes a number of related compounds, which have a methylated naphthoquinone ring structure in common, and which vary in the aliphatic side chain attached at the 3-position (Figure 22.6). Phylloquinone (vitamin K_1) contains four isoprenoid residues in its side chain, one of which is unsaturated. Menaquinones have side chains composed of a variable number of unsaturated isoprenoid residues; generally, they are designated as MK-n, where n denotes the number of isoprenoids.

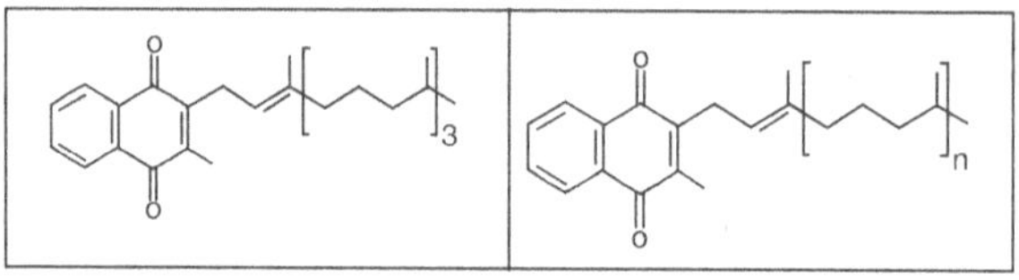

Figure 22.6   Structure of vitamin K

It is generally accepted that naphthoquinone is the functional group, so that the mechanism of action is similar for all K vitamins. However, differences may be expected with respect to intestinal absorption, transport, tissue distribution and bioavailability. These differences are caused by variance in lipophilicity of the various side chains and by different food matrices in which they occur.

PRODUCTION OF RIBOFLAVIN

Vitamin B_2 has been commercially produced by fermentation or chemical synthesis or a combination of both (Figure 22.7). Fermentation is the most recent and cost-effective method. However, manufacturers commonly use the mixed process. In this process, a four-step reaction sequence is used, starting from glucose. First, ribose is produced from glucose by fermentation. Then, a reaction with xylidine is used to convert ribose into riboside, which is then hydrogenated to produce ribamine, which is purified by crystallization. Subsequently, a reaction between ribamine and phenyl diazonium salt derived from aniline is started to produce phenylazoribitylamine. This compound is then crystallized, dried and converted into vitamin B_2 by cyclocondensation with barbituric acid.

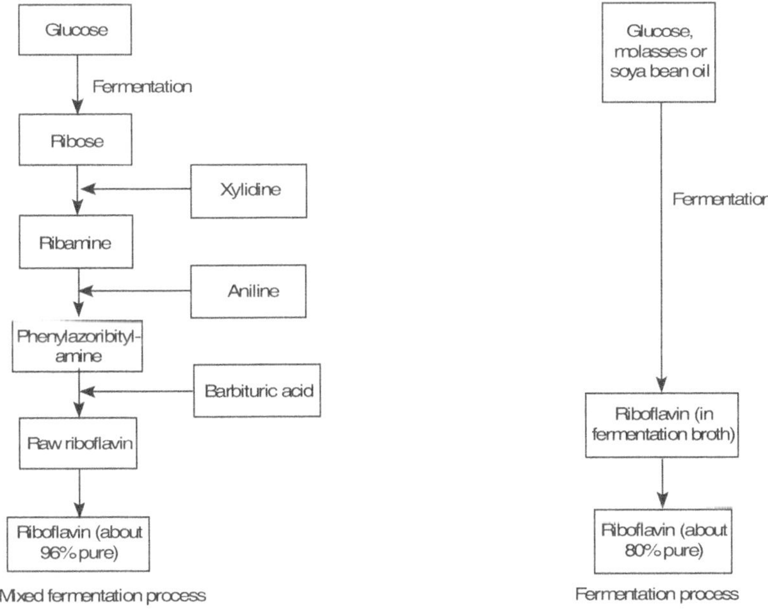

Figure 22.7   Production of vitamin B_2

Strains of Microorganisms

Although many strains of bacteria, fungi or yeasts can be used, each manufacturer may use a different organism. At present, the bacterium *Bacillus subtilis*, the fungi *Eremothecium ashbyii* and *Ashbya gossypii*, and the yeasts *Candida flareri* and *Saccharomyces cerevisiae* are commonly used in the production of vitamin B_2.

Raw Materials

Producers use different basic raw materials. Usually, vegetable oil, glucose or molasses, corn steep liquor and meat scraps are used as the sources of carbon (the significant "building block" for producing vitamin B_2). However, they vary greatly in price and carbon content. Further, enzymatically degraded collagen and lipids are used as energy sources together with corn steep liquor. The addition of peptones would give better results with bacterial and yeast strains.

Fermentation

Single-stage fermentation is commonly used, which saves substantial cost as compared to the multi-stage chemical process. The fermentation plant comprises simple mixing or stirring vessels and conventional purification systems.

Optimal fermentation conditions include aeration at about one third volume of liquid per minute and agitation at the rate of 1.0 hp/thousand litres of medium. Silicon and soybean oil are commonly used as foam control agents.

Productivity of Microorganisms

Productivity in biochemical processes is described in terms of conversion rate and space–time yield. Conversion rate is the proportion of carbon-source raw material that is converted into finished product. It dictates the amount of raw material required. Space–time yield is the amount of product made per unit volume per unit time. It determines the time duration required for fermentation and the size of vessels to be used. Conversion rate and space–time yield vary with microorganisms and raw materials used. For example, a slow organism may take many days to complete fermentation. In this case, the manufacturer will need numerous fermenting vessels. However, the productivity of organisms may be improved by natural mutation or by use of genetically modified organisms (GMOs). GM may give better results, but it adds a lot to the cost.

Recovery of Riboflavin

After separating the biomass, and evaporating and drying the concentrate, a product enriched with up to 80 per cent of riboflavin is obtained. For animal feed purpose, the pH is adjusted into 4.5 and the concentration of broth is reduced to 30%.

When a crystalline product is desired, the broth is heated at 121°C for one hour to achieve a soluble form of riboflavin. Using chemical treatment, riboflavin is crystallized and then dissolved in water or polar solvents. Further recovery of the product from polar solvents is done with acidification.

PRODUCTION OF VITAMIN C

Vitamin C is produced commercially from plants either by chemical synthesis, fermentation or by mixed fermentation method. Vitamin C may be produced in two ways: the Reichstein process or two-stage fermentation process. In the first step of either process, sorbitol is oxidized into sorbose by fermentation. Almost every manufacturer uses the same microorganism for fermentation. (Sorbitol is made by reducing glucose at a high temperature.) Neither the Reichstein process nor the two-stage fermentation process involves the use of GMOs.

Reichstein Process

The Reichstein process (Figure 22.8) is a mix of fermentation and chemical synthesis. In this process, sorbose is transformed into diacetoneketo-gluconic acid (DAKS) in a two-stage chemical process. The first step involves a reaction with acetone to produce diacetone sorbose, which is then oxidized using chlorine and sodium hydroxide to produce DAKS.

In the next step, DAKS is dissolved in a mix of organic solvents, and its structure is rearranged to form vitamin C using an acid catalyst. (This stage of the process is applicable to the two-stage fermentation process too.) As the last step in production, the crude vitamin C is purified by recrystallization. (This step is common for both processes.) Many technical and chemical modifications have now been made to the Reichstein process to optimize and shorten the various reaction routes.

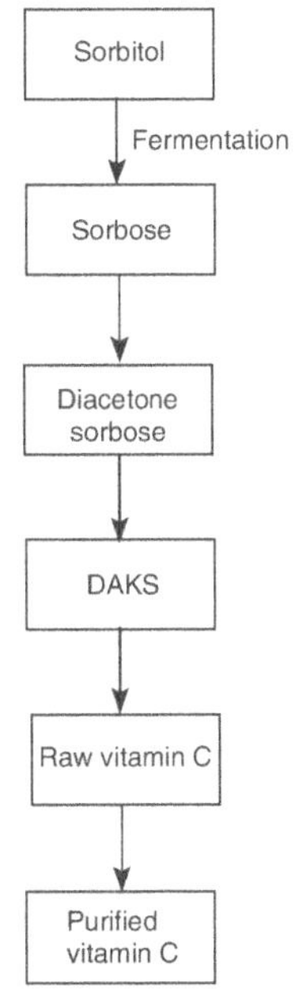

Figure 22.8 Production of vitamin C by Reichstein process

Consequently, each step now gives a yield of over 90 per cent. The overall yield of vitamin C from glucose is about 60 per cent. The Reichstein process uses considerable quantities of organic and inorganic solvents and reagents at various stages. These include acetone, sulphuric acid and sodium hydroxide. Although some of these compounds may be recycled, stringent environmental control is required, resulting in significant waste disposal overheads.

Two-stage Fermentation Process

This newer process was developed in China. In this process (Figure 22.9), the second fermentation replaces the chemical reactions used in the Reichstein method to produce DAKS. The second fermentation would result in an intermediate product called KGA. Finally, the last two steps as in the Reichstein process would complete the synthesis of vitamin C. The two-stage fermentation process makes lesser use of toxic solvents and reagents. Therefore, the cost of processing waste products is minimal.

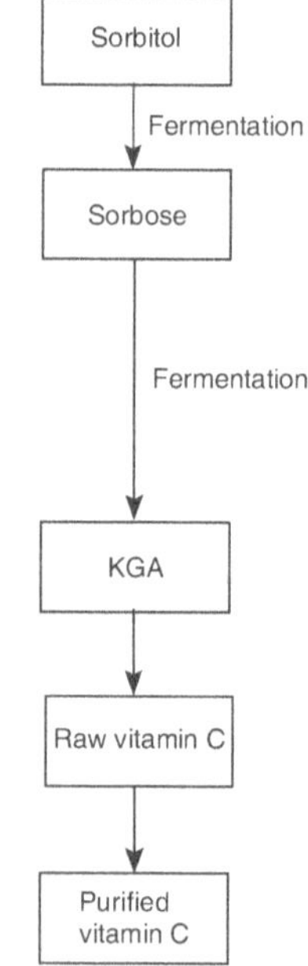

Figure 22.9   Production of vitamin C by fermentation process

Forms of Vitamin C

Usually, raw vitamins become unstable if affected by temperature, light, acidity or alkalinity. To prevent degradation, vitamins may be combined with other products, which develop into several chemical derivatives of vitamins.

Vitamin C is commercially available in many forms, including powder or granular forms and as chemical derivatives such as sodium ascorbate and calcium ascorbate. Sodium ascorbate is a preferred antioxidant as it would not discolour meat; moreover, it is almost identical to ascorbic acid in its biochemical properties. Calcium ascorbate is also commonly used in food supplements as a combined source of calcium and vitamin C.

REVIEW QUESTIONS

1. What are vitamins? Give the types of vitamins.

2. Explain in detail about types and importance of vitamins.

3. Describe in detail about production of riboflavin.

4. Explain in detail the production of vitamin C.

MICROBIAL POLYSACCHARIDES

XANTHAN

Xanthan gum is a natural polysaccharide and an important industrial biopolymer. It was discovered in 1950s at the Northern Regional Research Laboratories (NRRL) of the United States' Department of Agriculture. The polysaccharide B-1459 (xanthan gum) produced by the bacterium *Xanthomonas campestris* was extensively studied because of its properties that would allow it to supplement other known natural and synthetic water-soluble gums. Extensive research was carried out in several industrial laboratories during 1960s, culminating in semi-commercial production as Kelzan by Kelco. Substantial commercial production began in 1964. Today, the major producers of xanthan are Merck and Pfizer in the United States, RhoÃne Poulenc and Sanofi-Elf in France, and Jungbunzlauer in Austria.

Xanthan gum is a heteropolysaccharide with a primary structure consisting of repeated pentasaccharide units formed by two glucose units, two mannose units, and one glucuronic acid unit in the molar ratio 2.8 : 2.0 : 2.0. Its main chain consists of β-D-glucose units linked at the 1 and 4 positions. The chemical structure of the main chain is identical to that of cellulose. Trisaccharide side chains contain a D-glucuronic acid unit between two D-mannose units linked at the O-3 position of every other glucose residue in the main chain. Approximately one-half of the terminal D-mannose contains a pyruvic acid residue linked *via* keto group to the 4 and 6 positions with an unknown distribution. The D-mannose unit linked to the main chain contains an acetyl group at position O-6. The presence of acetic and pyruvic acids produces an anionic polysaccharide type. Chemical structure of xanthan is shown in Figure 23.1.

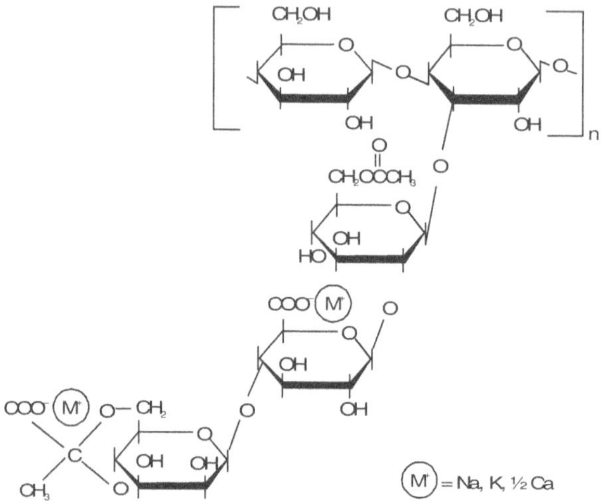

Figure 23.1 Chemical structure of xanthan

XANTHOMONAS

The trisaccharide branches appear to be closely aligned with the polymer backbone. The resulting stiff chain may exist as a single, double or triple helix, which interacts with other polymer molecules to form a complex. The molecular weight distribution ranges from 2×10^6 to 20×10^6 Da. This depends on the association between chains, forming aggregates of several individual chains. The variations of the fermentation conditions used in production are factors that can influence the molecular weight of xanthan. The polysaccharides produced by *Xanthomonas* are listed in Table 23.1.

Table 23.1 Composition of polysaccharides produced by *Xanthomonas*

Bacteria	D-Glucose	D-Mannose	D-Glucuronic acid	Pyruvate	Acetate
X.campestris	30.1	27.3	14.9	7.1	6.5
X.fragaria 1822	24.6	26.1	14.0	4.9	5.5
X.gummisudans 2182	34.8	30.7	16.5	4.7	10.0
X.juglandis 411	33.2	30.2	16.8	6.9	6.4
X.phaseoli 1128	30.9	28.6	15.3	1.8	6.4
X.vasculorum 702	34.9	30.2	17.9	6.6	6.3

Source: Kennedy and Bradshaw, 1984.

Xanthan gum is highly soluble in both cold and hot water, and this behaviour is related to the polyelectrolyte nature of the xanthan molecule. Xanthan solutions are highly viscous even at low polymer concentrations. These properties are useful in many industrial applications, especially in the food industry where xanthan is used as a thickener and to stabilize suspensions and emulsions. The thickening ability of xanthan solutions is related to its viscosity; however, a high viscosity resists the flow. Xanthan solutions are pseudoplastic, or shear thinning, and the viscosity decreases with increasing shear rate. The viscosity also depends on the temperature (both dissolution and measurement temperatures), biopolymer concentration, concentration of salts, and pH. The properties of xanthan are listed in Table 23.2.

Table 23.2   Properties of xanthan

Property	Value
Physical state	Dry, cream-coloured powder
Moisture (%)	8–15
Ash (%)	7–12
Nitrogen (%)	0.3–1
Acetate content (%)	1.9–6.0
Pyruvate content (%)	1.0–5.7
Monovalent salts (gL^{-1})	3.6–14.3
Divalent salts (gL^{-1})	0.085–0.17
Viscosity (cP)	13–35
(Melting point $T_M = 25°C$)	

APPLICATIONS OF XANTHAN GUM

Xanthan gum is widely used in foods, toiletries, oil recovery and cosmetics. It is non-polluting. Table 23.3 lists the various applications of xanthan gum.

In agriculture, xanthan has been used to improve the flow ability in fungicide, herbicide and insecticide formulations by uniformly suspending the solid component. The unique rheological properties of xanthan gum solution reduce drift and increase pesticide cling and permanence. Because of its ability to disperse and hydrate rapidly and give a good colour yield, xanthan is also used in jet injection printing.

Table 23.3 Industrial applications of xanthan gum

Applications	Concentration (% w/w)	Functionality
Salad dressings	0.1–0.5	Emulsion stabilizer, suspending agent, dispersant
Dry mixes	0.05–0.2	Eases dispersion in hot or cold water
Syrups, toppings, relishes, sauces	0.05–0.2	Thickener, heat stability and uniform viscosity
Beverages (Fruit and non-fat dry milk)	0.05–0.2	Stabilizer
Dairy products	0.5–0.2	Stabilizer, viscosity control of mix
Baked goods	0.1–0.4	Stabilizer, facilitates pumping
Frozen foods	0.05–0.2	Improves freeze, thaw stability
Pharmaceuticals (creams and suspensions)	0.1–1.0	Emulsion stabilizer, uniformity in dosage formulations
Cosmetic (denture cleaners, shampoos, lotions)	0.2–1.0	Thickener and stabilizer
Agriculture (additive in animal feed and pesticide formulations)	0.03–0.4	Suspension stabilizer, improved sprayability, reduced drift, increased cling and permanence
Textile printing and dyeing	0.2–0.5	Control of rheological properties of paste, preventing dye migration
Ceramic glazes	0.3–0.5	Prevents agglomeration during grinding
Slurry explosives	0.3–1.0	Thickens formulations, improves heat stability (in combination with guar gum)
Petroleum production	0.1–0.4	Lubricant or friction reduction in drill-hole
Enhanced oil recovery	0.05–0.2	Reduces water mobility by increasing viscosity and decreasing permeability

Recently, in the formulation of new generations of thermoset coatings, xanthan gum has been introduced to meet the challenges of producing environmental friendly products. In petroleum industry, xanthan gum is used in oil drilling, fracturing, pipeline cleaning, and fixing.

Xanthan gum has excellent compatibility with salt and is resistant to thermal degradation. Hence, it is useful as an additive in drilling fluids. The pseudoplasticity of its solutions would provide low viscosity at the drill bit where the shear rate is high, and high viscosity in the annulus where shear is low. Therefore, xanthan would serve a dual purpose by allowing faster penetration at the bit and suspending cuttings in the annulus. It is believed that when every barrel of oil is produced, double the amount of it is in unrecovered condition. So, the enhancement oil recovery (EOR) will be an important use of xanthan gum in future. The basic principle applied is to improve the separation of water and oil, thereby increase oil recovery. However, the quality of xanthan gum is of critical consideration because high impurities would make refining the oil more difficult. Xanthan gum is used in micellar-polymer flooding as a tertiary oil recovery operation. In this application, polymer-thickened brine is used to drive the slug of the surfactant through the porous reservoir rock to mobilize residual oil; the polymer prevents bypassing of the drive water through the surfactant band and ensures good area sweeping. In both applications, the function of polymers is to reduce the mobility of injected water by increasing its viscosity.

Other special applications that employ the xanthan gel are in removing rust, welding rods, wet slag, and cleaning other debris from gas pipelines. Many more applications of xanthan gum could be expected to be developed. These wide range of applications of xanthan gum may be attributed to its properties of (1) non-Newtonian behaviour, (2) high viscosity yield even at low concentrations (600–2000 ppm), (3) low sensitivity of viscosity to salinity changes, (4) resistance to mechanical degradation, (5) stable with respect to temperature (up to 90°C), and (6) a biodegradable material and hence environment friendly.

STRAIN OF *X. CAMPESTRIS* SP.

Xanthomonas is a genus of the Pseudomonaceae family. All organisms in this genus are plant pathogens. *Xanthomonas pathovars* infect a large selection of plants including crops, e.g. cabbage, alfalfa and beans. *Xanthomonas* cells occur as single straight rods, 0.4–0.7 mm wide and 0.7–1.8 mm long. The cells are motile, gram-negative, and they have a single polar flagellum (1.7 – 3 mm long). The microorganism is chemiorganotrophic and an obligate aerobe with a strictly respiratory type of metabolism that requires oxygen as the terminal electron acceptor. The bacterium cannot denitrify, and it is catalase-positive and oxidase-negative. The colonies are usually yellow, smooth and viscid. *Xanthomonas* sp. are able to oxidize glucose, and the Entner–Doudoroff pathway is predominantly used for glucose catabolism (the pentose phosphate pathway also occurs but uses only 8–16% of the total glucose consumed); both tricarboxylic acid and glyoxylate cycles are present. All the media employed for *X. campestris* growth are complex. The most commonly used are the yeast–mannitol medium and a semi-synthetic variant of the YM designated as YM-T. The growth is quite similar in both media, and the maximum biomass yields obtained are quite close for both; however, because of the two nitrogen sources present in YM-T, a diauxic growth pattern is obtained in this medium.

PRODUCTION OF XANTHAN

The culture environment and the operational conditions influence both the xanthan yield and the structure of xanthan produced. Some of these effects are discussed here.

Production Medium

To produce xanthan gum, *X. campestris* needs several nutrients, both micronutrients (e.g. potassium, iron and calcium salts) and macronutrients such as carbon and nitrogen. Glucose and sucrose are the most frequently used carbon sources. The concentration of carbon source affects the xanthan yield; a concentration of 2–4% is preferred. Higher concentrations of these substrates inhibit growth. Nitrogen, an essential nutrient, can be provided either as an organic compound or as an inorganic molecule. The C : N ratio used in production media is usually less than that used during growth.

Generally, lower concentrations of both nitrogen and carbon are suitable to produce the xanthan polymer. The best carbon sources are sugars (glucose and sucrose) and the best nitrogen source is glutamate at a concentration of 15 mM (higher concentrations inhibit the growth). Small quantities of organic acids (e.g. succinic and citric) when added to the medium enhance production. Nitrogen, phosphorous and magnesium influence growth whereas nitrogen, phosphorous and sulphur influence the production of xanthan. The optimal production medium composition is as follows: sucrose (40 gL^{-1}), citric acid (2.1 gL^{-1}), NH_4NO_3 (1.144 gL^{-1}), KH_2PO_4 (2.866 gL^{-1}), $MgCl_2$ (0.507 gL^{-1}), Na_2SO_4 (0.089 gL^{-1}), H_3BO_3 (0.006 gL^{-1}), ZnO (0.006 gL^{-1}), $FeCl_3 \cdot 6H_2O$ (0.020 gL^{-1}), $CaCO_3$ (0.020 gL^{-1}) and concentrated HCl (0.13 ml 0.13 ml/litre);

the pH is adjusted to 7.0 by adding NaOH. Figure 23.2 depicts the production process of xanthan.

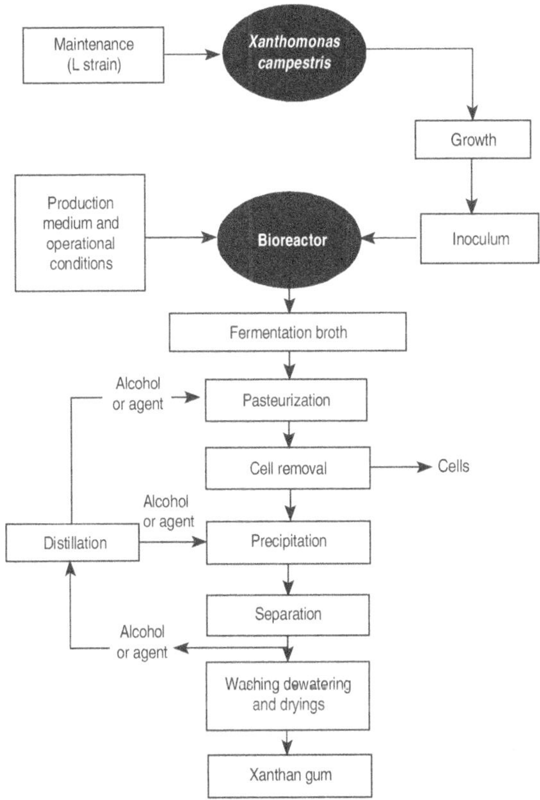

Figure 23.2   Flow chart of xanthan production

Inoculum Build-up

Because xanthan gum constitutes the bacterial capsule, its production is growth-associated. The aim of inoculum build is to increase cell concentration but it minimizes the production of xanthan because xanthan around the cells impedes mass transport of nutrients and extends the lag phase of growth. Suppressing xanthan production while building up the cell mass requires multiple stages of inoculum development.

The microorganism is transferred from a complex solid culture medium (usually YM agar) to a small volume (5 or 7 ml) of a complex liquid medium (usually YM), but the incubation is limited to 7 hours to prevent significant production of xanthan. This culture is transferred to 40–100 ml of a medium containing some inorganic salts (a semi-synthetic medium); the cells adapt to the new conditions that will be encountered in the production phase. The inoculum volume for the production fermenter is between 5 and 10% of the total broth volume in the vessel. The number of stages needed to build-up the inoculum while suppressing xanthan synthesis increases with the volume of the production bioreactor.

Temperature

Temperatures employed for xanthan production range from 25 to 34°C, but culture at 28°C and 30°C is quite common. The influence of temperature on xanthan gum production has been widely studied. It has been showed that 28°C is the optimal production temperature, and higher culture temperature increases xanthan production but lowers its pyruvate content. Some reports state that optimum process temperature is 33°C, proposing a temperature of 25°C for growth and 30°C for production. For high xanthan yield, a temperature between 31 and 33°C is recommended, but culture at 27–31°C is better at attaining a high pyruvate content in the gum. Some other reports conclude that the optimal temperature for xanthan production depended on the production medium used.

pH

Most authors agree that neutral pH is optimum for the growth of *X. campestris*. During xanthan production, the pH decreases from neutral pH to close to 5 because of acid groups present in xanthan. Some authors suggest that pH control is not necessary for this process, but others recommend control at neutral pH using alkalis such as KOH, NaOH and $(NH)_4OH$. A study of the pH effects showed that pH control enhanced cell growth but had no effect on xanthan production. When pH is controlled, xanthan production ceases once the stationary growth phase is attained and this effect is independent of the alkali used to control the pH. When pH is not controlled, the gum production continues during the stationary phase of growth.

Oxygen Mass Transfer Rate

Various types of bioreactors have been used to produce xanthan gum, but the sparged stirred tank is employed most frequently. In stirred reactors, the rate of oxygen mass transfer is influenced by the air flow rate and the stirrer speed. When stirred tanks are used, air flow rate is generally maintained at a constant value, usually 1 L/L/min. In contrast, the agitation speed used varies over a broad range. Some authors have used a constant speed, while others have varied the speed during fermentation.

When the stirrer speed is constant at < 500 rpm, the production of xanthan is reduced because oxygen mass transfer became limiting with increasing viscosity of the broth. When stirrer speed is held constant at > 500 rpm, the xanthan production is also poor because the cells were adversely affected by intense mechanical agitation. To deal with this problem, the stirrer speed is varied during culture from lower values (200–300 rpm) at initiation of the fermentation to higher values later on. Similar effects of excessive agitation have been reported for numerous other fermentations.

RECOVERY OF XANTHAN GUM

The recovery of xanthan from the fermentation broth is generally difficult and expensive. The final fermentation broth contains xanthan, cells and residual nutrients, and other metabolites. Because of a high xanthan concentration, the broth is highly viscous and difficult to handle. A high viscosity complicates biomass removal from the broth. In addition, mixing of the clarified broth with recovery reagents is power consuming because of viscosity. For processing, the broth is usually diluted at some stage of the process.

The specific purification method used is determined by the end user of the gum and the cost effect. The use of xanthan in enhanced oil-recovery requires removal of particles such as cells that could clog up the porous oil-bearing rock. Xanthan used as a food additive should be free of biomass and reagents used in the recovery process. Product specifications are less restrictive for uses such as textile processing. The main steps of the recovery process are deactivation and removal (or lysis) of microbial cells, precipitation of the biopolymer, dewatering, drying and milling. Processing must be done without degrading the biopolymer. The final product is usually a dry powder or a concentrated solution. Numerous methods have been developed to deactivate, lyse or remove cells from the broth. Chemical treatment (with alkali, hypochlorite, enzymes), mechanical treatment or thermal treatment is used. Chemical treatment at elevated pH can cause depyruvylation of the product. When enzymes are used, they must be removed from the medium and this adds to cost. Usually, the fermentation broth is pasteurized or sterilized to kill the cells. These thermal treatments enhance xanthan removal from the cells. Pasteurization of the fermentation broth at a high temperature often causes thermal degradation of the microbial exopolysaccharides. When the broth is treated under proper conditions (80–130°C, 10–20 minutes, pH 6.3–6.9), enhanced xanthan dissolution occurs without thermal degradation and disruption of cells. The increased temperature also reduces the viscosity of the broth to make removal of insolubles easy by centrifugation or filtration.

For highly viscous xanthan broths, viscosity reduction must precede filtration. Viscosity is reduced by dilution or heating. The fermentation broth is usually diluted with water, alcohol or mixtures of alcohol and salts in quantities lower than those needed for xanthan precipitation. The diluted and/or heated broth is filtered to remove the solids. Filtration is improved in the presence of alcohol. Xanthan in solution can be viewed as a hydrophilic colloid forming a true solution in water. Precipitation of polymer is achieved by decreasing the solubility of the dissolved colloid using methods such as addition of salts, water-miscible non-solvents, and concentration by evaporation. Recovery options that have been studied include precipitation with organic solvent such as ethanol and isopropyl alcohol (IPA); the use of mixtures of salts and alcohol; precipitation with trivalent or tetravalent salts; and use of ultrafiltration. The most common technique used for primary isolation and purification of polysaccharides is precipitation using water miscible non-solvents such as alcohols.

The cost of alcohol for recovery and the inevitable losses contribute significantly to the total cost of production. The knowledge of mechanisms controlling phase separation is useful for devising alternatives to alcohol precipitation and for determining the conditions under which alcohol usage can be minimized. The lower alcohols (methanol, ethanol, isopropanol) and acetone, which are non-solvents for the polysaccharide, can be added to the fermentation broth not only to decrease the solubility until phase separation occurs, but also to wash out impurities such as coloured components, salts and cells. The quantity needed depends on the nature of the reagent. Total precipitation of the gum is possible only when 3 volume of IPA or acetone is added per volume of the broth. If lower alcohols such as ethanol are used, 6 volume of alcohol is needed per broth volume.

The addition of salts in sufficient concentration causes precipitation or complex coacervation due to ion binding of the cations of the added salt to the ionized groups on the polyanion. This leads to charge reversal at the instance when all the available anionic groups are bound to a cation. Polyvalent cations such as calcium, aluminum and quaternary ammonium salts are especially effective in precipitation. Precipitation does not occur with monovalent salts such as sodium chloride. The addition of a non-solvent reagent promotes precipitation not only by decreasing the water affinity of the polymer, but also by enhancing the binding of the cations, which are present. Thus, xanthan precipitates with lesser amounts of reagents when alcohol and salt are used in combination. When xanthan is precipitated using a combination of salts and IPA, the quantity of the alcohol needed is lower than if only IPA is added. Alcohol volume is reduced when monovalent salts are used, but volume reduction is greater with divalent salts. However, the use of divalent cations leads to a less soluble xanthan salt as the final product.

The polysaccharide concentration in solution also influences the volume of the precipitating agent needed. When the polymer concentration in solution is increased, a smaller quantity of alcohol is needed for precipitating the biopolymer. Once the polymer is obtained as a wet precipitate, it is dried, milled and packed. The precipitate is dried in batch or continuous dryers under vacuum or with forced circulation of an inert gas. This prevents the combustion of organic solvent in the precipitate. Most commercial xanthans have a final moisture content of about 10%. After drying, the polymer can be milled to a predetermined mesh size to control dispersability and dissolution rates. Some commercial xanthan gums are only differentiated by mesh size. Care must be taken in milling so that excessive heat does not degrade or discolour the product. Finally, the packing used must be waterproof because xanthan is hygroscopic and subject to hydrolytic degradation.

COMMERCIAL PRODUCTION PROCESSES

Most commercial production of xanthan gum uses glucose or invert sugars, and most industries prefer batch instead of continuous fermentation for reasons of quality assurance and ease of control. A typical commercial production process starts with inoculums of *X. campestris* that are prepared in suitable fermentation medium in conventional batch processing using mechanically agitated vessels. The aerated culture that undergoes aerobic process is held at the following operating conditions: temperature approximately 28–30°C, pH 7, aeration rate higher than 0.3 (v/v), and specific power input for agitation higher than 1 kW/m^3. The fermentation process is carried out for about 100 hours to convert approximately 50% of the glucose into the final product that require a set of reactors of capacity ranging from 10–100 litres. The size of the inoculum preparation is based on the size of the production. During fermentation, the cells would grow exponentially resulting in rapid consumptions of nitrogen. After fermentation, the multi-step downstream processes would follow.

When industrial grade xanthan is produced, the post-fermentation process treatment may be started with pasteurization of the fermented broth to sterilize bacteria and to deactivate the enzymes. This process uses a large amount of alcohol to precipitate the xanthan gum, and the precipitated xanthan gum is sprayed dry or resuspended on water and then reprecipitated. When cell-free xanthan gum is required, cell centrifugation is facilitated by diluting the fermentation broth to improve cell separation. Cell separation by dilution process from highly viscous xanthan solution is a cost-intensive process. A favoured method is by adding alcohol and salt to improve precipitation by creating reverse effect charges. The xanthan gum obtained in wet solid form would undergo dewatering and washing to obtain the final purity required. Alcohol used for xanthan precipitation is recovered by distillation column. Washing may be carried out to improve the quality of the product, and hence it would entirely remove particulate matters such as cell debris, microgels, organic residues and pigments. Concentrated xanthan gum is then redissolved and washed with water or KCl to reduce viscosity, precipitated, and dewatered again until a product of satisfactory quality is achieved. Finally, the precipitated xanthan gum is spray-dried in batch or continuous driers. The dry xanthan gum is milled to the desired mesh size for control of disperse ability and the dissolution rate as well as to get the handling much easier. Figure 23.3 depicts the process involved in the commercial production of xanthan.

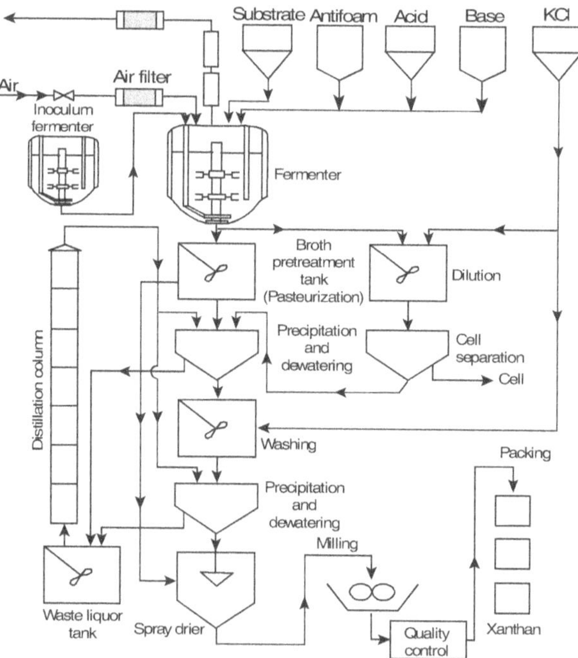

Figure 23.3 Flow chart of industrial production of xanthan

Batch versus Continuous Process

Although batch culture is commercially preferred, a problem in its operation is that the environment for cell growth keeps changing throughout the "growth cycle" and may cause adverse effects, such as toxic products, extreme pH and exhaustion of nutrients. On the other hand, in continuous culture, the growth medium is continuously supplied to the culture vessel, and extreme conditions will not occur because the medium is continuously diluted and removed from the vessel. Nevertheless, the continuous process gives a cost competitive system, and with suitable growth conditions considerable yields of polysaccharides can be maintained for more than 2000 hours. Thus, the continuous process could be a better choice. Although conventional methods can be improved by continuous fermentation, there is still a classic problem that the product may contain cells and cell debris, which contribute to lower filterability of the xanthan solution and limit its application. The production of cell-free xanthan gum is therefore desirable. In 1966, Esso Production Research Company, US found that continuous film fermentation reactions can be readily carried out by continuously depositing a suitable substrate on the surface of a rotating drum or moving belt, and applying a culture containing selected microorganisms to the film, and then continuously removing the fermentation product after sufficient residence time. Tests have shown that such a process makes it possible to use the substrate in higher concentrations, permits effective utilization of the substrate, reduces the time required for carrying out the fermentation reaction, minimizes variation in product quality, and simplifies the recovery of fermentation products.

REVIEW QUESTIONS

1. What is xanthan? Give the properties of xanthan.

2. Give some of the applications of xanthan.

3. Explain in detail about production and product recovery of xanthan.

4. How is the xanthan produced commercially? Explain the process.

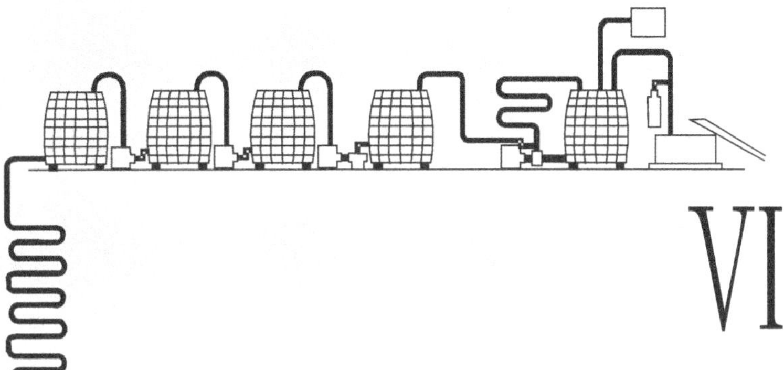

VI

AGRICULTURAL BIOPROCESSING

BIOFERTILIZERS

Biofertilizers are living fertilizer compounds of microbial inoculants or groups of microorganisms that can fix atmospheric nitrogen or solubilize phosphorus, decompose organic materials or oxidize sulphur in soil. They enhance the growth of plants, improve soil fertility, and reduce pollution.

A variety of nitrogen-fixing microorganisms are present in nature. They are broadly classified into three types:

- Symbiotic microorganisms (e.g. Legume-*Rhizobium*, Symbiosis)
- Asymbiotic or free living (e.g. *Azotobacter*, blue green algae)
- Associative symbiosis (e.g. *Azospirillum*)

In addition to nitrogen, these microorganisms supply a considerable amount of organic matter to the soil. Hence, biofertilizers or microbial inoculants may be defined as preparations that contain living or latent cells of efficient strains of nitrogen fixing, phosphorous-solubilizing or cellulolytic microorganisms.

TYPES OF BIOFERTILIZERS

The following are some important types of biofertilizers.

Rhizobium biofertilizers   In leguminous plants, the *Rhizobium* bacteria fix atmospheric nitrogen on the nodules formed on the roots of plants. These nodules are considered tiny nitrogen production factories in the plant. In the laboratory, a pure and efficient strain of *Rhizobium* can be multiplied on a suitable medium by shake technology or fermentation.

24

Azotobacter biofertilizers   *Azotobacters* are free-living microorganisms that grow in the rhizosphere. They fix atmospheric nitrogen non-symbiotically and make it available particularly to cereals. In addition, these bacteria produce substances that promote plant growth and yield. In the laboratory, a highly efficient strain of *Azotobacter* is grown either as shake culture or in a fermenter.

Azospirillum **biofertilizers**   The *Azospirillum* is called associative endosymbiont because it is having ability to fix the nitrogen with or without association of plants. Mostly the *Azospirillum* is having symbiotic association with grass and other similar plants. They fix atmospheric nitrogen and supply growth hormones and vitamins in host plants. These bacteria are commonly used in the preparation of commercial inoculants on a large scale.

Blue-green algal biofertilizers   Blue-green algae are considered as an important group of microorganisms capable of fixing atmospheric nitrogen non-symbiotically in the heterocyst cells in paddy. There is variation among different species of blue-green algae in the quality of nitrogen fixation. Efficient strains of blue-green algae are used in the preparation of inoculum. The production of inoculum in artificially controlled conditions is more expensive. On the other hand, the open air soil culture is most simple, less expensive and easily adaptable by farmers. This is based on the use of multi-strain inoculum of *Aulosira, Tolupothrix, Scytonema, Nostoc* and *Anabaena* as a starter culture.

Azolla biofertilizers   *Azolla* is a water fern in which the nitrogen-fixing blue-green algae Anabaena grows. It contains 2–3% nitrogen when wet and produces organic matter in soil. This type of biofertilizer is used all over the world. *Azolla* can be grown in a cool region. So, there is a need to develop a strain tolerant to high temperature and salinity and resistant to pests and diseases. However, the production technology can be easily adopted by paddy farmers. The only constraint in *Azolla* production is that it has to be kept growing on water, which is scarce in summer.

Phosphorus-solubilizer (Biofertilizer)   Phosphorus is one of the important elements required for plant growth and yield. This element is also essential for nodulation by *Rhizobium*. Phospho-microorganisms including bacteria and fungi are used commercially. Mycorrhizae have high potential of phosphorus accumulation in plants. There are two types of mycorrhizae, namely endo- and ectotrophics. Vesicular arbuscular (VA) mycorrhizae are most popular among the mycorrhizae.

Table 24.1 List of biofertilizers commonly produced in india

Name	Crops suited	Benefits	Remarks
Rhizobium	Legumes like pulses, groundnut, soybean	10–35% yield increase, 50–200 kg N/ha.	Fodders give better results. Leaves residual N in the soil.
Azotobacter	Soil treatment for non-legume crops including dry land crops	10–15% yield increase, 20–25 kg N/ha	Also controls certain diseases.
Azospirillum	Non-legumes like maize, barley, oats, sorghum, millet, sugarcane, rice etc.	10–20% yield increase	Enriches fodder response. Produces growth promoting substances. It can also be applied to legumes as co-inoculants.
Phosphate solubilizers	Applicable for all crops	5–30% yield increase	Can be mixed with rock phosphate.
Blue-green algae and *Azolla*	Rice/wet lands	20–30 kg N/ha; Azolla can give biomass up to 40–50 tonnes and fix 30–100 kg N/ha.	They produce growth promoting hormones; reduces soil alkalinity; can also be used as fish feed.
Mycorrhizae (VAM)	Many trees, some crops, and some ornamental plants	30–50% yield increase, enhances uptake of P, Zn, S and water.	Usually inoculated to seedlings.

MICROBIAL INOCULANTS

The activities of nitrogen fixation, mobilization of plant nutrients and degradation of lignocellulosic wastes are being carried out in soil by a large number of microorganisms. Artificially multiplied cultures of select microorganisms augment the natural recycling of organic resources. There are different types of microbial inoculants.

Nitrogen Fixers

1. *Symbiotic   Rhizobium,* inoculants for legumes

2. *Non-symbiotic*   For cereals, millets and vegetables

Bacteria

1. *Aerobic   Azotobacter, Azomonas, Azospirillum*

2. *Anaerobic   Clostridium, Chlorobium*

3. *Facultative anaerobes   Bacillus, Escherichia*

Blue-green algae   *Anabaena, Anabaenopsis, Nostoc*

ADVANTAGES OF BIOFERTILIZERS

Biofertilizers enhance the nutrient availability to crop plants and impart better health to plants and soil. Usually, biofertilizers do not create problems like salinity, alkalinity, soil erosion, etc. Vast areas of lands with low agricultural output can be turned into lands of substantial production.

PRODUCTION OF BIOFERTILIZERS

The process of manufacturing biofertilizers involves the following:

1. Selection of a suitable strain of the organism

2. Mass multiplication

3. Mixing the culture with carrier material

4. Packing

Culture Selection and Maintenance

The pure mother cultures of various strains are available in agricultural universities and regional biofertilizer labs. There are also international sources of supply like NifTAL, International Rice Research Institute, Phillipines (IRRI), etc. The mother culture of desired strain is obtained and is further subcultured and maintained purely for mass production by adopting standard techniques under the supervision of trained microbiologists.

Culture Augmentation

The culture is mass multiplied in two levels: (i) primary stage using shakers in flasks and (ii) secondary stage multiplication in fermenters. An important factor here is the preparation of growing medium. The composition of standard media is available for *Rhizobium* and other cultures. After the medium is prepared and sterilized, it is inoculated with 5% of starter cultures prepared from shake flask fermentation. The growing medium (broth) is continuously aerated by passing sterile air from compressors. After 3 or 4 days of fermentation, the broth will be ready for packing in a carrier material. The quality is tested by drawing samples at various stages.

Carrier Sterilization

While the broth is getting ready in the fermenter, the carrier material, which is usually the carbon source for the culture to survive, is sterilized in autoclaves and kept ready for mixing with broth. Peat is considered the best source of carrier material. However, it is costly, and lignite is used extensively in India. The carrier may be either sterilized in bulk or packed and sterilized.

Mixing and Packing

There are three methods commonly used.

1. In a non-sterile system, the broth is harvested from the fermenter into a sterilized carrier. Mixing is done manually under aseptic conditions and packed in polythene bags of desired quantity.

2. In a slightly upgraded system, the broth and sterilized carrier are mixed mechanically in a blender and the material is packed using semi-automatic machines.

3. In a completely sterile system, the carrier is taken in autoclavable polypropylene bags and pre-sealed and the broth from fermenter is directly injected into them with the help of dispensers. The injection hole is immediately sealed. The packets are kept in an incubation room for about a week.

AZOTOBACTER

These are free-living bacteria that grow well on a nitrogen-free medium. These bacteria utilize atmospheric nitrogen for cell protein synthesis. As soon as they die, this cell protein is mineralized in soil, thereby contributing towards the nitrogen availability of crop plants.

Azotobacters are known to fix on an average 10 mg of N/g of sugar in pure culture on a nitrogen-free medium. A maximum of 30 mg N fixed per gram of sugar was reported. However, *Azotobacter* poorly competes for nutrients in soil. Most efficient strains of *Azotobacter* would oxidize about 1000 kg of organic matter for fixing 30 kg of N/ha. This does not sound realistic for our soils, which have very low active carbon status. Besides, soils are inhabited by a large variety of other microbes, all of which compete for active carbon.

PRODUCTION OF *AZOTOBACTER*

Mother Culture

A pure growth of an organism on a small scale is called mother culture. The mother culture is always prepared in a conical flask of 500 or 1000 ml capacity, which is used for further production. In a one-litre conical flask, 500 ml of broth of nitrogen-free medium is added, and the flask is plugged with non-absorbent cotton and sterilized in an autoclave for 15–20 minutes at 75 lbs pressure. The flask is then inoculated with the mother culture aseptically with the help of inoculating needle. The flask is transferred to a shaker, and shaking is done for 72–90 hours so as to get optimum growth of bacteria in broth. The bacteria multiply by binary method (cell division). After 90 days, the number per millilitre comes to 100 crores. The starter culture or mother culture should be carefully done because the quality of the biofertilizer depends on how the mother culture is prepared.

Production on a Large Scale

Azotobacter may be multiplied on a large scale in a fermenter and a shaker. In the fermenter method, the medium is taken in a fermenter and sterilized. The pH of the medium is adjusted and 1% mother culture is added. The temperature and oxygen supply is adjusted so that a concentrated broth is made. The concentrated broth of the culture is then mixed with a carrier previously sterilized and the biofertilizer is prepared. Depending on the volume of production, a suitable fermenter is selected.

In shake method, a suitable medium is prepared and transferred to a conical flask of suitable capacity. The flask is sterilized in an autoclave at 15 lbs pressure for 15 minutes and then inoculated with 10-ml mother culture and kept in a shaker for 72–90 hours for multiplication. The broth is mixed with a suitable carrier previously sterilized. Finally, the biofertilizer is prepared, filled in plastic bags and stored.

Selection of the Carrier

A carrier is a substance that has high organic matter and water holding capacity and supports the growth of organisms. In order to make the biofertilizer easy to use, a suitable carrier is selected. Lignite coal, compost and peat soil are suitable carriers for *Azotobacter*. However, lignite is most suitable for this organism, because it is cheaper, keeps organism living for a longer period and does not lower the quality of biofertilizers.

Lignite is ground into a fine powder with a grinding machine. Its fineness should be 250–300 mesh. The pH of the carrier is adjusted to neutral by adding $CaCO_3$. Lignite is sterilized in an autoclave at 30 lbs pressure for 30 minutes. After this, the broth is mixed with lignite.

Sterilized galvanized trays are used for this. Lignite is transferred to these trays and broth is added (lignite 2: broth 1) and mixed properly. The trays are kept one above the other for 10–12 hours, allowing the organisms to multiply in the carrier. The mixture is then filled in plastic bags of 250 g or 500 g capacity. The plastic bags are properly sealed.

As per the Bureau of Indian Standards (BIS) standards, one gram of biofertilizer should have one crore cells of bacteria soon after it is prepared, and 15 days before expiry date, one gram of biofertilizer should have 10 lakh bacteria. If the biofertilizer is stored at 15–20°C, it will remain active for 6 months. However, at 0 to 4°C, the bacteria will remain active for 2 years. The storage periods are decided after testing the biofertilizer for storage conditions, temperature and humidity.

BLUE-GREEN ALGAE

The blue-green algae (BGA) are photosynthetic, prokaryotic microorganisms. Their main photosynthetic pigments are chlorophyll-*a*, carotenes, xanthophylls, together with phycobiliproteins, *c*-phycocyanin (blue) and *e*-phycoerythrin (red). Due to the presence of these pigments and mucilage, the colour of BGA ranges from dirty yellow through various shades of blue-green to brown or black.

Some blue-green algae can fix atmospheric nitrogen because they contain an O_2-sensitive enzyme nitrogenase. The term "algal biofertilizer" was coined in early sixties to represent blue-green algae that have the capacity to metabolize molecular nitrogen and add nitrogen content to the soil. The conversion of elemental nitrogen into ammonia was the monopoly of the heterocystous blue-green algae until Whyatt and Silvey reported in 1961 of nitrogen fixation by the unicellular algae *Gloeocapsa*. Since then, more than a dozen non-heterocystous genera of blue-green algae have been found to fix nitrogen.

Heterocysts

Some blue-green algae have a thick walled structure in their trichomes known as heterocysts. These specialized cells lack pigment system-II and, as such, there is no endogenous evolution of oxygen. The oxygen from air cannot diffuse through the thick wall. The anaerobic conditions so created inside the heterocysts keep nitrogenase active in them. Thus, under anaerobic conditions, only the heterocystous blue-green algae can fix nitrogen. Nitrogenase in the vegetative cells becomes active if the entire trichome is transferred to anaerobic conditions. The non-heterocystous blue-green algae are found in the subsoil region or soil–water interphase in the paddy field, where they add substantial amount of nitrogen.

Growth Promoting Effects of Blue-green Algae

In addition to contributing about 30 kg N/ha/season, the blue-green algae help in maintaining soil fertility by liberating growth-promoting substances like auxin and vitamins. They are photolithotrophic in nature. They can solubilize insoluble phosphates and improve the physical and chemical properties of soil. The blue-green algae have been found to synthesize and liberate biologically potent substances into the medium. The liberation of auxins, vitamin B_{12} and amino acids has been found maximum during the stationary phase of growth. These substances benefit crop growth and enable plants to utilize more nitrogen.

Photosynthesis by Blue-green Algae

The blue-green algae possess a property of simultaneously metabolizing both elemental nitrogen and carbon dioxide from the atmosphere. The process of photosynthesis in these organisms meets the entire energy requirement including the power needed for reducing nitrogen to ammonia. As such, these algae form a completely independent system and they do not depend on soil organic matter for energy supply. Organic matter is added to the soil as a consequence of algal growth. However, due to high nitrogen content and high rate of decomposition in water-logged conditions, no appreciable addition of organic matter to the soil is observed. The presence of organic matter in soil has been found to favour the growth of blue-green algae. This attributes to the increased availability of carbon dioxide for the process of photosynthesis.

The polysaccharide sheath present around the trichomes of algae binds the soil particles and increases the particle size. This improves aeration and water holding capacity of the soil. Some saprophytic algae like *Calothrix, Tolypothrix* and *Scytonema* grow on moist soil surface, forming a velvety layer that protects the soil from erosion.

PRODUCTION OF ALGAE

In general, there are four methods of algal production: 1) trough or tank method, 2) pit method, 3) field method and 4) nursery cum algal production method. The former two methods are essentially for farmers, and the latter two are for production on a commercial scale.

Trough Method

Shallow trays (2 m × 1 m × 23 cm) are prepared from galvanized iron sheets, or a permanent tank may be used. About 4 to 5 kg of river soil is spread in the tank with superphosphate and sodium molybdate. Soil is mixed with water in order to derive its nutrients. In high temperature area, the water level in the tray is maintained between 5 and 15 cm, whereas in a low temperature area, the level may be reduced. During bed preparation, the main nuisance is mosquitoes and insects. Furadon granules or any other suitable granules are mixed with the bed to prevent insects. The mixture is allowed to settle for 8–12 hours. After complete settlement of soil, the inoculum of BGA is seeded over the aqueous layer of the bed. The reaction of the soil should be neutral. If the soil is acidic, add $CaCO_3$. Within 10–15 days, the growth of blue-green algae as hard flakes will be visible on the surface of water/soil. The water level is reduced as a result of evaporation. By this way, water in the tray is allowed to evaporate and the algal flakes are allowed to dry. Once the soil is dried, the algal flakes are separated from soil and stored in plastic bags.

Pit Method

The pit method does not differ from trough method except that, instead of troughs or tanks, pits are dug in the ground. The pits are layered with thick polythene sheet to hold water. The rest of the process is the same as in trough method. This method is simple and less expensive to operate.

Field Scale Method

The field scale method is an improvement of trough method to produce BGA on a commercial scale. This method is more popular among farmers of south India.

- An area of 40 m^2 is demarcated from the field for algal production. Since algal production is envisaged immediately after crop harvest, no special preparation is necessary except that, if the soil is loamy, it should be well-puddled to facilitate water logging.

- Prepare a bund with soil to increase the water retaining capacity.

- Flood the area with water to a depth of 2.5 cm. In trough or pit methods, flooding is done only in the beginning, whereas in field scale method, flooding is repeatedly needed to keep the water standing.

- Apply superphosphate 12 kg/40m^2.

- To control insects and pests, apply carbofuran (3% granules) or furadon 250 g/40m^2.

- If the field has received algal application previously for two consecutive cropping seasons, no fresh algal application is required. Otherwise, apply composite algal culture 5 kg/40 m^2.

- In clayey soils, good algal growth shows in two weeks, while in loamy soils it takes three to four weeks.

- Once algae have grown and formed floating mats, they are allowed to dry in the field and the dry algal flakes are collected in gunny bags.

- One can continuously harvest algae from the same plot by reflooding the area and applying superphosphate and pesticides. However, the addition of algal inoculum may not be necessary.

- During summer months (April–June), the average yield of algae per harvest ranges from 16–30 kg/40m^2.

Nursery cum Algal Production

Some farmers produce algae in their nurseries. Alongside the plot used for nursery, an additional 40 m^2 is prepared for algal production, which may yield about 15–20 kg of algal material. This quantity of algal mass will be sufficient to inoculate one and a half hectares of land.

VERMICULTURE

Vermiculture is the art of rearing and maintaining earthworms for vermicompost preparation. Earthworms decompose waste organic materials and give out in a granular form called vermicompost, which includes cocoons and young earthworms. Different organic waste materials like crop residues, straws, leaves, animal waste, residues of green manure and household waste materials are used in vermicomposting. Generally 3200 species of earthworms occur in nature. Out of these, *Eisenia fetida* and *Eudrilus eugina* are used for preparing vermicompost.

Vermicompost may be prepared by two methods.

- ➲ Heap method
- ➲ Trench method

BENEFITS OF VERMICOMPOST

- ➲ Earthworms play a vital role in ploughing and fertilizing the soil and providing all the nutrients required by plants.

- ➲ They improve the soil structure, soil fertility, promote soil aggregation, encourage favourable soil reactions and enrich the nutrient status of soil, and in the process promote plant growth and improve the quality of the products.

- ➲ They make the soil porous.

- ➲ They help the soil achieve air, water and solids in proper ratio for maximum plant growth.

- ➲ They improve water infiltration rate and the soil's ability to absorb water.

- ➲ They neutralize soil pH.

- ➲ They stimulate microbial population. Innumerable free-living, nitrogen-fixing bacteria surround the earthworm's burrows.

PREPARATION OF VERMICOMPOST

➲ The base of the tree may be used as vermi-bed.

➲ About 5 or 10 kg of vermicastings is applied per tree.

➲ About 25 kg of any farm yard manure (FYM) is applied evenly on top of the vermicastings.

➲ Slashed weed available in the farm or any organic litter is mulched over on top.

➲ Moisture is applied regularly by watering.

➲ The earthworms subsequently consume the organic matter and turn it into rich vermicompost. Vermicompost is a biofertilizer rich in beneficial soil microorganisms. It contains all the essential plant nutrients like N, P and K.

Vermicomposting using Paddy Straw

Paddy straw has a wide C : N (80 : 1) organic material, is low in nitrogen and phosphorus but fairly rich in potassium. In the conventional method of composting, paddy straw takes 6–8 months for decomposition, resulting in poor compost. The conversion of organic materials into manure is chiefly microbiological and greatly influenced by the proportion of carbonaceous and nitrogenous materials present in organic wastes.

Microorganisms need carbon for cell structure formation and nitrogen for cellular protein synthesis. It was found that a C : N ratio of 30 : 1 or lower is desirable for efficient composting. Hence, organic materials poor in nitrogen should be made narrow by adding nitrogen in the form of a nitrogenous fertilizer. Superphosphate is generally added to fortify the phosphorous content of compost.

An experiment was conducted in Orissa taking red earthworm (*Eisenia fetida*) for decomposition of paddy straw in the presence of fertilizer sources. Dry paddy straw was chopped (3–4 cm), mixed with fresh cow dung slurry, and introduced into pots. Nitrogen in the form of calcium ammonium nitrate was applied to raise the N level to 2% and phosphorus as single superphosphate to raise the total P_2O_5 content to 0.2%. Watering was done to maintain the moisture content at 40–50%. After two weeks of preliminary decomposition, red earthworms were released at 10 adults per pot. The results showed neutral reaction of the compost masses indicating their suitability for soil application. The straw decomposition was 91%.

The C : N ratio decreased to 10 : 1 due to earthworm activity alone and further decreased to 8 : 1 when inoculated in the presence of N and P, showing better decomposition of a material with wide C : N ratio like paddy straw. The earthworm population was increased by up to 20 times.

REVIEW QUESTIONS

1. What are biofertilizers? Add notes on nitrogen-fixing organisms and types of biofertilizers.

2. Give the advantages and drawbacks of biofertilizers.

3. Write about the organisms used as biofertilizers.

4. Describe in detail about production of *Azotobacter.*

5. What are blue-green algae? How do they fix nitrogen?

6. Write in detail about algal cultivation.

7. What is vermiculture? Explain in detail about vermicomposting.

BIOPESTICIDES

Biopesticides are pest management tools based on beneficial microorganisms, nematodes and other safe, biologically active ingredients. Biopesticides effectively control insects, plant diseases and weeds.

Biopesticides fall into three major classes:

1. *Microbial pesticides*   Microbial pesticides consist of a microorganism (a bacterium, fungus, virus or protozoan) as the active ingredient. They can control different kinds of pests. However, each active ingredient is relatively specific for its target pest. For example, there are fungi that control certain weeds, and other fungi that kill specific insects.

   The subspecies and strains of *Bacillus thuringiensis* (Bt) are widely used in microbial pesticides. Each strain of this bacterium produces a different mix of proteins and specifically kills one or a few related species of insect larvae. While some Bt's control moth larvae found on plants, others are specific for larvae of flies and mosquitoes. Bt produces a protein that binds to the larval gut receptor, thereby causing the insect larvae to starve.

2. *Plant-incorporated-protectants (PIPs)*   PIPs are pesticidal substances produced by the plant itself due to genetic recombination. For example, the gene from a Bt pesticidal protein can be taken and introduced into the plant's own genetic material so that the plant produces the substance that destroys the pest.

3. *Biochemical pesticides*   Biochemical pesticides are naturally-occurring substances that control pests through non-toxic mechanisms. Conventional pesticides, in contrast, are generally synthetic materials

25

that kill or inactivate the pest. Biochemical pesticides include substances (e.g. insect sex pheromones) that interfere with mating as well as various scented plant extracts that attract insect pests. Some examples of biopesticides are given in Table 25.1.

Table 25.1   Examples of biopesticides

Type of control	Examples
Insect control	
Bacteria	*Bacillus thuringiensis, B. sphaericus, Bacillus popilliae, Serratia entomophila*
Viruses	Nuclear polyhedrosis viruses, granulosis viruses, baculoviruses
Fungi	*Beauveria* spp., *Metarhizium, Entomophaga, Zoopthora, Paecilomyces fumosoroseus, Nomuraea, Lecanicillium lecanii*
Protozoa	*Nosema, Thelohania, Vairimorpha*
Entomopathogenic nematodes	*Steinernema* spp., *Heterorhabditis* spp.
Weed control	
Fungi	*Colletotrichum gloeosporioides, Chondrostereum purpureum, Cylindrobasidium laeve*
Bacteria	*Xanthomonas campestris pv. poannua*
Plant disease control	
Fungi	*Ampelomyces quisqualis, Candida* spp. , *Clonostachys rosea, F. catenulate, Coniothyrium minitans, Pseudozyma flocculosa, Trichoderma* spp.
Competitive inoculants	*Bacillus pumilus, B. subtilis, Pseudomonas* spp., *Streptomyces griseoviridis*
Composts, soil inoculants	*Burkholderia cepacia*
Nematicides	
Nematode trapping fungi	*Myrothecium verrucaria, Paecilomyces lilacinus*
Bacteria	*Bacillus firmus, Pasteuria penetrans*
Mollusc parasitic nematode	*Phasmarhabditis hermaphrodita*

CHARACTERISTICS OF BIOPESTICIDES

Biopesticides are an important group of pesticides. Biopesticides, in general:

- have a narrow target range and a specific mode of action
- are slow acting
- have relatively critical application times
- suppress, rather than eliminate, a pest population
- have limited field persistence and short shelf life
- are safe to humans and environment
- present no residue problems

Microbial insecticides are another kind of biopesticides. They come from naturally-occurring or genetically altered bacteria, fungi, algae, viruses or protozoans. They suppress pests by:

- producing a toxin specific to the pest
- causing a disease
- preventing other microorganisms

An example of a microbial pesticide is *Bacillus thuringiensis* (Bt). *Bacillus thuringiensis* is a naturally-occurring soil bacterium that is toxic to the larvae of several species of insects. *Bacillus thuringiensis* may be applied to plant foliage or incorporated into the genetic material of crops. It is toxic to caterpillars (larvae) of moths and butterflies (Table 25.2). Several strains of Bt have been developed. These may be used against mosquitoes and black flies.

Table 25.2   Some biopesticides and their targets

Fungal species	Insect targets
Beauveria bassiana	Coleoptera, Homoptera, Lepidoptera and Diptera
Metarhizium spp.	Lepidoptera, Coleoptera, Homoptera and Orthoptera
Lagenidium giganteum	Mosquito larvae
Verticillim lecanii	Homoptera and Diptera
Nomuraea rileyi	Lepidoptera

ADVANTAGES OF BIOPESTICIDES

The government encourages the production and use of biopesticides. The manufacturers are given subsidy and financial assistance. The other reasons for their popularity are:

1. As compared to synthetic insecticides, biopesticides are easier to formulate.

2. Biopesticides are toxic to select groups of fungi. They does not pollute the environment and soil, leaves no residues on food and other edible materials.

DISADVANTAGES OF BIOPESTICIDES

There are several limitations in biopesticides. Some of them are given here.

1. Biopesticides have no stability. Because most agents in biopesticides are living, factors like temperature, moisture, pH, ultraviolet spectrum and soil conditions affect the mode of action and their efficacy.

2. Biopesticide agents face competition from other microbes. Once they autolyse, the effect is low. Soil and their factors also control the population of biopesticide agents.

3. Because of its selectivity in controlling insect-pests and pathogens, farmers do not give prominence to biopesticides. Moreover, biopesticides act slowly as compared to synthetic pesticides.

4. Biopesticides have very short shelf life. The distribution of biopesticides is also very erratic.

BIOPESTICIDE PRODUCTION

Biopesticides are produced by usual fermentation methods. Bacterial strains are cultivated by submerged fermentation, and fungal strains are cultivated by surface cultivation. The entomopathogenic organisms are cultivated by *in vivo* process.

Submerged Fermentation

The bacterial inoculum is developed by shake flask fermentation. Corn steep liquor and calcium carbonate are the common sources of nutrients. The production medium contains beet molasses, corn steep solids, cotton seed flour and calcium carbonate for development and sporulation of strains.

Surface Methodology

In surface methodology, the inoculum from the seed tank is absorbed on a carrier, usually a mixture of grains and inert substances like diatomaceous earth. The growth medium contains dextrose, corn steep, yeast hydrolysate and potassium phosphate with pH 7.0. After incubation, the seeds are immersed in a semi-solid medium containing dextrose, yeast hydrolysate, corn steep, potassium phosphate, sodium hydroxide and calcium chloride. Semi-solid fermentation is carried out in fermenters till high mycelial mass and spores form.

Solid state fermentation is usually done in metal bins instead of fermentation tanks. The solid medium contains wheat bran, perlite, soymeal, dextrose and calcium chloride. The pH is maintained between 6 and 7 and the moisture level is controlled.

In vivo Methodology

The entomopathogenic organisms are produced in living hosts. Caterpillars are the primary hosts. The saw fly larvae, gypsy moth larvae, boll worm larvae, cabbage looper larvae and mosquito larvae may also be used for cultivation of entomopathogenic organisms. The insect virus that can multiply in yeast and bacteria are produced by submerged fermentation techniques and cell lines. The viruses produced from cell lines have the same effect as the virus produced from the caterpillar.

REVIEW QUESTIONS

1. What is a biopesticide? Give its types.

2. Give examples for biopesticides.

3. Write about characters, advantages and disadvantages of biopesticides.

4. Explain in detail about biopesticide production.

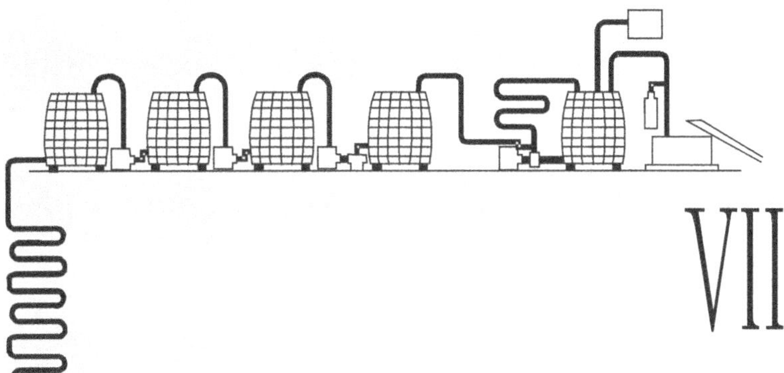

VII

ENVIRONMENTAL BIOPROCESSING

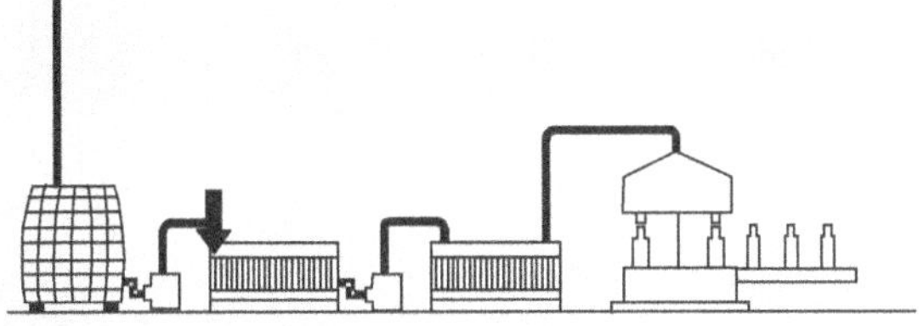

BIOREMEDIATION AND TRANSFORMATION

Bioremediation is the application of microorganisms or microbial processes or products in degrading contaminants from an area. On the other hand, bioaugmentation is the addition of microbes to a site. When a microbial population present at a site is capable of cleaning the site, the process is called natural attenuation. Natural attenuation is often slow. There are two methods of bioremediation, namely biostimulation and bioaugmentation. Biostimulation refers to modifying a site to enhance the growth of indigenous microbes already present in it. During biostimulation, nutrients (e.g. nitrates) are added and aeration is provided. This helps the existing population to breakdown the contamination quickly. The next step would be to add microbes to a deficient site (bioaugmentation). Bioaugmentation enables the use of specific microbes for specific contamination. Different microbes prefer different hydrocarbons, and bioaugmentation takes advantage of this property to quickly clean a site.

Bioremediation is helpful in cleaning up many organic wastes. The most hazardous wastes are listed in Table 26.1. However it has limitations. The extent of remediation is highly dependent on the toxicity and the level of contaminants, their ability to be biodegraded and the property of the soil. The types of contaminates generally targeted for bioremediation are non-halogenated volatile and semi-volatile organics and fuels. Sites with high metal concentrations (i.e., mercury), high chlorinated organics (compounds with many chlorine elements attached), and inorganic salts are hard enough to be cleaned with microbes. These compounds are toxic to microbes. There are, however, some plants that are capable of removing heavy metals from soils with little or no harm to plants. This process is called phytoremediation.

Table 26.1 Most hazardous waste types

Hazardous waste	Major contaminants
Petroleum hydrocarbons	Benzene, toluene, ethyl benzene and xylene (BTEX compounds) Aromatics Heterocyclic aromatics
Halogenated aliphatic compounds	Trichloroethylene,1,1,1-trichloroethane, tetrachloroethylene, *cis-trans*-1,2-dichloro- ethylene, 1,1-dichloroethane
Industrial waste lagoons and wood preservation	Creosote, polynuclear aromatics (or PAHs), Pentachlorophenol, nitrogen heterocyclic aromatics, aromatics
Chemical compounds and wastes	Acetone, acrylonitrile, animal fats and grease, anthracene, benzopyrene, t-Butanol, butylcellosolve, chrysene, 1,2-dichloroethane, dodecane, ethyl-acrylate, ethylbenzene, ethylene glycol, fatty amines, fluoranthene, gasoline, hexadecane, hexane, isopropyl acetate, methanol, methylene chloride, methylethyl ketone, methylmethacrylate, 2-Methylnaphthalene, monochlorobenzene, naphthalene, pentadecane, phenanthrene, phenols, polynuclear aromatic hydro-carbons (PAHs), pyrenes, stoddard solvent, styrene, tetrahydrofuran, trichloroethylene, 1-Tridecene

Bioremediation (except phytoremediation) usually involves a redox reaction that detoxifies the contaminants. The reaction occurs in limiting environmental conditions. However, the reaction can be expedited by improving the environmental conditions in one of the following ways:

Adding an electron acceptor   In some contaminants, electron donors are readily available, but they require electron acceptors to complete the redox reaction. Many organic compounds require electron acceptors to be added such as oxygen or nitrate.

Adding an electron donor   Sometimes, electron acceptors are present but a donor is absent. This is the case with contamination by nitrates or metals, which readily accept electrons. Adding a carbon source as an electron donor will enhance bioremediation.

Adding a limiting nutrient   Sometimes, lack of nutrients in the organism may hinder bioremediation. Fertilizers rich in nutrients like nitrogen and phosphorus can encourage bioremediation.

Increasing the bioavailability of the contaminant   In some cases, the contaminants may be inaccessible by the organism. Metals and organic compounds that are bound to sediments can be made bioavailable by the use of surfactants and chelates.

Stimulating the production of a specific enzyme   Bioremediation of certain organic molecules may be stimulated by bacterial enzymes. Certain substrates may be added to the environment for the bacteria to increase the productivity of the enzyme.

ADVANTAGES AND DISADVANTAGES OF BIOREMEDIATION

The use of intrinsic or engineered bioremediation processes may offer several advantages:

- Lower cost
- Contaminants are destroyed and contaminants converted into innocuous products
- Non-intrusive, potential for continued use
- Ease of implementation

However, there are disadvantages to bioremediation as well, which include:

- Difficult to control
- May cause other contamination problems
- May not reduce the concentration of contaminants to required levels
- Time consuming
- Requires extensive monitoring
- Lack of (hydraulic) control
- Difficult to predict effectiveness

IMPLEMENTING A BIOREMEDIATION PROCESS

Preliminary investigation is done on the site for geochemistry, environmental characteristics, soil conditions, hydrology, hydrogeology, concentration and distribution of contaminants.

- Identify the microbial reactions that can or have affected the concentration and distribution of contaminants.

- ⮞ A detailed site investigation is carried out to confirm the above.

- ⮞ Decide the bioremediation process appropriate to remediate the site. This is done in relation to the regulations and proposed use of the site.

- ⮞ Implement intrinsic or engineered bioremediation program, or investigate alternative options.

- ⮞ Monitor the site for efficacy of the program.

- ⮞ Modify the program on the bases of analyses and developments.

- ⮞ Document the entire process.

BIODEGRADATION AND METABOLISM

Biodegradation involves chemical transformations mediated by microorganisms that satisfy nutritional and energy requirements and, in turn, detoxify the immediate environment. It may also occur fortuitously such that the organism receives no nutritional or energy benefits.

1. Mineralization is complete biodegradation of organic materials into inorganic products. It often occurs through the combined activities of a group of microbes rather than a single microorganism.

2. Co-metabolism is partial biodegradation of organic compounds that occurs fortuitously. It does not provide energy or cell biomass to microorganisms. Co-metabolism can result in partial transformation that either serves as a carbon and energy substrate for microorganisms, as with some hydrocarbons, or can turn toxic to the transforming microbial cell, as with trichloroethylene (TCE) and methanotrophs.

Biodegradation reactions may be aerobic or anaerobic. Aerobic biodegradation involves the use of molecular oxygen (O_2), where O_2 (terminal electron acceptor) receives electrons transferred from an organic contaminant.

$$\text{Organic substrate} + O_2 \rightarrow \text{Biomass} + CO_2 + H_2O + \text{Other inorganics compounds}$$

Thus, the organic substrate is oxidized (addition of oxygen), and O_2 is reduced (addition of electrons and hydrogen) to water (H_2O). In this case, the organic substrate serves as the sources of energy (electrons) and cell carbon is used to build microbial cells (biomass). Some microorganisms (chemoautotrophic aerobes or lithotrophic aerobes) oxidize reduced inorganic compounds (NH_3, Fe^{2+}, or H_2S) to gain energy and fix CO_2 to build cell carbon.

$$NH_3 \text{ (or } Fe^{2+} \text{ or } H_2S) + CO_2 + H_2 + O_2 \rightarrow \text{Biomass} + NO_3 \text{ (or Fe or } SO_4^-)$$
$$+ H_2O$$

As a result of consumption of O_2 by aerobic microorganisms and slow recharge of O_2, the environment becomes anaerobic (lacking O_2), and mineralization, transformation and co-metabolism depend on microbial utilization of electron acceptors other than O_2 (anaerobic biodegradation). Nitrate (NO_3), iron (Fe^{3+}), manganese (Mn^{4+}), sulphate (SO_4) and carbon dioxide (CO_2) can act as electron acceptors if the organisms have the appropriate enzymes. The jet fuel constituents were observed to be biodegraded in the presence of NO as the electron acceptor. Iron and manganese are important microbial electron acceptors, with background concentrations in soils ranging from 20 to 3,000 mg/kg for manganese and 3.8 to 5.2 per cent for iron. An evaluation of the degradation of polycyclic aromatic hydrocarbons (PAHs) in aerobic and anaerobic environments was conducted based on thermodynamic principles. Biodegradation of pentachlorophenol (PCP) has been observed to increase in the presence of added Mn.

The halogenated compound may serve as a substrate for both aerobic and anaerobic microorganisms. The dehalogenation of the halogenated compounds may be taken place in two ways. In one way that is natural spontaneous removal of halogens. The other way is the removal of halogen groups from the compounds by the action of microbial enzymes through hydrolysis or reduction.

Table 26.2   Organisms commonly used in bioremediation

Type of contaminant	Genus
Petroleum	*Pseudomonas, Proteus, Bacillus, Penicillium, Cunninghamella*
Aromatic rings	*Pseudomonas, Achromobacter, Bacillus, Arthrobacter, Penicillium, Aspergillus, Fusarium, Phanerocheate*
Cadmium	*Staphylococcus, Bacillus, Pseudomonas, Citrobacter, Klebsiella, Rhodococcus*
Sulphur	*Thiobacillus*
Chromium	*Alcaligenes, Pseudomonas*
Copper	*Escherichia, Pseudomonas*

Table 26.3   Environmental factors critical to soil microbial activity

Environmental factor	Optimum levels
Oxygen	**Aerobic metabolism**—greater than 0.2 mg/L dissolved oxygen, minimum air-filled pore space of 10%.
	Anaerobic metabolism—less than 0.2 mg/L dissolved oxygen, O_2 concentration less than 1% air-filled pore space.
Nutrients	Sufficient nitrogen, phosphorus and other nutrients for microbial growth (suggested C : N : P ratio of 120 : 10 : 1)
Moisture	Unsaturated soil—25–85% of water holding capacity, –0.01 MPa; will affect oxygen transfer in soil (aerobic status); in saturated zone, water will affect transport rate of oxygen and therefore will affect the rate of aerobic remediation
Environment (pH)	5.5 to 8.5
Environment (redox)	**Aerobes and facultative anaerobes**—greater than 50 mV **Anaerobes:** less than 50 mV
Environment (temperature)	15–45°C (mesophilic)

METHODS OF BIOREMEDIATION

Generally two categories of bioremediation occur. They are *in site* or *in situ* treatment and *ex site* or *ex situ* treatment. The *in situ* bioremediation is carried out by augmentation of suitable microorganisms on the spot by creating favourable growth environment. The *ex situ* bioremediation differs from *in situ*. In *in situ* bioremediation, the polluted soil or water is removed from the polluted place and is treated in other place. After complete microbial treatment the treated soil or water will be reloaded at the same place.

SITE CHARACTERIZATION

Site characterization is the first step in bioremediation. A contaminated site generally consists of four phases:

1. solid (which has an organic matter component and an inorganic mineral component composed of sand, silt and clay)
2. oil (commonly referred to as non-aqueous phase liquid, or NAPL)
3. gas
4. aqueous (leachate or ground water)

These phases need to be characterized with regard to the extent and distribution of contamination as well as their potential exposure to human and environmental receptors. Each phase affects bioavailability, i.e., interactions with microorganisms and exposure to human and environmental receptors. In each phase, biological reactions result in the transformation of a parent chemical to CO_2, H_2O and inorganic species through the process of mineralization, or transformation to intermediates that persist or react with the soil components chemically and therefore alter the bioavailability of chemicals.

Evaluating the extent and distribution of contamination at a site will provide a basis to select specific bioremediation technologies. If contamination is widespread and low in concentration, natural attenuation (*in situ* treatment) may be feasible. Conversely, with high concentrations of contaminants, soil excavation and a confined treatment facility (CTF) or land treatment may be advisable.

The distribution of contaminant at the polluted site is determined by physical and chemical properties of contaminants and properties of the polluted site. The characters of the pollutant like leachability, volatilability and adsorbability capacity of the pollutant are characterized. Besides, the level of pollution in the layers of subsurface soil is also an important parameter to characterize the soil. These characters evaluate the bioremediation potential needed to remediate the polluted area.

IN SITU BIOREMEDIATION METHODS

The *in situ* techniques do not require excavation of contaminated soils. Hence they are less expensive, create less dust, and release lesser contaminants than *ex situ* techniques. It is also possible to treat a large volume of soil at once. In *in situ* techniques, however, may be slower than *ex situ* techniques, may be difficult to manage, and are most effective at sites with permeable (sandy or uncompacted) soil.

Bioventing

Bioventing systems (Figure 26.1) deliver air from atmosphere into the soil above the water table through injection wells placed on the ground where contamination exists.

The number, location and depth of wells depend on geological factors and engineering considerations.

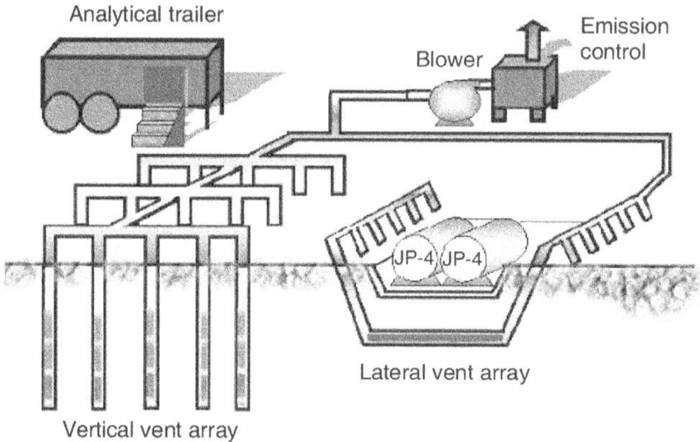

Figure 26.1 Bioventing system

Bacteria, fungi and other microorganisms are found in vast numbers in most soils, and significant numbers persist even at depths of several hundred feet. These microorganisms can metabolize a wide range of petroleum hydrocarbons. Petroleum serves as the electron donor and oxygen their preferred electron acceptor. The bacteria degrade petroleum hydrocarbons by converting them into energy, new cellular material, and oxidizing the rest to CO_2. Oxygen is used by the microorganisms to convert the organics while in the process it is reduced to water.

In most fuel contaminated sites, lack of oxygen is the only limiting factor for biodegradation. The microbial oxygen utilization rate exceeds the rate of oxygen diffusion in the soil, and therefore the oxygen supply is quickly exhausted. Anaerobic decomposition of the contaminants occurs at a rate usually slower than aerobic decomposition. Some chemicals, however, could not be degraded both aerobically and anaerobically. The goal of *in situ* bioventing is to supply the subsurface microorganisms with adequate supply of oxygen so that they will degrade the fuel as quickly as possible.

There are two ways to provide oxygen into the subsurface.

- ➲ air is directly injected into vent wells

- ➲ soil gas can be extracted from vent wells, creating a vacuum that draws oxygenated soil gas from the surrounding soils

Air injection can provide higher concentrations of oxygen to areas where contaminant concentrations are highest. It also pushes out volatile hydrocarbons into the neighbouring soils, thus increasing the contaminated area and allowing more microorganisms to assist in biodegradation. Air injection may extract volatile hydrocarbons from the subsurface, which requires emission monitoring at the soil surface. The injected oxygen stimulates microbial activity to degrade the contamination in the soil. The water table level is below the contaminant. If significant amounts of contaminant are found below the water table, dewatering may have to be performed initially.

Extracting soil gas draws off oxygen from clean soils into the contaminated soils at concentrations sufficient to allow for aerobic degradation of the fuel. However, this method cannot provide oxygen concentrations to the subsurface as high as with air injection.

Air sources   The most common sources of oxygen are atmospheric air, pure oxygen gas and hydrogen peroxide.

Atmospheric air is the preferred oxygen source since it is readily available in unlimited quantities at any site. However, sparging the groundwater with air (or pure oxygen) can only raise the dissolved oxygen level to moderate levels (8 to 40 mg/L). Hydrogen peroxide is infinitely soluble in water, dissociating into water and oxygen. However, hydrogen peroxide can be toxic to microorganisms at low concentrations and must be properly introduced into the subsurface environment.

Soil gas monitoring   Soil gas monitoring is necessary to ensure that vapour degradation is occurring. Monitoring points are created for pressure and soil gas monitoring. These points should be located in contaminated soils of levels greater than 1,000 mg/kg of total petroleum hydrocarbon. They are used to collect soil gas for carbon dioxide (a by-product of degradation) and oxygen analysis. Temperature monitoring is also completed by connecting thermocouples to monitoring points. The temperatures are usually constant at varying lateral distances but will change with depth.

Natural Attenuation

The term "natural attenuation" refers to a naturally-occurring process in soil and groundwater environments that act to reduce the mass, toxicity, mobility, volume or concentration of contaminants in those media. These *in situ* processes include biodegradation, dispersion, dilution, adsorption, volatilization and chemical or biological stabilization or destruction of contaminants.

Surface and subsurface soils possess different characteristics of natural attenuation. The movement of contaminant from subsurface soils diffuse into soil vapour and aqueous phase can be easily attenuated. The most higher molecular weight organic contaminant that persist in the matrix of subsurface soil is difficult to attenuate and degrade. These can be remediated by the events like releasing of fresh solvents to dissolve the contaminant, chemical and biochemical transformations and physical disturbances. But without proper exposure routes, this remediation may lead to increase in mobility and open exposure routes. Thus, the natural attenuation of subsurface soil may be considered a possible remedy on a site-specific basis.

In contrast, mobile contaminants in surface soils (typically at 1 to 2 feet) may have already degraded, volatilized or leached from the soil unless weathered free products have entrapped the contaminants. In addition, immobilized contaminants persist in surface soils for ages and thus, if undisturbed, are only slowly removed or not removed at all by natural attenuation. Unfortunately, wind or water erosion may mobilize the persistent or conserved contaminants through soil transport into surface waters. Bioaccumulation of contaminants and biomagnification in the watershed food chain has already produced widespread risk to human health and ecosystem. In addition, the resulting contaminated sediments are often uncontrollable, since natural events such as earthquakes and storms can resuspend or distribute contaminants over wide areas. Because of the risks, the contaminated surface soil needed more effort and effective containment.

For natural attenuation in soils, natural biotransformation processes such as dilution, dispersion, volatilization, biodegradation, adsorption and chemical reactions with soil materials act to reduce contaminant concentrations to acceptable levels. Natural attenuation may be considered for remediation of contaminants in soils if site-specific factors support its use, which include:

- protection of potential receptors during attenuation

- favourable geological and geochemical conditions

- documented reduction of degradable contaminant mass in the surface and subsurface soils

- confirmation in microcosm studies of contaminant cleanup

- for persistent or conserved contaminants, the control should be made during and after the attenuation.

Limitations

Factors that may limit the applicability and effectiveness of the process include:

- Extensive site characterization with modelling and long-term monitoring may be more expensive than active remediation.

- ➲ Toxicity of degradation and transformation products may exceed the levels of original contaminants.

- ➲ High risks occur at sites where geological characteristics such as fracture bed rock or Karst landscapes may prevent assessment of stable plume control for contaminants leached from soil.

- ➲ Contaminants may migrate (by erosion, leaching or volatilization) before they are degraded or transformed.

- ➲ Ground water at the site contaminated by the soil source will not be available for an extended period of time.

- ➲ Extensive free products, such as non-aqueous phase liquids, may have to be removed before natural attenuation can restore soil.

- ➲ Conservative metals can be temporarily immobilized where natural attenuation reestablished oxygenated soil condition.

EX SITU BIOREMEDIATION

Ex situ techniques can be faster, easier to control, and used to treat a wider range of contaminants and soil types than *in situ* techniques. However, they require excavation and treatment of the contaminated soil before and, sometimes, after bioremediation. *Ex situ* techniques include slurry-phase bioremediation and solid-phase bioremediation.

Slurry-phase Bioremediation

The contaminated soil is mixed with water and other additives in a bioreactor to keep the microorganisms, which are already present in the soil, in contact with the contaminants in the soil. Nutrients and oxygen are added, and conditions in the bioreactor are controlled to create optimum environment for the microorganisms to degrade the contaminants. Upon completion of the treatment, water is removed from the solids, which is disposed of or treated again if it still contains pollutants.

Slurry-phase biological treatment is a relatively rapid process compared to other biological treatment processes, particularly for contaminated clay. The success of the process is highly dependent on the specific soil and chemical properties of the contaminated material. This technology is particularly useful where rapid remediation is a high priority.

The bioreactor design is dependent on the contaminant to be remediated, the contaminated medium, and cost. The two major types of soil reactors are dry and slurry reactors. Dry reactors treat soil with no other amendments other than microbes and nutrients. Adequate moisture is maintained for microbial growth by sprinkler system. Physical mixing of the soil keeps it aerated. A liner may be fitted over the soil to collect vapours volatilizing from the soil. After the remediation process is complete, the soil may be transported to the desired location.

This system is only applicable in specific situations. Practically, only soils that are contaminated at a shallow depth are treated with a soil pile reactor. The formation of soil–microbe pellet hinders remediation by reducing the availability of pollutants to microbes. Slurry reactors have proven more effective and efficient against a wide range of pollutants. In a slurry reactor, soil is mixed with an equal or a greater amount of water, microbes and nutrients to form soil slurry. A slurry reactor is easier to maintain than dry reactors. It offers many advantages such as rapid treatment, reduced pellet formation, increased slurry homogenization, and increased bioavailability. Soil–water separation may become a problem, especially high clayey soils. Moreover, there is a need for wastewater treatment after the soil is dewatered.

Bioreactors for groundwater treatment (Figure 26.2) are usually fixed firm or some form of activated sludge reactors. Fixed firm reactors contain a high surface area for the medium, which supports microbial growth. Activated sludge reactors are aerated basins where microbes are mixed with wastewater and nutrients. Bioreactors may be operated in batch or steady state flow regimes.

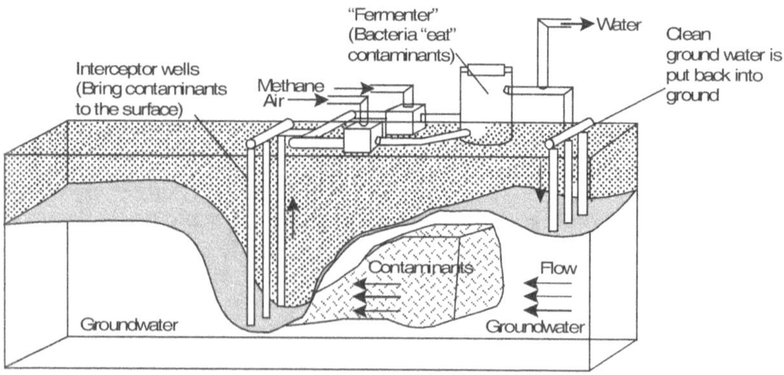

Figure 26.2   Slurry phase remediation of ground water

Bioreactors may be designed to operate aerobic and anaerobic processes (Figure 26.3).

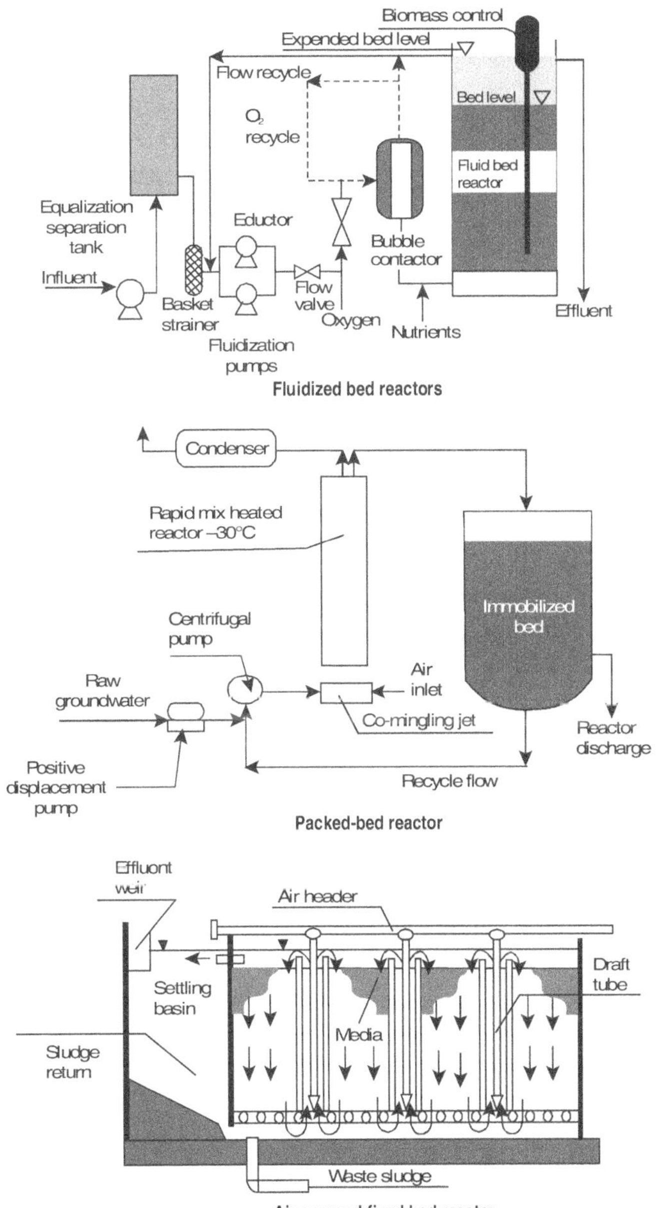

Figure 26.3   Some bioreactors used in remediation   (*Continues*)

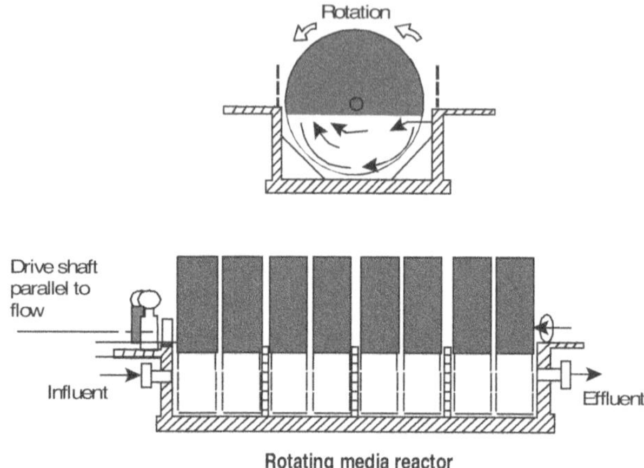

Figure 26.3 Some bioreactors used in remediation

Anaerobic degradation is useful in reducing highly halogenated compounds such as trichloroethylene. Conversely, aerobic degradation is effective against a wide range of pollutants. The use of anaerobic and aerobic processes in series offers a method to treat substances such as highly chlorinated organic pollutants that do not respond to conventional treatments. Anaerobic organisms can dechlorinate the substance to a point where aerobic organisms can completely degrade it.

SOLID-PHASE BIOREMEDIATION

Soil Biopiles

The contaminated soil is piled in heaps of several meters over an air distribution system. Aeration is provided through the soil pile with a vacuum pump. Moisture and nutrient levels are maintained at levels adequate for bioremediation. The soil heaps may be placed in enclosures. Volatile contaminants are easily pulled through the pile. Biopiles are also called biocells, bioheaps, biomounds and compost piles.

Biopiles are designed to optimize the conditions for aerobic bacteria to biodegrade organic contaminants. The effectiveness of a biopile system depends on many parameters, which may be grouped into three categories: 1) soil characteristics 2) constituent characteristics 3) climatic conditions.

The soil texture affects the permeability, moisture content and bulk density of the soil. Fine-grained soils are less permeable than coarse-grained soils.

Soils with lower permeability are more difficult to aerate but tend to retain moisture better than soils with higher permeability. However, lower permeability is usually associated with soils that clump together, making it difficult to distribute moisture, air and nutrients evenly. At times, the soil may need to be turned (or tilled) to promote continued biodegradation.

The soil usually contains a large number of diverse microorganisms including bacteria, algae, fungi, protozoa and actinomycetes. In well-drained soils, which are the most appropriate for biopiles, these organisms are generally aerobic. Of these organisms, bacteria are the numerous and biochemically active group, particularly at low oxygen levels. Although sufficient types and quantities of microorganisms are usually present in the soil for land farming, the recent applications of *ex situ* soil treatments include blending the soil with cultured microorganisms or animal manure.

Soil microorganisms require moist conditions for proper growth. Excessive soil moisture, however, restricts the movement of air through the subsurface, thereby reducing the availability of oxygen essential for metabolic processes.

The bacterial growth rate is a function of temperature. Soil microbial activity has been shown to significantly decrease at temperatures below 10°C. The activity of most bacteria useful in biodegradation of petroleum hydrocarbons diminishes at temperatures greater than 45°C. Within the range of 10°C and 45°C, the rate of microbial activity typically doubles for every 10°C rise in temperature. Because soil temperature varies with ambient temperature, there will be certain periods during the year when bacterial growth and, therefore, constituent degradation will diminish. When ambient temperatures return to the growth range, bacterial activity will be gradually restored.

The presence of very high concentrations of petroleum organics or heavy metals in soils can inhibit the growth and reproduction of bacteria responsible for biodegradation in biopiles. Conversely, very low concentrations of organic materials will result in diminished levels of microbial activity.

System Design

The height of biopiles varies between 3 and 10 feet. Additional land area is required to make the slope around the pile. The length and width of biopiles is generally not restricted unless aeration is to occur by manually turning the soils. In general, biopiles that will be turned should not exceed 6 or 8 feet in width.

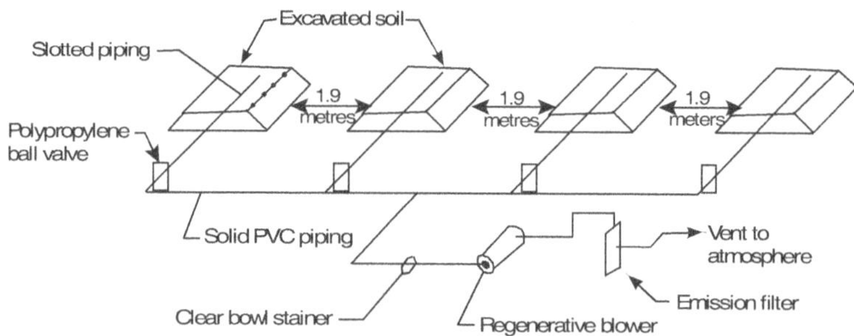

Figure 26.4 Design of a biopile

Biopiles (Figure 26.4) are typically constructed in 'lifts'. Blended soil is mounded up to a few feet, and aeration and moisturizing piping is laid prior to the addition of the next lift. This process is repeated until the pile reaches a desired height.

Blending the soil may involve the addition of 1) manure to augment microbial population and provide additional nutrients, 2) soil amendments (e.g. gypsum) and bulking materials (e.g. sawdust or straw) to ensure that the biopile medium has a loose or divided texture, and 3) chemicals to adjust the soil pH within 6 or 8, with a value of 7 (neutral) being optimal.

Periodically, moisture must be added to the biopiles because soil may become dry as a result of evaporation, especially during aeration operations. Excessive accumulation of moisture may occur within biopiles in areas of high precipitation or poor drainage. Volatile constituents tend to evaporate from the biopiles during extraction or injection, rather than being biodegraded by bacteria, and the containment of vapours may be done by covering the biopile and installation of collection piping. If air is passed through the aeration pipe, volatile constituent vapours will pass into the extracted airstream which may be treated. In some cases, it may be acceptable to re-inject the extracted vapours back into the soil pile for additional degradation. However, the vapours may need to be treated (typically through carbon adsorption). To prevent possible leaching of contaminants from the biopile into the underlying groundwater, biopiles may require to be constructed on top of an impermeable liner. Leachate that drains from the biopile may be collected for treatment and disposal. The advantages and disadvantages of biopiles are listed in Table 26.4.

Table 26.4   Advantages and disadvantages of biopiles

Advantages	Disadvantages
Relatively simple to design and implement.	Concentration reductions (>95%) and constituent concentrations (<0.1 ppm) are very difficult to achieve.
Short treatment times: usually 6 months to 2 years under optimal conditions.	May not be effective for high constituent concentrations (>50,000 ppm total petroleum hydrocarbons).
Relatively cost competitive.	Presence of significant heavy metal concentrations (> 2,500 ppm) may inhibit microbial growth.
Effective on organic constituents with slow biodegradation rates.	Volatile constituents tend to evaporate rather than biodegrade during treatment.
Requires less land area than landfarms.	Requires a large land area for treatment.
Can be designed to be a closed system; vapour emissions can be controlled.	Vapour generation during aeration may require treatment prior to discharge.
Can be engineered to be potentially effective for any combination of site conditions and petroleum products.	May require bottom liner in case of leaching from the biopile.

COMPOSTING

Biodegradable waste is mixed with a bulking agent such as straw, hay or corn cobs to make it easier to deliver optimum levels of air and water to the microorganisms. The three common designs in composting are static pile composting (compost is formed into piles and aerated with blowers or vacuum pumps), mechanically agitated in-vessel composting (compost is placed in a treatment vessel where it is mixed and aerated), and windrow composting.

Biofiltration

Biofiltration refers to the use of a biologically active material to remove contaminants from a waste stream. Compost has been shown to be an excellent medium for biofiltration for two reasons: First, the compost product has a high exchange capacity, which translates into a high absorptive capacity. The waste products undergo physical absorption onto the compost particles and water surfaces surrounding it. Compost has a large moisture holding capacity. This is useful in biofiltration of volatile organic compounds (VOCs) and odorous gases. In a study by Yang and Allen (1994), compost was evaluated for its hydrogen sulphide removal rate, which was demonstrated to be over 99.9% when the compost water content was between 30 and 62%. There are two possible ways of metabolizing an odorous gas molecule. For one, the gas molecule may diffuse to the surface of the compost medium, from where it is absorbed. Then, it enters the water phase where microbial action takes place. Another possibility is that the gas molecule enters directly into the microscopic water phase.

The second reason that compost is an excellent medium for biofiltration is that it supports a large heterogeneous microbial population that could metabolize numerous natural and xenobiotic compounds. Microbial degradation is the second method of filtration. The compost inherently has a generous amount of carbon to be utilized as a food source. The microbial activity is also affected by the moisture content of the medium. The microbial action may involve metabolism, utilization or oxidation of organic compounds. The compost supports an indigenous population of bacteria, fungi and actinomycetes.

Table 26.5 Organisms that degrade organic compounds under aerobic conditions

Organic compound	Organism	Organic compound	Organism
2-Chlorobenzoic acid	*Pseudomonas*	4-Chlorophenol	*Mycobacterium, Alcaligenes, Flavobacter*
4-Chlorobenzoic acid	*Arthrobacter*	Pentachlorophenol (PCP)	*Arthrobacter, Flavobacter, Pseudomonas, Coryneform*
3,5-Dichlorobenzoic acid	*Psuedomonas*	2,4-Dichlorophenoxyacetic acid (2,4-D)	*Psuedomonas, Azotobacter*
3-Chlorobenzene	*Pseudomonas*	2,4,5-Trichlorophenoxyacetic acid (2,4,5-T)	*Pseudomonas*
1,4-Dichlorobenzene	*Pseudomonas*	1,1,1-Trichloro-2,2-bis (*p*-chloro -phenol) ethane (DDT)	*Escherichia, Pseudomonas, Aerobacter, Clostridium, Proteus, Fusarium, Mucor, Cylindrotheca, Nocardia Streptomyces, Phanerochaete*
1,4-Dichlorobenzene chlorobenzene	*Alcaligenes*	Polychlorobiphenyl (PCB)	*Alcaligenes, Acinetobacter, Pseudomonas*
3-Chlorophenol	*Nocardia*		

BIOTRANSFORMATION AND REMEDIATION OF COMPOUNDS

Hydrocarbons

Hydrocarbons (Figure 26.5) are introduced into the environment through various sources including incomplete combustion of fossil fuels, accidental discharge, disposal of petroleum products and other organic wastes, incineration of refuse and wastes, and various industrial processes. They may also be produced as a result of natural processes including forest fires and volcanic eruptions. Certain hydrocarbons, known as polycyclic aromatic hydrocarbons (PAHs), are found in the soils at wood preservation and gas manufacturing plants.

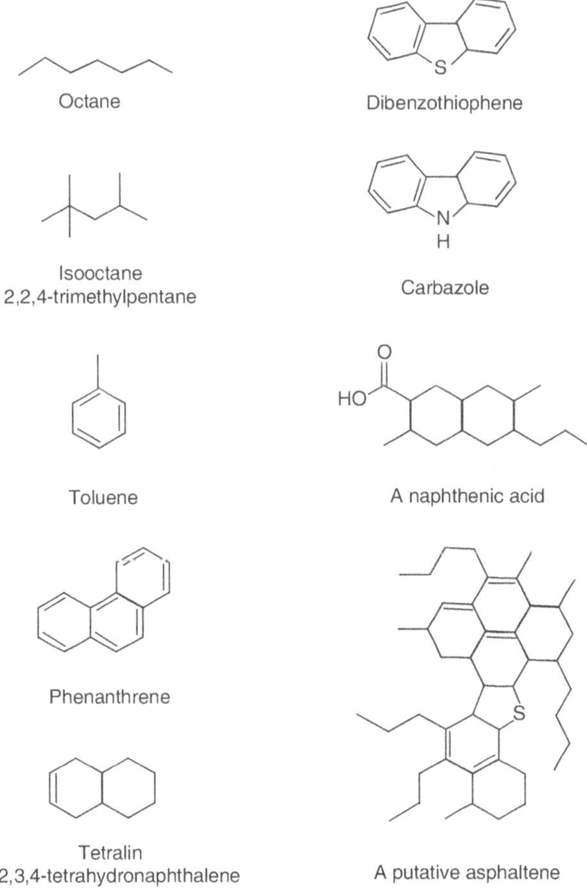

Figure 26.5 Chemical structure of hydrocarbons

The inertness of high molecular weight coupled with low solubility in water and strong lipophilic character lead to high accumulation levels of hydrocarbons. Some hydrocarbons (such as benzene) are carcinogenic. Industrial oil spills may destroy entire ecosystems. Excess hydrocarbons in aqueous environments can deplete the supply of oxygenating molecules, leading to anoxic conditions. Most crude oils contain hydrocarbons ranging in size from methane to molecules with hundreds of carbons. When crude oils reach the surface environment, they are biodegraded usually under aerobic conditions.

A numerous genera of bacterial and fungal species are able to degrade hydrocarbons aerobically. The hallmark of an oil-degrading organism is its ability to insert oxygen atoms into hydrocarbons. Once a hydrocarbon possesses carboxylate or alcohol functionality, it is invariably a readily degradable compound. Sometimes, hydrocarbons may be oxidized by bacteria under anaerobic conditions, where the oxygen probably comes from water. Limited hydrocarbon biodegradation has been shown under sulphate-, nitrate, carbon dioxide- and ferric iron-reducing conditions.

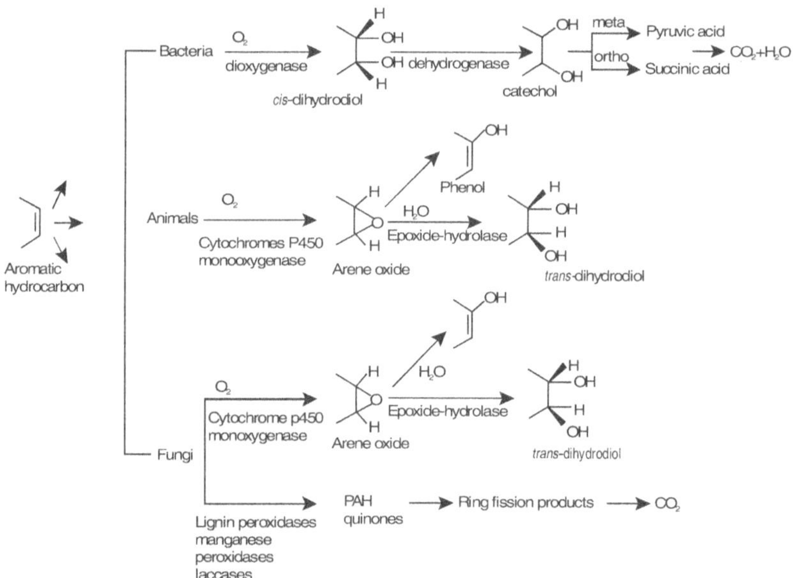

Figure 26.6 Biotransformation pathway of hydrocarbons

For example, slow biodegradation of larger hydrocarbon molecules will occur in an anaerobic environment that exists in lake sediments. However, anaerobic biodegradation is so slow as compared to aerobic biodegradation and photo-oxidation (degradation of hydrocarbons by sunlight and oxygen). Therefore, most bioremediation processes focus on utilizing aerobic processes combined with abundance of sunlight and oxygen. The biotransformation pathway of hydrocarbons are shown in Figure 26.6.

Bioremediation of Hydrocarbons

Bioremediation of hydrocarbons depends on their locations in the environment. Hydrocarbon vapours in air can be filtered using biofilters that utilize organisms indigenous to the filter material or provided by soil or commercial inoculum. In an ocean, chemical dispersants can be used to stimulate biodegradation by increasing the surface area of the oil available for microbial attachment and, perhaps, providing nutrients to stimulate microbial growth.

Oil can penetrate the surface gravel on shorelines, creating a situation that is easily remedied by aerobic bacteria (the gravel is highly permeable and, therefore, conducive to oxygen availability). However, nitrogen and phosphorous are limiting nutrients in this type of environment, and so addition of nutrients may be necessary for a significant period of time.

In areas that lack oxygen (e.g. marshes and mangroves), aeration is likely to be most effective. "Pump and treat" refers to a groundwater purification technique that involves bringing contaminated water to the surface, removing free product by flotation, and re-injecting clean water into the aquifer. As in other environments, adding oxygen and nutrients can stimulate the degradation of residual hydrocarbons that are not extracted by pumping. "Bioslurping" involves pumping the liquid contaminant out while drawing in air to stimulate aerobic degradation. Fertilizer nutrients are frequently added at air injection wells.

In any contaminated site, biodegradation can be enhanced through the addition of oxidizing molecules such as oxygen, nitrate, ferric iron, sulphate and CO_2. In contaminated soil, oxygen is a limiting nutrient, and soil tilling is widely practised. Bioventing can also be a successful remedy, wherein air is injected and extracted through wells in order to strip volatiles and provide oxygen to indigenous organisms. Plants can be used to help deliver air to soil microbes. The other methods include biopiles, composting and slurry bioreactors.

Halogenated Organic Solvents (HOS)

Halocarbons, both halogenated compounds and solvents, are widespread pollutants in air, water and soil; they are recalcitrant molecules resistant to mineralization because the carbon–halogen bond is very stable. The stability and chemical inertness make many halogenated compounds suitable for many industrial processes. However, their degradation is slow once they are released into the environment. In general, the more the halogens on a molecule, the slower the degradation. The focus of bioremediation of halocarbons involves exploiting the ability of organisms to mineralize these toxic contaminants into benign molecules.

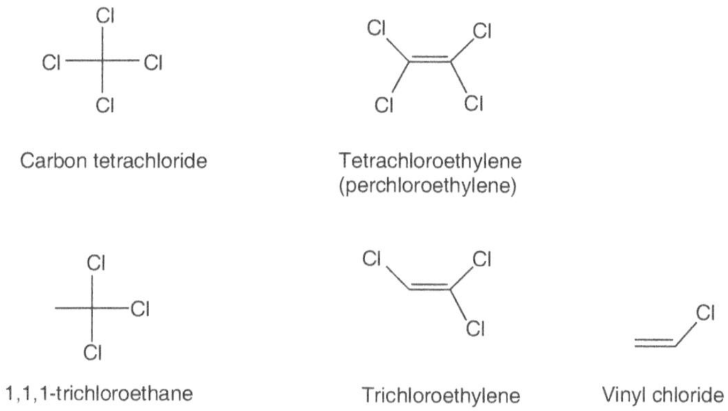

Carbon tetrachloride

Tetrachloroethylene
(perchloroethylene)

1,1,1-trichloroethane

Trichloroethylene

Vinyl chloride

Figure 26.7   Chemical structure of halogenated organic solvents

Halogenated organic solvents (Figure 26.7) have a wide range of applications including metal processing, electronics, dry cleaning and paint, paper and textile manufacturing. They have the potential to contaminate almost every aspect of the environment, particularly water and soil. Since halogenated solvents are generally denser than water, they tend to sink and accumulate in groundwater sources (they are also known as dense non-aqueous phase liquids [DNAPLs]) Acute overexposure to halogenated solvents causes serious health consequences including nervous system damage, heart failure and increased chance of cancer. At lower doses, halogenated solvents are considered potential carcinogens and have been shown to cause cancer in laboratory animals.

Halogenated solvents may be degraded by microorganisms in both aerobic and anaerobic conditions. The anaerobic process proceeds by reductive dechlorination (Figure 26.8) of one chlorine atom at a time. Halogenated solvents have been shown to be the sole carbon source for many anaerobic organisms.

Under aerobic conditions, halogenated solvents can be mineralized by bacteria; however, the bacteria do not use them directly for growth (they are co-metabolized). The process relies on the presence of additional growth substrates, and the halogenated solvents are co-oxidized as part of the

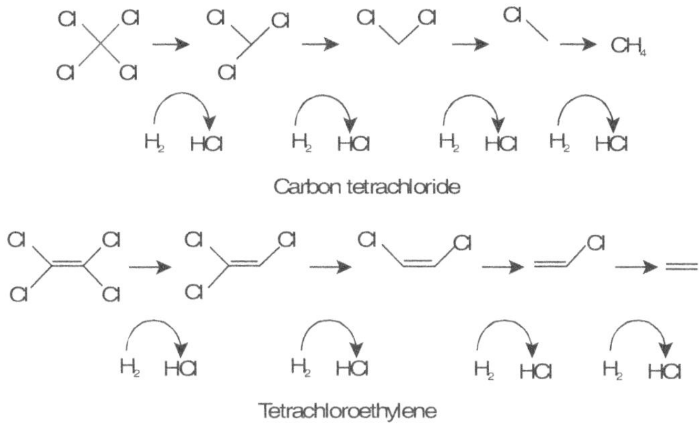

Carbon tetrachloride

Tetrachloroethylene

Figure 26.8   Reductive dechlorination of carbon tetrachloride and tetrachloroethylene

metabolism process. The oxidation process (Figure 26.9) is catalysed by two types of enzymes that utilize O_2: monooxygenases and dioxygenases. Monooxygenases insert one oxygen atom into the halogenated solvent and reduce the other to water. Dioxygenases insert both atoms into the halogenated solvent.

Monooxygenase

Dioxygenase

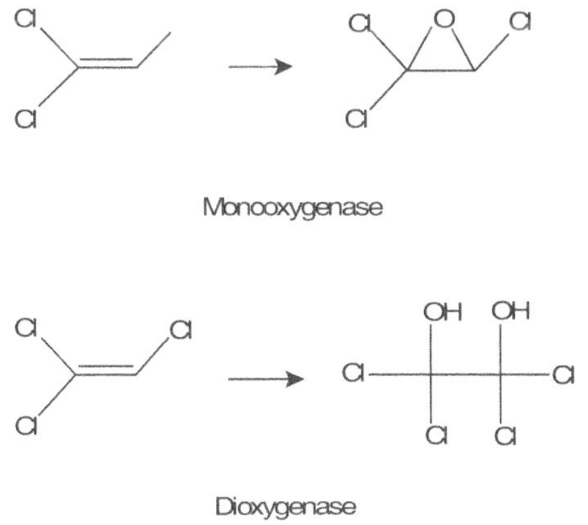

Figure 26.9   Aerobic oxidation of tetrachloroethylene

Co-metabolism is a phenomenon that occurs when a compound is transformed by a microorganism, but the organism does not grow on the compound nor derive any energy, carbon or other nutrients from the transformation. Co-metabolic transformation occurs when an enzyme of a microorganism growing on compound A recognizes compound B as substrate and then transforms it into a product. The transformation is limited, because the next enzyme of the organism that should attack in sequence has a higher specificity and does not recognize the product of B as a substrate. It is a dead-end transformation without benefit to the organism.

Bioremediation of HOS

Biofilters may be used to treat air and water waste from industrial processes. Contaminants in the airstream are passed through a porous medium (usually compost or soil) and partitioned out into a thin film of water, where they can be biodegraded by microorganisms. While using biofilters, it is important to cultivate a crop of microorganisms capable of degrading the contaminants. This process is highly dependent on temperature, humidity conditions, rate of flow, and the components of the waste stream, which affect the health of microorganisms. The advantage of using a biofilter is that the hazardous materials are completely destroyed. Biofilters have been shown to be effective in removing halogenated organic solvents such as dichloromethane from waste streams.

In situ treatment   Halogenated organic solvents tend to accumulate on fractured bedrock at the bottom of aquifers, which is a difficult location to treat. Bioremediation is advantageous over other methods for reasons of low capital costs, no secondary waste, destruction of contaminants, low maintenance, minimal site disturbance, and decreased working hazards. Subsurface anaerobic microbes use chlorinated solvents as a source of energy with dissolved hydrogen as the electron acceptor. Other microbes co-metabolize compounds when more than one substrate or carbon source is available. Nutrients or fertilizers and additional microorganisms may be added (bioaugmentation) to the site to enhance biodegradation. Hydrogen release compounds (HRC) may be used to promote anaerobic dechlorination at contaminated sites. When injected into the site, HRC provides a slow-time release of dissolved hydrogen, which serves as the electron donor, while the chlorinated solvent is the electron acceptor in microbe metabolism. The reductive dechlorination process for tetrachloroethylene (TCE):

$$C_2Cl_4 + H_2 \rightarrow C_2HCl_3 + H^+ + Cl$$

$$C_2HCl_3 + H_2 \rightarrow C_2H_2Cl_2 + H^+ + Cl$$

$$C_2H_2Cl_2 + H_2 \rightarrow C_2H_3Cl + H^+ + Cl$$

$$C_2H_3Cl + H_2 \rightarrow C_2H_4 + H^+ + Cl$$

Halogenated Organic Compounds (HOC)

Halogenated compounds (Figure 26.10) have been widely used in different industrial processes. However, they present a danger to human health and include toxins and potential carcinogens such as dioxins, pesticides and PCBs. One example of a halogenated organic compound is the pesticide DDT, which can bioaccumulate in animal fat tissue, disrupt hormone function, and damage ecosystems. Polychlorinated biphenyls (PCBs) is another type of halogenated organic compounds that have wide industrial applications as coolants, lubricants, plasticizers and dyes. PCBs has been shown to cause cancer in laboratory studies. It may damage neurological, immune system, reproductive system and other organs in humans.

Halogenated compounds belong to a class of molecules known as persistent organic pollutants (POPs) that tend to biodegrade very slowly. It was thought that no natural sources of halogenated compounds were present in the environment, and hence no organisms had evolved to exploit them. However, it has been recently discovered that organisms as well as volcanic eruptions can produce these compounds, and that natural production of chlorinated phenols may be greater than anthropogenic sources. Since these compounds may have existed for millions of years, many strains of bacteria have evolved to break down halogenated compounds, opening up the possibility for its bioremediation.

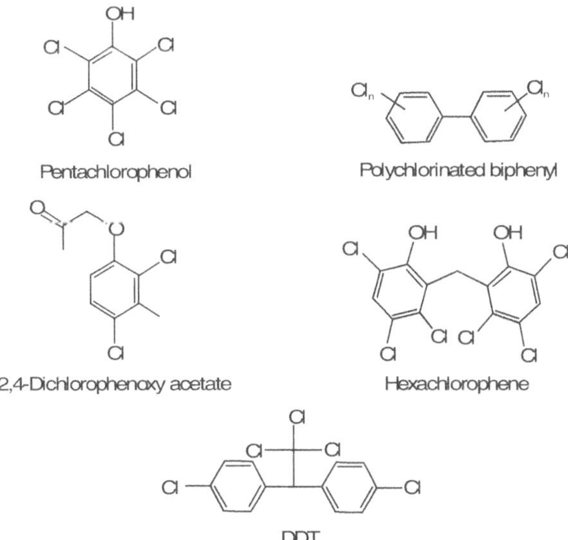

Figure 26.10 Halogenated compounds

Biodegradation of halogenated compounds depends on the ability of the bacterial dehalogenase enzymes to catalyse the cleavage of carbon-chlorine bonds. There are seven known mechanisms of enzyme-catalysed dehalogenation: reductive, oxygenolytic, hydrolytic and thiolytic dehalogenation, intramolecular nucleophilic displacement, dehydrohalogenation, and hydration. Reductive dehalogenation in methanogenic conditions as well as aerobic degradation with monooxidase of persistent compounds, such as DDT and polychlorinated biphenyls, has been observed.

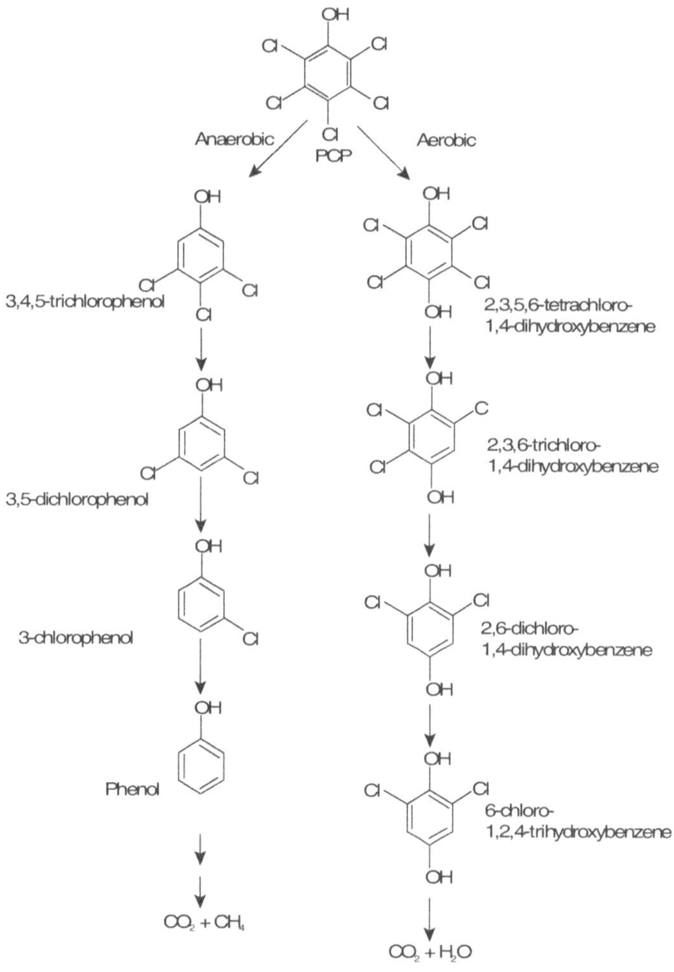

Figure 26.11   Microbial degradation pathway of polychlorophenol (PCP)

Polychlorophenol (PCP) may be degraded aerobically by bacteria with monooxygenase enzyme and anaerobically through reductive dechlorination (Figure 26.11). Anaerobic degradation relies on methanogenic or sulphidogenic microbial consortia in the presence of additional carbon source. These microorganisms derive energy from halogenated compounds (in contrast to haloalkyl degraders that simply co-metabolize contaminants). They are substrate-specific in the types of chemicals they degrade.

Bioremediation of HOC

Bioaugmentation   The site is seeded with a culture of microorganisms that can mineralize the contaminants. Bacteria and fungi are good at metabolizing hazardous chemicals. Bioengineering may discover many new strains of bioremediating microbes.

Phytoremediation   Plants or algae may be used to remove, transfer, stabilize or destroy contaminants in soil, sediment and groundwater. Enhanced rhizosphere biodegradation involves microbes in the soil or groundwater immediately surrounding plant roots. Phytoextraction or phytoaccumulation describes the process by which the plant accumulates contaminants in its tissues (which may be removed by harvesting the plant). Phytodegradation occurs when the plant can metabolize the contaminant into a less harmful form. Phytostabilization occurs when the plant produces chemicals that can effectively immobilize contaminants in the soil.

HERBICIDES AND PESTICIDES

When pesticides, herbicides and fungicides are applied in correct quantities, they are usually biodegradable and that they do not accumulate in the food chain. However, when these chemicals (Figure 26.12) are introduced into the environment in large quantities, they present a contamination problem. Furthermore, some non-chlorinated pesticides and herbicides may contaminate the water supply and pose a potential health hazard. Contamination may also result from heavy rainfall that washes away pesticide and herbicide residue from newly-treated fields into streams. Bioremediation strategies may be employed to mineralize these compounds at contaminated sites.

Contamination in drinking water can cause serious consequences, even to those who do not actually drink such water. The problem can quickly jump from phytoplankton feeding on contaminated water to the humans (phytoplankton → clams → fish → humans). Thus, the managing of storage and spills of toxins such as pesticides and herbicides is critical.

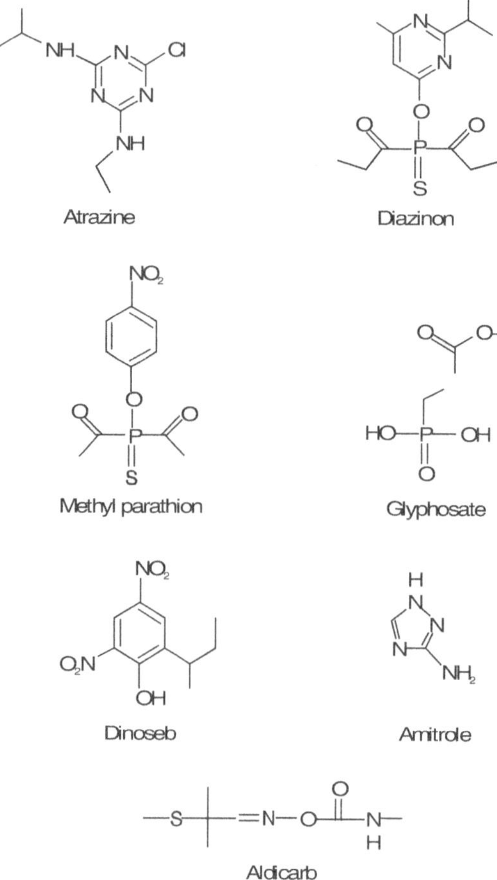

Figure 26.12   Structures of chemicals commonly used as pesticides and herbicides

It is critical to find a biological process that accelerates the process of biodegradation of pesticides because their accumulation can cause serious problems throughout the food chain. Atrazine, a common herbicide, is resistant to aerobic degradation due to its triazine ring (Figure 26.13). Thus, it increases the possibility for long-term contamination effects in the environment. Atrazine contamination can cause serious repercussions, even at seemingly low concentrations. Twenty pounds of atrazine evenly dispersed throughout an aquifer is sufficient to contaminate 10 million litres of water at a level of 100 ppb (parts per billion). This level is significantly higher than the accepted contamination level in drinking water, which is 3 ppb.

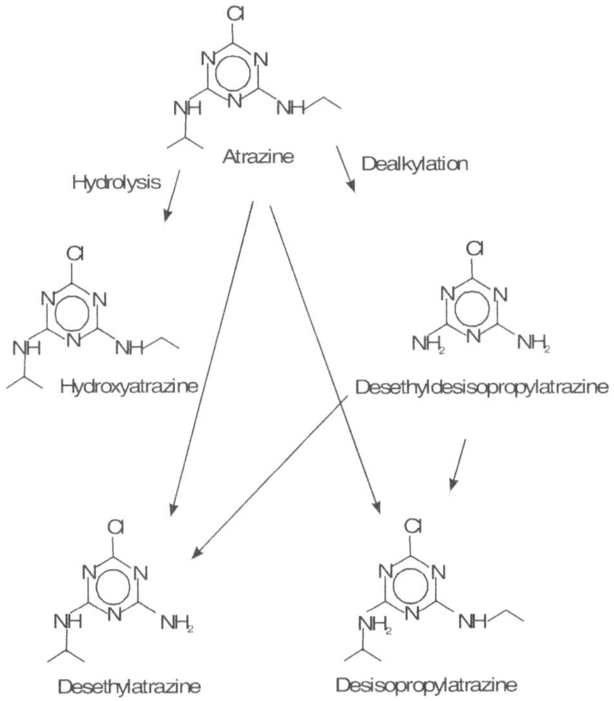

Figure 26.13   Structures of atrazine and its metabolites *via* three different pathways (biological dealkylation, chemical hydrolysis and biological hydrolysis)

BIOREMEDIATION OF PESTICIDES

Chemical Degradation of Pesticides

One way of bioremediation is to chemically denature the pesticides using alkaline water or soil. Carbamates and organophosphates are generally susceptible to this type of degradation. For the organophosphate diazinon, a chemical most stable at a pH of 7 with a half-life of 10 weeks, a pH of 5 will reduce the half-life to 2 weeks.

Microbial Degradation of Pesticides

This process usually occurs in the soil, whereby the pesticides are broken down into a less active, less toxic compounds by fungi, bacteria and other microorganisms. There are many conditions that strongly influence the effectiveness of the reaction, such as moisture, temperature, aeration, pH, and the amount of organic matter. If the pesticide is applied more to a given area, rapid microbial degradation is more likely to occur as repeated applications can stimulate a build-up of microorganisms.

Inoculation of soil with active organisms can spur bioremediation of contaminated soil. Herbicide degradation tends to occur faster in cultivated soil, suggesting the possibility for rhizosphere enhanced phytoremediation. In groundwater, atrazine is an s-triazine herbicide that controls weeds by inhibiting photosynthesis. In soil, atrazine usually breaks down within several months. However, once atrazine enters the water supply, it is much more recalcitrant. Atrazine has been shown to cause liver, kidney and heart damage in animal tests. Bioremediation of contaminated water may be achieved by adding atrazine metabolizing species to the water supply. However, complete degradation of atrazine may require several communities of microorganisms, as no one organism has the ability to metabolize atrazine completely.

Nitrogen Compounds

Nitrate (NO_3^-) and ammonia (NH_3) contamination occurs as a result of agricultural run-off of fertilizers or animal waste. It is highly soluble and can easily percolate through the soil to contaminate aquifers. Excess nitrate poses a problem since it leads to eutrophication in streams, which can cause adverse health effects in humans. Nitrogen contamination may also exist in the form of nitroaromatic compounds such as nitrobenzene, nitrophenol, nitrotoluene (Figure 26.14) and nitrobenzoate, which are used in the manufacturing of pesticides, explosives, dyes, pharmaceuticals and plastics.

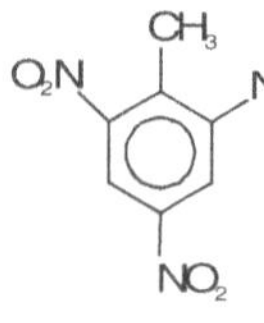

Figure 26.14   Chemical structure of trinitrotoluene (TNT)

Nitrate ingestion can cause blue-baby syndrome (methaemoglobinaemia) when nitrate-reducing bacteria in the intestine produce nitrite which bind with haemoglobin in blood. This inhibits the red blood cells from transporting oxygen to tissue cells. Nitroaromatics are highly toxic as they are mutagenic and carcinogenic.

Under natural circumstances, nitrogenous pollutants like ammonia and nitrates are bioremediated by nitrifying and denitrifying bacteria. Ammonia is aerobically converted to nitrate, and nitrate is anaerobically converted to nitrogen gas. *Nitrosomonas* oxidize ammonia to nitrite, *Nitrobacter* oxidize nitrite to nitrate and the denitrifying bacteria like *Pseudomonas aeruginosa* anaerobically reduce nitrate to nitrogen gas.

This process balances the influx of fixed nitrogen with the outflow of nitrogen gas. However, due to anthropogenic nitrogen input (fertilizer and waste), fixed nitrogen is often added to soil and groundwater faster than it can be converted to N_2, thus leading to nitrogen contamination. Therefore, the strategies for bioremediating nitrogenous compounds should enhance the capabilities of nitrifying and denitrifying bacteria.

While ammonia and nitrate are the main nitrogenous pollutants, bacteria can assist in the degradation of nitroaromatics as shown in the reaction pathway for nitrobenzene metabolism (Figure 26.15).

Metals

A variety of metal contaminants exist in groundwater, surface water and soils resulting from industrial and agricultural activity. Toxic metals such as lead, mercury, cadmium, arsenic, chromium and uranium can cause damage to human health and the environment. The most prevalent form of metal pollution is acid mine drainage. This occurs when the mining of coal and metal ores exposes metals and radionuclides to the atmosphere, allowing them to be oxidized by certain bacteria (e.g. *Thiobacillus ferrooxidans*). For example, pyrite (often exposed in coal mining) can be oxidized to iron hydroxide and sulphuric acid.

$$FeS_2 + \frac{15}{4} O_2 + \frac{7}{2} H_2O \rightarrow Fe(OH)_3 + 2H^+ + 2HSO_4$$

The effluent produced, known as acid mine drainage, is highly toxic since it is acidic and contains toxic metals. Such effluents contaminate the nearby watersheds, killing much of the aquatic life. Streams that contain acid mine effluent are often discoloured due to $Fe(OH)_3$ or "yellow-boy" contamination.

Unlike organic compounds, metals cannot be broken down into non-toxic components. However, biological organisms can reduce their toxicity through processes such as chelation and precipitation.

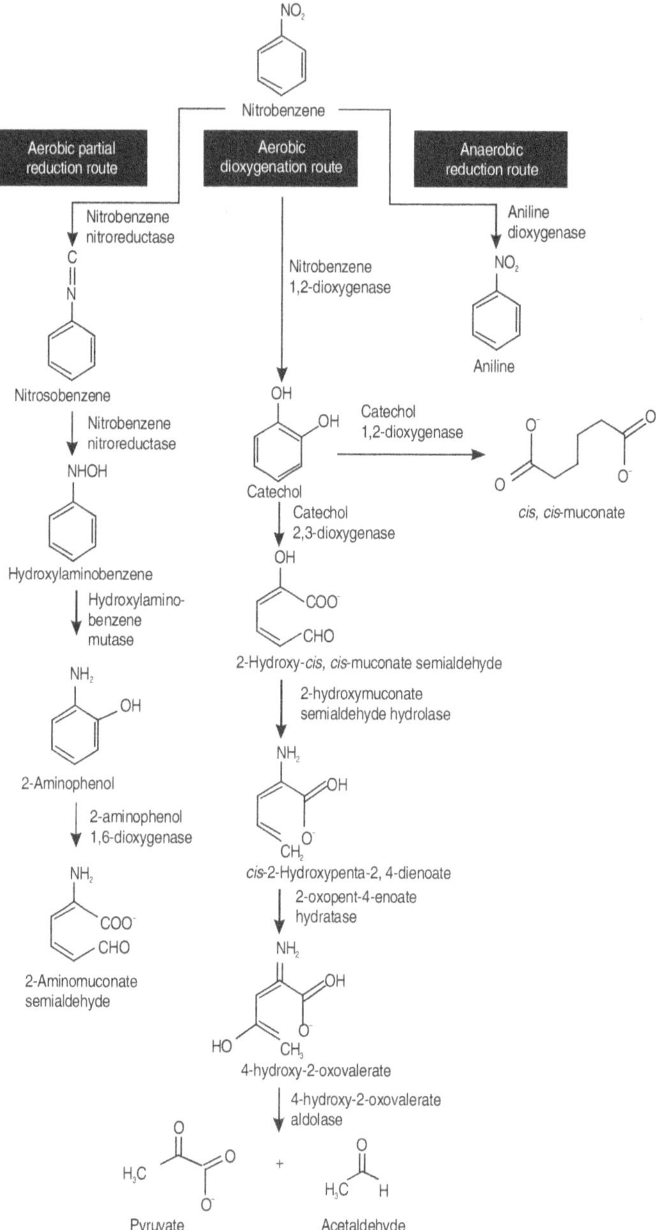

Figure 26.15 Microbial metabolism of nitrogen compounds

Metal Chelation

Chelates are compounds that contain many ligand sites that attach to the metal. The tight bond of the chelate renders the metal non-toxic. Chelates like mugineic acid readily bind to metals (Figure 26.16).

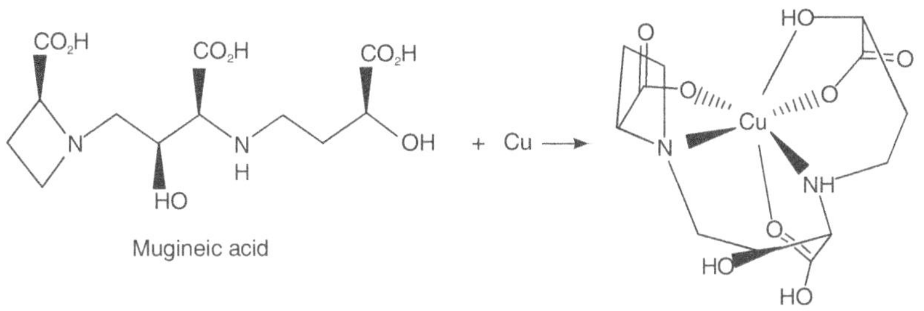

Figure 26.16   Chelation of mugineic acid with metal

Many organisms including plants and bacteria use chelates to aid the absorption and transportation of essential metal nutrients. Moreover, their binding property is often used to stabilize toxic metals as well. Thus, chelates are critical compounds for bioremediation, especially in phytoremediation. Other chelates including metallothionens and phytochelatins facilitate metal transportation in organisms. While metallothionens exist in many kinds of organisms, phytochelatins are produced only by certain species of plants.

Metal Precipitation

Metals in acid mine effluents can be removed as solid precipitates by anaerobic bacteria. The bacteria reduce the oxidation state of the metal (redox reaction), which usually causes it to form a harmless solid precipitate. For instance, iron reducing bacteria like *Geobacter*, which normally gain energy by reducing Fe (III) to Fe (II), can reduce U (VI) to U (IV) too. The reduced form of the metal forms a non-toxic solid precipitate.

Sulphur reducing bacteria such as *Desulfovibrio* may also be used in bioremediation, but the chemistry requires an extra step.

First, the bacteria reduce the sulphate producing hydrogen sulphide.

$$SO_4^{-2} + 2CH_2OH^+ \rightarrow H_2S + H_2O + CO_2$$

Then, hydrogen sulphide reacts with metals to form a sulphide that precipitates out of the effluent. Additionally, some bicarbonate (HCO_3^-) produced along with the CO_2 in the sulphur redox reaction acts to neutralize the acid in the effluent. With the acidity reduced and the metals now existing as a harmless precipitate, the effluent is effectively remediated. Although precipitation could effectively remediate these toxic metals, the environmental conditions are not always suitable for this process. Sometimes, an electron donor may be missing to reduce the metal. In this case, additional nutrients are supplemented to provide an electron source. In some situations, there is a more suitable electron acceptor such as oxygen.

REVIEW QUESTIONS

1. What is bioremediation and bioaugmentation?

2. Write about advantages and disadvantages of bioremediation.

3. Give the relation between metabolism and bioremediation.

4. Describe in detail about *in situ* bioremediation?

5. Explain about bioventing.

6. What is natural attenuation.

7. Give a detailed account on *ex situ* bioremediation.

8. What are biopiles? Explain.

9. What is biotransformation?

10. Describe in detail about biotransformation of hydrocarbons.

11. Explain in detail about biotransformation of HOS and HOC.

12. Explain in detail about biotransformation of herbicides and pesticides.

BIOLOGICAL WASTE TREATMENT

It is well known in biology that microorganisms need some resources for their maintenance and growth. The general idea of biological waste-water treatment is to use the right microorganisms that transform undesirable substrate (for example nitrates used by intensive agriculture) into biomass. In this way, the pollution present in soluble state in water is concentrated into biomass. This biomass is filtered; the water is now free from pollution. The resulting biomass may then be used to improve soil fertilization.

27

$$\text{Biomass} + \text{Substrate} \rightarrow \text{Biomass}$$

MATERIALS THAT NEED TREATMENT

Many materials should not be discharged into water bodies without treatment. They are given below.

1. *Pathogens*   Many disease-causing viruses, parasites, and bacteria are present in waste water and enter from almost anywhere in the community. These pathogens often originate from people and animals who are infected with or are carriers of a disease. Gray water and black water from typical homes contain enough pathogens to pose a risk to public health. Other likely sources in communities include hospitals, schools, farms, and food processing plants.

2. *Organics*

   Non-biodegradable   Living organisms feed on but cannot degrade these materials. Therefore, non-biodegradables persist in the environment, and as they move up in food chains, many of them concentrate in the fat of animals.

Biodegradable   Organisms whose growth is limited by their carbon source will proliferate if large quantities of the source are suddenly made available. For example, sulphate (SO_4^{2-}) reducers are limited by carbon. Therefore, if a carbon source is discharged into the environment, these organisms multiply. Moreover, sulphate reduction generates a toxic gas, hydrogen sulphide (H_2S). In addition, as these organisms grow, their predators and other organisms that live on their decayed materials also grow and deplete the oxygen supply. Consequently, other aerobic organisms such as fish starve for oxygen and die. This phenomenon is known as eutrophication.

Table 27.1   Notable water-borne pathogens

Pathogenes	Disease caused
Bacterial	
Salmonella typhi	The causative agent of typhoid fever.
Vibrio cholerae	The causative agent of cholera. The acidic gastric (stomach) juices provide some protection against this organism. Infection requires 10^9 organisms in an acidic environment but only 10^4 organisms in neutral environments. This disease strikes children more often than adults because the stomachs of children are generally less acidic.
Protozoan	
Giardia lamblia	The causative agent of giardiasis, a severe form of gastroenteritis. This organism has a cyst phase that is not killed by chlorine (Cl_2). Cases of contamination occur especially in relatively unpopulated areas. If the filtration step of water treatment is omitted, such cysts may contaminate the drinking water.
Entamoeba histolytica	It may cause ulcers and abscesses in the liver and lungs. If untreated, brain damage may occur.
Viral	
Hepatitis A	It is a single-stranded RNA virus that infects the liver. The resulting liver inflammations are usually not fatal.
Poliovirus	It is a single-stranded RNA virus. A worldwide vaccination program has made infections by this virus extremely rare.

3. *Phosphate*   The growth of organisms such as algae and cyanobacteria, which fix their own carbon and nitrogen, is limited by phosphate. Thus, if a phosphate source is introduced into the environment, these primary producers will proliferate and eutrophication may occur.

4. *Nitrogen sources* (NH_4^+, NO_3^-, NO_2^-)   The introduction of nitrogen sources into the environment will allow nitrogen-limited organisms to

grow, initiating eutrophication. Fertilizers are rich in such nitrogenous materials.

The main point to note is that untreated, nutritious materials, when released into the environment, may favour the growth of certain microorganisms, upset the normal balance of ecosystems, and kill many larger creatures that require oxygen to live.

There can be three stages of sewage treatment. The primary treatment physically removes particulate material with screens in a settling pond. The secondary treatment uses microorganisms to reduce the level of organic matter; the amount of residual organic matter is quantified as the biochemical oxygen demand. Tertiary treatment includes processes that remove inorganic nutrients, such as phosphate and nitrate, from the waste water. Most sewage plants use only primary and secondary treatment of waste water.

PRIMARY TREATMENT

Preliminary treatment of waste includes screening, grinding, grit removal, floatation, equilization and flocculation (Figure 27.1).

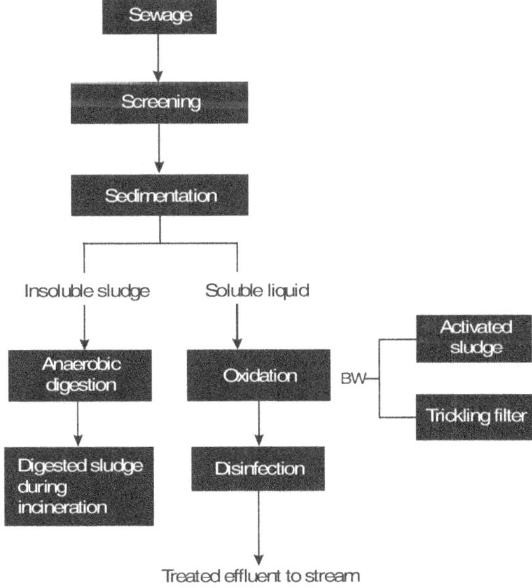

Figure 27.1   Flow diagram of waste treatment

Screens, grinders and grit removal are provided for the protection of other equipment in the treatment plant. Air floatation and flocculation aid in the removal of suspended solids and oil in the primary clarifier and reduce the biological loading on secondary treatment processes. Prechlorination or preaeration may be required to prevent odour problems and to eliminate septic conditions where waste water abnormally runs through a long course to the plant. Equalization structures are used to dampen diurnal flow variations and to equalize flows to the treatment facilities.

RACKS AND SCREENS

Racks and screens (Figure 27.2) are mainly employed to remove larger particles from the waste. Raw sewage enters the treatment plant through two influent pumping stations. At this pumping station, large screens with vertically placed bars remove large debris like rags, sticks and paper. After being screened, the waste water is pumped into a parallel bank of grit basin tanks.

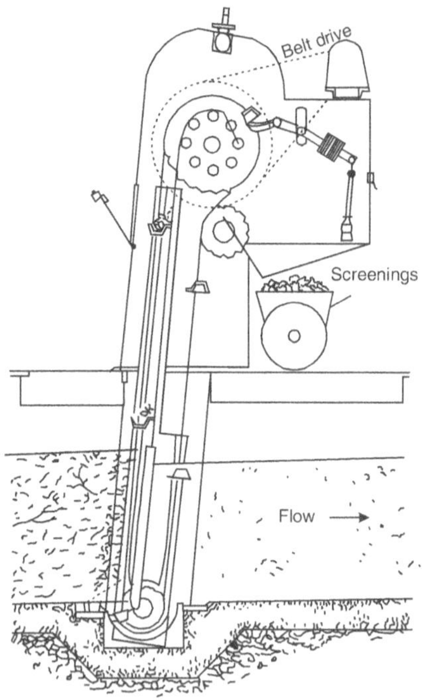

Figure 27.2   Basic design of racks and screen

The primary function of coarse screening is protection of downstream facilities rather than effective removal of solids from the plant influent. All screens used in sewage treatment plants or in pumping stations may be classified as follows.

1. Trash racks, which have a clear opening between bars of 1½ to 4 inches and are usually cleaned with hand, by means of a hoist, or with a power-operated rake.

2. Standard, mechanically cleaned bar screen with clear openings from ½ to 1½ inches.

3. Fine screen with openings of ¼ inch or smaller.

Design Basis

Screens are positioned such that they are readily accessible. An approach velocity of 2 feet per second, based on average flow of waste water through the open area, is required for manually cleaned bar screens. For a mechanically cleaned screen, the approach velocity will not exceed three feet per second at maximum flow.

Bar spacing   Clear openings of 1 inch are usually satisfactory for bar spacing, but 0.5 to 1.5 openings may also be used. The standard practice will be to use $0.32" \times 2"$ bars up to 6 feet in length and $0.38" \times 2"$ or $0.38" \times 5"$ bars up to 12 feet in length. The bar should be long enough to extend above the maximum sewage level by at least 9 inches.

Size of screen channel   The maximum velocity through the screen bars, based on maximum normal daily flow, will be two feet per second. For wet weather flows or periods of emergency flow, a maximum velocity of three feet per second will be allowed. This velocity will be calculated on the basis of the screen being entirely free from debris. To select the proper channel size, knowing the maximum storm flow and the maximum daily normal flow, the procedure is as follows: the sewage flow (million gallons per day) multiplied by the factor 1.547 will give the sewage flow (in cubic feet per second). This flow (in cubic feet per second) divided by the efficiency factor obtained from Table (27.2) will give the wet area required for the screen channel. The minimum width of the channel should be 2 feet and the maximum width should be 4 feet. As a rule, it is desirable to keep the sewage in the screen channel as shallow as possible in order to keep down the head loss through the plant; therefore, the allowable depth in the channel may be a factor in determining the size of the screen. In any event, from the cross-sectional area in the channel, the width and depth of the channel can be readily obtained by dividing the wet area by the depth or width, whichever is the known quantity.

Table 27.2 Efficiency of bar spacing

Bar size (inches)	Opening size (inches)	Efficiency
1/4	1	0.800
5/16	1	0.768
3/8	1	0.728
7/16	1	0.696
1/2	1	0.667

Velocity check   Although screen channels are usually designed on the basis of maximum normal flow or maximum storm flow, it is important to check the velocities, which would be obtained through the screen for minimum or intermediate flows. The screen will be designed so that, at any period of flow, the velocities through the screen do not exceed 3 feet per second under any flow condition.

GRIT CHAMBER

The primary purpose of grit chambers (Figure 27.3) is to protect pumps and other mechanical equipment. They may not be required if surface runoff is excluded from the sanitary sewer system; however, silting occurs through improper joints, broken manholes, and other openings in the system even without contributing to surface runoff. Grit chambers will be located ahead of pumps and commuting devices. Coarse bar racks will be placed ahead of mechanically cleaned grit-removal facilities. There are two types of grit chambers: horizontal-flow and aerated. The first attempt to reduce the velocity of the waste water flow is by placing the grit chambers. This velocity will carry most of the smaller organic particles through the chamber and will tend to resuspend those that settle but will allow the heavier inorganic grit to settle out. In recent years, aerated grit chamber has been more widely used because introducing oxygen into the waste water early in the treatment process is beneficial and there is minimal head loss through the chamber; however, the increased operational and energy costs must be evaluated.

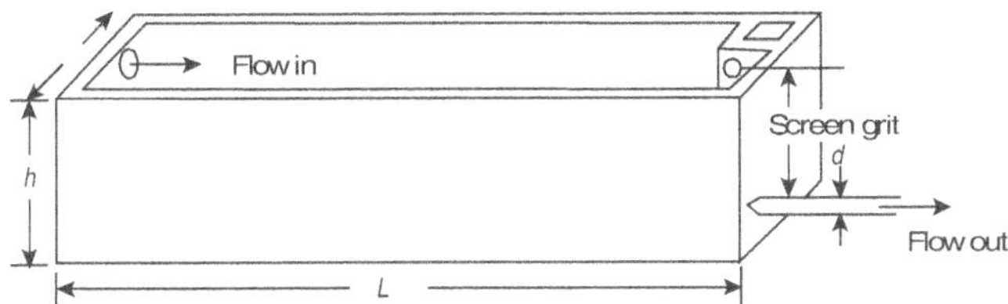

Figure 27.3   Basic design of grit chamber

Ideally, the desired flow rate and retention time would be known, and then height (h) relative to pipe, and (L) could be calculated. The h relative to pipe is important because it determines how high the opening of the outlet pipe should be placed for the given flow rate. The width of the grit chamber, w, is fixed throughout the calculations. These parameters are all linked to the desired flow rate and retention time as explained below.

First, input the desired flow rate, "Q", and retention time, "t". The desired flow rate is linked to the outlet velocity, v, as follows:

$$v = \frac{Q}{\dfrac{\pi}{d^2/4}}$$

Once the outlet velocity is known, it is then possible to solve for h relative to pipe, hp, using the equation:

$$hp = \frac{v^2}{2g}$$

Now that hp is known, you can use the desired retention time to determine the height, h, and length, L, of the grit chamber:

$$L = \frac{Qt}{hw}$$

It is important to note that Qt should be converted to units of m^3. Also, the width of the grit chamber, w, should be chosen by the user. The only two unknown variables left are h and L. Through trial and error, keep changing the values until you obtain suitable h and L values for the grit chamber. In addition, there may be limiting factors, such as space availability, at the site that would affect your choices for w, h and L.

Horizontal-flow Grit Chamber

This grit chamber will be designed for a controlled velocity of 1 foot per second (at the average rate of flow) in order to prevent settling of organic solids (at low rates of flow) and scouring (at high rates of flow). The velocities at these conditions will not vary more than 10% from the design velocity. Theoretically, the channel of the grit chamber should be 50 per cent length of the grit chamber to allow for turbulence and outlet distribution. The floor of the chamber will be far enough below the weir crest to make allowance for the accumulation of about 2.5 cubic feet of grit every 10 days per million gallons of waste water. This depth allowance will be not less than 2½ inches. The channels will be as narrow as possible without causing serious submergence of the crown of the inlet sewer. The effluent channel will be designed so that objectionable shooting velocities are not produced and submergence of the weir crest by tail water will not exceed permissible limits.

To size a grit chamber with rectangular cross-section, first determine the horizontal surface area using the following equation:

$$A = QX$$

where,

A = horizontal surface area (sq. ft.),

Q = flow rate (cfs) and

X = settling rate of grit particle (sec./ft.).

Assume a grit chamber of width 2 feet for flows less than 1 million gallons per day, and 2 to 4 feet for flows between 1 and 2 million gallons per day. For flows greater than 2 million gallons per day, divide the area calculated above by the assumed width to obtain the length of the channel. At a flow velocity of 1 foot per second, the depth of flow can be determined by equation

$$D = \frac{L}{VX}$$

where,

D = depth of flow (ft.)

L = length of channel (ft.)

V = flow velocity (fps) and

X = settling rate (sec./ft.).

The rectangular cross-section channel will be used in treatment plants with capacity less than 2 million gallons per day. The parabolic channel is costly and best suited for larger plants. The square grit chambers are sized as rectangular cross-section channel chambers; horizontal area is the basis of the design. These units are best suited for flows greater than 2 million gallons (1 gallon = 3.75 litres) per day.

Aerated Grit Chamber

When waste water flows into an aerated grit chamber, the grit will settle at rates dependent on the size, specific gravity, and velocity of roll in the tank. The variable rate of air diffusion is a method of velocity control, which can be easily adjustable to different field conditions. The following design criteria are to be used.

Air rates   The air system should be designed to provide 8 cubic feet per minute per foot of grit chamber length. The design should allow the air rate to be controlled over a range.

Detention time   The chamber should be designed to have a detention time of 3 minutes at the maximum flow rate.

Geometry   The inlet and outlet should be placed to prevent short-circuiting in the chamber. In addition, the inlet should introduce the waste water directly into the circulation pattern caused by the air. The outlet should be at a right angle to the inlet with a baffle. A length to width ratio of 4 : 1 should be used.

Coagulation and Flocculation

Coagulation of waste water may be accomplished with any of the common water coagulants including lime, iron and aluminum salts, and synthetic polymers. The choice is based on suitability for a particular waste, availability and cost of the coagulant, and sludge treatment and disposal considerations. For example, iron is sometimes available free as a waste product in the form of pickling liquor, and its presence in sludge presents no specific problems for anaerobic digestion. Lime generally provides good clarification, a rapidly settling sludge, and permits the use of a simple method of recovery that also ensures destruction of most sewage solids in the resulting sludge.

An important aspect of waste water purification and clarification is the removal of suspended colloidal material and as much of the solutes as possible. To remove colloidal material, a floc forming chemical is needed. Most floc forming chemicals are tied up with peripheral chemicals, which cause the resulting effluent to have higher total dissolved solids (TDS) content. If this water is being reused, the high TDS levels contribute to a high final water usage cost.

Perhaps the floc chemical could be put into contaminated water in its ionic form, without adding the peripheral chemicals. This should be ideal because the colloidal contaminants and soluble contaminants that would react with the floc chemical could be removed. The result would be clarified water with reduced TDS. Considerable research has been undertaken to evaluate methods by which these objectives could be obtained. One possible approach is treatment through electrochemistry.

Electrochemistry is defined as the use of direct current to cause sacrificial electrode ions to move into an electrolyte and remove undesirable contaminants either by chemical reaction and precipitation or by causing colloidal materials to coalesce and then be removed by electrolytic floatation. The electrochemical system has proven to be able to cope with a variety of waste waters. These waters include paper mill waste, metal plating, tanneries, canning factories, steel mill effluent, slaughter houses, chromate, lead and mercury-laden effluents as well as domestic sewage. These waters may be reduced to clear, clean, odourless and reusable water. Impingement diffusers, jet diffusers or helix-type diffusers may also be used for air flocculation. Mechanical flocculation is achieved by revolving or reciprocating paddles, radial-flow turbine impellers, or draft tubes. In typical situations, mechanical aerators (vertical draft tube type) will be used in single tanks arranged for cross flow.

Mechanical and air flocculation units for domestic waste water will be designed for 30 minutes of flocculation detention time followed by a clarifier with surface settling rates of 800 gallons per day per square foot. The air requirement for flocculation is 0.1 standard cubic feet per gallon at 30 minutes of detention time. To ensure proper agitation, air will be supplied at 2.5 standard cubic feet per minute per linear foot of tank channel. The number and size of air diffusers are determined by dividing the total air requirement by optimum air diffusion rate per unit (4.0 cubic feet per minute per square foot of porous diffuser area). Water depths vary from 8 to 13 feet. In rectangular tanks, the ratio of length to width will be 3 : 1. For mechanical flocculation, revolving paddles may be either horizontal or vertical. Peripheral-paddle speed should be kept in the range of 1.0 to 3.0 feet per second to minimize deposition and yet avoid destruction of the flocs.

Sedimentation

Sedimentation is one of the primary treatment of waste. These can be achieved by sending waste water through the sedimentation tanks. These tanks may be rectangular, circular or square, but all operate on the same principle of collecting the settled solids by slow-moving scrapers to the point of removal.

In rectangular tanks, the waste water enters at one end and flows horizontally to the other end. The scrapers (flights) are attached by their ends to two parallel chains that pass over sprockets. The flights move slowly along the tank floor, pushing the settled solids to a sludge hopper at the end of the tank. At the same time, the partially submerged flights, in their return path, push the floating solids, grease and oils (scum) to a trough at the end of tank.

In centre-feed circular tanks, the waste water enters at the centre and flows radially and generally horizontally to the periphery. The circular tanks have scraper arms attached to a central motor-driven shaft. The bottom of the tanks is sloped towards the centre and the scrapers move the settled solids to a sludge hopper at the centre. Skimmer arms, if present, are attached to the central shaft at the surface for the collection of floating solids, grease and oils (scum).

In square tanks, the waste water enters at the centre and flows to the four sides. The scraper mechanism is similar to that in the circular tanks. The major difference is that the rigid arms of the scraper mechanism are equipped with pivoted corner blades that reach out into the four corners of the tank and move the solids in these areas to the path of circular mechanism.

Plant Operation

The collection and removal of sludge from the sedimentation tank, as well as correct maintenance procedures, are important factors in successful plant operation. The mechanical collection equipment can be run intermittently but is most often run on a continual basis. This prevents excess accumulation of solids on the tank bottom and reduces the load on the collector mechanisms and thereby helps to prevent equipment damage. Solids left in the tank bottom too long will gasify and rise to the surface; therefore, sludge should be removed from the tank as often as necessary. This, in turn, is dependent on waste stream characteristics, volume of flow and sludge quality. The sludge removal schedule must be worked out for each plant by observations and tests, keeping in mind that the objective is to remove sludge at the proper rate, proper concentration and with proper quality for the receiving process unit. Concentrating the sludge reduces the volume of water being pumped and contributes to pumping efficiency. There are, however, pump design limitations to consider. In addition, scum and grease should be removed daily from the tank surface.

Physical Observations

Physical observations at the water surface of each clarifier can provide the operator with valuable information regarding clarifier efficiency and related plant status. The observed data can dictate the operational control necessary to achieve optimum unit efficiency and subsequent plant performance. These results should be confirmed through laboratory testing.

Observations to be made daily and the results recorded:

1. Floating solid accumulations should be noted, if present, whether existing as clumps or as dispersed accumulation.

2. Formation of gas bubbles.

3. Absence or presence of odours.

4. Distribution of flow to multiple tanks.

5. Distribution of flow over weirs.

6. Loss of solids in the effluent.

Other factors that should occasionally be checked are elevation and placement baffles.

Laboratory Control

Evaluation of primary treatment is dependent on laboratory analysis. The frequency of testing and the "range" of test results will vary from plant to plant. Waste stream characteristics, environmental conditions and in-plant operation will affect test data at each facility.

Table 27.2 Laboratory tests commonly associated with primary tanks

Parameter	Type of sample	Location
BOD	Composite	Influent Effluent
Suspended solids	Composite	Influent Effluent
Settleable solids	Grab	Influent Effluent
pH	Grab	Influent Effluent
Temperature	Grab	Influent

The comparison of influent versus effluent test results provides information required to calculate and evaluate clarifier efficiency.

SECONDARY TREATMENT

In normal waste water, the total solids may be classified as organic or inorganic in origin. In terms of the size of solids, the distribution is approximately 30% suspended, 6% colloidal and 65% dissolved solids. We know that the function of primary treatment is to remove as much of the suspended solids as possible. Primary treatment utilizes clarifiers or settling tanks, which remove settleable organic and inorganic solids from waste water. The effluent from primary treatment therefore contains mainly colloidal and dissolved organic and inorganic solids. Recent effluent standards and water quality standards require a greater degree of removal of organics from waste water than can be accomplished by primary treatment. This additional removal of organics can be accomplished by secondary treatment. Secondary treatment consists of the biological treatment of waste water by utilizing many different types of microorganisms in a controlled environment (Figure 27.4).

In biological treatment of waste waters, a mixed population of microorganisms utilizes the colloidal and dissolved organics found in the effluent from primary treatment. In consuming these organics, the microorganisms utilize a part of the organic substances to obtain energy needed for their life activities. When the oxidation of organics occurs in the presence of dissolved oxygen, the end products include carbon dioxide, water, sulphates, nitrates and phosphates. The remainder part of consumed organics is used as building blocks in a series of synthesis (reproduction) reactions that result in an increase in the population of microorganisms. Therefore, the colloidal and dissolved organics originally present in waste water have been transformed in part into a stable form, such as carbon dioxide, and part into a viable biological mass. This biochemical reaction is active in all biological treatment processes. The biological mass must subsequently be separated from waste water to ensure a proper degree of treatment within effluent and water quality standards. If this biological mass is not properly removed from the waste stream, usually by final clarification, treated effluent quality will be less due to the highly suspended solids (SS) and biological oxygen demand (BOD).

In activated sludge process, the microorganisms are dispersed throughout the water phase. While in trickling filters or biodiscs, the microorganisms are attached to a fixed surface forming a biological film. In either process, the microorganisms carry out the treatment and therefore all precautions must be taken to assure a favourable environment for their life cycle.

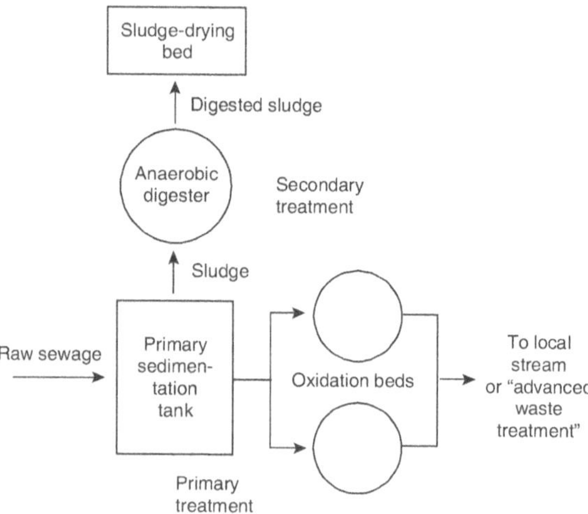

Figure 27.4 Schematic diagram of primary and secondary treatment of waste

AEROBIC DIGESTION

Aerobic digestion of waste is a natural biological degradation and purification process in which bacteria that thrive in oxygen-rich environments break down and digest the waste. During oxidation process, pollutants are broken down into carbon dioxide, water, nitrates, sulphates and biomass (microorganisms). By operating the oxygen supply with aerators, the process can be significantly accelerated. Of all the biological treatment methods, aerobic digestion is widely used throughout the world. Aerobic bacteria demand oxygen to decompose dissolved pollutants. Large amounts of pollutants require large quantities of bacteria; therefore the demand for oxygen will be high. The biological oxygen demand (BOD) is a measure of the quantity of dissolved organic pollutants that can be removed in biological oxidation by the bacteria. It is expressed in mg/l. The chemical oxygen demand (COD) measures the quantity of dissolved organic pollutants that can be removed in chemical oxidation, by adding strong acids. It is expressed in mg/l. The BOD/COD gives an indication of the fraction of pollutants in the waste water that is biodegradable.

Advantages of Aerobic Digestion

Aerobic bacteria are very efficient in breaking down waste products. Aerobic treatment usually yields better effluent quality than that obtained in anaerobic processes. The aerobic pathway (Figure 27.5) also releases a substantial amount of energy. A portion is used by the microorganisms for synthesis and growth of new microorganisms.

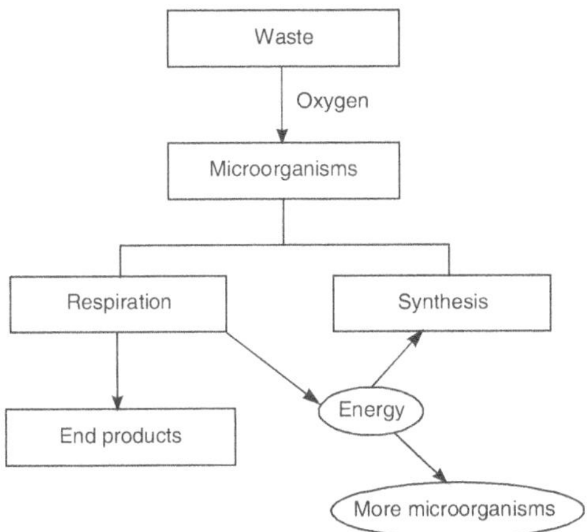

Figure 27.5   Pathway of aerobic digestion of waste

AEROBIC DECOMPOSITION

It is a biological process in which organisms use available organic matter to support biological activity. The process uses organic matter, nutrients and dissolved oxygen, and produces stable solids, carbon dioxide, and more organisms. Microorganisms that can only survive in aerobic conditions are known as aerobic organisms. In sewer lines, the sewage becomes anoxic if left for a few hours and becomes anaerobic if left for more than 1½ days. Anoxic organisms work well with aerobic and anaerobic organisms. The facultative organisms and anoxigenic organism have the same concept.

TYPES OF AEROBIC TREATMENT OF WASTE

ACTIVATED SLUDGE

The activated sludge process was developed in England in 1914 and was so named because it involved the production of an activated mass of microorganisms capable of aerobically stabilizing the organic content of the waste. Activated sludge is probably the most versatile of the biological treatment processes capable of producing an effluent with any desired BOD. The process thus found wide applications among domestic waste water and industrial waste water-treatment.

Like the trickling filter, activated sludge is a biological contact process where bacteria, fungi, protozoa and small organisms, such as rotifers and nematode worms, are commonly found. Bacteria are the most important group of microorganisms for they are the ones responsible for the structural and functional activity of activated sludge floc. All types of bacteria make up activated sludge. The predominate type of bacteria present will be determined by the nature of organic substances in the waste water, the mode of operation of the plant, and the environmental conditions present in the process.

Fungi are relatively rare in activated sludge. When present, most of the fungi tend to be of the filamentous form, which prevents good floc formation and therefore decreases the efficiency of the plant. A high carbohydrate waste, unusual organic compounds, low pH, low dissolved oxygen concentrations, and nutrient deficiencies stimulate fungal growth. The other forms of microorganisms present in activated sludge play a minor role in the actual stabilization of organics in waste water.

Microorganisms in the Activated Sludge Process

Activated sludge process is a treatment technique in which waste water and reused biological sludge full of living microorganisms is mixed and aerated. The biological solids are then separated from the treated waste water in a clarifier and returned to the aeration process or wasted. The microorganisms are mixed thoroughly with the incoming organics. As they grow and are mixed with air, the individual organisms clump together (flocculate). Once flocculated, they settle more readily in the secondary clarifier. The activated sludge is constantly growing and produce more than can be returned for use in the aeration basin. Some of this sludge must, therefore, be wasted to a sludge handling system for treatment and disposal. The volume of sludge returned to the aeration basins is normally 40 to 60% of the waste-water flow; the rest is wasted. A fixed growth of microorganisms occurs on synthetic media similar to trickling filters. By means of sludge recalculation, a population of suspended growth microbes is developed in addition to the fixed growth on the media.

Basic Design of Activated Sludge Equipment

The basic activated sludge process (Figure 27.6) consists of several interrelated components:

- ➲ An aeration tank where the biological reactions occur.

- ➲ An aeration source that provides oxygen and mixing.

- ➲ A tank, known as the clarifier, where solids settle and are separated from treated waste water.

- ➲ A collecting means for the solids either to return them to the aeration tank (returned activated sludge, RAS), or to remove them from the process (waste activated sludge, WAS).

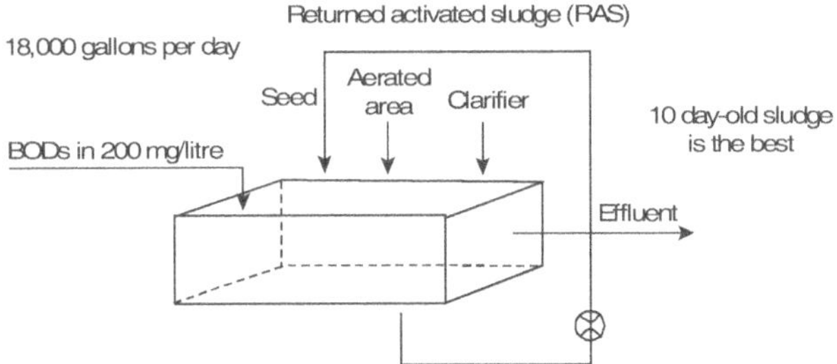

Figure 27.6 Principle diagram of activated sludge process

Primary settling tanks   The oldest and most widely used form of water and waste-water treatment uses gravity settling to remove particles from water. The tanks may be round, square or rectangular in shape. Sedimentation takes place in the primary settling tanks and is relatively simple and inexpensive. Particulates suspended in surface water can range in size from 10^{-1} to 10^{-7} mm in diameter, the size of fine sand and small clay respectively. Turbidity or cloudiness in water is caused by those particles larger than 10^{-4} mm, while particles smaller than 10^{-4} mm contribute to the water's colour and taste. Such very small particles may be considered for treatment purposes, because it will not settle properly and dissolve more than higher particles.

Water-containing particulate matter flows slowly through a sedimentation tank and is detained long enough for the larger particles to settle to the bottom before the clarified water leaves the tank over a weir at the outlet end. Particles that have settled to the bottom of the tank are removed manually or by mechanical scrapers. Detention time is typically 3 hours in tanks that are 3–5 m deep.

Aeration tanks   The waste water flows into an aeration chamber usually constructed of steel, polystyrene, fibreglass or concrete. The aeration chamber normally provides 6–24 hours retention time for the waste water. The contents of the aeration tank are referred to as mixed liquor, and the solids are called mixed liquor suspended solids (MLSS). The latter includes inert materials as well as living and dead microbial cells. In the aeration tank, microorganisms are kept in suspension for 4 to 8 hours by mechanical mixers and/or diffused air, and their concentration in the tank is maintained by the continuous return of settled biological floc from the secondary settling tank to the aeration tank.

Final settling tanks (Settling chamber or secondary clarifier)   Like primary tanks, final tanks may be rectangular or circular, and occasionally square, but they provide longer detention (2 hours) and lower overflow rates (30–50 m^3/m^2 per day). The final settling tank receives the overflow of the aeration chamber. When the sludge settles to the bottom of the tank, it is still active, and it is possible to remove more BOD from waste water. Returning the activated sludge to the aeration chamber on a continuous basis maintains and increases the microorganism concentration in the aeration chamber. This is a key factor to increase BOD removal from the waste water. The sludge will continue to build up. Occasionally, some of the sludge should be drained to keep the effluent from deteriorating. The cleaner water at the top of the settling chamber overflows through openings at the top of the chamber. It can then be treated for reuse or discharged. For less than 24-hour retention period in the aerobic process, BOD concentration should not exceed 2,000 mg/l on the effluent.

Types of Activated Sludge Process

Plug flow   The plug flow is a simple way of modelling for fluid flow where no back mixing is allowed. Waste water is routed through a series of channels constructed in the aeration basin.

- The waste water flows through as a plug and is treated as it finds its way through the tank.

- As the waste water goes through the system, BOD and organic concentrations are greatly reduced.

- Variations to this method include adding return sludge and/or decreasing the amounts at various locations along the length of the tank; waste water BOD is reduced as it passes through the tank, and thus air requirements and the number of bacteria required decrease accordingly.

Complete mix   Waste water may be immediately mixed throughout the entire contents of the aeration basin (mixed with oxygen and bacteria).

- ⮑ This is the most common method used today.

- ⮑ Since waste water is completely mixed with bacteria and oxygen, the volatile suspended solids' concentration and oxygen demand are the same throughout the tank.

Contact stabilization

Contact stabilization is a conventional activated sludge process. In contact stabilization, two aeration tanks are used. One tank is for separate reaeration of the return sludge for at least four hours before it is permitted to flow into the other aeration tank.

- ⮑ Microorganisms consume organics in the contact tank.

- ⮑ Raw waste water flows into the contact tank where it is aerated and mixed with bacteria.

- ⮑ Soluble materials pass through bacterial cell walls, while insoluble materials stick to the outside.

- ⮑ Solids settle out later and are wasted from the system or returned to a stabilization tank.

- ⮑ Microbes digest organics in the stabilization tank and are then recycled back to the contact tank because they need more food.

- ⮑ Detention time is minimized, and so the size of the contact tank can be smaller.

- ⮑ Volume requirements for the stabilization tank are also smaller because the basin receives only concentrated return sludge; there is no incoming raw waste water.

- ⮑ Often no primary clarifier before the contact tank due to the rapid uptake of soluble and insoluble food.

Extended aeration

This is the same as complete mix, with just longer aeration.

- ⮑ Used to treat industrial waste water containing soluble organics that need longer detention times.

- ⮑ Longer detention time in the aeration tank; it provides equalization to absorb sudden/temporary shock loads.

⮌ Less sludge is generally produced because some of the bacteria are digested in the aeration tank.

Other modifications

Oxidation ditch   Similar to plug flow but uses a circulator aeration basin.

Step feed   Microbes gradually feed organics in a step feed mode at multiple points along the tank plug flow.

Tapered aeration   Air flow rate to the aeration basin may be tapered along the length of the basin.

Kraus process   A portion of the return sludge flow from the secondary clarifier is mixed with the anaerobically digested sludge and the digester supernatant before being combined with the return sludge stream and recycled back to the aeration basin. Denser anaerobically digested sludge settles rapidly and is added to the lighter secondary sludge to add weight and improve settling.

Process Design Consideration

The activated sludge process is usually employed following primary sedimentation. The waste water contains some suspended and colloidal solids and, when agitated in the presence of air, the suspended solids form nuclei on which biological life develops and gradually build up to larger solids that are known as activated sludge.

Activated sludge with its living organisms has the property of absorbing or adsorbing colloidal and dissolved organic matter. The biological organisms utilize the absorbed material as food and convert it to inert insoluble solids and new bacterial cells. Much of this conversion is a step-by-step process. Some bacteria attack the original complex substances to produce simpler compounds as their waste products. Other bacteria use the waste products to produce still simpler compounds, and the process continues until the final waste products can no longer be used as food for bacteria.

The generation of activated sludge or floc in waste water is a slow process, and the amount so formed from any volume of waste water during its period of treatment is small and inadequate for rapid and effective treatment of the waste water. Such concentrations are built up by collecting the sludge produced from each volume of waste water treated and reusing it in the treatment of subsequent waste water flows. The sludge so reused is known as returned sludge. This is a cumulative process so that eventually more sludge has been produced and is available to maintain a viable biological population of organisms to treat the incoming wastes. The surplus or excess activated sludge is then permanently removed from the treatment process and conditioned for ultimate disposal.

The activated sludge must be kept in suspension during its period of contact with the waste water being treated by some method of agitation. The activated sludge process, therefore, consists of the following steps:

- ⟳ Mixing the activated sludge with the waste water to be treated (mixed liquor).

- ⟳ Aeration and agitation of this mixed liquor for the required length of time.

- ⟳ Separation of the activated sludge from the mixed liquor in the final clarification process.

- ⟳ Return the proper amount of activated sludge for mixture with the waste water.

- ⟳ Disposal of the excess activated sludge.

Mixing the Activated Sludge with the Waste water to be Treated

The first step in the activated sludge process is to bring the microorganisms in contact with the organics of the waste water. This is generally accomplished by the rapid mixing of the return sludge with the waste water at the inlet of the aeration tank. In some cases, small mixing chambers may be provided.

Aeration and Agitation of Mixed Liquor

Aeration serves at least three important functions: 1) mixing the returned sludge with the effluent from primary treatment, 2) keeping the activated sludge in suspension, and 3) supplying oxygen to the biochemical reactions necessary for the stabilization of the waste water. The theoretical oxygen requirement can be computed by knowing the BOD of the waste, the amount of organisms wasted from the system per day, and the degree of treatment (whether a nitrified effluent is required). For practical purposes, enough air should be added to the waste to maintain at least 2 mg/l of dissolved oxygen under all conditions of loading in all parts of the aeration tank. The air requirements for good treatment can be satisfied either with a diffused air system or by mechanical aerators.

Table 27.3 Standard operating procedures—aeration and dissolved oxygen control

Procedure	Frequency	Method	Range	Condition	Probable cause	Response
Check DO level	Every 2 hours	DO meter or Iodometric method*	Normally 1 to 3 mg/L	High	Too much aeration	Decrease aeration
				Satisfactory		Continue monitoring
				Low	Too little aeration	Increase aeration
Check uniformity of aeration pattern in aeration tank	Daily	Visual observations	Uniform mixing and roll pattern and air bubble disbursement	Dead spots	Improper distribution of air	Perform DO profiles and balance air distribution with header valves
				Uneven roll	Improper distribution of air	
				Localized boiling	Diffuser malfunction	Pull and check for plugged diffuser
Check air requirement (diffused aeration)	Daily	Calculation SCF/LB COD or LB BOD removed		High	Poor O_2 transfer or nitrification	Check uniformity of aeration check for nitrification
				Satisfactory		Continue monitoring
				Low	Inaccurate DO, COD, or BOD measurement	Recalibrate DO meter check lab analysis
Check air requirement (mechanical aeration)	Monthly	Calculation lbs O_2/lb COD or LB BOD removed		High	Low loading	Reduce number of units in operation check for adequate mixing
				Satisfactory Low loading capacity	Low loading capacity	Improve primary treatment, increase number of units in operation

In a diffused air system, air under low pressure, generally not more than eight to ten pounds, is supplied by blowers and forced through various types of porous material in plates or tubes installed near the bottom of the aeration tank. As the air is discharged into the liquid phase, it breaks up into fine bubbles, thus increasing the surface area across which oxygen diffuses from air into the waste water. The plates or tubes are so located in the aeration tank that a rotary motion is imparted to the waste-water mixture, resulting in a considerable amount of air being absorbed from the atmosphere. Diffuser plates may be composed of fused crystalline alumina or high-silica sand. They are set in containers usually made of reinforced concrete. Diffuser tubes are usually made of a similar material or corrugated stainless steel pipe with multiple outlets and wrapped with saran (vinylidine chloride polymer plastic) twisted cord. These are suspended in the aeration tank in sections and can be disconnected above the waste-water surface and removed for cleaning or renewal. When installed on swing joint connections so that they may be brought to the surface of the tank, they are known as "swing diffusers".

To prevent clogging of the diffuser plates or tubes, the air supplied to them should be filtered to remove dust, oil or other impurities, and the piping should be of non-corrosive material. There are a number of types of filters available based on different principles used.

Mechanical aerators are generally of two types—surface and turbine. Surface aerators consist of submerged or partially submerged impellers, which are centrally mounted in the aeration tank. They agitate the waste water vigorously, entraining air in the waste water and causing a rapid change of the air–water interface. Another type of surface aerator, more popular in Europe, consists of a paddle wheel or brush, partly submerged in the waste water, revolving on a horizontal axis. Air is absorbed by surface contact and by droplets thrown through the air by the paddle mechanism.

Turbine aerators are usually violent agitators which will give the upflow air entrainment of the solution. A draft tube may be utilized to control the flow pattern of the circulating liquid within the aeration tank. The draft tube is a cylinder with flared ends mounted concentrically with the impeller, and extending from just above the floor of the aeration tank to just beneath the impeller.

Surface aerators are rated in terms of their oxygen-transfer rate or pounds of oxygen per hour at standard conditions. They normally have efficiency from 2 to 4 lb O_2/h.p/h while turbine aerators vary from 2 to 3 lb O_2/h.p/h.

In the activated sludge process, the sludge accomplishes the major part of the removal of BOD from the waste water being treated in a relatively short period of aeration. It takes, however, a much longer time for the sludge to assimilate the organic matter that it has absorbed. During this time, an aerobic environment must be maintained. To effect the most complete treatment of waste water and the most economical operation in the conventional activated sludge process, an aeration detention time of six to eight hours has been found to be adequate for diffused air aeration, and nine to twelve hours for mechanical aeration. Substantially shorter periods are used in some of the modifications of the conventional process. These shorter aeration periods generally result in lowering the quality of the plant effluents.

Separation of Activated Sludge from the Mixed Liquor

The function of the secondary clarifier is to separate the activated sludge solids from the mixed liquor. These solids represent the colloidal and dissolved solids that were originally present in the waste water. In the aeration unit, they were incorporated into the activated sludge floc, which are settleable solids. The separation of these solids, a critical step in the activated sludge process, is accomplished in the secondary or final settling tanks. These tanks are similar in design to the mechanically cleaned primary sedimentation tanks but with a surface settling rate not exceeding 800 gallons per square foot per day.

The cycle of sludge removal from the secondary tanks is much more important than with primary tanks. Some sludge is being removed continuously to be used as returned sludge in the aeration tanks. The excess sludge must be removed before it loses its activity because of the death of the aerobic organisms resulting from lack of oxygen at the bottom of the tank. Anaerobic sludge in the final clarifier can cause "rising sludge". This should not be confused with a bulking sludge. Rising sludge is a result of denitrification and septicity. It is possible, where facilities are available, to reactivate return sludge in separate reaeration tanks before addition to the waste water. However, it is much wiser to retain the activity of the sludge by prompt withdrawal from the tank.

Return Sludge Requirements

The purpose of return sludge is to maintain a concentration of activated sludge in the aeration tank sufficient for the desired degree of treatment. Ample return sludge pump capacity should be provided since the return sludge volume may range from 10 to 50% of the volume of waste water being treated and sometimes more. For a conventional plant, the percentage is usually between 20 and 30. The best concentration must be determined for each plant by trial operation and should be carefully maintained by controlling the proportion of return sludge. The maximum concentration is limited by the air supply and waste water load. If solids area is allowed to build up, the air and food requirements will exceed those available and an upset will occur.

Advantages and Disadvantatages of Activated Sludge Process

Advantages

⮺ Highly diverse, the sludge plants can be run in various sizes.

⮺ Removes organics

⮺ Oxidation and nitrification achieved

⮺ Biological nitrification without adding chemicals

⮺ Biological phosphorus removal

⮺ Solids/liquids separation

⮺ Stabilization of sludge

⮺ Capable of removing ~ 97% of suspended solids

Disadvantages

⮺ Does not remove colour from industrial wastes and may increase the colour through formation of highly coloured intermediates through oxidation

⮺ Does not remove nutrients, and so tertiary treatment is necessary

⮺ Problem of getting well settled sludge

⮺ Keeps high biomass concentration in aeration tanks, allowing it to be performed in technologically acceptable detention times

TRICKLING FILTERS

A trickling filter (Figure 27.7) consists of a bed of highly permeable media on whose surface a mixed population of microorganisms is developed as a slime layer. The word "filter" in this case is not correctly used because there is no straining or filtering action involved. The passage of waste water through the filter causes the development of a gelatinous coating of bacteria, protozoa and other organisms on the media. With time, the thickness of the slime layer increases, preventing oxygen from penetrating the full depth of the slime layer. In the absence of oxygen, anaerobic decomposition becomes active near the surface of the media. The continual increase in the thickness of the slime layer, the production of anaerobic end products next to the media surface, and the maintenance of a hydraulic load to the filter eventually cause sloughing of the slime layer to form. This cycle is continuously repeated throughout the operation of a trickling filter. For economy and to prevent clogging of the distribution nozzles, trickling filters should be preceded by primary sedimentation tanks equipped with scum collecting devices.

Primary treatment ahead of trickling filters makes available the full capacity of the trickling filter for use in the conversion of non-settleable, colloidal and dissolved solids to living microscopic organisms and stable organic matter temporarily attached to the filter medium and to inorganic matter carried off with the effluent. The attached material intermittently sloughs off and is carried away in the filter effluent. For this reason, trickling filters should be followed by secondary sedimentation tanks to remove these sloughed solids and to produce a relatively clear effluent.

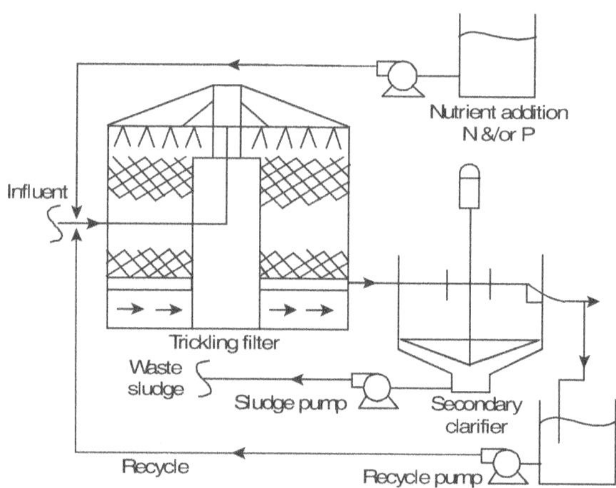

Figure 27.7   Trickling filter set-up

Construction and Design

The primary factors that must be considered in the design of trickling filters (Figure 27.8) include: 1) the type of filter media to be used, 2) the type and dosing characteristics of the distribution system, and 3) the configuration of the under drain system.

Filter media   The ideal filter medium is a material that has a high surface area per unit volume, is low in cost, has a high durability, and does not readily clog. The choice of filter medium is more often governed by the material locally available, which may include field stone, gravel, broken stone, blast furnace slag and anthracite stones. Stones less than one inch in diameter do not provide sufficient pore space and may therefore result in plugging of the media and ponding. Using of 2.5 inch of stones is now considered the minimum diameter. Large-diameter stones tend to avoid ponding situations but limit the surface area per unit volume available for the slime layer to grow. An upper size limit of about 4 inches is therefore recommended.

Distribution system The rotary distributor has become standard for the trickling filter process because of its reliability and ease of maintenance. It consists of a hollow vertical centre column carrying two or more radial pipes or arms, each of which contains a number of nozzles or orifices for discharging the waste water onto the bed. All these nozzles point in the same direction at right angles to the arms and the reaction of the discharge through them causes the arms to revolve. The necessary reaction is furnished by a head of 18" to 24". The speed of revolution will vary with the flow rate, but it should be in the range of one revolution in 10 minutes or less for a two-arm distributor. A dosing tank should be provided for standard rate of trickling filters. It should shut off the flow when the level of the solution reaches the minimum level which is required to revolve the arms at particular speed.

A clearance of 6 to 9 inches should be allowed between the bottom of the distributor arm and top of the bed. This will permit waste streams from the nozzles to spread out and cover the bed uniformly, and it will also prevent ice accumulation from interfering with the distributor motion during freezing weather.

Fixed spray nozzles were used when trickling filters were first developed. The nozzles were attached to pipes laid in the filter medium and were fed intermittently from a siphon controlled dosing tank. By this method, waste water is applied to the filter for short periods of time. Between applications, the filter has rest periods while the dosing tank is filling. Many types and shapes of nozzles were developed and the siphon dosing tank was designed to attain best possible uniform distribution of waste water over the entire surface of the filter. However, the distribution is usually not uniform and there are areas on the filter where very little waste water is sprayed.

In addition, because a number of nozzles are used for the distribution of waste, clogging and increased operational and maintenance problems were encountered.

Underdrain system   The underdrain system in trickling filters serves two purposes: 1) to carry the waste water passing through the filter and the sloughed solids from the filter to the final clarification process, and 2) to provide for ventilation of the filter to maintain aerobic conditions. Underdrains are specially-designed vitrified clay blocks with slotted tops that admit the waste water and yet support the media. The blocks are laid directly on the filter floor, which is sloped toward the collection channel at a 1 to 2 per cent gradient. Since the underdrains also provide ventilation for the filter, it is desirable that the ventilation openings total at least 20% of the total floor area. Normal ventilation occurs through convection currents caused by a temperature difference between waste water and ambient air temperature. In deep filters or heavily loaded filters, there may be some advantage in force ventilation.

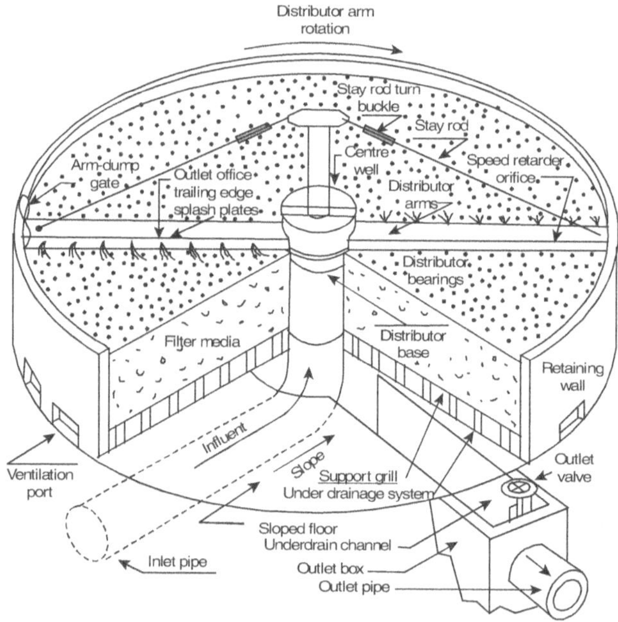

Figure 27.8   Parts of a trickling filter

Types of Trickling Filters

Trickling filters are classified by hydraulic or organic loading as high-rate or low-rate. The organic load on a filter is the BOD content in pounds (1 pound is 0.4536 kg) applied to the filter. This is usually expressed as pounds of BOD per day per 1000 cubic feet of filter medium or pounds of BOD per day per acre. The hydraulic load, including recirculation flow if used, is the gallons of flow per acre of filter surface per day.

Low-rate filters are relatively simple treatment units that normally produce a consistent effluent quality even with varying influent strength. Depending upon the dosing system, waste water is applied intermittently with rest periods that generally do not exceed five minutes at the designed rate of waste flow. With proper loadings, the low-rate trickling filter, including primary and secondary sedimentation units, should remove from 80 to 85 per cent of the applied BOD. While there is some unloading or sloughing of solids at all times, major unloadings usually occur several times a year for comparatively short periods of time. High-rate filters are usually characterized by higher hydraulic and organic loadings than low-rate filters. Higher BOD loading is accomplished by applying a larger volume of waste per acre of surface area of the filter.

One method of increasing the efficiency of a trickling filter is to incorporate recirculation. Recirculation is a process by which the filter effluent is returned to and reapplied onto the filter. Recycling of the effluent increases the contact time of the waste with the microorganisms and also helps to "seed" the lower portion of the filter with active organisms.

When recirculation is used, the hydraulic loading per unit area of filter media is increased. As a result, higher flow velocities will usually occur causing a more continuous and uniform sloughing of excess growths. Recirculation also helps to minimize problems with ponding and restriction of ventilation. Recirculation may be continuous or intermittent. Return pumping rates may either be constant or variable. Sometimes, recycling is practiced during periods of low flow to keep the distributors in motion, to prevent the drying of filter growths, and to prevent freezing during colder temperatures. Also, recirculation in proportion to flow may be utilized to reduce the organic strength of the incoming wastes, and to smooth out diurnal flow variations.

Recirculation can be accomplished by various other techniques. Some of which are as follows:

Biofilter Biofilter (Figure 27.9) is a high-rate filter, usually 3 to 4 feet in depth, employing recirculation at all times. Recirculation involves bringing the effluent of the filter or of the secondary sedimentation tank back through the primary settling tank. The secondary settling tank sludge is usually very light and can be continually fed back to the primary settling tank where two types of sludges are collected together and pumped to the digester.

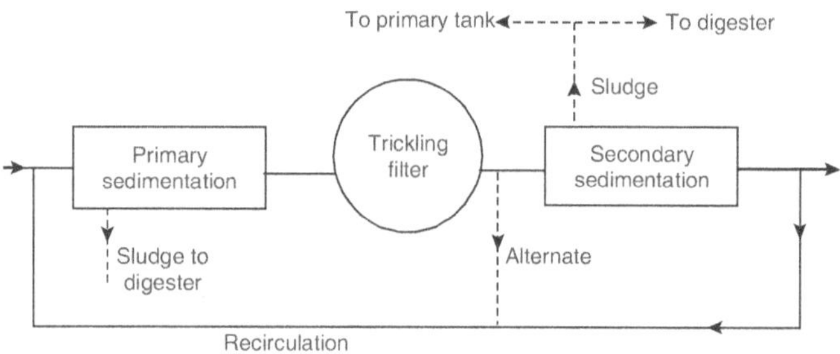

Figure 27.9 Flow diagram of a biofilter

Accelo-filter An accelo-filter recirculates unsettled effluent from the filter back to the inlet of the filter distributor. It is used for both low-rate and high-rate filters, the former being applicable if a well-nitrified effluent is required (Figure 27.10).

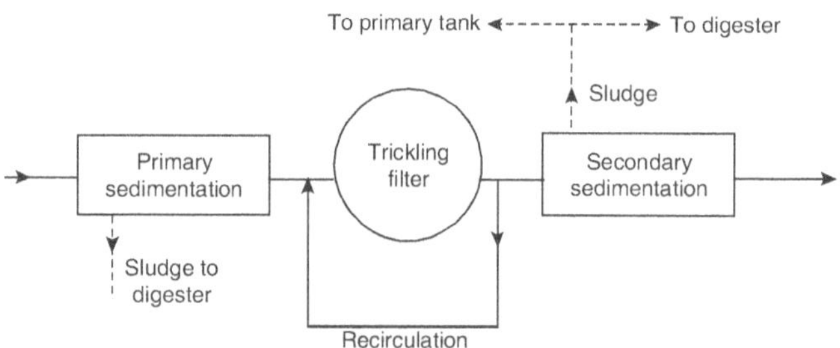

Figure 27.10 Design of an accelo-filter

Aero-filter An aero-filter (Figure 27.11) distributes the waste water by maintaining a continuous rain-like application of the waste water over the filter bed. For small beds, distribution is accomplished by a disc distributor revolving at a high speed of 260 to 369 rpm set 20" above the surface of the filter. For large beds, a large number of revolving distributor arms, 10 or more, tend to give more uniform distribution. These filters are always operated at a rate in excess of 10 million gallons per acre of surface area per day.

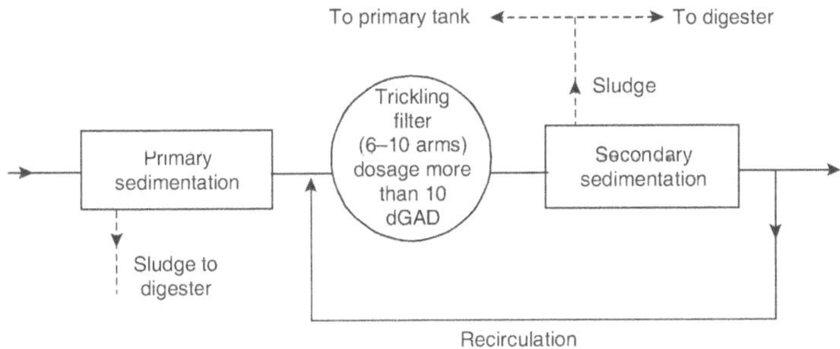

Figure 27.11   Design of aero-filter

High-rate trickling filters, including primary and secondary sedimentation, should remove from 65 to 85 per cent of the BOD in waste water under normal operation. Recirculation should be adequate to provide continuous dosage at a rate equal to or in excess of 10 million gallons per acre per day. As a result of continuous dosing at such high rates, some of the solids accumulated on the filter medium are washed off and carried away with the effluent continuously. High-rate trickling filters have been used advantageously for pretreatment of industrial wastes, unusually strong waste waters. When so used, they are called "roughing filters". With these filters, the BOD loading is usually in excess of 110 pounds of BOD per 1000 cubic feet of filter medium.

Generally, most organic wastes can be successfully treated by trickling filtration. Food processing, textile, fermentation and some pharmaceutical process wastes are amenable to trickling filtration.

Some industrial waste waters cannot be treated by trickling filtration if they contain excessive concentration of toxic materials, such as pesticide residues, heavy metals, and high acidic and alkaline wastes. Since organisms growing on these media are temperature dependent, climatic changes will affect the filter's performance. The organism's metabolic rate increases with warmer weather. Therefore, higher loadings and greater efficiencies are possible in warmer temperatures and climate if aerobic conditions can be maintained in the filter.

ROTATING BIOLOGICAL CONTACTORS (RBC)

The rotating disc process, presently popular in Europe, is used to treat municipal and industrial waste waters. The process operates as a fixed-film biological reactor. A biological film or biomass grows on the surface of a series of discs mounted on a shaft and placed in a tank conforming to the general shape of the discs. The discs are usually made of plastic or some other non-corrosive durable material. The discs are slowly rotated while one-third to one-half immersed in the waste water being treated. The rotation rate is usually 2–4 rpm. As the disc rotates, the microbial population attached to it forms a film and later becomes several millimetres thick. The rotation of the disc brings the attached biologic film into contact with the waste water for removal of organics, and then with the atmosphere for absorption of oxygen. Excess biomass generated during the cycle is stripped off by shearing forces exerted as the discs rotate.

The process has shown to be effective in the treatment of waste water with removal up to 95 per cent for a disc surface loading of 3–5 Kilo/sq.ft. of surface area. In addition, the process is inherently stable under conditions of fluctuating hydraulic and BOD loads. The use of simple mechanical components for the process hardware results in very low maintenance cost. Maintenance is largely limited to greasing of bearings and inspecting the chains and sprockets for wear and slack. The low disc speed used in the biodisc process achieves sufficient mixing and aeration while consuming relatively little power.

Like other biological treatment processes, this process does have its limitations. Organic and hydraulic shock loads as well as toxic discharges will tend to decrease process efficiency.

Moreover, since the metabolic rate of the biomass is temperature dependent, process efficiency would be expected to decrease during colder temperatures. Therefore, RBC's are usually covered or enclosed. The rotating disc process is normally designed on the basis of hydraulic loading. At a specific hydraulic loading rate, there is a specific percentage of BOD reduction. Thus, the percentage of BOD reduction is mainly a function of the hydraulic loading rate and hydraulic detention time. One inherent limitation of the RBC is the lack of operational flexibility. Once the plant is designed and installed, little can be done to alter the operation.

In activated sludge processes, flexibility and adjustments must be incorporated in the design to meet changing conditions. On the other hand, the lack of flexibility in RBC's (no alternative flow schemes or modes of operation, etc.) gives the process simplicity and stability. As long as the discs and media keep rotating and the hydraulic loading remains within the design capacity, the RBC's will function properly.

Intermittent Sand Filters

An intermittent sand filter is a specially prepared bed of sand on which effluents from primary treatment or from trickling filters or secondary settling tanks may be applied intermittently by using troughs or perforated pipe distributors. The effluent from the filter is removed through an under-drainage system.

Bed construction   The filter bed should have a depth of clean sand of at least 24 inches overlying clean, graded gravel placed in at least three layers around the underdrains and to a depth of at least six inches over the top of the underdrains. The sand itself should have an effective size of 0.3 to 0.6 millimetres and have a uniform coefficient of not more than 3.5. The centre to centre spacing of underdrains should not be more than 10 feet.

Capacity   When treating a primary effluent from normal sewage, the rate of application on the filter should not exceed 125,000 gallons per acre per day, which should be reduced if the sewage is strong. With trickling filter and secondary settling tank effluent, the loading should not exceed 500,000 gallons per acre per day.

The intermittent sand filter is a true filter that strains out and retains fine suspended solids, and acts as an oxidizing unit. A major portion of straining and oxidation is carried out at or near the surface of the sand. Straining results from the fine nature of the sand medium with small voids and from the biological slime growth of organisms on the surface of the sand. Oxidation is carried out, as in all secondary treatment devices, by the living aerobic microorganisms that develop primarily at the surface, forming a slime layer, but also extending into the sand medium.

Operation   It is important that the filter be allowed to empty itself and obtain a supply of fresh air at intervals. This is accomplished by intermittent dosing of the sewage onto the filter. Sewage is applied two to six times a day in quantities sufficient to cover the surface down through the sand, and air is drawn in from the surface. Sand filters are constructed in two or more units and used in rotation. Eventually, the slime layer on the surface causes the top layer of the sand to become clogged, which necessitates the removal of the top layer of sand in order to put the unit back into efficient operation.

Pooling   Pooling should not be allowed to develop on the beds because it tends to produce septic action, obnoxious odours and an effluent of poor quality. Pooling indicates that cleaning is necessary. The surface of the beds should be kept on a level to afford uniform distribution of sewage, and weeds, grass, etc. should not be allowed to grow on the beds.

During winter, the open air beds should be ridged every two feet in order to hold the ice off the main body of sand. An alternative method is that of raking up small four to six inch piles of the top surface every three feet over the surface of the bed. The bed should be cleaned and levelled as early as possible in the spring. Where natural sand percolation beds are used, it is sometimes advisable to prepare shallow beds after careful removal of all organic deposits. Such harrowing is not advisable with underdrained beds because of possible damage to the under-drainage system.

Efficiency and use A well-operated intermittent sand filter will give a clear, sparkling stable effluent, almost completely oxidized and nitrified. The overall removal of 95 per cent or more of BOD and suspended solids in the raw sewage can be expected. This exceeds the performance of other accepted secondary treatment processes.

However, as compared with other sewage treatment processes, this process requires large areas of land and high costs of construction and maintenance. The use of filters of this type is restricted to situations where the volume of sewage to be treated is small, or where an exceptionally high grade of plant effluent is necessary. They have been used effectively for additional treatment of secondary treatment effluents. Modifications in sand filters have been proposed in the past but not proven practical for treatment of large volumes of sewage.

AERATED LAGOONS AND STABILIZATION PONDS

Aerated Lagoons

Aerated lagoons are relatively shallow lagoons in which waste water is added at a single point either at the edge or middle of the lagoon, and the effluent is removed from another point. The retention time is a function of the per cent removal of BOD. It may vary from 6 to 18 days as the removal of BOD from domestic waste water varies from 75 to 90 per cent. Oxygen is supplied by means of surface aerators or by diffused aeration units. The action of the aerators maintains the solids of the lagoon in suspension. Depending on the degree of mixing, lagoons may be operated as either aerobic or aerobic–anaerobic systems.

In aerobic lagoons, all biological solids are in continual suspension, and stabilization of the organics occurs under aerobic conditions. In the case of aerobic–anaerobic lagoon, a large portion of solids settles to the bottom of the lagoon. As solids build up, a portion will undergo anaerobic decomposition. Therefore, stabilization in this case occurs partly under aerobic conditions and partly under anaerobic conditions.

Stabilization Ponds

A stabilization pond or "oxidation pond" is a shallow earthen basin of controlled shape, which is designed for treating waste waters from small communities or industrial plants. The ponds are usually 2 to 4 feet deep, although much deeper ponds have been used quite successfully. Stabilization ponds have been applied singly as part of a treatment scheme or as the sole process providing the complete treatment.

The process involves two major steps in the decomposition of organic matter in waste water. The carbonaceous matter is first oxidized by aerobic microorganisms with the formation of carbon dioxide and inorganic forms of nitrogen and phosphorous. These inorganic forms are then used by algae in their photosynthetic reactions. One of the end products of photosynthesis is oxygen, which becomes available to aerobic microorganisms. As a result of the reactions in the ponds, the organics in waste water are partly oxidized and partly converted to algal cells, which may be harvested and used in animal feed as a protein source. Therefore, the treatment of waste water with the production of a useful by-product is possible in stabilization ponds.

Most stabilization ponds are designed for loadings of one acre per 400 persons and 50 pounds of BOD per acre per day or 15 pounds of BOD per acre per day with detention periods generally greater than 30 days. The natural soil in which they are located should be fairly impervious so that seepage will not materially affect the surface level of the waste water in the pond. These ponds can be constructed and operated with low cost.

Performance of the Ponds

Treatment efficiencies that can be expected with ponds vary more than most other treatment devices. Some of them are:

Physical factors   This includes the type of soil, surface area, depth, wind action, sunlight, temperature, short-circuiting and inflow variations.

Chemical factors   It includes organic material, pH, solids, concentration and nature of waste.

Biological factors   It includes the type of bacteria, type and quantity of algae, activity of organisms, nutrient deficiencies and toxic concentrations.

Operation and Maintenance

Most problems that arise from ponds are due to operator's neglect and poor housekeeping practices. Some operation and maintenance problems associated with ponds are:

Scum control   Scum is usually present in springs when water warms up and biological activity resumes. It can promote the growth of blue-green algae, which can give rise to disagreeable odours. If scum is allowed to accumulate, it can cut off sunlight to the pond, which would affect the dissolved oxygen produced from photosynthesis.

Odour control   Odours are usually associated with overloading and poor housekeeping. Most odours occur during spring warm-up when biological activity resumes. The reduction of odours can be accomplished with the use of surface aerators or by adding chlorine. Sometimes, sodium nitrate may be used as a supplemental source of oxygen.

Weed control   Weeds are mostly objectionable because they can promote mosquito breeding and scum accumulation. Aquatic weed roots may puncture pond linings and hinder pond circulation.

Insect control   Mosquitoes will breed in sheltered areas of standing water where there is vegetation or scum to which female mosquitoes can attach their eggs. Mosquitoes are a nuisance and may pose public health problems. There are many minute, shrimp-like predators that feed on algae, usually during warmer months, which may clear the pond of algae, reducing the dissolved oxygen content in the pond and producing noxious odours. This condition is normally temporary. Effective insect control is usually accomplished by good housekeeping practices and with the use of insecticides.

ANAEROBIC SLUDGE DIGESTION

This process employs microorganisms to convert industrial waste water into readily disposable digested sludge. Anaerobic digestion is a bacterial process that breaks down organic materials in the absence of oxygen. It is generally run in closed tanks. Usually, the biomass consisting of sewage or processing wastes is mixed with water and fed into the digester without air. The waste stream generally contains fats, oils and greases. These waste streams are produced in processes involving the manufacture of detergents and soaps as well as within the petrochemical industry. Municipal sewage treatment facilities also use sludge digesters. The generalized equation for anaerobic sludge digestion is:

Organic matter + Combined oxygens $\longrightarrow$ Anaerobic microbes + New cells + Energy for life processes + CH_4 + CO_2 + Other gases

Combined oxygens consist of CO_3^{2-}, SO_4^{2-}, NO_3^{1-} and PO_4^{3-}.

Anaerobic digestion occurs in four steps (Figure 27.12).

Hydrolysis   Complex organic matter is decomposed into simple soluble organic molecules using water to split the chemical bonds between the substances.

Fermentation or acidogenesis   Chemical decomposition of carbohydrates by enzymes, bacteria, yeasts or moulds in the absence of oxygen.

Acetogenesis   The fermentation products are converted into acetate, hydrogen and carbon dioxide by acetogenic bacteria.

Methanogenesis   It is formed from acetate and hydrogen/carbon dioxide by methanogenic bacteria.

Acetogenic bacteria grow in close association with methanogenic bacteria during the fourth stage of the process. The reason for this is that the conversion of the fermentation products by acetogens is thermodynamically possible only if the hydrogen concentration is kept sufficiently low. This requires a close relationship between both classes of bacteria.

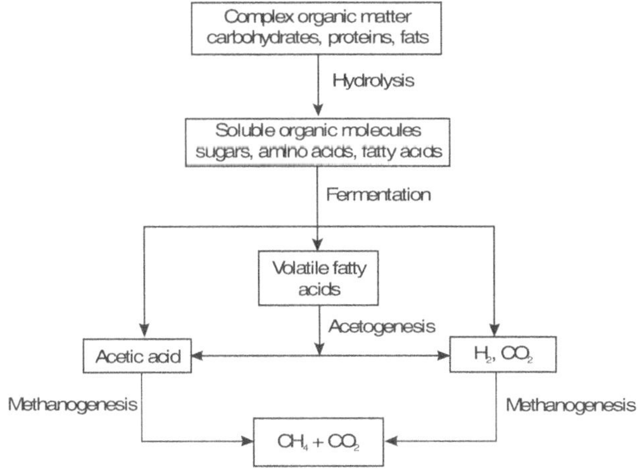

Figure 27.12   Pathway of anaerobic digestion

The anaerobic treatment can take place only under strict anaerobic conditions. It requires specific adapted biosolids and particular process conditions, which are considerably different from those needed for aerobic treatment.

Advantages of Anaerobic Digestion

Waste water pollutants are transformed into methane, carbon dioxide and smaller amounts of biosolids. The biomass growth is much lower as compared to those in aerobic processes. They are much more compact than aerobic biosolids.

The benefits of anaerobic digestion include:

⮕ Odour reduction

⮕ Reduction in the BOD of treated effluent by up to 90%, reducing the risk of water contamination

⮕ Improved nutrient application control, because up to 70% of nitrogen in the waste is converted to ammonia, the primary nitrogen constituent of fertilizers

⮕ Reduced the pathogens such as virus, protozoa and other disease causing pathogens due to anaerobic condition

⮕ It is used as a potential to generate electricity

ANAEROBIC DECOMPOSITION

It is a biological process in which decomposition of organic matter occurs without oxygen. Anaerobic decomposition involves two processes. Firstly, facultative acid forming bacteria use organic matter as food source to produce volatile (organic) acids, gases such as carbon dioxide and hydrogen sulphide, stable solids and more facultative organisms. Secondly, anaerobic methane formers use volatile acids as food source to produce methane gas, stable solids and more anaerobic methane formers. The methane gas produced by the process can be used as a fuel. The methane former works slower than the acid former; therefore the pH has to stay constant consistently, slightly basic, to optimize the creation of methane. Sodium bicarbonate is added to keep it basic.

ANAEROBIC REACTORS FOR WASTEWATER TREATMENT

Conventional digesters, such as sludge and continuous stirred tank reactors (CSTR), have been used worldwide in sewage treatment plants for stabilization of activated sludge and sewage solids. In recent times, the emphasis has shifted to high rate biomethanation systems that are based on the concept of sludge immobilization techniques.

Fixed Film Reactor

In stationary fixed film reactors (Figure 27.13), the cells are deliberately attached to large-sized solid support. The reactor has a biofilm support structure (media) for biomass immobilization; waste-water distribution system for uniform distribution of waste water above or below the media; and facilities for effluent draw-off and recycle (if required). Fixed film reactors offer distinct advantages such as simplicity of construction, elimination of mechanical mixing, better stability at higher loading rates, and capability to withstand large toxic loads. In addition, these reactors can tolerate sudden organic shock loads at constant hydraulic loading and recover normal performance within a few days if alkalinity is high enough to maintain the pH above 6.2. The reactors can process different waste streams with little compromise in capacity and can adapt readily to changes in temperature. This is important for installations where waste water characteristics change rapidly. The reactor start-up will be very quick after a period of starvation (one or two days to reach maximum capacity after three weeks of starvation).

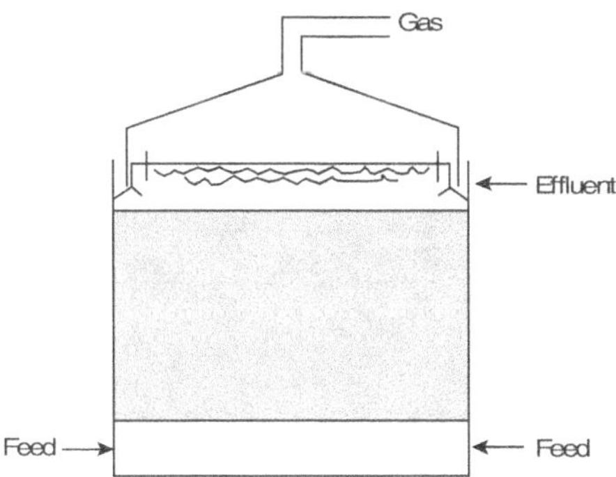

Figure 27.13 Fixed film reactor

The main limitation of this design is that the reactor volume is relatively high as compared to other high-rate processes due to the volume of the media. Another common problem associated with stationary fixed film reactors is clogging due to non-uniform growth of biofilm thickness and/or high suspended solid concentration in waste water. Non-uniform growth and consequent clogging occurs especially at the entry of influent. Some measures to combat this problem include recirculation of effluent and gas for developing a relatively thin film and sloughing of biomass; provision for relatively thin layer of media near the load entering area to accumulate excess biofilm; and improvement in the flow distribution system to avoid very low liquid velocity.

Activated carbon, polyvinylchloride (PVC), hard rock particles, and ceramic rings are the various types of film support that have been tried. Reactor configuration and operation (upflow or downflow mode of operation) have a marked effect on the performance of the reactor. With wastes containing large amounts of hard-to-digest suspended solids, recirculation aids in degradation since it keeps the solids in suspension.

UPFLOW ANAEROBIC SLUDGE BLANKET (UASB) REACTOR

The UASB technology is used extensively for a large number of different types of industrial effluents. The system uses sludge granules as a means of achieving high mean cell residence time (MCRT), thereby achieving highly cost-effective designs. UASB processes have found a variety of applications in recent years in the treatment of high-strength and low or medium-strength waste water and a variety of other substrates. The process has been successfully applied to waste water generated from distilleries, food processing units, tanneries, etc. in addition to municipal waste water. A major advantage in the technology is that it requires less investment as compared to an anaerobic filter or a fluidized bed system. Figure 27.14 shows the structure of an upflow anaerobic sludge blanket reactor.

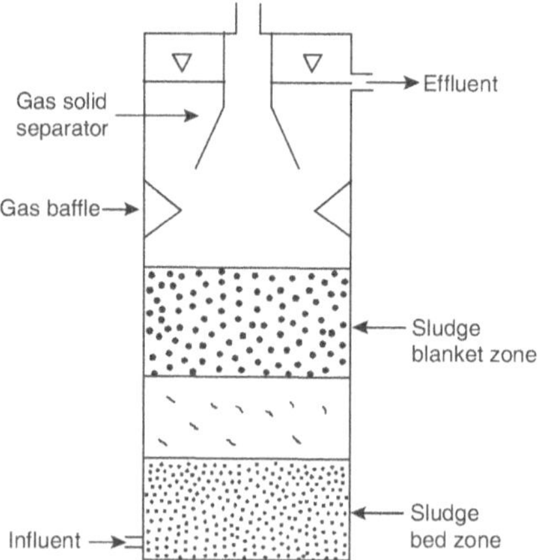

Figure 27.14   Upflow anaerobic sludge blanket reactor

Notable disadvantages include 1) long start-up period, 2) requires sufficient amount of granular seed sludge for faster start-up, and 3) significant wash-out of sludge during the initial phase of the process. A UASB reactor essentially consists of a gas-solid separator (to retain the anaerobic sludge within the reactor), an influent distribution system, and effluent draw-off facilities.

Anaerobic Fluidized Bed Reactor

In an anaerobic fluidized bed, the medium for bacterial attachment and growth is kept in the fluidized state through application of high upflow velocities, normally achieved by recycling the effluent. Increasing recycling allows the process to tend towards the operational characteristics of a completely mixed system. The design of a fluidized bed reactor (Figure 27.15) thus consists of a waste water distributor, a media support structure, media, head space, and effluent draw-off and recycle facilities.

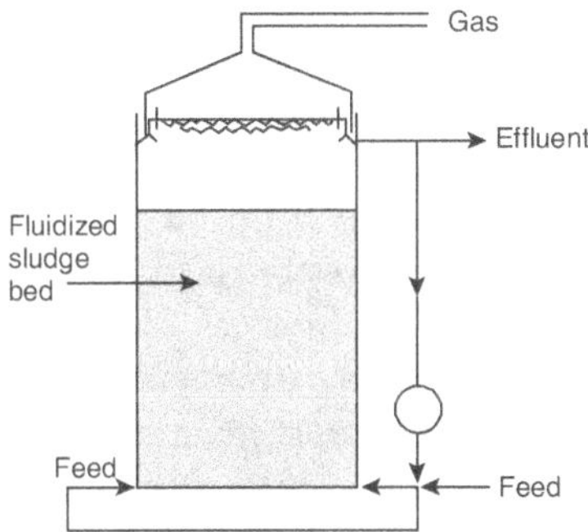

Figure 27.15   Anaerobic fluidized bed reactor

The thickness of the biofilm is controlled by bed regeneration and by the size and density of the inert media in combination with upflow velocity. Excess sludge can be removed from the top of the fluidized bed where the biofilm tends to have maximum thickness. Fluidized bed technology is more effective than anaerobic filter technology as it favours the transport of microbial cells from the bulk to the surface and thus enhances the contact between the microorganisms and the substrate. These reactors have several advantages over anaerobic filters. They are less likely to clog and have a low hydraulic head loss combined with better hydraulic circulation. They are able to operate at lower retention times and/or higher loading rates and have a greater surface area per unit of reactor volume. Finally, the capital cost is lower due to reduced reactor volumes. However, the recycling of effluent may be necessary to achieve bed expansion. Thus, the stationary packed-bed technology is adequate for the treatment of easily biodegradable waste water or where high COD removal is not required, while the fluidized bed technology is especially suitable for treatment of hazardous wastes with recalcitrant compositions.

Covered Anaerobic Lagoons

A covered lagoon digester is the simplest anaerobic digestion system. It typically consists of an anaerobic combined storage and treatment lagoon, an anaerobic lagoon cover, an evaporative pond for the digester effluent, and a gas treatment and/or energy conversion system. Figure 27.16 shows a typical schematic of a floating covered anaerobic lagoon. Covered lagoon digesters typically have a hydraulic retention time (HRT) of 40 to 60 days. HRT is the amount of time a given volume of waste remains in the treatment lagoon. A collection pipe leading from the digester carries the biogas to either a gas treatment system, such as a combustion flare, or to an engine or generator or boiler that uses biogas to produce electricity and heat. Following the treatment, the digester effluent is often transferred to an evaporative pond or to a storage lagoon prior to land application. The climate affects the feasibility of using covered lagoon digesters to generate electricity. Engine or generator systems typically do not produce sufficient heat to maintain the temperature high enough in covered lagoon digesters in winter to sustain consistently high biogas production rates. Using propane or natural gas to provide additional heat for the lagoon contents is typically not economically viable. Without additional heat, most covered lagoon digesters produce less biogas in colder temperatures, and little or no gas below 3.9°C. As a result, covered lagoon digesters are most appropriate for use in warm climates if biogas is to be used for energy or heating purposes.

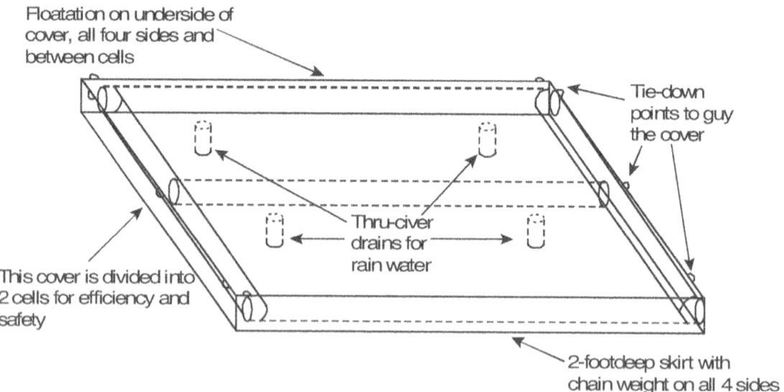

Figure 27.16 Principle diagram of anaerobic lagoon construction

Table 27.4   Odour control strategies for anaerobic lagoons

Odour control technology	Odorous gas emissions reduction (%)
Feed processing/additives	
Grinding feed	5–12
Wet-feeding hogs (3 : 1 water to feed)	23–31
Reducing sulphur-containing amino acids	49-63
Adding fibre (soybeans, hulls) to diet	Up to 68
Biofilters	50
Solids separation	50-60
Soil injection of waste upon land application	50–80 (land application odours only)
Surface aeration	Up to 85
Aerobic cap	Up to 90
Lagoon additives	Up to 90
Lagoon covers	80–90
Anaerobic digestion	80–90
Composting	Up to 100 for well-managed systems

Complete Mix Digesters

Complete mix digester systems consist of a mix tank, a complete mix digester, and a secondary storage or evaporative pond (Figure 27.17). The mix tank may be an above-ground tank or concrete in-ground tank that is fed regularly from the mix tank. Waste is stirred in the mix tank to prevent solids from settling prior to being fed to the digester. Complete mix digesters with in-ground lagoons often employ covers similar to those used in covered lagoon digesters. In the digester, a mix pump circulates the waste slowly around the heater to maintain a uniform temperature. Hot water from an engine/generator or cogeneration water jacket or boiler is used to heat the digester.

Complete mix digesters have an HRT of 15 to 20 days, which means that they can reduce the overall lagoon volume required for waste storage and treatment. This makes complete mix digesters comparable to covered lagoon digesters in cost, despite the increased complexity of stirring, mixing and plumbing components. In addition, biogas production rates, and therefore heat and electricity production, are greater and more consistent in complete mix digesters than in covered lagoons. Like covered lagoon systems, the digester effluent from complete mix digesters is frequently stored in evaporative ponds or storage lagoons.

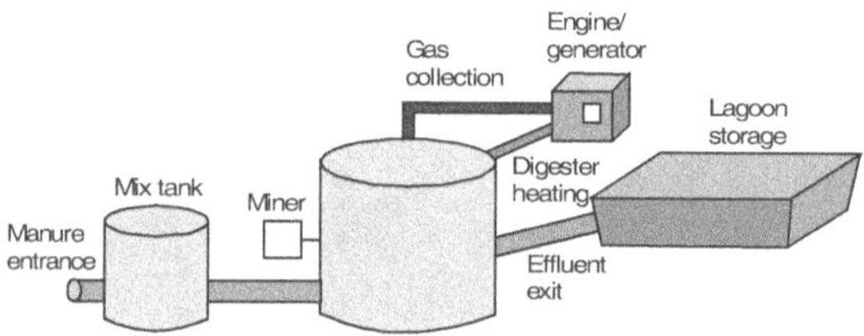

Figure 27.17   Schematic diagram of a complete mix digester

Hybrid Reactor

A hybrid reactor (Figure 27.18) is an improved version of the UASB system and combines the merits of the upflow sludge blanket and fixed film reactors. This reactor offers strong resistance to disturbances and large fluctuations in loading rate. Since most of the microbes adhere firmly to the support media, any change in fermentation conditions would only temporarily affect the microbes. Even at a very high upflow velocity of the effluent, the biofilm is not washed away. A successful phase separation (in terms of acidogenic and methanogenic phases) can be achieved with this type of design. It is possible to maintain the desired pH conditions for both acidogens and the methanogens. Thus, not only does the hybrid reactor provide the best environment for the growth of both the organisms, it also overcomes the limitations usually experienced with the UASB system. Hybrid reactors can be used for a wide variety of industrial effluents. The inert matrix material increases the retention of granular sludge and prevents the washout of microbial population. By choosing a suitable highly porous packing material with a large specific surface, the adhesion of microbes can be greatly improved and the concentration of activated sludge in the reactors can be considerably enhanced. The rate of mass transfer is also higher owing to the increased contact time between the feed and microbes. Since the material that immobilizes microbes can capture most of the sludge when the slurry passes through the reactor, the loss of sludge is minimized. Apart from the advantages of simplicity in operation and design, the hybrid reactor works out to be more economical than fixed-bed system at the industrial scale.

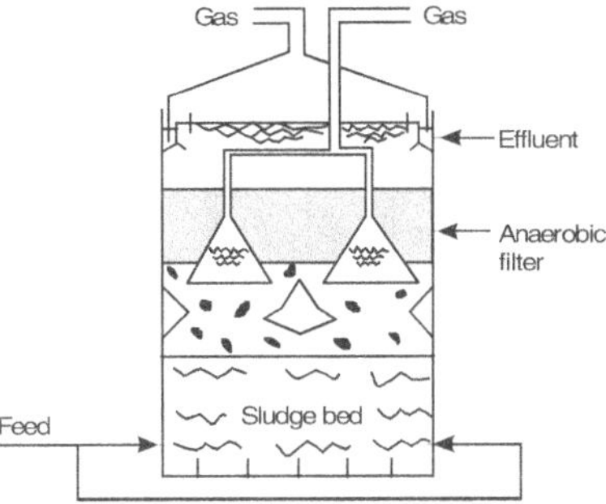

Figure 27.18   Hybrid anaerobic reactor

TERTIARY TREATMENT

Tertiary treatment methods purify the waste water further until it is drinkable. The following chemical and physical methods are used:

1. *Non-biodegradable organics*

   i. Activated charcoal can bind non-specifically to many of the dissolved organics and remaining particulates, thus easily removable from water.

   ii. Chlorine (Cl_2) oxidizes and precipitates organics that can be removed after they settle.

2. *Phosphate*   Selective precipitation removes phosphate from the solution as insoluble salts. These precipitates, which also carry along most remaining microorganisms, are allowed to settle and then removed.

   i. Addition of calcium chloride ($CaCl_2$) leads to the formation of calcium phosphate ($Ca_3(PO_4)_2$), which comes out of the solution.

   ii. Addition of ferric chloride ($FeCl_3$) leads to the formation of ferric phosphate ($FePO_4$), which comes out of the solution.

3. *Nitrogen*   Treatment with a strong base (e.g. NaOH) converts the ammonium ions (NH_4^+) to ammonia gas (NH_3) that bubbles out of the solution.

$$NH_4^+ + OH^- \longrightarrow NH_3 \text{ (gas)} + H_2O$$

DISINFECTION OF WASTEWATER

Primary, secondary and even tertiary treatment cannot by expected to remove 100 per cent of the incoming waste load, and many organisms still remain in the waste water stream. While most of these microorganisms are not pathogenic, pathogens must be assumed to be potentially present. Thus, whenever waste water effluents are discharged to receiving waters, which may be used for water supply, swimming or shell fishing, the reduction of bacterial numbers is a very desirable goal.

Disinfection is treatment of the effluent for destruction of all pathogens. Another term to this is sterilization. Sterilization involves the destruction of all microorganisms. While disinfection indicates the destruction of all disease causing microorganisms, no attempt was made in waste water treatment to obtain sterilization. However, disinfection procedures applied to waste waters will result in a substantial reduction of microbes to a safe level.

In general, disinfection can be achieved by a variety of physical or chemical methods that are capable of destroying microorganisms under certain conditions. Physical methods might include, for example, heating to boiling, incineration, or irradiation with X-rays or ultraviolet rays. Chemical methods might theoretically include the use of strong acids, alcohols, or a variety of oxidizing chemicals or surface active agents (such as special detergents). However, the treatment of waste waters for the destruction of pathogens demands the use of practical measures that can be used economically and efficiently on large volumes of waste waters, which need to be treated to various degrees.

In the past, waste water treatment practices have principally relied on the use of chlorine for disinfection. The prevalent use of chlorine has come about because chlorine is an excellent disinfecting chemical and has been available at reasonable cost. However, the rising cost of chlorine coupled with the fact that chlorine, even at low concentrations, is toxic to fish and other biota as well as the possibility that potentially harmful chlorinated hydrocarbons may form have made chlorination less favoured in waste water treatment. The increased use of ozone (ozonation) or ultraviolet light as a disinfectant is a distinct possibility in waste water disinfection. Both ozone and ultraviolet light leave no toxic residual. Ozone can, in addition, raise the dissolved oxygen level of water. However, ozone has just recently begun to compete economically with chlorination. Ultraviolet light too appears effective and economically competitive with chlorination as a disinfectant.

The use of chlorine and ozone as chemical disinfectants and their disinfecting properties and actions will be considered individually. However, since chlorine continues to be used extensively as a disinfectant, we will mainly be concerned with the principles and practice of chlorination.

CHLORINATION

Chlorination is the application of chlorine to waste water to accomplish some definite purpose. The purpose of chlorination may not always be disinfection and may, in fact, involve odour control or some other objective. Chlorine may be applied in two general ways: gaseous and liquid. In general, the effective chemicals in the form of chlorine will either destroy microbes or act against odour that is caused by the microbes, etc. Gaseous forms of chlorine are first dissolved in water prior to addition to the waste water stream, while liquid forms of chlorine (called hypochlorites) are available in the form of water soluble salts. Because chlorine gas generally costs lesser than hypochlorites, it is normally used in treatment plants. The application of chlorine is usually controlled by special devices called chlorinators or chlorinizers.

Reactions of chlorine in waste waters   When chlorine is mixed with pure water, it immediately dissolves, forming hypochlorous acid and then hypochlorites:

The two forms of chlorine (hypochlorous acid and hypochlorite ion) are called "free" residual chlorine, as opposed to the reaction products ("combined" residual chlorine) of chlorine with other compounds. Free residual chlorine is a more effective disinfecting agent than combined residual chlorine, and generally hypochlorous acid is a much more effective disinfectant than hypochlorite ion. In waste waters, free residual chlorine is seldom detected and chlorine is usually found in the combined residual form.

Chlorine is an extremely active oxidizing chemical that will react with many substances in waste waters. If a small amount of chlorine is added to waste waters, it will react rapidly and is thus consumed. Chlorine first reacts readily with substances like hydrogen sulphide, ferrous iron, manganese or thiosulphates, which may have their origin from industrial wastes. However, almost any "reducing" compound capable of reacting will react with chlorine (an oxidizing compound). If all of the chlorine is consumed in these reactions, no disinfection will result.

Chlorine generally reacts in a prescribed order, first with inorganic reducing compounds. The addition of more chlorine would result in reactions of chlorine with the organic matter that is present. This forms chloroorganic compounds, which have little or no disinfecting action. A further addition of a little more chlorine will react with ammonia or other nitrogenous compounds to produce chloramines or other combined forms of chlorine, which can have disinfecting action but not as effective as free chlorine.

The continued addition of chlorine will result in the destruction of chloramines and the formation of free chlorine. While chlorine is seldom applied to this level in waste water treatment, the addition of chlorine in sufficient dosages to where free chlorine can form is called "breakpoint" chlorination.

For effective chlorine disinfection, both sufficient chlorine dosages as well as contact time are necessary. Generally, both these factors must be worked out experimentally. Several other factors may affect the effectiveness of chlorination. Among the principal factors are bacterial numbers, pH, temperature and contactor. In "pure" systems, "bacterial kill" at a particular chlorine dosage is directly related to the number of bacteria present when chlorine is first added. The pH will affect the form of chlorine present; generally, at neutral pH hypochlorous acid, the most effective form of chlorine, is favoured. Temperature affects the speed with which chemical reactions take place, and colder temperatures are less favourable for disinfection. Proper contacting or mixing or agitation is necessary to make sure that the chlorine applied contacts or reaches the vital parts of the microbial cell.

The precise mechanism of the disinfecting action of chlorine is not fully known. However, chlorine is capable of undergoing a wide variety of reactions and would probably react with the microbial cell at several levels. At high concentrations, massive oxidation takes place, which affect the membranes and all organic components. At lower concentrations, chlorine probably affects the vital protein systems as well as the membranes. From the point of view of waste water treatment, the mechanism of action of chlorine is much less important than its effects as a disinfecting agent.

The quantity of reducing substances, both organic and inorganic, in waste waters varies, and so the amount of chlorine that has to be added to waste water for different purposes will vary. The amount of chlorine used up by organic and inorganic reducing substances is defined as chlorine demand. Chlorine demand is equal to the amount of chlorine added minus that remaining as combined chlorine after a period of time, which is generally 15 minutes. This relationship can be expressed as:

Chlorine demand = Applied chlorine dose – Chlorine residual

It is important to note that disinfection is carried out by that amount of chlorine remaining after the chlorine demand has been satisfied. This quantity is defined as residual chlorine and expressed as mg per litre. For example, a chlorinator is set to feed 50 lbs of chlorine per 24 hours and the waste water flow is at the rate of 0.85 mg/dl. The chlorine measured after 15 minutes contact is 0.5 mg/L. Thus, the chlorine feed or dose is

$$\frac{50\,\text{lbs}}{0.85\;\text{mgd}} = 59\,\text{lbs/mg/dl} = \frac{59}{8.34\,\text{lbs/gal}} = 7.1\;\text{mg/L}$$

Chlorine dose	7.1 mg/L
Chlorine residual	0.5 mg/L
Chlorine demand	6.6 mg/L

OZONATION OF WASTEWATERS

Ozone (O_3) is the triatomic form of oxygen, which means it is composed of three oxygen atoms. Under normal conditions, ozone is unstable and quickly decomposed to the more stable gaseous oxygen, O_2. Because ozone is unstable, it cannot be stored safely, and so it must be produced at the point of application.

Ozone can be generated by passing oxygen, or air containing oxygen, through an area having an electrical discharge or spark. You may have observed a clean smell in the air after a thunder and lightning storm. The clean smell is most likely that of ozone formed by lightning bolts passing through the atmosphere.

To generate a sufficient quantity of ozone for a waste water treatment plant, ozonators are developed to discharge the ozone in halo manner. Ozonators have two large-area metal electrodes separated by a dielectric and an air gap. An alternating electric current is applied to the electrodes to create an electrical discharge. At the same time, air or oxygen is passed through the air gap. As the air or oxygen flows through the air gap, the electrical discharge converts a portion of oxygen to ozone. The dielectric is necessary to spread the electric discharge over the entire electrode area and avoid producing an intensive single arc.

A by-product from the corona discharge is the generation of a large amount of heat. The oxygen or air flow in the air gap is not sufficient to cool the electrodes. Since high temperatures cause ozone to very rapidly decompose to oxygen, it is necessary to provide a cooling system for the electrodes. At present, two types of cooling systems are used: 1) air-cooled and 2) water-cooled.

The formation of oxides of nitrogen also takes place in the corona discharge. Oxides of nitrogen react with water to form nitric acid, which would attack the materials inside the ozonator. To avoid this problem and extend the useful life of the ozonator, the air or oxygen flowing through the air gap in the ozonator should be moisture free. This can be accomplished by cooling the compressed gas to remove the moisture before allowing it to enter the air gap.

The concentration of the ozone leaving the ozonator is approximately 1 to 2% by weight and is applied to the waste water to be disinfected. As with chlorination, the effectiveness of disinfection is dependent on the concentration of the disinfectant, thorough mixing and contact time. To satisfy the mixing and contact time requirements, three general types of contactors are generally used: 1) packed bed, 2) sparged column and 3) sparged column with mixing. The design of most efficient contactor will vary among the treatment plants.

Advantages and Disadvantages of Ozonation

The advantages of ozonation include:

1. eliminates odours

2. reduces oxygen demanding matter, turbidity and surfactants

3. removes most colours, phenolics and cyanides

4. increases dissolved oxygen

5. production of no significant toxic by-products

6. increases suspended solids reduction

7. high capital cost

8. high electric consumption

9. highly corrosive, especially with steel or iron and even oxidizes neoprene

To minimize the disadvantages of ozonation in waste water treatment plants, some innovations have been developed. The use of oxygen feed instead of ozone into air gap improves the for contactors. These innovations are increasing the effectiveness of the system and the cost effect is low because the used oxygen is recycled and reused. Normally, the cold ozone feed system requires about 6 to 9 kWh/lb whereas the cold oxygen feed system requires only about 2.5 to 3.5 kWh/lb.

Disinfection Using Ozone

Ozone is thirteen times more soluble in water than oxygen. When introduced into waste water, very little disinfection occurs initially. Ozone is usually rapidly consumed, satisfying the ozone demand of inorganic salts and organic matter dissolved in the waste water. The disinfecting properties of ozone come into play only after the ozone demand is satisfied. When the demand is satisfied, ozone brings about disinfection 3100 times faster than chlorine, as research data point out. It has also been found that disinfection occurs within contact times of 3 to 8 seconds. Typical ozone dosages needed to reach the disinfection stage vary with the quality of the effluent. Dosages between 5 and 15 mg/L are commonly cited for disinfection of secondary waste water effluents. Ozone also exhibits excellent viricidal properties at these dosages but with longer contact time of about 5 minutes. It has also been found that any residual ozone in the effluent of the contactor disappears in a matter of seconds outside the contactor.

Other Uses of Ozone in Waste-Water Treatment

1. Ozone has the ability to remove solids from waste water by oxidation and physical floatation. Foam develops when waste water is ozonated. It has been found that this foam traps a significant amount of solids and nutrient material such as phosphates and nitrates.

2. pH has been found to increase very slightly because of ozonation. This is probably the effect of carbon dioxide being driven out of the solution by the gas feed in the ozone contactor.

3. The colour and turbidity are reduced by the addition of ozone. This is brought about by chemical oxidation of the substance.

4. Some minor nitrification occurs, but not at levels high enough to consider ozonation as an effective nitrification process.

ULTRAVIOLET LIGHT FOR DISINFECTION

Ultraviolet (UV) light is an electromagnetic radiation with a wavelength of approximately 200–400 nm. These wavelengths are outside the region normally detected by the human eye, and therefore are considered invisible.

Ultraviolet light has been used since the early 1900s in Europe for the disinfection of municipal water supplies. UV light is currently receiving renewed attention as a method of waste water disinfection. A major reason for this renewed interest is that UV light adds nothing to the waste water during the disinfection process.

UV light disinfects by altering the DNA of the bacterial cells that are exposed. It has been found that UV radiation with a wavelength of approximately 254 nm is most efficient for disinfection purposes.

In waste water treatment plants, UV light is produced by low pressure mercury lamps. These lamps, which provide radiation of 253.7 nm, are usually housed in specially fused quartz sleeves. Glass sleeves cannot be used as they can absorb UV light with great efficiency. This would result with no or little UV light reaching the waste water; therefore, very little disinfection would occur. The quartz sleeves serve as an electrical insulator by preventing the waste water from contacting the electrical portion of the lamp, and also as a temperature buffer so that the bulb may remain at its optimum operating temperature for maximum efficiency.

In a usual configuration, the mercury lamps are housed in a closed unit. This is done for safety and to promote complete mixing. The closed unit contains baffles, which direct the waste water flow so that the microorganisms will spend a maximum amount of time close to the ultraviolet source.

For the UV disinfection process to be effective, the UV radiation must be directed on the bacteria. The UV unit therefore attempts to expose all bacteria to the radiation at a reasonably close range (i.e., 1.4 inch). To achieve disinfection, most bacteria require 6000 to 13000 microwatt per seconds of exposure. Commercially available UV disinfection units can provide in excess of 30000 microwatt per seconds.

The three main disadvantages of UV are: 1) high cost of operation, 2) anything that can prevent the UV light from reaching the bacteria will prevent an effective kill, and 3) UV light tends to ionize compounds and break them apart (e.g. nitrate could become nitrite in UV light, causing toxic effects on the effluent). Suspended solids, slime growth, turbidity, and colour are some of the factors that have an adverse effect on UV disinfection.

Better design of the disinfection units has attempted to solve some of these problems. For example, the disinfection unit may have automatic wipers for the quartz tubes to control slime build-up.

REVIEW QUESTIONS

1. What is biological waste treatment? Why is it needed?

2. Give a detailed account on steps involved in biological waste treatment.

3. Explain about grit chamber and its types.

4. Explain in detail about anaerobic digestion in waste treatment.

5. Write in detail about aerobic digestion in waste treatment.

6. Explain activated sludge process.

7. Give an account on trickling filter.

8. Write about RBC.

9. What is aerated lagoon?

10. Give a detailed account on different reactors for anaerobic digestion of waste.

11. Write elaborately about tertiary treatment of waste.

12. Give the role of ozone in waste treatment? Add a note on advantages of ozonation.

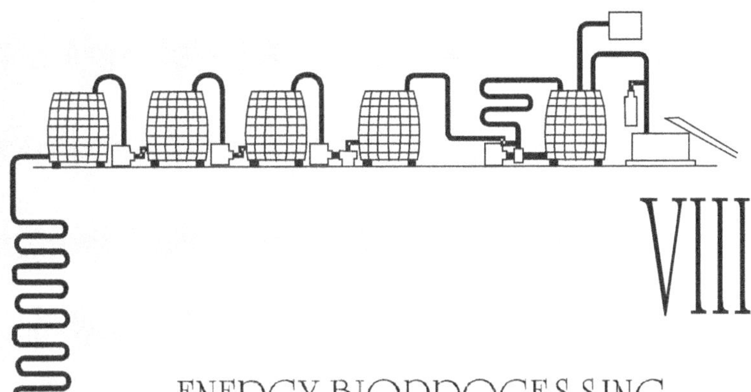

VIII

ENERGY BIOPROCESSING

BIOGAS PRODUCTION

INTRODUCTION

When organic waste is stored in the absence of air, microbial degradation is started, where biogas is derived. The process of anaerobic digestion runs at its optimum at a temperature range of 25 to 38°C (mesophilic conditions), but also up to 55°C in the thermophilic range, however temperature increases due to metabolic activity of microbes. The gas thus produced contains 55 to 70% methane, 30 to 45% carbon dioxide and trace gases. One cubic metre of biogas has the energy equivalent of 0.6 l of fuel oil or 6.36 kW/h.

BIOGAS POTENTIALS

Technical Energy Potential of Biogas Production

Landfill gas, which is recovered from landfill sites, sewer gas from water purification plants as well as biogas from organic household waste make only a small fraction of the available potential for biogas production. However, at the moment, much of the energy from these biological gases is derived from these sources (landfill and sewer gases).

Technically from organizational and economical point of view, over half the excrements from farm animals could be used energetically in farm or centralized biogas plants.

Unrecorded Biogas Potential

In communities, in landscape maintenance, in food processing and canteens organic waste produced increases, which can't be easily disposed off *via* the traditional ways of composting, dumping, animal feeding or further processing. This is due to unfavourable environment, inadequate place to form the land filling and chance to spread animal diseases.

Biogas from Waste of Food Processing

Organic wastes from the food processing industry as well as left over food from canteens were disposed and rarely came into further processing. The energy potential of organic waste from industry is difficult to estimate, because the digestibility of some substrates is not yet examined enough. On the other hand, the recycling paths and the availability of substrates can change very quickly. Experts reckon that the available amount for anaerobic digestion will reach 25% in the long run on these kind of substrates.

Biogas from Grass and Green Cuttings

In some of the countries like Germany, agricultural land has been taken out of production through EC-regulations. On this area of land, the cultivation of crops such as grass, maize, fodder beets and other energy crops are already grown and used for energy production through anaerobic digestion, most of the natural energy producing companies have utilized these crops for energy production. The recycled slurry was used as compost fertilizer for cultivable land.

PRESENT USES

The advantages of biogas technology are not only the CO_2-neutral generation of energy (e.g. electricity and heating) but also the avoidance of odours, methane- and nitrous oxide emissions, saving of fertilizer and chemical sprays, reduction of landfill area, and protection of ground water. The aim of this protocol to tackle greenhouse gases can be effectively supported. The treatment of household waste water, organic residues from food processing and organic household waste offers a wide range of applications for agricultural biogas plants. This is conforming to the requirements of a sustainable waste recycling system.

In many countries, the produced biogas is mostly utilized to promote heat and electricity generation plants. The digested slurry is spread on to the fields as an odourless, organic fertilizer. The contribution of biogas technology in agriculture is still marginal, but continuously growing. At present in Germany, the biogas plants are contributing to generate 2–3 % of the total electricity production from renewable energies.

BIOWASTE TREATMENT PLANTS

The anaerobic digestion of biowaste from source separated household waste has a high energy potential as compared to the aerobic treatment in a composting plant (Table 28.1).

Table 28.1   Comparison between anaerobic digestion and composting of biowaste

	Anaerobic digestion	Composting
Energy	Production (300–600 kWt/h)	Consumption (20–100 kWt/h)
Sanitation	Guaranteed under consideration of legal standards	Guaranteed under consideration of legal standards
Emissions	Low (odours, ammonia)	High (odours, ammonia, methane, nitrous oxide, hydrogen sulphate, germs)
N-fertilizing effect	Fast	Slow
Unsuitable substrates	Tree and bush cuttings	Half liquid substrates with no structure (fatty residues, kitchen waste, slaughter house waste, food processing waste) biowaste without structure can only be composted after wood or other structured biowaste has been added

Very few places are using anaerobic digestion for treatment of their half liquid and liquid biowaste and food residues. Most installations are on a large-scale demonstration and pilot stage, but there is a growing interest in the technology especially in central Europe and Asia (China, India) for tackling the huge waste problems of the sprawling cities and metropolitan areas. At the moment, it is still possible to dump or incinerate the waste, but more and more environmental and energetic problems arise. When there is a source separation of biowaste, then composting is the traditional way of treatment, but liquid and half liquid organic wastes cause emission problems, which lead to additional costs of a closed off treatment process. Even being more expensive, anaerobic digestion systems can offer a cost effective and environmentally sound alternative considering energy production for own consumption and selling to the grid. Often the combination of digestion and composting makes sense, where substrates with different moisture contents are separately treated and the energy production makes the operation completely energy self-sufficient.

ANAEROBIC TREATMENT OF SEWAGE SLUDGE AND LANDFILL GAS UTILIZATION

The community and industrial applications of anaerobic digestion have a relatively high degree of use at present in developed countries, where higher environmental standards have been applied for many years. Waste water treatment plants can be found in almost every town, and bigger installations also have an anaerobic digester. The utilization of sewer gas in cogeneration sets is only done in installations, where several thousands of inhabitants are connected to the sewage system and a considerable amount of sewer gas is produced. Sewer gas as well as landfill gas is already produced through the necessary installations. Therefore, industrialized countries already use 30% of these gases.

INTERACTIONS BETWEEN VARIOUS MICROBIAL GROUPS

Microbial diversity in biogas digesters is as great as that of rumen wherein 17 fermentative bacterial species have been reported to play an important role in the production of biogas. Furthermore, it is the nature of the substrate that determines the type and extent of the fermentative bacteria present in the digester. It has been reported that the presence of proteolytic organisms is higher in cow dung-fed digesters and other animal waste-fed digesters. However, while cow dung-fed digesters supported higher amylolytic microorganisms, poultry waste-fed digesters showed higher proteolytic population. Among fermentative organisms, *Bacteroides succinogens, Butyrivibrio fibrisolvens, Clostridium cellobioparum, Ruminococcus albus* and *Clostridium* sp. are predominant. It has been observed that a clear differentiation existed in the type of cellulolytic bacterial distribution in rumen and biogas digester. In rumen, *Ruminococcus* sp. alone accounted for 60% of the total population, whereas in the biogas digester, the predominant species belonged to the genera *Bacteroides* and *Clostridium* rather than the genus *Ruminococcus*. Besides, *Ruminococcus flavefaciens, Eubacterium cellulosolvens, Clostridium cellulosolvens, Clostridium cellulovorans, Clostridium thermocellum, Bacteroides cellulosolvens* and *Acetivibrio cellulolyticus* are some of the other predominant fermentative bacteria present in cattle dung-fed digesters.

Most of these bacteria adhere to the substrate prior to extensive hydrolysis. While the digester slurry contained higher cellulolytic population, the outlet of the digester recorded the least cellulolytic population. Furthermore, the particulate-bound cellulolytic bacteria were the predominant group in the slurry of the digester. It has been observed that, out of the total cellulolytic population of 42×10^4 ml^{-1} of slurry, the particulate-bound bacteria accounted for 34×10^4 ml^{-1} of slurry. Furthermore, the particulate-bound bacteria predominated up to 20th day of initiation of biogas digester. It is also known that the particulate-bound bacteria showed direct relation to the biogas yield from the digester. Figure 28.1 shows the nutrient utilization of biogas-producing bacterial population.

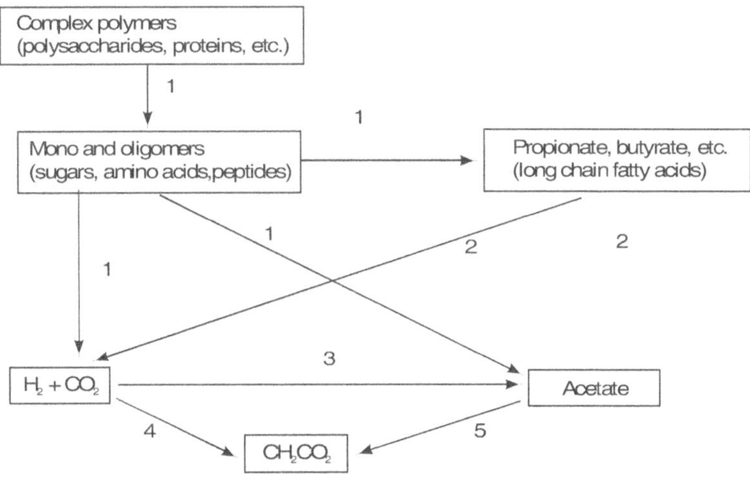

Figure 28.1   Nutrient utilization of biogas-producing bacterial population

Table 28.2   Reactions occurring in biogas digester

Representative reactions	Productions
Glucose + 3 H$_2$O	3 CH$_4$ + 3 HCO$_3$ + 3 H$^+$
Glucose + 4 H$_2$O	2 CH$_3$COO$^-$ + 2 HCO$_3$ + 4 H$^+$ + 4 H$_2$
CH$_3$COO$^-$ + H$_2$O	CH$_4$ + HCO$_3$ + H$^+$
4 H$_2$ + HCO$_3$ + H$^+$	CH$_4$ + 3 H$_2$O
4 H$_2$ + 2 HCO$_3$ + H$^+$	CH$_3$COO$^-$ + 2 H$_2$O
Butyrate + 2 H$_2$O	2 CH$_3$COO + H$^+$ + 2 H$_2$
Propionate + 3 H$_2$O	CH$_3$COO$^-$ + HCO$_3$ + H$^+$ + 3 H$_2$
Benzoate + 7 H$_2$O	3CH$_3$COO$^-$ + HCO$_3$ + 3 H$^+$ + 3 H$_2$

Though a variety of products are formed by the action of fermentative bacteria, volatile fatty acids are the primary products of carbohydrate fermentation in biogas digesters, as they are in rumen. The partial pressure of hydrogen can influence the products of carbohydrate metabolism. The partial pressure of hydrogen can be maintained either by hydrogen oxidizing methanogens or sulphate reducing bacteria. However, in biogas digesters, the action of former organisms is preferred, resulting in methane as the end-product. Under these conditions, oxidation of NADH and the conversion of hexose to acetate, H_2 and CO_2 by fermentation occurs, yielding 4 ATP molecules per hexose molecule by glycolysis or acetyl phosphate pathway. But, under conditions of higher partial pressure of hydrogen, the formation of more reduced products results in the following order: propionate, butyrate, ethanol and lactate. Also, the fermentation of hexose either to ethanol or lactate yields only 2 ATP per hexose molecule by glycolysis, depriving thereby methanogens of the substrate (acetate) needed for its growth and activity. Though less in number, hydrogen-producing acetogenic bacteria are one of the important groups in biogas digesters. These organisms oxidize fatty acids that are longer than acetate to acetate, and thereby release energy from the substrate in the form of methane. It has been isolated from *Syntrophobacter wolinii*, which beta-oxidizes propionate to acetate, and isolated *Syntrophomonas wolfei*, which beta-oxidizes C_4 to C_7 fatty acids. But, these reactions are favourable only if the hydrogen partial pressure is below 10^{-3} atmosphere. However, for propionate oxidation, the partial pressure has to be still lower. Some reports suggest that the accumulation of propionic acid and lauric acid resulted in the acidification of digester, which inhibited further methane formation in castor-oil-cake-fed digesters. Furthermore, the propionate is toxic to methanogens even at a concentration of 8 mM. Subsequently, the development of consortia comprised of syntrophic co-cultures in association with hydrogen-utilizing methanogens, which stabilized methane production from castor-oil-cake-fed digester. The example for co-culture system is the isolate of propionate degrading bacteria associated with *M. formicicum* isolated from cattle dung digester. Though these organisms occurred at a pH < 6.0 and below 45°C, methanogenesis is observed at pH > 6.5 and temperature above 40°C.

Reductive halogenation of various halogenated aliphatic and aromatic compounds is the next important process by fermentative group of organisms. These reactions are thermodynamically feasible and can support growth. Anaerobic degradation of aromatic compounds also depends on hydrogen-consuming bacteria. Most of the bacterial strains oxidize benzoate to acetate with the help of obligate co-culturing system and with either methonogen or sulphate reducing bacteria. In the digester, phenol at 1000 ppm concentration inhibits the gas production up to six weeks, which increases with prolonged incubation. However, a mixed culture consortia enriched with phenol utilized the phenol immediately, showing an increase in gas production without any lag phase. In anaerobic degradation of aromatic compounds, the benzoic acid was the major intermediary product, which was further metabolized to acetate and butyrate.

The hydrogen-consuming acetogenic bacteria are the minor groups involved in fermentative reactions in biogas digesters. The activity in the formation of acetate by reduction of CO_2 in cattle dung-fed digesters accounted for less than 5% of the total acetate formed. However, a substantial activity by these groups of organisms should definitely increase methane formation, as acetate is the preferred substrate for *Methanosarcina barkeri*, the predominant methanogen in biogas digester in cattle waste-fed digester. However, *Methanobacterium formicicum* is the predominant methanogenic bacteria in cattle dung-fed digesters, followed by *M. ruminantium*. Methanogens possess very limited metabolic repertoire, using only acetate or C1 compounds (H_2 and CO_2, formate, methanol, methylamines or CO), with methane being the end-product of the reaction. Of the methanogenic genera, *Methanosarcina* sp. and *Methanosaeta* sp. form methane by aceticlastic reaction. While the apparent Km for methane formation from acetate for *Methanosaeta* sp. was under 1 mM, for *Methanosarcina* sp., it was 3–5 mM. Therefore, while faster-growing *Methanosarcina* sp. are predominant in high-rate, shorter-retention digesters wherein acetate concentration is higher, *Methanosaeta* sp. are predominant in low-rate, slow-turnover digesters.

Both carbon dioxide-reducing and aceticlastic-methanogens play an important role in maintaining the stability of the digester. The failure in a biogas digester can occur if carbon dioxide-reducing methanogens fail to keep pace with hydrogen production. While apparent Km for hydrogen consumption in methanogenic environments is near 10^{-2} atmosphere, hydrogen consumption must be below 10^{-3} atmosphere for oxidation of fatty acid to acetate, and thus the carbon dioxide-reducing methanogens in a biogas digester are greatly undersaturated for hydrogen. Despite this, hydrogen in biogas digester can build up rapidly to levels inhibitory to methanogenesis due to failure of the activity of hydrogen-scavenging organisms, shifting the fermentation products away from acetate. Moreover, the failure of aceticlastic methanogens to keep up with acetic acid production results in the accumulation of fatty acids, resulting thereby in the failure of the digester.

THE BIOGAS PLANT

The biogas plant (Figure 28.2) consists of two components: a digester (or fermentation tank) and a gas holder. The digester is a cube-shaped or cylindrical waterproof container with an inlet into which the fermentable mixture is introduced in the form of liquid slurry. The gas holder is normally an airproof steel container that, by floating like a ball on the fermentation mix, cuts off air to the digester (anaerobiosis) and collects the gas generated. In one of the most widely used designs, the gas holder is equipped with a gas outlet, while the digester is provided with an overflow pipe to lead the sludge out into a drainage pit.

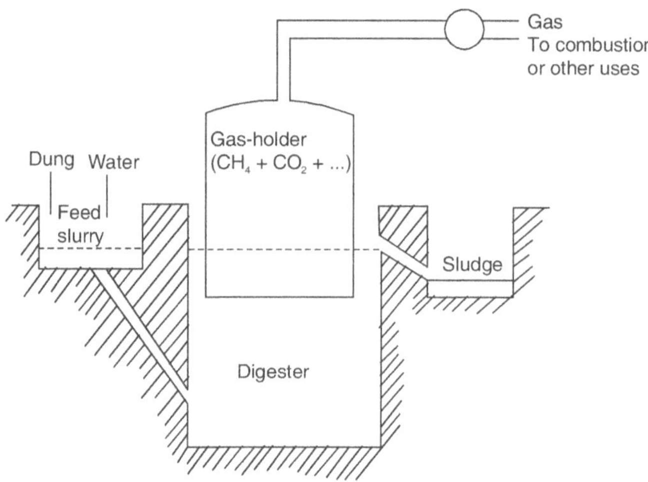

Figure 28.2   The ideal biogas apparatus

For the construction of a biogas plant, the important criteria are: 1) the amount of gas required for specific uses, and 2) the amount of waste material available for processing. There are several guidelines for consideration in the designing of batch (periodic feeding) and continuous (daily feeding) compartmentalized and non-compartmentalized biogas plants that are of either vertical or horizontal type. Digester reactors are constructed from brick, cement, concrete and steel. In Indonesia, where rural skills in brick making, brick laying, plastering and bamboo craft are well established, clay bricks have successfully replaced cement blocks and concrete. In areas where the cost is high, the 'sausage' or bag digester appears to be ideal. The digester is constructed of 0.55 mm thick hypalon laminated with neoprene and reinforced with nylon. The bag is fitted with an inlet and an outlet made from PVC. In rural areas, the whole installation is completed in a matter of minutes. A hole in the ground accommodates a bag, which is filled two-thirds with waste water. The production of gas fully inflates the bag, which is weighted down and fitted with a compressor to increase gas pressure. An Indonesian model biogas tank is shown in Figure 28.3.

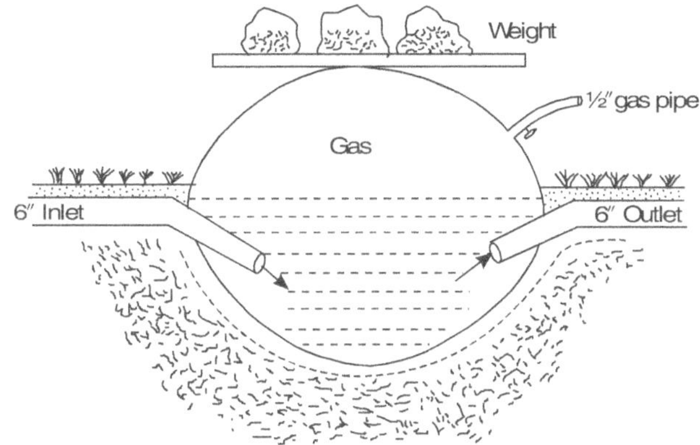

Figure 28.3   Ideal biogas tank in Indonesia

ENVIRONMENTAL AND OPERATIONAL CONSIDERATIONS

Raw Materials

Raw materials may be obtained from a variety of sources—livestock and poultry wastes, night soil, crop residues, food-processing and paper wastes, and materials such as aquatic weeds, water hyacinth, filamentous algae, and seaweed. Different problems may be encountered with each of these wastes with regard to collection, transportation, processing, storage, residue utilization, and ultimate use. Agricultural residues such as spent straw, hay, cane trash, corn and plant stubble, and bagasse need to be shredded in order to facilitate their flow into the digester reactor as well as to increase the efficiency of bacterial action. Succulent plant material yields more gas than dried matter does, and hence materials like brush and weeds need semi-drying. The storage of raw materials in a damp confined space for over ten days initiates anaerobic bacterial action that, though causing some gas loss, reduces the time for the digester to become operational.

Influent Solids Content

If fermentation materials are too dilute or too concentrated, it would result in low biogas production and insufficient fermentation activity, respectively.

Experience has shown that the raw-material (domestic and poultry wastes and manure) ratio to water should be 1 : 1, i.e., 100 kg of excrete to 100 kg of water. In the slurry, this corresponds to a total solids concentration of 8–11% by weight.

Loading

The size of the digester depends upon the loading, which is determined by the influent solids content, retention time and the digester temperature. Optimum loading rates vary with different digesters and their sites of location. Higher loading rates have been used when the ambient temperature is high. In general, the literature is filled with a variety of conflicting loading rates. In practice, the loading rate should be an expression of either (a) the weight of total volatile solids (TVS) added per day per unit volume of the digester, or (b) the weight of TVS added per day per unit weight of TVS in the digester. The latter principle is normally used for smooth operation of the digester.

Seeding

The common practice involves seeding with an adequate population of both the acid-forming and methanogenic bacteria. Actively digesting sludge from a sewage plant constitutes ideal "seed" material. As a general guideline, the seed material should be twice the volume of the fresh manure slurry during the start-up phase, with a gradual decrease in the amount added over a three-week period. If the digester accumulates volatile acids as a result of overloading, the situation can be remedied by reseeding, or by the addition of lime or other alkali.

pH

Low pH inhibits the growth of methanogenic bacteria and gas generation and is often the result of overloading. A successful pH range for anaerobic digestion is 6.0–8.0; efficient digestion occurs at a pH near neutrality. A slightly alkaline state is an indication that pH fluctuations are not too drastic. Low pH may be remedied by dilution or by the addition of lime.

Temperature

With a mesophilic flora, digestion proceeds best at 30–40°C; with thermophiles, the optimum range is 50–60°C. The choice of temperature to be used is influenced by climatic considerations. In general, there is no rule of thumb, but for optimum process stability, the temperature should be carefully regulated within a narrow range of the operating temperature. In warm climates, with no freezing temperatures, digesters may be operated without added heat. As a safety measure, it is common practice either to bury the digesters in the ground on account of the advantageous insulating properties of the soil, or to use a greenhouse covering. Heating requirements and, consequently, costs, can be minimized through the use of natural materials such as leaves, sawdust, straw, etc., which are composted in batches in a separate compartment around the digester.

Nutrients

The maintenance of optimum microbiological activity in the digester is crucial to gas generation, and consequently is related to nutrient availability. Two of the most important nutrients are carbon and nitrogen, and a critical factor for raw material choice is the overall C/N ratio. Domestic sewage and animal and poultry wastes are examples of N-rich materials that provide nutrients for the growth and multiplication of anaerobic organisms. On the other hand, N-poor materials like green grass, corn stubble, etc. are rich in carbohydrate substances that are essential for gas production. Excess availability of nitrogen leads to the formation of NH_3, the concentration of which inhibits further growth. Ammonia toxicity can be remedied by low loading or by dilution. In practice, it is important to maintain, by weight, a C/N ratio close to 30 : 1 for achieving an optimum rate of digestion. The C/N ratio can be judiciously manipulated by combining materials low in carbon with those that are high in nitrogen, and vice versa.

Toxic Materials

Wastes and biodegradable residue are often accompanied by a variety of pollutants that could inhibit anaerobic digestion. Potential toxicity due to ammonia can be corrected by remedying the C/N ratio of manure through the addition of shredded bagasse or straw, or by dilution. Common toxic substances are the soluble salts of copper, zinc, nickel, mercury and chromium. On the other hand, salts of sodium, potassium, calcium and magnesium may be stimulatory or toxic in action, both manifestations being associated with the cation rather than the anionic portion of the salt. Pesticides and synthetic detergents may also be troublesome to the process.

Stirring

When solid materials not well shredded are present in the digester, gas generation may be impeded by the formation of a scum that is comprised of these low-density solids that are enmeshed in a filamentous matrix. In time, the scum hardens, disrupting the digestion process and causing stratification. Agitation can be done either mechanically with a plunger or by means of rotational spraying of fresh influent. Agitation, normally required for bath digesters, ensures exposure of new surfaces to bacterial action, prevents viscid stratification and slowdown of bacterial activity, and promotes uniform dispersion of the influent materials throughout the fermentation liquor, thereby accelerating digestion.

Retention Time

Other factors such as temperature, dilution, loading rate, etc. influence retention time. At high temperature, biodigestion occurs faster, reducing the time requirement. A normal period for the digestion of dung would be two to four weeks.

KINETICS OF ANAEROBIC FERMENTATION

Several kinetic models have been developed to describe the anaerobic fermentation process. Monod showed a hyperbolic relationship between the exponential microbial growth rate and substrate concentration. In this model, the two kinetic parameters, namely microorganism growth rate and half velocity constant, are deterministic in nature, and these predict the conditions of timing of maximum biological activity and its cessation. This model can be used to determine the rate of substrate utilization (r_s) by the equation:

$$ r_S = q_{max} - \frac{S_x}{K} + S, $$

where,

S = limiting substrate concentration,

K = half constant,

x = concentration of bacterial cells, and

q_{max} = maximum substrate utilization rate.

The above equation is applicable for low substrate concentration. However, for high substrate concentration, the equation is rewritten as:

$$ r_S = q_{max} \cdot x $$

The Monod model suffers from the drawback that one set of kinetic parameters is not sufficient to describe the biological process both for short-

and long-retention times, and that kinetic parameters cannot be obtained for some complex substrates. To alleviate the limitations of the Monod model while retaining its advantages, Hashimoto developed an alternative equation, which attempts to describe the kinetics of methane fermentation in terms of several parameters. According to this equation, given below, for a given loading rate S_o/q, daily volume of methane per volume of digester depends on the biodegradability of the material (B_o) and kinetic parameters μm and K.

$$r_V = \left(B_o \times \frac{S_o}{q} \right) \cdot \left\{ 1 - \left(\frac{K}{q}\mu m - 1 + K \right) \right\}$$

where,

r_v = volumetric methane production rate, $l\ CH_4\ l^{-1}$digester d^{-1},

S_0 = influent total volatile solids (VS) concentration, $g\ l^{-1}$,

B_o = ultimate methane yield, $l\ CH_4\ g^{-1}$ VS added as q,

q = hydraulic retention time d^{-1},

μm = maximum specific growth of microorganism d^{-1} and

K = kinetic parameter, dimensionless.

PRODUCTION OF METHANE

Methane fermentation is a versatile biotechnology capable of converting almost all types of polymeric materials to methane and carbon dioxide under anaerobic conditions. This is achieved as a result of consecutive biochemical breakdown of polymers to methane and carbon dioxide in an environment in which a variety of microorganisms, which include fermentative microbes (acidogens); hydrogen-producing, acetate-forming microbes (acetogens); and methane-producing microbes (methanogens), harmoniously grow and produce reduced end-products. Anaerobes play important roles in establishing a stable environment at various stages of methane fermentation.

Methane fermentation offers an effective means of pollution reduction, superior to that achieved *via* conventional aerobic processes. Although practiced for decades, anaerobic fermentation has only recently been focused for its use in the economic recovery of fuel gas from industrial and agricultural surpluses.

The biochemistry and microbiology of anaerobic breakdown of polymeric materials to methane and the roles of various microorganisms involved are discussed here. The recent progress in the molecular biology of methanogens is reviewed, new digesters are described and improvements in the operation of various types of bioreactors are also discussed.

Microbial Consortia and Biological Aspects of Methane Fermentation

Methane fermentation is the consequence of a series of metabolic interactions among various groups of microorganisms (Figure 28.4). The description of microorganisms involved in methane fermentation is based on an analysis of bacteria isolated from sewage sludge digesters and from the rumen of some animals. The first group of microorganisms secretes enzymes that hydrolyse polymeric materials to monomers, such as glucose and amino acids, which are subsequently converted to higher volatile fatty acids, H_2 and acetic acid. In the second stage, hydrogen-producing acetogenic bacteria convert the higher volatile fatty acids produced, e.g. propionic and butyric acids, to H_2, CO_2 and acetic acid. Finally, the third group, methanogenic bacteria, converts H_2, CO_2 and acetate to CH_4 and CO_2.

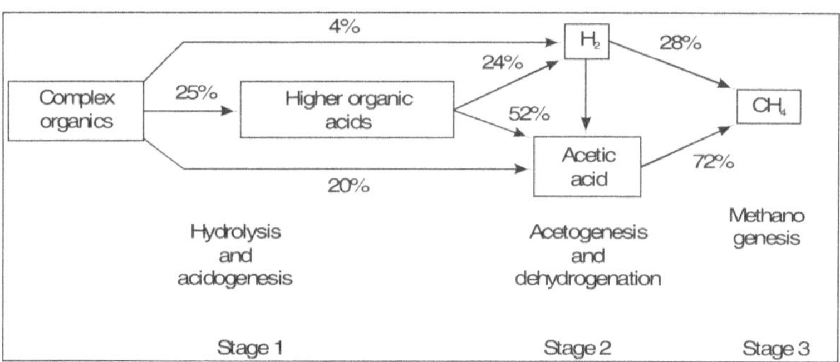

Figure 28.4 Stages of methane production

Hydrolysis and Acidogenesis

Polymeric materials such as lipids, proteins and carbohydrates are primarily hydrolysed by extracellular hydrolases excreted by microbes present in stage 1. Hydrolytic enzymes (lipases, proteases, cellulases, amylases, etc.) hydrolyse their respective polymers into smaller molecules, the primarily monomeric units, which are then consumed by microbes. In methane fermentation of waste waters containing high concentrations of organic polymers, the hydrolytic activity relevant to each polymer is of paramount significance, in that polymer hydrolysis may become a rate-limiting step for the production of simpler bacterial substrates to be used in subsequent degradation steps.

Lipases convert lipids to long-chain fatty acids. A population density of $10^4 - 10^5$ lipolytic bacteria per ml of digester fluid has been reported. Clostridia and micrococci appear to be most of the extracellular lipase producers. The long-chain fatty acids produced are further degraded by β-oxidation to produce acetyl-CoA.

Proteins are generally hydrolysed to amino acids by proteases secreted by *Bacteroides, Butyrivibrio, Clostridium, Fusobacterium, Selenomonas* and *Streptococcus*. The amino acids produced are then degraded to fatty acids, such as acetate, propionate, and butyrate, and to ammonia as found in *Clostridium, Peptococcus, Selenomonas, Campylobacter* and *Bacteroides*.

Polysaccharides such as cellulose, starch, and pectin are hydrolysed by cellulases, amylases and pectinases. The majority of microbial cellulases are composed of three species: (a) endo-α-1,4-glucanases, (b) exo-β-1,4-glucanases, (c) cellobiase or β-glucosidase. These enzymes act synergistically on cellulose, effectively hydrolysing its crystal structure to produce glucose. Microbial hydrolysis of raw starch to glucose requires amylolytic activity, which consists of 5 amylase species: (a) a-amylases that endocleave α-1-4 bonds, (b) β-amylases that exocleave α-1-4 bonds, (c) amyloglucosidases that exocleave α-1-4 and α-1-6 bonds, (d) debranching enzymes that act on α-1-6 bonds, and (e) maltase that acts on maltose liberating glucose. Pectins are degraded by pectinases, including pectinesterases and depolymerases. Xylans are degraded with α^2-endo-xylanase and α^2-xylosidase to produce xylose.

Hexoses and pentoses are generally converted to C_2 and C_3 intermediates and reduced to electron carriers (e.g. NADH) *via* common pathways. Most anaerobic bacteria undergo hexose metabolism *via* the Emden–Meyerhoff–Parnas pathway (EMP), which produces pyruvate as an intermediate along with NADH. Pyruvate and NADH, thus generated, are transformed into fermentation endo-products such as lactate, propionate, acetate and ethanol by other enzymatic activities, which vary tremendously with microbial species.

Thus, in hydrolysis and acidogenesis, sugars, amino acids and fatty acids produced by microbial degradation of biopolymers are successively metabolized by groups of bacteria and are primarily fermented to acetate, propionate, butyrate, lactate, ethanol, carbon dioxide and hydrogen .

Acetogenesis and Dehydrogenation

Although some acetate (20%) and H_2 (4%) are directly produced by acidogenic fermentation of sugars and amino acids, both products are primarily derived from the acetogenesis and dehydrogenation of higher volatile fatty acids. Obligate H_2-producing acetogenic bacteria are capable of producing acetate and H_2 from higher fatty acids. Only *Syntrophobacter wolinii*, a propionate decomposer and *Syntrophomonos wolfei*, a butyrate decomposer, have thus far been isolated, since H_2 produced severely inhibits the growth of these strains. The use of co-culture techniques incorporating H_2 consumers such as methanogens and sulphate-reducing bacteria may, therefore, facilitate elucidation of the biochemical breakdown of fatty acids.

H_2 production by acetogens is generally energetically unfavourable due to high free-energy requirements ($\alpha''G° > 0$). However, with a combination of H_2-consuming bacteria, co-culture systems provide favourable conditions for decomposition of fatty acids to acetate and CH_4 or H_2S ($\alpha''G° < 0$). In addition to the decomposition of long-chain fatty acids, ethanol and lactate are also converted to acetate and H_2 by an acetogen and *Clostridium formicoaceticum*, respectively. The effect of the partial pressure of H_2 on the free energy is associated with the conversion of ethanol, propionate, acetate and H_2 or CO_2 during methane fermentation. An extremely low partial pressure of H_2 (10^{-5} atm) appears to be a significant factor in propionate degradation to CH_4. Such a low partial pressure can be achieved in a co-culture with H_2-consuming bacteria as previously described.

Methanogenesis

Methanogens are physiologically united as methane producers in anaerobic digestion. Although acetate and H_2/CO_2 are the main substrates available in the natural environment, formate, methanol, methylamines and CO are also converted to CH_4.

Table 28.3   Reactions involved in fatty acid catabolism by *Syntrophomonas wolfei*

Even-numbered

$$CH_3CH_2CH_2COO^- + 2H_2O \rightarrow 2H_3COO^-2H_2 + H^+$$

$$CH_3CH_2CH_2CH_2CH_2COO^- + 4H_2O \rightarrow 3CH_3COO^- + 4H_2 + H^+$$

$$CH_3CH_2CH_2CH_2CH_2CH_2CH_2COO^- + 6H_2O \rightarrow 4CH_3COO^- + 6H_2 + 3H^+$$

Odd-numbered

$$CH_3CH_2CH_2CH_2COO^- + 1\,H_2O \rightarrow CH_3CH_2COO^- + CH_3COO^- + 2\,H_2 + H^+$$

$$CH_3CH_2CH_2CH_2CH_2CH_2COO^- + 4\,H_2O \rightarrow CH_3CH_2COO^- + 2\,CH_3COO^- + 4\,H_2 + 2H^+$$

Branched-chained

$$CH_3CHCH_2CH_2CH_2COO^- + 2\,H_2O \rightarrow CH_3CHCH_2COO^- + CH_3COO^- + 2H_2 + H^+$$
$$\hspace{4em}|\hspace{14em}|$$
$$\hspace{4em}CH_3\hspace{13em}CH_3$$

Table 28.4 Free-energy changes for reactions involving anaerobic oxidation in pure cultures or in co-cultures with H_2-utilizing methanogens or *Desulfovibrio* spp.

Equations	$\alpha"G^{\circ'}$ (kJ/reaction)
Proton-reducing (H_2-producing) acetogenic bacteria	
A. $CH_3CH_2CH_2COO^- + 2H_2O \rightarrow 2\ CH_3COO^- + 2H_2 + H^+$	+48.1
B. $CH_3CH_2COO^- + 3H_2O \rightarrow CH_3COO^- + HCO_3^- + H^+ + 3H_2$	+76.1
H_2-using methanogens and desulfovibrios	
C. $4H_2 + HCO_3^- + H^+ \rightarrow CH_4 + 3\ H_2O$	−135.6
D. $4H_2 + SO_4^{2-} + H^+ \rightarrow HS^- + 4\ H_2O$	−151.9
Co-culture of 1 and 2	
A + C2 $CH_3CH_2CH_2COO^- + HCO_3^- + H_2O \rightarrow 4\ CH_3COO^- + H^+ + CH_4$	−39.4
A + D2 $CH_3CH_2CH_2COO^- + SO_4^{2-} \rightarrow 4\ CH_3COO^- + H^+ + HS^-$	−55.7
B + C4 $CH_3CH_2COO^- + 12H_2 \rightarrow 4\ CH_3COO^- + HCO_3^- + H^+ + 3\ CH_4$	−102.4
B + D4 $CH_3CH_2COO^- + 3\ SO_4^{2-}\ 4\ CH_3COO^- + 4\ HCO_3^- + H^+ + 3\ HS^-$	−151.3

Table 28.5   Energy yielding reactions of methanogens

Reaction	G° (kJ/mol substrate)
$CO_2 + 4\ H_2 \rightarrow CH_4 + 2H_2O$	−130.7
$HCO_3^- + 4\ H_2 + H^+ \rightarrow CH_4 + 3\ H_2O$	−135.5
$CH_3COO^- + H^+ \rightarrow H_4 + CO_2$	−37.0
$CH_3COO^- + H_2O \rightarrow CH_4 + HCO_3^-$	−32.3
$HCOO^- + H^+ \rightarrow 0.25\ CH_4 + 0.75\ CO_2 + 0.5\ H_2O$	−36.1
$CO + 0.5\ H_2O \rightarrow 0.25\ CH_4 + 0.75\ CO_2$	−52.7
$CH_3OH \rightarrow 0.75\ CH_4 + 0.25\ CO_2 + 0.5\ H_2O$	−79.9
$CH_3NH_3^+ + 0.5\ H_2O \rightarrow 0.75\ CH_4 + 0.25\ CO_2 + NH_4^+$	−57.4
$(CH_3)_2NH_2^+ + H_2O \rightarrow 1.5\ CH_4 + 0.5\ CO_2 + NH_4^+$	−112.2
$(CH_3)_2NCH_2CH_3H^+ + H_2O \rightarrow 1.5\ CH_4 + 0.5\ CO_2 + {}^+H_3NCH_2CH_3$	−105.0
$(CH_3)_3NH + 1.5H_2O \rightarrow 2.25\ CH_4 + 0.75\ CO_2 + NH_4^+$	−170.8

Since methanogens, as obligate anaerobes, require a redox potential of less than −300 mV for growth, their isolation and cultivation was somewhat elusive due to technical difficulties encountered in handling them under completely O_2-free conditions. However, as a result of a greatly improved methanogen isolation techniques developed by Hungate, more than 40 strains of pure methanogens have now been isolated. Methanogens can be divided into two groups: H_2/CO^{2-} and acetate-consumers. Although some of the H_2/CO_2 consumers are capable of utilizing formate, acetate is consumed by a limited number of strains, such as *Methanosarcina spp.* and *Methanothrix spp.* (now *Methanosaeta*), which are incapable of using formate. Since a large quantity of acetate is produced in the natural environment, *Methanosarcina* and *Methanothrix* play an important role in the completion of anaerobic digestion and in accumulating H_2, which inhibits acetogens and methanogens. H_2-consuming methanogens are also important in maintaining low levels of atmospheric H_2.

DEVELOPMENTS IN BIOREACTOR TECHNOLOGY

Methane fermentation has been used since 1900 for treating excess sludge discharged from sewage-treatment plants. This technology has since been developed to treat waste waters, such as those derived from alcohol distillation, antibiotic production, and baker's yeast manufacture. However, the conventional system requires a long hydraulic retention time (HRT) due to factors such as low microbial concentration and instability against environmental shocks. Research and development efforts have been directed at retaining a high density of useful microorganisms in order to achieve rapid and effective treatment with the objective of improving the conventional system. To this end, considerable technological developments in microbial floc formation and in microbial adhesion onto carrier materials, which retain cells in the reactor, have been made. For the former purpose, the upflow anaerobic sludge blanket (UASB) has proven useful, while for the latter, the upflow anaerobic filter process (UAFP) and anaerobic fluidized-bed reactor (AFBR) have been developed. In all of these newly developed processes, however, acidogenesis may occur more frequently than methanogenesis, leading to the accumulation of inhibitory products such as volatile fatty acids. Two-phase anaerobic digestion processes have been developed in order to resolve this problem.

Two-phase Methane Fermentation Processes

Novel bioreactors for methane fermentation such as the UASB, UAFP and AFBR experience inherent problems when operated at high COD loads,

Table 28.6   Operational methane-generating bioreactors and their current applications

| Reactor type | Location | Waste characteristics | | | | Design | | Medium |
		Type	COD (mg/l)	Flow (m³/day)	HRT (day)	COD load (kg/m³/day)	COD removal (%)	
UASB	USA							
	La Crosse, Wis.	Brewery	2,500	23,000	0.2	14.1	86	
	Caribou, Me.	Starch	22,000	910	2.0	11.0	85	
	Plover, Wis.	Potato	4,300	3,000	0.7	6.0	80	
	Europe							
	Netherlands	Sugar	17,000	3,000	1.0	13.3	94	
	Netherlands	Sugar beet	3,000	6,960	0.2	14.5	85	
	Netherlands	Starch	7,700	1,750	0.1	8.0	85	
	Switzerland	Potato	2,260	2,210	0.3	8.3	80	
	Germany	Sugar beet	7,500	2,400	0.6	12,0	86	
	Netherlands	Alcohol	5,330	2,090	0.3	16.0	90	
	Netherlands	Sugar	17,000	3,000	1.0	13.3	94	

(Contd.)

Table 28.6   (Continued)

Reactor type	Location	Waste characteristics				Design		
		Type	COD (mg/l)	Flow (m³/day)	HRT (day)	COD load (kg/m³/day)	COD removal (%)	Medium
UAFP	Spokane Washington	Starch gluten	8,800	490	0.9	3.8	64	Graded rock 2.5–7.6 cm
	Vernon, Texas	Guar gum	9,140	823	1.0	16.0	60	9 cm Pall rings
	San Juan Puerto Rico	Rum distillery	95,000	1,325	7.0–8.0	8.9	75	Synthetic ("Vinyl core")
	Bishop, Texas	Chemical	12,000	3,785	1.5	9.6	80	9 cm Pall rings
	Pampa, Texas	Chemical	14,400	3,785	1.5	10.4	90	9 cm Pall rings
AFBR	Birmingham, Ala.	Soft-drink bottling waste	6,900	380	0.3	9.6	77	0.6 mm E.S. sand
	Midwest	Soy-processing waste	9,000	-	<1.0	13.0	-	0.4 mm E.S. sand

due to the fact that the overall growth rate of acidogenic bacteria proceeds faster (10-fold) than that of methanogenic bacteria. When this occurs, inhibitory products such as volatile fatty acids and H_2 accumulate in the reactor, slowing down the entire process. In order to overcome this, two-phase processes consisting of acidogenic and methanogenic fermentations have been investigated. In one full scale two-phase system can produce 3–13 Kg/m^3/day of biogas with 65–80% methane content and 70–97% COD removal in first phase of acidogenic fermentation. In another example, a two-phase system consisting of a complete stirred reactor for the first phase and a UASB for the second phase was constructed. When this system was applied in the treatment of alcohol distillery waste (COD =10,000 mg/l) at HRTs of 16–72 hours in the first phase, and 14 hours in the second phase, 84% COD removal and 92% BOD removal were accomplished. A two-phase system consisting of a UAFP for the first phase and a horizontal AFP for the second phase has also been proposed, with which it should be possible to treat sewage waste water (COD 800 to 2,600 mg/l) at HRTs of 2–5.5 hours with a high methane content (~90%). In addition, since SS in waste water greatly influences the performance of the UASB or UAFP, an acidogenic fermentation first phase in combination with a UASB or UAFP second phase is useful in reducing the SS which enters the second phase.

PRODUCTION OF HYDROGEN

Hydrogen gas is seen as a future energy carrier by virtue of the fact that it is renewable, does not evolve the "greenhouse gas" CO_2 in combustion, liberates large amounts of energy per unit weight in combustion, and is easily converted to electricity by fuel cells. Biological hydrogen production has several advantages over hydrogen production by photoelectrochemical or thermochemical processes. Biological hydrogen production by photosynthetic microorganisms, for example, requires the use of a simple solar reactor such as a transparent closed box with low energy requirements. Electrochemical hydrogen production *via* solar battery-based water splitting requires the use of solar batteries with high energy requirements.

Low conversion efficiencies of biological systems can be compensated for by low energy requirements and reduced initial investment costs. Moreover, in laboratory experiments, light energy conversion efficiency as high as 7% has been obtained using the photoheterotrophic process. The basic characteristics of biological hydrogen production and experiments designed to improve the feasibility of biological hydrogen production, particularly through the use of photosynthetic microorganisms, are described in this section.

Table 28.7   Feed-stocks for methane production

Feedstock	Hydraulic retention time	Organic loading rate (kg VS m⁻³ d⁻¹)	Methane yield (m³kg⁻¹ VS)
Fruit wastes	8	3.8	0.030
	16	3.8	0.250
	20	3.8	0.320
	24	3.8	0.190
	16	5.7	0.110
	16	7.6	0.110
	16	9.5	0.420
Tomato processing waste	24	4.3	0.420
Banana peeling			
0.088 mm size	NA	NA	0.408
0.40 mm size	NA	NA	0.409
1.0 mm size	NA	NA	0.396
6.0 mm size	NA	NA	0.374
30 × 10 m size	NA	NA	0.291
Cauliflower waste			
0.088 mm size	NA	NA	0.423
0.40 mm size	NA	NA	0.423
1.0 mm size	NA	NA	0.423
Ipomoea sp.			
0.088 mm size	NA	NA	0.429
0.40 mm size	NA	NA	0.427
1.0 mm size	NA	NA	0.421
6.0 mm size	NA	NA	0.413
30 × 10 mm size	NA	NA	0.387
Bermuda grass			
0.088 mm size	NA	NA	0.226
0.40 mm size	NA	NA	0.228
1.0 mm size	NA	NA	0.214
6.0 mm size	NA	NA	0.205
30 × 10 mm size	NA	NA	0.137

(Contd.)

Table 28.7   (Continued)

Feedstock	Hydraulic retention time	Organic loading rate (kg VS m^{-3} d^{-1})	Methane yield (m^{-3} kg^{-1} VS)
Wheat straw			
0.088 mm size	NA	NA	0.249
0.40 mm size	NA	NA	0.248
1.0 mm size	NA	NA	0.241
6.0 mm size	NA	NA	0.227
30 × 10 mm size	NA	NA	0.162
Paddy straw			
0.088 mm size	NA	NA	0.365
0.40 mm size	NA	NA	0.367
1.0 mm size	NA	NA	0.358
6.0 mm size	NA	NA	0.347
30 × 10 mm size	NA	NA	0.241
Ipomoea fistulora stem	NA	NA	0.361
0.4 mm size	NA	NA	0.182
6.0 mm size	NA	NA	0.181
Gliricidia maculata	5	4.95	0.034
Parthenium hysterophorus	10	2.48	0.117
	20	1.24	0.115
	40	0.62	0.101
Ageratum, partially decomposed	NA	NA	0.241
Pistia stratiotes	NA	NA	0.361
Salvinia sp.	NA	NA	0.242
Azolla pinnata	NA	NA	0.132
	NA	NA	0.117
Lemna minor	NA	NA	0.106
Lantana camera	NA	NA	0.236
Night soil (NS)			
100% NS	NA	NA	0.500
50% NS + 50% Cow dung	NA	NA	0.350
100% Cow dung	NA	NA	0.130

Biophotolysis of Water by Microalgae and Cyanobacteria

Microalgae are primitive microscopic plants living in aqueous environments. Cyanobacteria, formerly known as blue-green algae, are now recognized as bacteria since the anatomical characteristics of their cells are prokaryotic (bacterial type). Microalgae and cyanobacteria, along with higher plants, are capable of oxygenic photosynthesis according to the following reaction:

$$CO_2 + H_2O \rightarrow 6[CH_2O] + O_2$$
organic compounds

Photosynthesis consists of two processes: light energy conversion to biochemical energy by a photochemical reaction, and CO_2 reduction to organic compounds, such as sugar phosphates, through the use of this biochemical energy by Calvin-cycle enzymes. Under certain conditions, however, instead of reducing CO_2, a few groups of microalgae and cyanobacteria consume biochemical energy to produce molecular hydrogen. Hydrogenase and nitrogenase enzymes are both capable of hydrogen production.

Hydrogenase-dependent Hydrogen Production

The green algae, *Scenedesmus*, produce molecular hydrogen under light conditions after being kept under anaerobic and dark conditions. A 25 to 30% sugar concentration may be obtained regardless of the sugar concentration of the initial saccharified solution. Figure 28.8 shows the process involved in the hydrogenase-dependant hydrogen production.

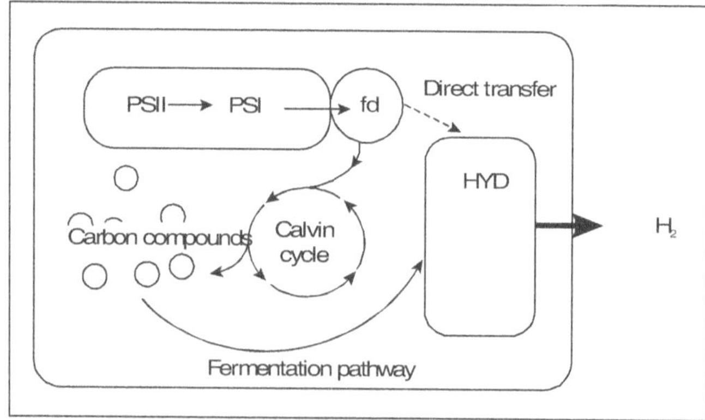

Figure 28.8   Hydrogenase-mediated hydrogen production

Hydrogenase, the enzyme responsible for this hydrogen production, catalyses the following reaction:

$$2H^+ + 2X_{reduced} \rightarrow 6\,H_2 + 2X_{oxidized}$$

The electron carrier, X, is thought to be ferredoxin. Since ferredoxin is reduced with water as an electron donor by the photochemical reaction, green algae are theoretically water-splitting microorganisms. The mechanisms involved in hydrogen production have determined that the reducing power (electron donation) of hydrogenase does not always come from water, but may sometimes originate intracellularly from organic compounds such as starch. The contribution of the decomposition of organic compounds to hydrogen production is dependent on the algal species concerned and on culture conditions. Even when organic compounds are involved in hydrogen production, an electron source can be derived from water, since organic compounds are synthesized by oxygenic photosynthesis. The reason for hydrogenase inactivity in green algae under normal photosynthetic growth conditions is unclear. Hydrogenase is thought to become active in order to excrete excess reducing power under specific conditions, such as anaerobic conditions.

Green algae were known as light-dependent, water-splitting catalysts, but the characteristics of their hydrogen production were not practical for exploitation. Hydrogenase is too oxygen-labile for sustainable hydrogen production: light-dependent hydrogen production ceases within a few to several minutes since photosynthetically produced oxygen inhibits or inactivates hydrogenases. A continuous gas flow system designed to maintain low oxygen concentrations within the reaction vessel was employed in basic studies, but has not been found practically applicable.

Greenbaum and co-workers reported very high (10 to 20%) efficiencies of light conversion to hydrogen, based on PAR (photosynthetically active radiation which includes light energy of 400–700 nm in wavelength). These authors recently reported what may represent a "short circuit" of photosynthesis, whereby hydrogen production and CO_2 fixation occurred by a single photosystem (photosystem II only) of a *Chlamydomonas* mutant.

Green algae are applicable in another method of hydrogen production. The *Scenedesmus* produces hydrogen gas not only under light conditions, but also under dark anaerobic conditions fermentatively with intracellular starch as a reducing source. Although the rate of fermentative hydrogen production per unit of dry cell weight is less than that obtained through light-dependent hydrogen production, hydrogen production is sustainable due to the absence of oxygen, on the basis of experiments conducted on fermentative hydrogen production under dark conditions. In the hydrogen production in light or dark cycle, CO_2 is reduced to starch by photosynthesis in the daytime (under light conditions), and the starch thus formed is decomposed to hydrogen gas and organic acids and/or alcohols under anaerobic conditions during night time (under dark conditions). The technological merits of this include: oxygen-inactivation of hydrogenase can be prevented through maintenance of green algae under anaerobic conditions; night time is used effectively; temporal separation of hydrogen and oxygen production does not require gas separation for simultaneous water-splitting; and organic acids and alcohols can be converted to hydrogen gas by photosynthetic bacteria under light conditions.

Nitrogenase-dependent Hydrogen Production

The nitrogen-fixing cyanobacterium, *Anabaena cylindrica*, produces hydrogen and oxygen gas simultaneously in an argon atmosphere for several hours. Nitrogenase is responsible for nitrogen-fixation and is distributed mainly among prokaryotes, including cyanobacteria, but does not occur in eukaryotes, under which microalgae are classified. Molecular nitrogen is reduced to ammonium with the consumption of reducing power and ATP. Nitrogenase-dependant hydrogen production process is shown in Figure 28.9. The reaction is substantially irreversible and produces ammonia:

$$3H_2 + N_2 \rightarrow NH_3$$

However, nitrogenase catalyses proton reduction in the absence of nitrogen gas (i.e., in an argon atmosphere).

$$2H^+ + 2e^- \rightarrow H_2$$

$$4ATP \rightarrow 4(ADP + Pi)$$

Hydrogen production catalysed by nitrogenase occurs as a side reaction at a rate of one-third to one-fourth that of nitrogen-fixation, even in a 100% nitrogen gas atmosphere.

Nitrogenase itself is extremely oxygen-labile. Unlike the case of hydrogenase, however, cyanobacteria have developed mechanisms for protecting nitrogenase from oxygen gas and supplying it with energy (ATP) and reducing power. The most successful mechanism is the localization of nitrogenase in the heterocysts of filamentous cyanobacteria. Vegetative cells (ordinary cells) in filamentous cyanobacteria carry out oxygenic photosynthesis. Organic compounds produced by CO reduction are transferred into heterocysts and are decomposed to provide nitrogenase with reducing power. ATP can be provided by PSI-dependent and anoxygenic photosynthesis within heterocysts. Investigations into prolongation and optimization of hydrogen production revealed that the hydrogen-producing activity of cyanobacteria was stimulated by nitrogen starvation.

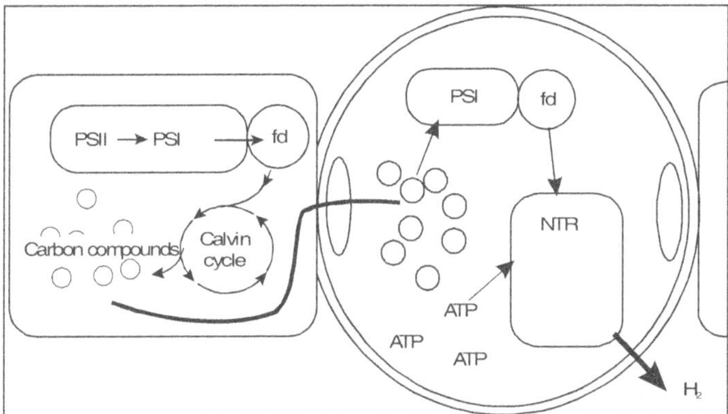

Figure 28.9 Nitrogenase-mediated hydrogen production in Heterocystous cyanobacteria

The presence and physiological roles of hydrogenases in nitrogen-fixing cyanobacteria remain controversial, but uptake hydrogenase appears to consume and reuse hydrogen gas, resulting in a decrease in net hydrogen production. The aerobic hydrogen production by a nitrogen-fixing *Anabaena* sp. believed to be an uptake hydrogenase-deficient strain. After being cultured for 12 days, the strain accumulated approximately 10% hydrogen and 70% oxygen gas within the gas phase of the vessel by the nitrogenase side reaction, even in the presence of air. *Anabaena cylindrica*, a nitrogen-starved culture of the cyanobacteria, was continuously sparged by argon-based gas, while the hydrogen content of the effluent gas was measured. The average conversion efficiency over a period of one month (combustion energy of hydrogen gas produced by cyanobacteria or incident solar energy into the photobioreactor area) was approximately 0.2%.

HYDROGEN FROM ORGANIC COMPOUNDS

Hydrogen Production by Photosynthetic Bacteria

Photosynthetic bacteria undergo anoxygenic photosynthesis with organic compounds or reduced sulphur compounds as electron donors. Some non-sulphur photosynthetic bacteria are potent hydrogen producers, utilizing organic acids such as lactic, succinic and butyric acids, or alcohols as electron donors. Since light energy is not required for water oxidation, the efficiency of light energy conversion to hydrogen gas by photosynthetic bacteria is in principle much higher than that by cyanobacteria. Hydrogen production by photosynthetic bacteria is mediated by nitrogenase activity, although hydrogenases may be active for both hydrogen production and hydrogen uptake under some conditions. Miyake and Kawamura demonstrated a maximum energy conversion efficiency (combustion energy of hydrogen gas produced/incident light energy) of 6 to 8% using *Rhodobacter* sp. in laboratory experiments.

Combined Photosynthetic and Anaerobic Bacterial Hydrogen Production

Generally, anaerobic bacteria produce hydrogen and organic acids by metabolizing the sugar. But the organisms are incapable to proceed further to break down organic acids. The conversion of organic acids into hydrogen is achieved by the cumulative activity of photosynthetic bacteria and anaerobic bacteria. Theoretically, one mole of glucose can be converted to 12 moles of hydrogen through the use of photosynthetic bacteria capable of capturing light energy in such a combined system. From a practical point of view, organic wastes frequently contain sugar or sugar polymers. It is not, however, easy to obtain organic wastes containing organic acids as the main components. The combined use of photosynthetic and anaerobic bacteria should potentially increase the likelihood of their application in photobiological hydrogen production. Figure 28.10 shows the processes involved in hydrogen producing reactions.

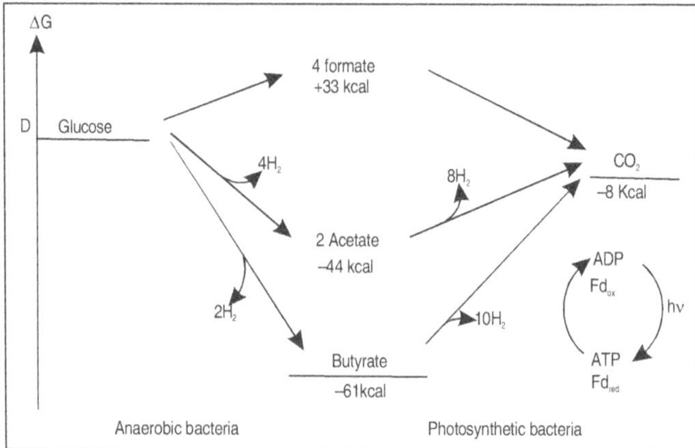

Figure 28.10   Free energy changes in hydrogen-producing reactions by anaerobic and photosynthetic bacteria

PRODUCTION OF GOBAR GAS FROM COW DUNG

Cow dung gas is 55–65% methane, 30–35% carbon dioxide, with some hydrogen, nitrogen and other traces. Its heating value is around 600 British thermal unit (BTU) per cubic foot. Natural gas consists of around 80% methane, yielding a BTU value of about 1000. Biogas may be improved by filtering it through limewater to remove carbon dioxide; iron fillings to absorb corrosive hydrogen sulphide; and calcium chloride to extract water vapour after the two former processes. Cow dung slurry is composed of 1.8–2.4% nitrogen (N_2), 1.0–1.2% phosphorus (P_2O_5), 0.6–0.8% potassium (K_2O) and 50–75% organic humus. About one cubic foot of gas may be generated from one pound of cow manure at around 28°C. This is enough to cook a day's meal for 4–6 people in India. About 1.7 cubic metres of biogas equals one litre of gasoline. The manure produced by one cow in one year can be converted to methane equivalent to over 200 litres of gasoline. Gas engines require about 0.5 m³ of methane per horsepower per hour. Some care must be taken with the lubrication of engines using solely biogas due to the "'dry' nature of the fuel and some residual hydrogen sulphide.

Fermentation

There are two basic types of organic decomposition: aerobic (in the presence of oxygen), and anaerobic (in the absence of oxygen). All organic material, both animal and vegetable, can be broken down by these two processes, but the products of decomposition will be quite different in both cases. Aerobic decomposition (fermentation) will produce carbon dioxide, ammonia and some other gases in small quantities, heat in large quantities, and a final product that can be used as a fertilizer. Anaerobic decomposition will produce methane, carbon dioxide, some hydrogen and other gases in traces, very little heat, and a final product with a higher nitrogen content than is produced by aerobic fermentation.

Anaerobic decomposition is a two-stage process as specific bacteria feed on certain organic materials. In the first stage, acidic bacteria dismantle the complex organic molecules into peptides, glycerol, alcohol and simpler sugars. When these compounds have been produced in sufficient quantities, a second type of bacteria starts to convert these simpler compounds into methane. These methane producing bacteria are particularly influenced by the ambient conditions, which can slow down or halt the process completely, if they do not lie within a fairly narrow level.

Acidity

Anaerobic digestion will occur best within a pH range of 6.8 to 8.0. More acidic or basic mixtures will ferment at a lower speed. The introduction of raw material will often lower the pH (make the mixture more acidic). Digestion will stop or slow dramatically until the bacteria have absorbed the acids. A high pH will encourage the production of acidic carbon dioxide to neutralize the mixture again.

Carbon–Nitrogen Ratio

The bacteria responsible for the anaerobic process require both elements, as do all living organisms, but they consume carbon roughly 30 times faster than nitrogen. Assuming all other conditions are favourable for biogas production, a carbon–nitrogen ratio of about 30 : 1 is ideal for the raw material fed into a biogas plant. A higher ratio will leave carbon still available after the nitrogen has been consumed, starving some of the bacteria of this element. These will in turn die, returning nitrogen to the mixture, but slowing down the process. Too much nitrogen will cause this to be left over at the end of digestion (which stops when the carbon has been consumed) and reduce the quality of the fertilizer produced by the biogas plant. The correct ratio of carbon to nitrogen will prevent loss of either fertilizer quality or methane content.

Temperature

Anaerobic breakdown of waste occurs at temperatures between 0°C and 69°C, but the action of the digesting bacteria will decrease sharply below 16°C. The production of gas is most rapid between 29°C and 41°C or between 49°C and 60°C. This is due to the fact that two different types of bacteria multiply best in these two different ranges, but the high temperature bacteria are much more sensitive to ambient influences. A temperature between 32°C and 35°C has proven most efficient for stable and continuous production of methane. Biogas produced outside this range will have a higher percentage of carbon dioxide and other gases than within this range.

Percentage of Solids

Anaerobic digestion of organics will proceed best if the input material consists of roughly 8% solids. In the case of fresh cow manure, this is the equivalent of dilution with roughly an equal quantity of water.

Basic Design of a Biogas Plant

The central part of an anaerobic biogas plant is an enclosed tank known as the digester. This is an air-tight tank filled with organic waste slurry where the gas is generated and collected. Design differences mainly depend on the type of organic waste to be used as raw material, the temperatures to be used in digestion and the materials available for construction.

Systems intended for the digestion of liquid or suspended solid wastes (for example, cow manure) are mostly filled or emptied using pumps and pipe work. A simpler version is to gravity feed the tank and allow the digested slurry to overflow the tank. This has the advantage of being able to consume more solid matter as well, such as chopped vegetable waste, which would block the pump very quickly. This provides extra carbon to the system and raises the efficiency. Cow manure is nitrogen rich, which can be improved by the addition of vegetable matter.

Biogas plants constructed above ground must be made of steel to withstand the pressure within, and it is generally simpler and cheaper to build the digester below ground. This also makes gravity feed of the system much simpler. However, the maintenance is much simpler for systems built aboveground, and a black coating will help provide some solar heating. This should make it clear that biogas is a practical application and use of a waste product.

Continuous Feeding (Mostly Liquids)

The complete anaerobic digestion of cow manure takes about eight weeks at normally warm temperatures. One-third of the total biogas will be produced in the first week, another quarter in the second week and the remainder of biogas production will be spread over the remaining six weeks. Gas production can be accelerated and made more consistent by continuously feeding the digester with small amounts of wastes. This will preserve the nitrogen level in the slurry for use as fertilizer. If such a continuous feeding system is used, then it is essential to ensure that the digester is large enough to contain all the material that will be fed through in a whole digestion cycle. One solution is to use a double digester, consuming the waste in two stages, with the main part of the biogas (methane) being produced in the first stage and the second stage finishing the digestion at a slower rate, but still producing another 20% or so of the total biogas.

Batch Feeding (Mostly Solids)

There are biogas systems that are designed to digest solid vegetable waste alone. Since plant solids will not flow through pipes, this type of digester is best used as a single batch digester. The tank is opened, the slurry is removed for use as fertilizer, and the new charge is added. The tank is resealed and ready for operation. Dependent on the waste material and operating temperature, a batch digester will start producing biogas after two to four weeks, which slowly increase and drop off after three or four months. Batch digesters are, therefore, best operated in groups, so that at least one is always producing useful quantities of gas.

Most vegetable matter has a much higher carbon–nitrogen ratio than dung has, and so some nitrogen producers (preferably organic) must generally be added to the vegetable matter, especially when batch digestion is used. However, vegetable matter produces about eight times as much biogas as manure, so the quantity required is much smaller for the same biogas production. A mixture of dung and vegetable matter is hence ideal in most ways, with a majority of vegetable matter to provide the biogas and the valuable methane contained in it.

Stirring

Some method of stirring the slurry in a digester is always advantageous, if not essential. If not stirred, the slurry will tend to settle out and form a hard scum on the surface, which will prevent the release of biogas. This problem is much greater with vegetable waste than with manure, which will tend to remain in suspension and have better contact with the bacteria as a result. Continuous feeding causes fewer problems in this direction, since the new charge will break up the surface and provide a rudimentary stirring action. If some form of heating is needed for the biodigester, this will also provide some circulatory action, which will tend to stir the contents.

Temperature Control

In hot regions, it is relatively easy to simply shade the digester to keep it in the ideal range of temperature, but cold climates present more of a challenge.

The first action is, naturally, to insulate the digester with straw or wood shavings. A layer about 50–100 cm thick coated with a waterproof covering is a good start. If this still proves to be insufficient in winter, then heating coils may have to be added to the biogas digester. It is relatively simple to keep the digester at the ideal temperature if hot water, regulated with a thermostat, is circulated through the system. Usually, it is sufficient to circulate the heating for a couple of hours in the morning and again in the evening. Naturally, the biogas produced by the digester can be used for this purpose. The small quantity of gas wasted on heating the digester will be more than compensated for by the greatly increased biogas production.

Gas Collection

The biogas in an anaerobic digester is collected in an inverted drum. The walls of the drum extend down into the slurry to provide a seal. The drum is free to move to collect the gas produced. The weight of the drum provides the pressure on the gas system to create flow. The biogas flows through a small hole in the roof of the drum. A non-return valve here is a valuable investment to prevent air being drawn into the digester, which would destroy the activity of the bacteria and provide a potentially explosive mixture inside the drum. Larger plants may need counterweights of some sort to ensure that the pressure in the system is correct. The drum must obviously be slightly smaller than the tank, but the difference should be as small as possible to prevent loss of gas and dipping of the drum.

REVIEW QUESTIONS

1.   What is biogas?

2.   Give the principle and structure of biogas plants.

3.   What are the operational conditions of biogas plants?

4.   Describe in detail about production of methane.

5.   Explain methanogenesis.

6.   Describe in detail about biohydrogen production.

7.   Explain in the detail about Gobar gas production from cow dung.

BIOFUELS

29

INTRODUCTION

The use of alcohol as a motor fuel is nothing new. In 1872, when Nikolaus Otto invented the internal combustion engine, gasoline was not available. Ethyl alcohol at 180–190 proof was the fuel then. The automobile T Ford was designed to run on crude gasoline, alcohol or any combination of the two. Alcohols (ethanol, in particular) make excellent motor fuels. The reason why alcohol fuel has not been fully exploited is that, gasoline has been cheap and easy to produce. However, crude oil is getting scarce, and the price difference between alcohol and gasoline is getting narrower.

Currently, there is a need to develop alternative sources of energy so that dwindling reserves of crude oil and other fossil fuels may be conserved. As Edward Teller pointed out: "No single prescription exists for a solution to the energy problem. Energy conservation is not enough. Petroleum is not enough. Coal is not enough. Nuclear energy is not enough. Solar and geothermal energy are not enough. New ideas and developments will not be enough by themselves. Only the proper combination of all of these will suffice."

Alcohol fuel is by no means a panacea. If all of the surplus agricultural products were converted to ethanol, it would supply less than 5% of our motor fuel needs. The possibility of converting cellulose residues to ethanol and general biomass to methanol falls short of 10% of our present needs. However, even a 5 or 10% may be very important because it can be renewed each year, and each gallon of alcohol produced will save a gallon of oil.

ALCOHOL FUEL

Alcohol is an excellent alternative motor fuel. However, it is not suitable for home use or for stationary power requirements. The production of alcohol consumes energy, which depends on the feedstock (raw material) and the efficiency of the distillation process. A small operation would expend 30,000–40,000 Btu per gallon of ethanol. The real advantage of alcohol is that it can be burned with little or no modification.

ALTERNATIVE FUELS

Methanol (wood alcohol) is used as an alternative fuel. Methanol is a viable substitute for gasoline, and it can be produced from a wide variety of renewable biological resources. However, methanol is not as easy to produce on a small scale. The simplest method of producing methanol is by destructive distillation (pyrolysis) of wood. The process involves heating the wood residues in a dry distillation apparatus and collecting methanol at the other end. The process requires relatively simple equipment. The problem is that, along with methanol, a considerable amount of impurities are produced, which include acetone, acetic acid and other substances. These by-products are difficult and expensive to remove. If left in methanol, they will quickly corrode the engine. In simple words, the small-scale production of methanol by destructive distillation requires a large plant and energy necessary to remove the impurities.

Methanol may be produced from other renewable resources, such as hydrogen, carbon monoxide, cellulose and biomass. However, these methods are feasible on a very large scale. Methane gas can be considered a good motor fuel. It may be generated, for example, by the action of bacteria on manure. Methane has a very low heat value (energy content per unit of weight) and engine conversion is necessary. Methane is better suited to stationary power requirements than for use as a motor fuel.

Natural gas, propane and butane are other possible motor fuels. However, they cannot be considered as renewable resources. Much research has been done to separate water into hydrogen and oxygen in order to use hydrogen as a fuel. Besides alcohol and methane, there seems to be no suitable alternative fuel that can be produced from renewable resources. Vehicles that operate on electricity need to be developed more. So now, ethanol is the best motor fuel from renewable resources.

BASIC FUEL THEORY

Chemical Composition

Alcohol and gasoline, despite the fact that they are from different chemical classes, are remarkably similar. Gasoline is a mixture of hydrocarbons. Hydrocarbons are chemical substances composed exclusively of carbon and hydrogen atoms. Hydrocarbons are a very large chemical class containing many thousands of substances. Most fuels such as coal, gasoline, kerosene, butane and propane are chiefly hydrocarbons. The simplest member of this group is methane, which consists of a single carbon atom and four hydrogen atoms, followed by ethane with two carbons and six hydrogen atoms. Propane has three carbons and butane has four (Figure 29.1). These substances are gases under ordinary conditions. However, as we add more carbons to the hydrocarbon molecule, the chemicals turn into liquids such as pentane, hexane, heptane and octane. As we continue, the molecules become more complex, and the substances get progressively oilier, waxier and finally solid.

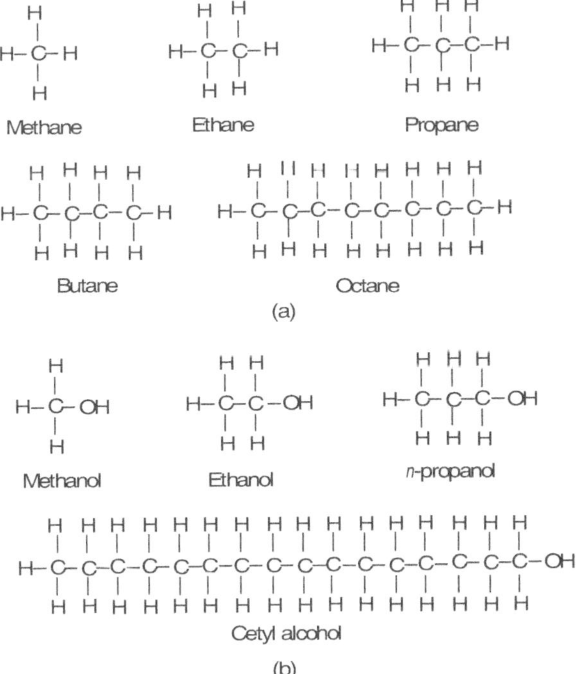

Figure 29.1   Chemical structures (a) hydrocarbons (b) alcohols

Alcohols may be considered as hydrocarbons in which one of the hydrogen atoms has been replaced by a hydroxyl group, which consists of a hydrogen atom bonded to an oxygen atom (Figure 29.1). Thus, methane becomes the simplest alcohol, i.e., methanol; ethane becomes ethanol; propane becomes propanol, and so on. There are many alcohols of ever increasing complexity.

Combustion Properties

One of the most important properties of a fuel is the amount of energy it releases when burnt. Octane, which represents an ideal gasoline, contains no oxygen. In contrast, all alcohols contain an oxygen atom bonded to a hydrogen atom in the hydroxyl radical. When alcohol is burnt, the hydroxyl combines with a hydrogen atom to form a molecule of water. Thus, oxygen contained in alcohol contributes nothing to the fuel value.

Table 29.1   Physical properties of alcohol and gasoline

	Typical regular gasoline	Octane	Methyl alcohol	Ethyl alcohol
Chemical formula	Complex	C_6N_{18}	CH_3OH	C_2H_5OH
Molecular weight	Complex	116	32	66
Heating value (Btu/lb)				
High value	20,250	20,570	9.710	12.740
Low value	19.000	19.000	8,640	11,550
Latent heat of vaporization (Btu/lb)	140	161	674	361
Specific gravity	0.745	0.702	0.746	0.796
Stoichiometric ratio	15.1	15.1 : 1	6.65 : 1	9 : 1
Boiling temperature (°C)	37–204	124.5	64.7	78.5
Octane number (Research)	80	100	106	106
Energy of stoichiometric mixture (Btu/ft³)	96.8	96.6	96.5	96.7

The relative atomic weights of atoms involved are: hydrogen, 1; carbon, 12; and oxygen, 16. Since methyl alcohol has an atomic weight of 32, half the molecule cannot be burned, and so it does not contribute any fuel value. Methanol has less than half the heat value (expressed in Btu/lb) of gasoline. Ethanol, with 35% oxygen, is slightly better with 60% of the heat value of gasoline.

If the heating values of methyl and ethyl alcohol were considered, they would appear poor choices as motor fuels. However, latent heat of vaporization and anti-knock values make alcohol fuels superior to gasoline.

A certain amount of air is required for complete combustion of a fuel. When the quantity of air and the quantity of fuel are exactly balanced, the fuel–air mixture is said to be stoichiometrically correct. The stoichiometric ratio for gasoline is 15 : 1, i.e., 15 pounds of air for each pound of gasoline. The figures for methyl and ethyl alcohol are 6.45 : 1 and 9 : 1, respectively. Practically, this means that to burn alcohol effectively, the fuel jets in the carburettor must provide 2.3 pounds of methanol or 1.66 pounds of ethanol for each 15 pounds of air. An interesting fact is that, if we provide the correct stoichiometric mixture and then compare on the basis of energy (in Btu's) contained in each cubic foot of different fuel or air mixtures, the fuels are almost identical: gasoline 94.8 Btu per cubic foot; methanol 94.5 and ethanol 94.7. This means that gasoline and alcohol have equal "volumetric efficiency" when burned in a correctly adjusted engine. Table 29.1 provides the physical properties of alcohol.

Volatility

Another important quality in a motor fuel is volatility, i.e., the ability to be vaporized. As previously noted, methyl alcohol contains less than half the heat value of gasoline and ethyl alcohol contains only about 60%. The next higher alcohol, propyl alcohol with three carbon atoms, contains only 26.6% oxygen and, thus, about 74% of the heat value of gasoline. It is apparent that the more complex the alcohol, the closer its heat value comes to that of gasoline. Cetyl alcohol, for example, contains only about 6.6% oxygen and, thus, has about 90% of the heat value of gasoline. However, this alcohol is a solid wax. It cannot be conveniently vaporized and mixed with air in an engine. In considering alcohol fuels, however, a compromise must be made between heat value and volatility.

Latent heat of vaporization is closely related to volatility. When a liquid is at its boiling point, a certain amount of additional heat is needed to change the liquid to a gas. This additional heat is called latent heat of vaporization, expressed in Btu/lb. This effect is the principle behind refrigeration and the reason that water evaporating from your skin feels cool. Gasoline has a latent heat of about 140 Btu/lb; methanol, 474 Btu/lb; and ethanol, 361 Btu/lb. In an engine, vaporization of the gasoline fuel or air mixture results in a temperature drop of about 4°C. Under similar conditions, the temperature drop for ethyl alcohol will be more than twice that of gasoline, and for methanol the drop will be over three times as great. These temperature drops result in a considerably greater mass density of the fuel entering the engine for alcohol as compared to gasoline. The result is a greatly increased efficiency for alcohol fuels. At a given pressure, the amount of space a gas occupies is directly proportional to temperature. For example, if one pound of a gas fits into a certain container at a given pressure and the temperature is cut in half, the container will now hold two pounds of the gas at the same pressure. In an engine, a stoichiometric mixture of methanol and air would be over three times colder than the same gasoline or air mixture. This means that there is now over three times (by weight) as much methanol in the cylinder. Even though methanol has only half the heat value of gasoline, the net gain in volumetric mass efficiency is over three times. So, for example, if the gasoline or air mixture in a given engine cylinder produces 100 Btu on each stroke, the same engine would produce 150 Btu per stroke with methanol. This power gain due to increased volumetric mass efficiency is the primary reason for the popularity of methyl alcohol as a racing fuel. With ethanol, the effect is not so high, but the greater heat value partially offsets the lower latent heat. Above all, the power increase with alcohol fuels considerably mitigates the liability of low heat value.

However, the increased cooling due to latent heat sometimes creates a problem in the engine. Once vaporized, a certain amount of heat is required to keep the fuel safe from condensing before it reaches the cylinder. The engine should be designed so as to provide heat to accomplish this. Due to its greater latent heat, alcohol requires more heat than gasoline. This is one of the reasons why racing engines have multiple carburettors. The shorter the distance the fuel travels to the cylinder, the lesser the chance of condensation and fuel distribution problems. On a practical level, most engines that have been converted to use alcohol supply enough heat once they are warmed up. The main problem with high performance racing engines is in starting a cold engine.

Octane Ratings

If a certain fuel is burned in an engine in which the compression ratio is gradually increased, a point will be reached when the fuel will detonate prematurely. This is because heat is generated as gas is compressed. If the explosive fuel–air mixture in an engine cylinder is compressed enough, the resulting heat will cause it to detonate. Since gasoline engines are designed so that the mixture is detonated by the spark plug at the beginning of the downward movement of the piston following the compression stroke, pre-ignition or "knock" occurring during the compression stroke is undesirable. Indeed, severe knock can quickly overstress and destroy an engine.

Since greater compression ratios mean increased power per stroke and greater efficiency, the ability of a fuel to resist premature detonation is a desirable quality. The octane numbers assigned to fuels are based on the pure hydrocarbon (octane), which is considered to be 100. At the other end of the scale, *n*-heptane is considered to have an octane rating of zero. The octane number of an unknown fuel is based on the percentage volume of a mixture of octane and *n*-heptane that matches it in pre-ignition characteristics. In practice, these tests are conducted in a special test engine with variable compression. Alcohols have a relatively high anti-knock or octane rating and the ability to raise the octane ratings of gasolines considerably when they are mixed. The effect is greatest on the poorer grades of gasoline. A 25% of ethanol prepared in gasoline (40 octane number) will have a net increase of 30 point than normal ethanol. This increase is one of the major advantages of "gasohol". The ability to increase octane rating means that: 1) a lower (therefore cheaper) grade of gasoline can be used to obtain a fuel with a certain octane rating; and 2) the use of traditional pollution producing anti-knock additives, such as tetraethyl lead, can be eliminated. The addition of about 10–15% ethanol to unleaded gasoline raises the octane rating enough so that it can be burned in high compression engines that could not use unleaded fuel previously. Ethanol was the original gasoline additive for increasing the octane rating. The term "ethyl" used to describe a high-test gasoline comes from ethyl alcohol, not tetraethyl lead. Figure 29.2 shows the octane increase of alcohol or gasoline blends.

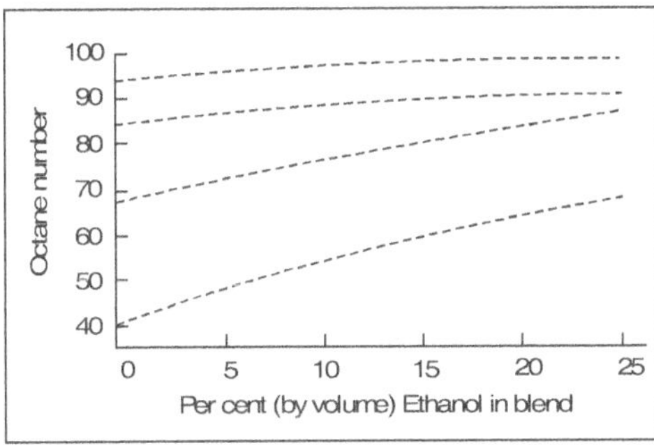

Figure 29.2   Octane increase of alcohol or gasoline blends

Water Injection

During World War II, the military made extensive use of water injection in high performance piston aircraft engines. Later, water injection was used by both civilian and military jet aircraft to provide extra thrust, principally on takeoff. Even today, water injection systems are installed in automobiles. The fact is that these systems actually increase power. The latent heat of vaporization for gasoline is about 140 Btu/lb and for ethanol is 361 Btu/lb. Water has a latent heat of about 700 Btu/lb. Therefore, if a little water is injected into the carburettor in the form of an ultra-fine mist, the latent heat of water will cool the charge and increase volumetric efficiency. In addition, when the charge is fired in the cylinder, the water will turn to high-pressure steam and provide additional power due to the pressure exerted by the steam. However, there are limits to the amount of water that can be injected. Too much will cause excessive cooling and misfiring. The use of water injection in a gasoline fuelled engine requires a separate metering and injection system because water and gasoline do not mix. Ethanol and water, however, do mix and so the benefits of water injection can be improved by adding the desired amount of water to alcohol in the fuel tank.

Exhaust Composition

In theory, a hydrocarbon fuel when burned should produce only water and carbon dioxide as exhaust gases. Carbon dioxide, of course, is completely non-poisonous. However, such ideal combustion rarely occurs even in the most perfectly adjusted engine. What is actually produced is a large amount of poisonous carbon monoxide and other complex (and undesirable) emissions like sulphur, lead or phosphorus. Pure alcohol when burned under ideal conditions produces, in theory, only carbon dioxide and water. However, in practice, varying amounts of carbon monoxide are produced but much lower than with gasoline. In addition, alcohol fuel will contain no sulphur and additives. Pure alcohol fuels burn extremely clean.

Since alcohol burns considerably clean, the amount of emission is proportional to the amount of alcohol in the blend. Pure alcohol, as an anti-pollution fuel, would easily meet all emission requirements. With alcohol blends, the main advantage would be in the use of ethanol to replace lead and other undesirable compounds used to raise the octane number.

ENGINE PERFORMANCE WITH STRAIGHT ALCOHOL

Having discussed a few basic factors that influence the performance of fuels in an engine, let us now examine some actual engine tests. Figure 29.3 is a plot of ethyl alcohol as compared to gasoline. "Mean effective pressure" in the graph is a direct indication of the power produced. The increased mean effective pressure (MEP) of alcohol at all mixture ratios is the most noticeable difference between the two fuels. The increase in MEP is mainly due to the greater volumetric efficiency that results from the high latent heat of vaporization of ethanol and the resulting greater mass density of the fuel–air mixture.

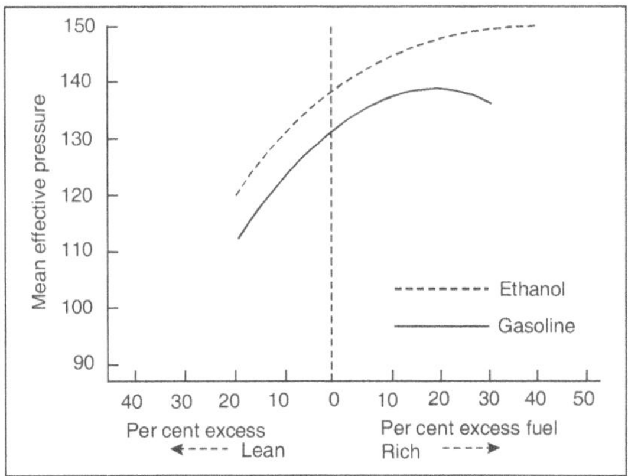

Figure 29.3   Engine performance of ethanol vs. gasoline

Note that the MEP of ethanol increases with mixtures having up to 40% excess fuel, whereas for gasoline, the maximum pressure is reached at 20% excess fuel. Also, it should be noted that the rich mixtures necessary to obtain maximum MEP. are accompanied by incomplete burning of the fuel and the resultant lowering of the overall thermal efficiency. The lean limits for alcohol and gasoline, therefore, are the same, and both fuels develop maximum thermal efficiency at about 15% excess air. With mixtures leaner than 15%, both fuels lose thermal efficiency.

Figure 29.4 compares engine horsepower and air–fuel ratios for ethanol and gasoline in a six-cylinder engine. The fuels in this case were 190 proof (95%) ethanol and regular gasoline having a specific gravity of 0.745. In the tests, air was supplied to the intake manifold at a constant 37°C, and the carburretor needle valve was adjusted to provide the desired fuel–air ratios. The 2/3 and 1/3 loads were established by adjusting the throttle to give the same manifold pressure for both fuels.

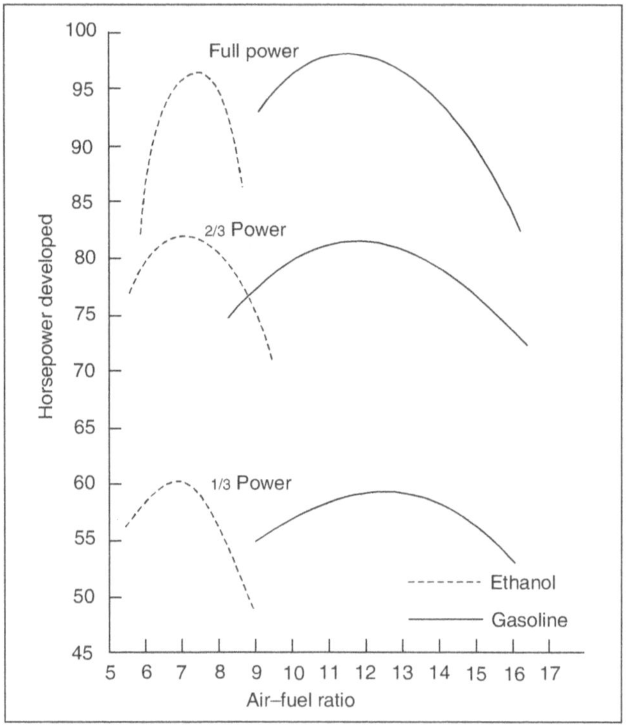

Figure 29.4   Horsepower comparison of ethanol vs. gasoline

The smaller air–fuel ratios for ethanol in comparison with gasoline are evident. In this test with the air supplied at the same temperature for both fuels, the correct fuel–air mixture should produce about 2% more power from gasoline than ethanol. However, alcohol, with its greater latent heat, requires more manifold heat to remain completely vaporized. In another test where this additional heat was supplied, the correct alcohol–air mixture gave 8.6% more power with ethanol. Note also that the test depicted in Figure 29.4 was run with alcohol that contained 5% water. This benefit of water injection probably inflated the alcohol power results to a certain degree. However, the main point illustrated is that the two fuels are remarkably similar in performance in a correctly adjusted engine.

ENGINE PERFORMANCE WITH ALCOHOL BLENDS

Although alcohol blends can be made from both ethanol and methanol, the primary interest seems to be in the direction of ethanol. Methanol and gasoline have limited miscibility (mixability) while ethanol and gasoline can be mixed in all proportions. Ethanol can be more readily made from renewable resources. In addition, ethanol is a slightly superior motor fuel.

A major advantage of blends is that up to a certain concentration (between 10 and 20%), they can be used with absolutely no modification of the engine. Many studies on how the various blends affect engine performance are contradictory. Some tests claim slightly better emissions, others claim no significant change. In relation to power output, the tests are equally ambiguous. However, the overall conclusion is that, in terms of fuel economy, emissions and performance, there is no real difference.

Another major advantage of alcohol blends is its ability to raise the anti-knock quality of the gasoline with which it is mixed. This means that lower grades of gasoline can be used to obtain a fuel with desired octane rating, and the use of pollution producing additives can be eliminated. This is a significant advantage from the economic standpoint. Furthermore, as previously mentioned, it is possible to raise the octane rating of unleaded gasoline so that it can be used in engines that previously required high-test leaded gasoline. Alcohol blends do have a minor drawback. The presence of even small amounts of water in the blend will cause a portion of the alcohol and gasoline to separate. At room temperature, less than 1% water can do the damage. As the temperature is lowered, amounts as small as 0.01% can cause separation. However, various substances such as benzene (benzol), acetone and butyl alcohol can be added to the blend to increase water tolerance. Closed fuel systems, now in use, prevent moisture from forming inside the gas tank. Moreover, alcohol blends have been extensively used over the past 50 years.

UTILIZATION OF ALCOHOL FUELS

Alcohol fuels may be utilized in three basic ways: as a blend with gasoline; as a straight unblended fuel; or as alcohol–water mixture in an injection system. Each method has both advantages and disadvantages.

Alcohol Blends

Alcohol blends have the advantage that up to 25% concentration of alcohol may be used without modification to the engine. The actual concentration usually varies with each engine type, but generally a four-cylinder engine will tolerate a stronger blend than a six or eight. Small single-cylinder engines, such as lawn mowers, can be run on pure alcohol by merely adjusting the mixture control screw. Even with larger engines, slight modification such as adjusting the carburettor and, perhaps, advancing the timing a little may allow the use of blends in the 25–40% range.

The disadvantage is that the alcohol used must be perfectly dry. The highest concentration of alcohol that can be achieved by ordinary methods is 190 proof or 95%. In order to blend the alcohol with gasoline, the remaining 5% water must be removed. However, drying the alcohol requires a separate manufacturing step and the expenditure of additional energy.

Pure Alcohol

The advantages of burning relatively pure (80–95%) alcohol are several. First of all, the fuel can be produced for less than the cost of gasoline. Secondly, by leaving 5–15% water in alcohol, it gives the benefits of water injection. The only disadvantage is the expense of modifying the engine to burn alcohol. The principal engine modification is the enlargement of the carburettor jets.

In addition, there may be a problem of cold starting. Alcohol has a higher latent heat of vaporization than gasoline and requires more manifold heat to keep the mixture in the vapour state. With most engines, this problem can be overcome by installing a higher temperature thermostat because the engine would run fine as soon as it is warmed up. However, the engine will be difficult to start, especially in cold weather. The easiest solution to this problem is to start the engine on gasoline and, after it has warmed up, switch to alcohol. To accomplish this, a small gasoline tank may be installed under the hood and a selector valve is fitted to the motor. It is desirable to replace the automatic choke with a manual control. Moreover, switching back to gasoline prior to shutting the engine down will aid in restarting. A more complex solution to this problem would be to install a priming pump and manifold heater glow plugs similar to those found in diesel engines. Other alternatives are to preheat the fuel or squirt an easily volatilized liquid, such as pentane, into the carburettor. The addition of 8% pentane directly to the alcohol in the fuel tank will solve the starting problem. Another problem related to latent heat is that of fuel distribution. Larger engines are more likely to encounter this problem. There is insufficient heat to keep the fuel vaporized and some of it liquefies before it reaches the outer cylinders. This causes misfires and poor performance. A simple solution is insulating the intake manifold or installing a higher temperature thermostat. Heating the fuel before it enters the carburettor helps in the same way as heating the intake air. The ultimate solution is to install multiple carburettors and a short-path manifold.

Although most engines can be easily converted to use alcohol, each engine is different. Alcohol engine modification is a relatively "rediscovered" field. Turbocharged engines present no special conversion problems once the jets, etc., have been enlarged. Alcohol and turbochargers work well together.

Diesel Engines

Diesel engines can be run on pure alcohol. The main problem, however, is the lubrication of the injectors. This can be solved by the addition of 5–20% vegetable oil (or other suitable lubricant) to alcohol. It is also possible to make a diesel "gasohol" with up to 80% alcohol. Since alcohol and oil cannot mix when water is present, both alcohol and oil must be anhydrous. Engines may require adjustment of the metering pump for optimum performance. Diesel engines, especially turbocharged diesels, may be run with alcohol–water injection systems.

ALCOHOL INJECTION

Alcohol injection is the third alternative for the utilization of alcohol fuel. It is similar to water injection except that alcohol or an alcohol–water mixture is injected into the engine. Since the water–alcohol injection mixture ratio can be as low as 50%, the first run product from a still can be used. This is a considerable saving because most of the energy used in alcohol production is expended in distillation to obtain 95% alcohol. Another advantage is that engines with an injection system will retain complete dual fuel capability. Finally, alcohol injection can be used with fuel-injected, turbocharged and diesel engines. A basic injection system is shown in Figure 29.5.

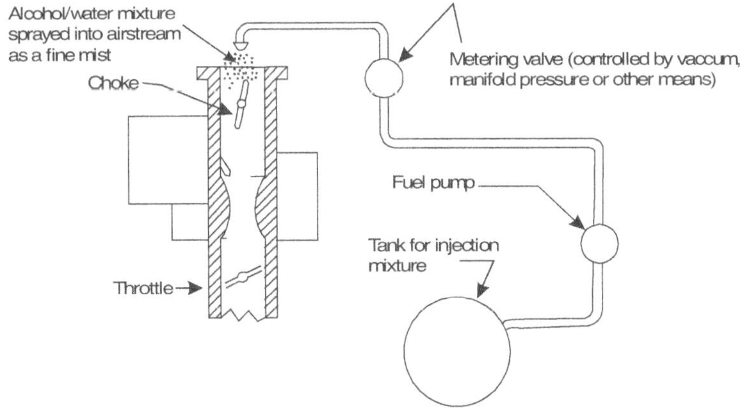

Figure 29.5   Basic injection system

The alcohol–water mixture is filled in separate tank. During feeding, the mix is passed with little pressure through nozzle located at the carburettor. The mixture is metered into the carburretor airstream where it mixes with air and is taken into the engine. There are many ways of metering the alcohol–water mixture. It can be done by mechanically linking a metering valve to the throttle.

Other methods include using combinations of vacuum and or manifold pressure. The metering system should work in parallel with the throttle in any system. That is, the flow of alcohol–water mixture should increase as the load increases.

Figure 29.6 illustrates a similar system for turbocharged engines. This is an extremely simple system. The alcohol–water tank is pressurized by bleed air from the compressor on the turbocharger. The mixture is metered into the turbocharger airstream by an orifice.

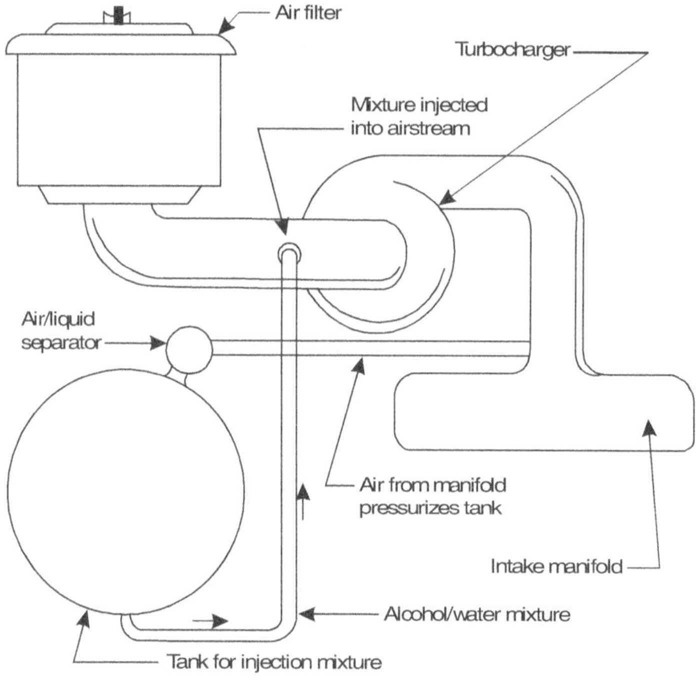

Figure 29.6   Injection of turbo engines

The size of the orifice is determined based on engine requirement. The metering system operates with the turbocharger. As the boost increases, more pressure is supplied to the tank and, thus, more mixture is drawn into the engine. On a diesel tractor rated at 125 HP and consuming 8.5 gallons of fuel per hour, the injection system produced the same power with only six gallons of diesel fuel and two gallons of a 50% alcohol–water mixture. This is an overall saving of 6% in fuel consumption and a saving of almost 30% in diesel fuel. Other benefits include trouble-free automatic operation, increase in power, lowering of engine operating temperatures, and prolonged engine life.

REVIEW QUESTIONS

1. What is biofuel?

2. Write about alcohol fuels and alternate fuels.

3. What is fuel theory? Explain.

4. Give the combustion properties of some biofuels.

ETHANOL

Ethanol or ethyl alcohol, CH_3CH_2OH, is one of the most exotic synthetic oxygen-containing organic chemicals. It has a unique combination of properties as a solvent, a germicide, a beverage, an antifreeze, a fuel, a depressant, and a versatile chemical intermediate for other organic chemicals (Table 30.1).

Under ordinary conditions, ethanol is a volatile, flammable, clear, colourless liquid. Its odour is pleasant and has a characteristic taste. The physical and chemical properties of ethanol are primarily dependent upon the hydroxyl group. This group imparts polarity to the molecule and gives rise to intermolecular hydrogen bonding. In liquid state, hydrogen bonds are formed by the attraction of hydroxyl hydrogen of one molecule and hydroxyl oxygen of a second molecule. The effect of this bonding is to make liquid alcohol behave as though it were largely dimerized. This behaviour is analogous to that of water, which is more strongly bonded and appears to exist in liquid clusters of more than two molecules.

The chemistry of ethanol is largely that of the hydroxyl group, namely reactions of dehydration, dehydrogenation, oxidation and esterification (Figure 30.1). The hydrogen atom of the hydroxyl group can be replaced by an active metal, such as sodium, potassium and calcium, to form a metal ethoxide (ethylate) with the evolution of hydrogen gas.

30

Table 30.1   Properties of ethanol

Property	Value
Normal boiling point	78.32 °C
Critical temperature	243.1 °C
Density	0.7893 d_4^{20}, g/ml
Heat of combustion at 25°C	29676.69 J/g
Auto-ignition temperature	793.0 °C
Flammable limits in air	
Lower	4.3 vol%
Upper	19.0 vol%

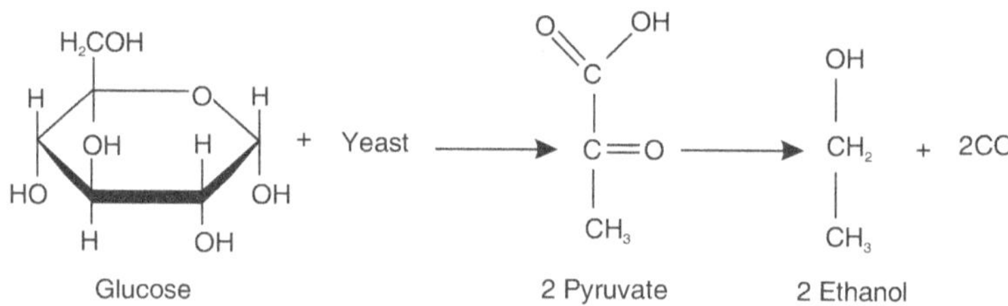

Figure 30.1   Metabolic pathway of ethanol production

INDUSTRIAL ETHANOL AND ITS USES

Industrial ethanol may be categorized into three classes: solvent, chemical intermediate and fuel. In each of these applications, fermentation ethanol must compete with petroleum derived ethanol as well as with alternative commodities. The extent of ethanol use and the mix of fermentative vs. petroleum derived ethanol vary substantially across the world.

The fermentation route is more prevalent in less industrialized nations. Alcohol production can begin in relatively small plants, which serve as a first step towards a diversified chemical industry. Today, India and Brazil are expanding their chemical industries based on fermentation ethanol. Brazil is actively marketing its ethanol technology to other less industrialized nations. Fermentative ethanol has remained important in Belgium, Italy, France and the Netherlands, and is growing in Japan.

Uses of Ethanol Solvent

Ethanol ranks second (to water) as an industrial solvent. Solvent applications include resins, pharmaceuticals, cosmetics, household cleaning products and industrial solvents; 50–55% of industrial (non-fuel) alcohol use is for solvent applications.

Ethanol Fuel

Fuel use in internal combustion engines is the fastest growing application of fermentative ethanol. The original Otto engine was developed to use anhydrous ethanol fuel. Ethanol use in blends of up to 20 vol% with gasoline was widespread during World War II.

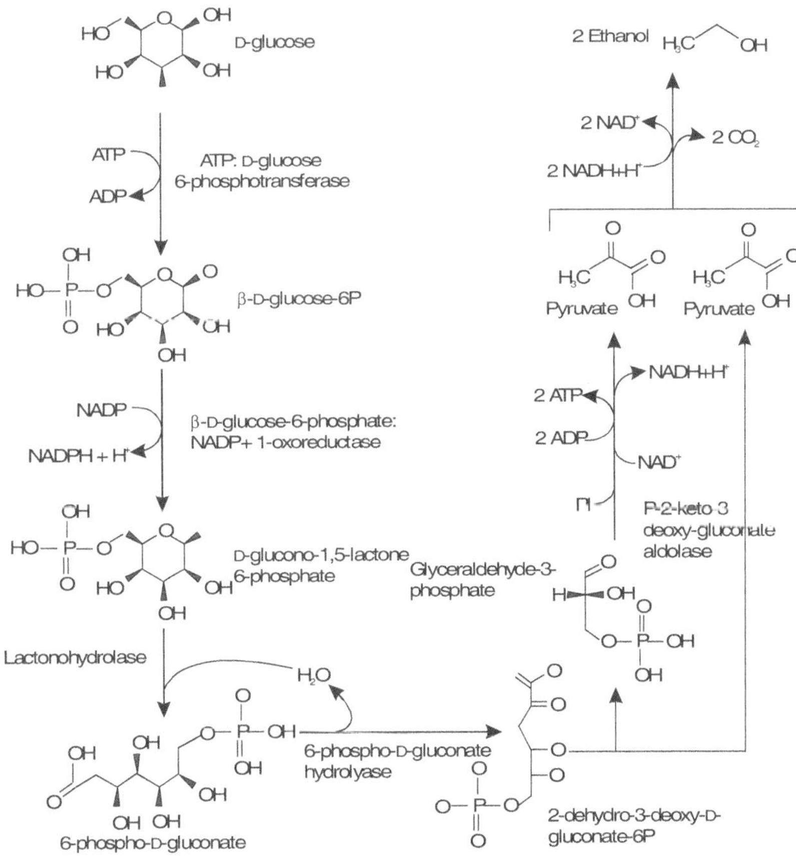

Figure 30.2   Ethanol biosynthesis

As early as 1931, Brazil legislated the use of 5 vol% alcohol blend with gasoline as a means to utilize sugar refinery molasses and to stabilize sugar prices.

Ethanol Use as Chemical Intermediate

Many important chemicals can be derived from ethanol. A new process allows the production of a hydrocarbon mixture similar to high-octane gasoline using a shape-selective zeolite catalyst. Therefore, it is possible to produce practically all petroleum derived chemicals from ethanol.

In US, butadiene production for synthetic rubber led to tremendous growth in the ethanol fermentation industry during World War II. After the war, petroleum replaced ethanol as the raw material for butadiene. Acetaldehyde became the major intermediate based on ethanol in 1963. A direct route to acetaldehyde from ethylene was developed, and acetaldehyde production from ethanol was reduced in 1975. The trend to replace ethanol as a chemical intermediate has continued, and the major chemical products from ethanol were esters, amines and ethers.

ORGANISMS FOR ETHANOL FERMENTATION

Bacteria

A great number of bacteria are capable of producing ethanol. Many of these microorganisms, however, generate multiple end products in addition to ethyl alcohol. These include other alcohols (butanol, isopropylalcohol, 2,3-butanediol), organic acids (acetic acid, formic acid and lactic acid), polyols (arabitol, glycerol and xylitol), ketones (acetone) and various gases (methane, carbon dioxide and hydrogen).

Microbes that are capable of yielding ethanol as the major product (i.e., a minimum of 1 mol ethanol produced per mol of glucose utilized) are shown in Table 30.2.

Many bacteria (e.g. Enterobacteriaceae, Spirochaeta and Bacteroides) metabolize glucose by the Embden–Meyerhoff pathway. Briefly, this path utilizes 1 mol of glucose to yield 2 mol of pyruvate, which is then decarboxylated to acetaldehyde and reduced to ethanol. Besides, the Entner–Doudoroff pathway is an additional mean of glucose consumption in many bacteria. When *Z.mobilis* is cultivated in a media which contains glucose, fructose, and sucrose, it was found that the glucose uptake is high while compared with uptake of other sugars.

Table 30.2   Bacterial species that produce ethanol

Mesophilic organisms	Ethanol produced (in mM) per mM glucose metabolized
Clostridium sporogenes	4.15
Clostridium indolis	1.96
Clostridium sphenoides	1.8
Clostridium sordelli	1.7
Zymomonas mobilis	1.9
Spirochaeta aurantia	1.5
Spirochaeta stenostrepta	0.84
Spirochaeta litoralis	1.1
Erwinia amylovora	1.2
Leuconostoc mesenteroides	1.0
Streptococcus lactis	1.0
Sarcina ventriculi	1.0

Table 30.3   Kinetic parameters for growth of *Zymomonas mobilis* strain ZM4 in batch culture with different carbon substrates (initial concentration 250 g/L)

Kinetic parameter	Glucose	Fructose	Sucrose
Specific growth rate, μ (h^{-1})	0.18	0.10	0.14
Specific substrate consumption rate, q_s (g/g/hr)	11.3	10.4	10.0
Specific ethanol production rate, q_P (g/g/hr)	5.4	5.1	4.6
Cell yield, Y(g/g)	0.015	0.009	0.014
Ethanol yield, $Y_{p/s}$ (g/g)	0.48	0.48	0.46
Ethanol yield (% of theoretical)	117	94.1	90.2
Maximum ethanol concentration (g/L)	0	119	89
Time period of calculation of maximum rates (hr)	0–19	0–28	0–15

Yeast

The organisms of primary interest in industrial fermentation of ethanol include *Saccharomyces cerevisiae, S. uvarum, Schizosaccharomyces pombe* and *Kluyueromyces* sp. Under anaerobic conditions, yeast metabolize glucose to ethanol primarily by way of the Embden–Meyerhoff pathway. The overall net reaction involves the production of 2 moles each of ethanol, but the yield attained in practical fermentation does not usually exceed 90–95% of theoretical. This is partly due to the requirement for some nutrients to be utilized in the synthesis of new biomass and other cell maintenance related reactions. A small concentration of oxygen must be provided to the fermenting yeast as it is a necessary component in the biosynthesis of polyunsaturated fats and lipids. Typical amounts of O_2 maintained in the broth are 0.05 – 0.10 mm Hg oxygen tension. The relative requirements for nutrients not utilized in ethanol synthesis are in proportion to the major components of the yeast cell. These include carbon, oxygen, nitrogen and hydrogen. To a lesser extent, quantities of phosphorus, sulphur, potassium and magnesium must be provided for the synthesis of minor components. Minerals (e.g. Mn, Co, Cu, Zn) and organic factors (amino acids, nucleic acids and vitamins) are required in trace amounts. Yeast is highly susceptible to ethanol inhibition. A concentration of 1–2% (w/v) is sufficient to retard microbial growth, and at 10% (w/v) alcohol, the growth rate of the organism is nearly halted.

Selection of Yeast Strain

Yeast strains are generally chosen from among *Saccharomyces cerevisiae, S. ellysoideus, S. carlsbergensis, S. fragilis* and *Schizosaccharomyces pombe*. For whey fermentation, *Torula cremoris* or *Candida pseudotropicalis* is used. Yeasts are carefully selected for its 1) high growth and fermentation rate, 2) high ethanol yield, 3) ethanol and glucose tolerance, 4) osmotolerance, 5) low pH fermentation optimum, 6) high temperature fermentation optimum, 7) general hardiness under physical and chemical stress. High growth and fermentation rate allows the use of smaller fermentation equipment. Ethanol and glucose tolerance allows the conversion of concentrated feeds to concentrated products, reducing energy requirements for distillation and stillage handling. Osmotolerance allows the handling of relatively "dirty" raw materials such as blackstrap molasses with its high salt content. Osmotolerance also allows the recycling of a large portion of stillage liquids, thus reducing stillage handling costs. Low pH fermentation combats contamination by competing organisms. High temperature tolerance simplifies fermenter cooling. General hardiness allows yeast to survive the ordinary stress of handling (such as centrifugation) as well as stresses arising from a plant upset.

RAW MATERIALS FOR ETHANOL PRODUCTION

Ethyl alcohol may be produced from three basic types of raw materials or "feedstock". The three basic types of feedstock are:

Saccharine   It is a sugar containing material in which the carbohydrate (actual substance from which the alcohol is made) is present in the form of simple, directly fermentable, six and twelve carbon–sugar molecules such as glucose, fructose and maltose. Sugarcane, sugar beet, fruit (fresh or dried), citrus molasses, cane sorghum, whey and skim milk are rich in saccharine.

Fruits   Fruits commonly used in ethanol production along with their average sugar content may be as follows: grapes, 15.0%; bananas, 13.8%; apples, 12.2%; pineapples, 11.7%; pears, 10.0%; peaches, 7.6%; oranges, 5.4%; prickly pear, 4.2%; watermelon, 2.5%; and tomatoes, 2.0%. For achieving 75% extraction with apples, for example, the total fermentable material would be about 9% of the original weight. On this basis, a ton of apples would yield about 13 gallons of alcohol. Assuming an 80% extraction with grapes, a ton should yield about 17 gallons. With watermelons and a 90% extraction, a ton would yield only 3 or 3.5 gallons. In all the above cases, the percentage of fermentable material in the extracted juice is low enough so that dilution is unnecessary and undesirable. To ferment these materials, the juice need to be adjusted to the proper pH (between 4.8 and 5.0) and yeast is added at the rate of 2 pounds per 1000 gallons of mash. To provide proper nutrients to the yeast, backslopping of about 20–25% by volume is desirable.

The materials may be simply crushed or pulped instead of extracted in a press. This way the total sugar content is available for fermentation.

Molasses   Beet or cane molasses are excellent sources of alcohol. They contain 50–55% fermentable sugar; a ton should yield between 70 and 80 gallons of alcohol. Molasses with sugar content above 15–20% will need to be diluted. Since molasses is usually low in nutrients necessary for proper yeast growth, backslopping may be of particular benefit. Up to 50% stillage (by volume) is recommended. Moreover, molasses is naturally alkaline, and acid will be needed in addition to stillage to obtain proper pH value.

Cane sorghum   Cane sorghum is a good source of alcohol because it has about 14% fermentable sugar content. The main drawback in using this material is that the extraction requires heavy-duty shredding and pressing equipment. An alternative process is to shred the stalks as much as possible and dissolve the sugar by heating with a minimum amount of water. The process must be repeated several times to retrieve most of the sugar. Note that, in this type of process, two extractions of one gallon each are better than one extraction of two gallons. A conventional 65% extraction should yield about 13–14 gallons of alcohol per ton. Acidification may be necessary and backslopping to about 25% can be tolerated.

Sugar beet   Sugar beet is an excellent material for ethanol production. It contains about 15% sugar, 82% water, and the rest as various solids. The juice may be extracted in a press, or the beet may be crushed and fermented. Because beet contains a small amount of starch, the addition of small quantities of malt (1–2% by weight) or enzyme will greatly improve the alcohol yield. The adjustment of pH is, of course, necessary, and backslopping in the range of 20–25% is desirable. A ton of beet should produce 20–25 gallons of alcohol.

Sugar corn wastes   Stalks from sugar corn contain 7–15% sugar. The stalks are shredded and extracted in a manner similar to sugar cane or sorghum stalks. A relatively efficient operation should yield 8–18 gallons of alcohol per ton of materials. As usual, backslopping to 20–25% and acidification are necessary.

Processing Steps Specific to Saccharine Materials

Saccharine materials require the least processing from among other ethanol feedstock. Molasses and other sugar-containing syrups need to be diluted and pH adjusted prior to fermentation. Other materials, including the fruits, need to be crushed or extracted to make the sugar readily accessible to the yeast enzymes.

Prior to fermentation, saccharine materials are usually put through an extraction process to separate the sugar-containing juice from the rest of the materials. Extraction is usually done in a press. However, extraction is not absolutely necessary. The materials may be simply crushed to expose the juices to the fermentation process. However, in most distillation equipment, the solids will have to be removed prior to transferring the juice to the still.

There are certain problems associated with extraction or crushing. Extraction in a press usually leaves an appreciable amount of fermentable material behind. Typically, only 75% of the sugars can be extracted from apples, and about 80% from grapes. One technique in press extraction that can be used to increase yield is to take the residue from the first pressing, soak it in a minimum amount of water to dissolve the sugar, and press it again. However, this method creates additional dilution, which lowers the alcohol content of the fermented mash (called beer) and requires more energy and time in the distillation process. For example, if a fruit juice contains 10% sugar, the final alcohol concentration going to the still will be about 5%. More water used to wash additional sugar from the residue will further dilute the final concentration. The lower the alcohol concentration, more water must be removed from alcohol during distillation. However, in many instances, the greater total amount of alcohol gained justifies the additional dilution.

Crushing the material instead of extracting it in a press leaves more sugar available for fermentation, although the material usually must be strained prior to distillation. However, some juice will remain in the residue, and the only solution is to wash it with a little water. If a simple pot still is used, filtering the residue is not absolutely necessary as long as the still pot is cleaned after each run.

Materials like sweet corn stalks and sugar cane require heavy hydraulic presses to effectively extract the juice. The alternative process is to shred the material and heat it with as little water as possible to dissolve the sugar. Note that, to obtain complete recovery of the sugar, the process must be repeated several times. However, a point may be reached where dilution offsets the amount of sugar released and some compromise must be made. Note that two extractions of one gallon each will dissolve more sugar than a single two-gallon extraction.

STARCHY MATERIALS

Materials that contain more complex carbohydrates, such as starch and inulin, can be broken down into simpler six and twelve carbon sugars by hydrolysis with acid or by the action of enzymes. This process is called malting. Such materials include corn, grain sorghum, barley, wheat, potatoes, sweet potatoes, jerusalem artichokes, cacti, manioc and arrowroot.

Grains   Grains must be milled, diluted, cooked and converted prior to fermentation. They contain large amounts of potentially fermentable materials. The average content of convertible starch and sugar in typical grains are barley, 50%; maize, 66%; oats, 50%; rye, 59%; sorghum seed, 67%; and wheat, 65%. Alcohol yield per ton is dependent on how completely the starches are converted to fermentable sugar; usually, it should be between 70 and 100 gallons.

After milling, the grain must be diluted and cooked. In an average, dilution yields between 56 and 64 gallons per 100 pounds of grain, depending on moisture and starch content. The method of cooking with minimal water and adding the balance prior to conversion has the dual advantage of reducing the amount of energy needed for cooking and shortening the cooling time. Pre-malting with malt or enzymes is generally desirable.

Cooking is accomplished by heating the diluted and pre-malted mash to a slow boil and holding at this temperature for 30–60 minutes. Generally, mash can be sufficiently cooked when it is soft and mushy. The mash is cooled to 62°–65°C and malt slurry is added. Malt slurry consists of 2.5 pounds of dried or green malt per gallon of water.

On a weight/weight basis, corn or wheat will require 8–10 pounds of malt per 100 pounds of grain. Rye will require 10–12 pounds of malt for 100 pounds of grain. Other grains fall somewhere in between. Malt slurry is stirred constantly during conversion. For wheat, the conversion will complete in 5–15 minutes. Corn will require about 30 minutes, and rye between 30 and 60 minutes.

When the conversion is complete, the mash is cooled to 21°–23°C and yeast slurry is added. Note that most grain mashes have an acceptable low pH and need no adjustment. Backslopping should be limited to 20–25%. The following is the general procedure for converting corn with Miles Laboratories enzymes. The procedure for other materials and enzymes will differ slightly. After milling, the grain is partially diluted (slurried) at a ratio of 35 gallons of water per 100 pounds of grain. The pH is adjusted to above 5.5 with an optimum range between 6.0 and 6.5. Pre-malting or liquefaction is accomplished by the addition of 0.3 ounces of Taka-Therm enzyme.

The mash is slowly heated. Gelatinization will begin at about 65°C and the mash will rapidly thicken. Constant stirring is necessary at this point. At about 71°C, the liquefying action of the enzyme will begin. Heating may be more rapid after the liquefying action of the enzyme begins to take effect. After the mash reaches 93–100°C, an additional 1.3 ounces of Taka-Therm enzyme is added. After the mash has been held at a slow boil for 20–30 minutes, an additional 33 gallons of water is added to complete dilution and cool the mash. When the mash has cooled to 57–60°C, the pH is adjusted to 4.2 with an acid and an enzyme Diazyme L-100 is added at a ratio of 4 ounces per 100 pounds of grain. The enzyme completes the conversion in about 30 minutes and, after cooling to 21–26°C, the mash is fermented in the usual manner.

Jerusalem artichokes   Jerusalem artichokes are an abundant source of alcohol because they contain between 16 and 18% of fermentable material. In addition, the starches present can be converted without the use of malt or enzymes by cooking for a sufficient length of time. A ton should yield about 25 gallons of alcohol. To prepare artichokes for fermentation, they should be crushed to pulp and cooked for 2–3 hours. If the starch test indicates that some starch is still present unconverted, conversion with small amounts of malt or enzyme may be needed. Shorter cooking times are possible if a greater amount of malt or enzyme is used. For example, a 30-minute cooking time should be sufficient for conversion using 3–6% malt or the equivalent amount of enzyme. Dilution is not necessary because the root usually contains 79–80% water. After cooking, the pH is adjusted and fermentation is carried out in the usual manner.

Table 30.4   Average yield of ethanol per ton

Material	Gallons*	Material	Gallons*
Wheat (all varieties)	85.0	Yams	27.3
Corn	84.0	Potatoes	22.9
Buckwheat	83.4	Sugar beets	22.1
Raisins	81.4	Figs, fresh	21.0
Grain sorghum	79.5	Jerusalem artichokes	20.0
Rice, rough	79.5	Pineapples	15.6
Barley	79.2	Sugarcane	15.2
Dates, dry	79.0	Grapes (all varieties)	15.1
Rye	78.8	Apples	14.4
Prunes, dry	72.0	Apricots	13.6
Molasses, blackstrap	70.4	Pears	11.5
Sorghum cane	70.4	Peaches	11.5
Oats	63.6	Plums (non-prunes)	10.9
Figs, dry	59.0	Carrots	9.8
Sweet potatoes	34.2		

* 1 Gallon = 3.785 litres

Potatoes Potatoes contain between 15 and 18% fermentable material and are a traditional source of alcohol. On the average, a ton of potatoes will yield about 22–25 gallons of alcohol. Damaged or sprouted potatoes are not undesirable, and the use of sprouted potatoes will reduce the amount of malt or enzyme required for conversion. Commercially, potatoes are usually cooked with steam under pressure. Alternatively, potatoes should be shredded or cut up and placed in the cooker with as little water as possible and cooked in steam until the potatoes are reduced to a soft mash. Pre-malting to reduce viscousness is a definite advantage. Usually only 3–4 pounds of malt per 100 pounds of potatoes is required. After cooking, the mash is cooled to the conversion temperature. The mash must be constantly stirred during conversion, which may take about 15–20 minutes. Because cooking and conversion times may vary due to indifference in starch content, the use of starch test is recommended. Once converted, the pH should be checked and the mash is fermented in the usual manner.

Sweet potatoes Sweet potatoes contain in average about 22% starch and 5–6% sugar in a total of 27–28% fermentable material. A ton should yield up to 40 gallons of alcohol. Sweet potatoes are cooked and converted in a manner similar to potatoes with the exception that some dilution is necessary because they contain only about 66% water.

Processing Steps Specific to Starchy Materials

Starchy materials fall into two main categories: those in which starch is encased or protected by grain hulls (e.g. grains) and others where starch is more readily available (e.g. potatoes). Milling or grinding the material to expose the starch is necessary for the former group. Alternatively, a certain amount of cooking and conversion of starch to sugar may be necessary prior to fermentation. There are two basic methods of conversion. The first uses malt or an extract of enzymes contained in malt and the second uses dilute acid in a process called acid hydrolysis.

Manufacturing alcohol from saccharine feedstock generally requires 1) extraction or crushing to make the sugars available to the yeast enzymes during fermentation, 2) dilution may be required for certain materials, 3) fermentation, and 4) distillation. Starchy materials usually undergo the processes of: 1) milling to remove grain kernels, 2) dilution, 3) cooking to dissolve and gelatinize the starch, and 4) conversion of starch to fermentable sugars by malting or acid hydrolysis in addition to the steps of fermentation and distillation. Cellulose materials are similar to starchy materials and that they must be converted prior to fermentation. Starchy materials generally require milling, cooking and conversion prior to fermentation. However, materials like potatoes and sweet potatoes do not require milling, and artichokes do not require conversion.

Milling Grains and other starchy materials must be milled to expose the starch to the cooking, conversion and fermentation processes. Ideally, the material is ground as fine as possible without producing an excessive amount of flour. Fine (flour) particles are difficult to remove if the material must be filtered prior to distillation. However, if a simple pot still is used, the presence of fine flour particles is not objectionable.

A large amount of flour can make the mash too viscous (thick) and hard to handle. This is objectionable. However, viscousness may be tolerated if we use a simple pot still. Otherwise, pre-malting can solve the problem. The grain may be milled in any grain-milling equipment.

Cooking Cooking is necessary for starchy materials. The objective is to dissolve the water-soluble starches and then, as much as possible, gelatinize them. In commercial operations, cooking is always done with steam under pressure and usually in a continuous process. Water is usually boiled at 100°C at sea level and at a lower temperature as altitude increases. With pressure cooking equipment, higher temperatures and shorter cooking times can be achieved. At 150 pounds pressure, for example, grain starches can be cooked in less than six minutes.

Because a lot of energy is needed to boil the water in the cooking process, it is best to cook with as little water as possible. After cooking, water may be added to dilute the mash to optimum concentration. If water is added after cooking and prior to conversion, cooling time may be saved. Most grains can be cooked with as little as 15–20 gallons of water per load. When cooking with minimum water, however, stirring the mash may give advantage. Otherwise, lumping and burning may occur. New methods of cooking are being developed, which combines milling and cooking into one operation without the use of water. The process uses heat generated by friction in the milling process to simultaneously cook the grain.

Conversion Conversion of starch to fermentable sugars can be accomplished either with the use of malt or extracts of enzymes contained in malt, or with acid in a process called acid hydrolysis. Each method is discussed separately in detail.

Malting When seeds are moistened and allowed to sprout, certain enzymes (amylases) are produced in them, which have the ability to convert starch to a form of fermentable sugar called maltose. All cereal grains produce these enzymes. Barley produces much enzymes and is usually the most economical to use.

Dry or ground barley malt is commercially available, or one may produce it from the grain. However, undried or green malt will not keep unless dried.

In converting starch to sugar, malt enzymes exert two forms of chemical activity: liquefaction and saccharification. The intensity of these reactions depends on the temperature of the mash. The liquefying power is greatest at 70°C. It begins to weaken at 79°C and ceases below 93°C. The saccharifying (sugar making) power is strongest between 48°C and 54°C and is destroyed completely at 79°C. Both these actions are desirable. The conversion process usually begins after the cooked mash is allowed to cool to about 65°C. The material is held at this temperature for a certain length of time and then allowed to cool to the optimum fermenting temperature.

In average, half to 1 pound of dried malt is required for each 10 pounds of grain. Dried malt is mixed with warm water at a ratio of 2.5 pounds per gallon to form a slurry. After an hour, when the slurry cools to proper temperature, it is mixed with the mash. Because barley malt is expensive, it may be used as little as possible. A little of the converted mash is taken and filtered in a cheesecloth or similar material. The filtered liquid is placed in a white dish and several drops of a solution composed of 5 g of potassium iodide and 5 g of iodine crystals in 250 ml of distilled water is added. Any blue colour produced indicates the presence of unconverted starch. If the test indicates no blue colour, the next trial should be run with lesser malt.

Pre-malting During the cooking process, the starch in the grain is gelatinized. When the mash is cooled, it may become too thick to be stirred and handled during the malting process. The technique of pre-malting solves this problem by taking advantage of the liquefying property of malt prior to conversion. To pre-malt, simply add about 10% of the total malt weight to the mash prior to cooking. It gives sufficient liquefaction to the mash and facilitates handling during subsequent operations. It also helps to prevent thermal destruction of the malt enzymes and hence reduces the production of undesirable by-products. After cooking, the remaining 90% of malt slurry is added and conversion is continued as usual.

Preparation of Malt

In a basic process of making malt, any grain may be used; however, barley is by far the best. Malt is simply sprouted grain. The basic requirements for sprouting are moisture, warmth and darkness. Grains may be sprouted in any container of choice. The container, if small, should have small holes in the bottom and, for larger containers, a valve is provided with a screen or mesh to allow water to drain but retain the material inside.

The grain is soaked for 8–12 hours or until the kernels can be crushed and the inside is soft; it may take considerably longer time for corn. The water is drained out. Thereafter, warm water is sprinkled over the grain several times a day to keep the grain moist but not wet. If the grain is too wet, it may rot. Generally, sprouting generates some heat. The optimum temperature for sprouting is 26°C, but most enzymes seem to be produced at about 15°C. When sprouting in large containers, the grain should not get too warm. If it does, the grain should be spread out on a concrete floor in a dark place and the sprouting is continued. Small containers will not give the problem of too much heat. Sprouting will take about four days. The malt is ready when the sprout is about a half inch long.

The malt will have to be crushed in a mill before use. It is also possible to crush the malt in an ordinary blender or food processor. Fresh, undried malt is called green malt. It must be dried because it may rot if stored wet.

Enzyme conversion   Malt enzymes are commercially available from several manufacturers. The use of enzyme extracts is usually superior to malt. Enzyme extracts are usually cheaper, and they generally produce better results and higher yields.

The three basic types of commercially available enzymes are alpha, beta and gluco amylases. Alpha amylases split the starch molecules randomly to produce a type of sugar called dextrose. Beta amylases act similarly to produce maltose. Together, these two enzymes can convert about 85% of the starch to fermentable sugar. Gluco amylases can reduce the remaining starches. The use of all three can achieve total conversion of the starch.

Acid Hydrolysis   Starch (and cellulose) can be converted to fermentable sugars by the action of acid. This process is relatively simple, but it requires acid proof equipment and high temperatures. For these reasons, it is not recommended for small-scale production. Basically, dilute mineral acid (usually sulphuric acid) is added to the grain slurry prior to cooking at a concentration of 1–4% as calculated on a weight/weigh basis. The mash is then cooked at a temperature of about 176°C.

Cooking and conversion of starch can take place simultaneously. The mash is immediately neutralized with calcium hydroxide (or some other base) and fermented in the usual manner. High temperatures essential to this process are obtained by the use of pressure cooking. The steam pressure required is about 150 pounds per square inch. Together with the necessity for acid-proof equipment, this process is considered unsuitable for small-scale use. However, the process is excellent for large operations because cooking and fermentation times are short and the method is readily adaptable to continuous operations.

Mash cooling   Malting is conducted at a temperature of about 62–65°C. Fermentation is commenced at an optimum temperature of 21–26°C. Between the two steps, the mash must be cooled. One of the biggest problems affecting alcohol yield is bacterial contamination of the mash before or during fermentation. The chief protection is the control of acidity in mashing and fermentation operations. However, even with perfect pH control, bacterial infections can set in. This happens mostly during the cooling stage between mashing and fermentation. If bacterial contamination becomes a problem, the only solution is to shorten the cooling time as much as possible. The cooling coil is the best solution for a long term; however, if the problem occurs occasionally, as during the summer months, a plastic bag full of ice may be suspended in the mash.

CELLULOSE MATERIALS

Wood, wood waste, paper, straw, corn stalks, corn cobs and cotton contain a material that can be hydrolysed with acid, enzymes, or otherwise converted into fermentable sugars called glucose.

Processing Steps Specific to Cellulose Conversion

Cellulose feedstock, which includes corn stalks, wood, straw and cotton, are potentially good sources of alcohol. For example, a ton of old newspapers would yield up to 70 gallons of alcohol. Cellulose materials are extremely cheap. Cellulose may be converted with enzymes or acid hydrolysis. Nova Laboratories produces special enzymes called Cellulase and Cellobiase 250 L for conversion of cellulose to fermentable glucose. The acid process involves either strong acid and relatively low temperatures, or weak acid and high temperatures. The strong acid process has a disadvantage that glucose may be destroyed almost as fast as it is formed unless the contact time with acid is brief. The weak process requires acid proof, pressure cooking equipment as described earlier. However, for obvious reasons, these methods are not recommended on a small scale.

The main problem with cellulose as an ethanol feedstock is getting the cellulose itself. In a plant, cellulose is encased in a substance called lignin.

Lignin is the substance that gives wood its strength. To get the cellulose, the lignin must be dissolved away. The paper industry uses substances like sulphur dioxide, calcium bisulphite, sodium sulphate, sodium sulphide and sodium hydroxide (lye) to dissolve lignin. Concentrated mineral acid, may also be used to dissolve lignin. Unfortunately, a strong acid, as it dissolves lignin, may destroy the glucose.

Figure 30.3 depicts the processes involved in the production of ethanol from cellulose material. Commercial processes are being developed to convert cellulose into alcohol with the use of strong acid without destroying cellulose.

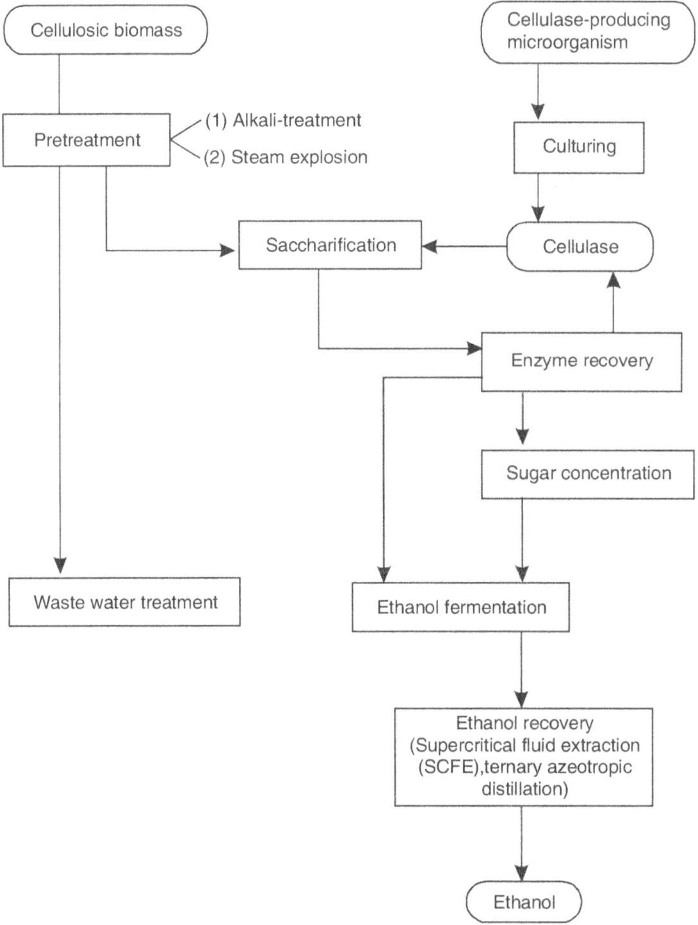

Figure 30.3   Production of ethanol from cellulose material

However, the process is complicated and not economically feasible on a small scale. The only alternative to dissolving lignin is to reduce the cellulose material to a fine material by powdering, grinding or pressing so that at least some cellulose may be recovered. The yield of cellulose is directly proportional to how finely the starting material is. Other cellulose materials are, however, easier to process except those with high lignin content. Newspapers are almost pure cellulose and can be easily converted by enzyme or acid process. Remember, in order for a plant to produce cellulose, it must first produce glucose, which is the sugar we are trying to extract. Therefore, plants that are processed wet and green have the advantage of having fermentable sugar already available. These materials can be fermented without conversion and are considered low-yield saccharine feedstock.

Multiple enzyme treatment   Usually, materials used in the production of ethanol contain some cellulose. Saccharine materials might benefit from a separate step of cellulose conversion. Starchy materials could have the cellulose enzymes added during conversion in addition to starch enzymes. Depending on the amount of cellulose available, this procedure could dramatically increase yields.

FERMENTATION

Batch Fermentation

Batch fermentation begins with the production of an active yeast inoculum. It may be either by conventional serial growth method or by rapid semi-aerobic method. Aseptic techniques are used throughout. In serial growth method, a pure culture inoculum from an agar slant is used to seed a laboratory shake flask. At the peak of growth (12–24 hours) the culture is used to seed a succeeding culture 30–35 times larger. This is generally repeated three times through laboratory level cultivation and three times though working-plant-level cultivation to produce 2–3% of inoculum for primary fermentation. The inoculum is grown on a medium similar to the final fermentation mash to minimize the acclimatization time in the final fermenter, but a higher level of growth nutrients may be used to produce a higher cell density.

Cylindro-conical Nathan vessels are preferred for fermentation because they promote better circulation and allow thorough drainage. For very large plants, sloped-bottom, cone-roof tanks of volume up to 1 million litres are used, and these large vessels are often agitated by carbon dioxide evolution during fermentation.

After emptying and rinsing from the previous batch, the mash at 13–17 wt% sugar is pumped into the fermenters. Once 20% full, the inoculum is added to allow growth during the remainder of the filling cycle, which may last 4–6 hours. The temperature of fermentation is regulated by circulating cool water through submerged coils, or passing the mash through external heat exchanges, or spraying the vessel walls with cool water (adequate for small fermenters only). The feed is generally introduced at 0–1°C, and the temperature is allowed to gradually rise as heat is evolved. The temperature thus varies from 30–35°C during the initial period (which is optimal for yeast growth). Cooling prevents the temperature from exceeding 2–4°C, which is optimal for ethanol production. The temperatures may be modified depending on the yeast strain used. The pH is set initially between 4.5 and 5.5 and decreases slowly, generally holding at 4.0 or above. This is important for fermentation of grain mashes with simultaneous dextrin hydrolysis, as many amylase enzymes are rapidly denatured at lower pH.

The rate of ethanol production is the product of specific (per cell) productivity and concentration of cells. Initially, the rate of production is quite low but, as the number of yeast cells increases, the overall rate increases and with rapid carbon dioxide evolution, the beer appears to boil. After 20 hours, maximum ethanol productivity is reached. The effect of reduced sugar concentration and ethanol inhibition then becomes important. The fermentation continues at a decreasing rate until, at 36 hours, 94% of the sugar is utilized and a final ethanol content of 69 g per litre is achieved. The average volumetric ethanol productivity over the course of fermentation is 1.9 g/L/h. Fermentation time may vary depending on the yeast strain and substrate. The Hawaiian and Cuban blackstrap molasses usually requires 36 hours for fermentation. Molasses from Java may require as long as 72 hours. The fermentation time may be reduced when molasses are clarified. Grain fermentation requires 40–50 hours to allow complete residual dextrin conversion.

The Biostil Process

The Biostil process (Figure 30.4) is a modification of the continuous fermentation process in which fermentation and distillation are closely coupled and a very high stillage backsetting rate is used. Fermented beer is continually cycled (through a centrifuge for yeast recycle) to a small rectifying column from where ethanol is removed. Most part of ethanol depleted beer (with residual sugars and non-fermentables) is recycled to the fermenters. The yeast cell density is maintained at 500 billion cells per liter. The fermented ethanol concentration can be maintained at any desired non-inhibitory level by adjustment of the beer cycle rate. A large liquid recycle provides an internal dilution so that highly concentrated feeds can be processed. Liquid flows external to the recycle loop are greatly reduced. Less water is consumed and a more concentrated stillage is produced. The flow rate capacities of most auxiliary equipment can be substantially reduced. Extensive heat exchange is incorporated into the beer cycle loop to maintain energy efficiently. As bacteria are not well separated by the centrifuge, the cycling through a hot distillation stage provides continuous pasteurization and infection risk is reduced. As with the Melle–Boinet process, ethanol yield can be increased by cell recycling. When ethanol inhibition is overcome, the Biostil process is inhibited by the build-up of toxic non-fermentable feed components and fermentation by-products. In pilot plant studies, the stillage flow could be reduced by a factor of 20 for concentrated cane syrup, but only a factor of 3 for blackstrap molasses with its high non-fermentable content. The process can also be applied to prehydrolysed starch feeds.

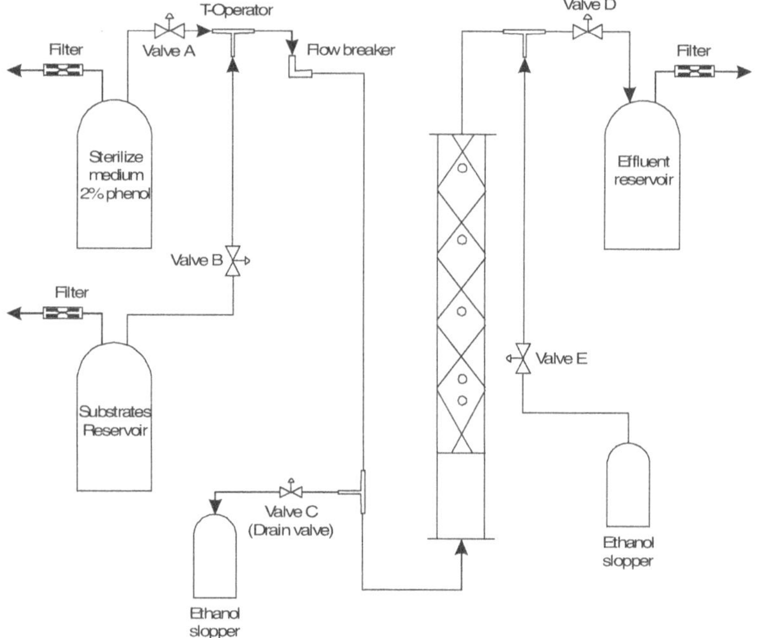

Figure 30.4   Schematic diagram of ethanol continuous fermentation with immobilized cells

Product Removal During Fermentation

Ethanol fermentation is a good example of product accumulation inhibiting the microbial culture. Most strains of yeast have a much slower alcohol production rate when ethanol reaches about 10 per cent, and wine or saki strains that achieve over 20 per cent by volume of ethanol are very slow. A system known as the Vacuferm for removal of alcohol by distillation as such it is formed.

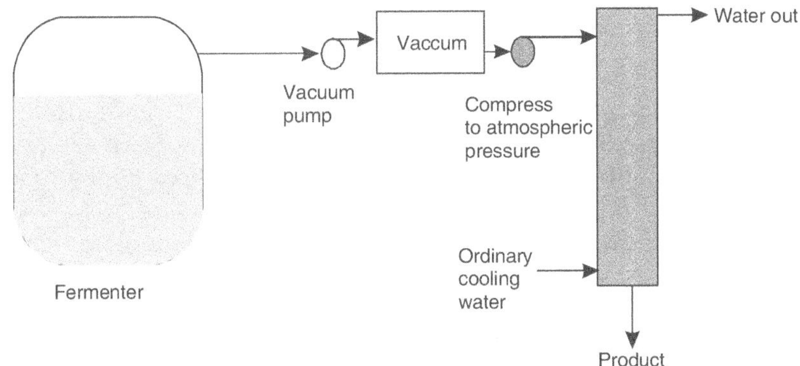

Figure 30.5   Production and distillation of alcohol

The vacuum may be adjusted to the vapour pressure of the alcohol–water solution at the fermentation temperature (30–40°C). Volumetric productivity is far better than that of a conventional fermenter, but there is a disadvantage of having to recompresses large volumes of vapour so that alcohol can be condensed with normal cooling water instead of expensive cold brine. Furthermore, the large amounts of carbon dioxide generated by fermentation are evacuated and recompressed along with alcohol and water vapours.

The fermenter operates at normal pressure so that carbon dioxide escapes; broth is circulated through the flash pot for vaporization of ethanol. Although this system may seem to have attractive energy economy because metabolic heat is removed as the vapour flashes, the initial investment and cost of pumping the vapours are high. Operating ethanol fermentation at higher temperatures with thermophilic organisms has better economics in terms of milder vacuum and less recompression because of the higher vapour pressure. Other ways of overcoming inhibition by the product are extraction from the fermentation broth with an immiscible solvent and/or operating with very dense cultures so that low productivity per cell is compensated with more cells.

PRODUCT RECOVERY

Alcohol recovery is energy intensive, typically accounting for more than 50% of the total fermentative ethanol plant energy consumption. When heat from burning of raw material residues (such as bagasse) is not sufficiently available, it constitutes a significant operating cost. Depending on the design of the recovery system, recovery equipment cost generally makes up 6–12% of the total plant capital investment.

Three-column Barbet System

For higher quality alcohol, the three-column Barbet system is commonly employed. In this system, the stripped beer is first purged of most aldehydes in a purifying column which then feeds a relatively dilute alcohol stream to a stripper-rectifier. The evaporated alcohol is condensed by the cooling rectifier which is present below the head of the common rectifier. A small purge of additional aldehydes from the head of the rectifier is recycled to the purifying column. Fusel oils are tapped from the lower plates of the rectifier, and in some cases an additional fusel oil concentrating column is employed. The energy consumption for a four-column system is 4.1 kg of steam per litre of 94 wt% alcohol and the product may be contaminated with traces of aldehydes.

Three-column Othmer System

In the 1930s, an alternative three-column Othmer system incorporating vapour reuse methods to reduce steam consumption was developed. A third aldehyde stripping column is added to the basic two-column design. The produced alcohol is rectified through the cooling condenser and is then fed to the top of the aldehyde column. Aldehyde carrying vapours from the head of the aldehyde stripper are returned to the rectifying column and a technical grade alcohol product is tapped from the rectifier reflux to purge the aldehydes as in the simple two-column design. The main product is virtually free of aldehydes and is of higher quality than the Barbet system product.

Heat for alcohol vapour boil-up in the aldehyde stripper reboiler is provided by condensation of a portion of the stripper vapour product, which is then returned to the stripper as reflux to reduce the stripper tray requirement. This reuse of heat increases energy efficiency, and steam requirement is not increased over the simple two-column design (2.4–3.0 kg per litre of 94 wt % product).

ANHYDROUS ETHANOL PRODUCTION

Ethanol without water is called anhydrous ethanol.

Azeotropic Distillation

Azeotropic distillation is the prevalent method for ethanol dehydration. Water is only slightly soluble in non-polar benzene, while ethanol is freely soluble. Vapour pressure of water over the benzene/ethanol/water–liquid mixture is thus greatly enhanced. In distillation, the ethanol–water mixture is fed at a midpoint in the column while an organic-rich phase (75 mol% benzene decanted from the cooled distillate) is used as reflux at the column head. Benzene concentrates in the liquid phase on the plates above the feed, enhancing the volatility of water and forcing it into the rising vapour phase. Ethanol concentrates benzene/ethanol/water ternary azeotrope (24% ethanol, 54% benzene, 22% water) is taken as the column head product. Benzene is stripped from ethanol, and anhydrous ethanol is recovered from the azeotropic column reboiler. The distillate of the azeotropic column, when cooled and decanted, separates into two phases —the organic-rich phase used for reflux and a water-rich phase (35% ethanol, 4% benzene, 61% water), which is treated in the benzene recovery column. This column produces the benzene concentrated azeotrope as head product. The dilute alcohol can be reconcentrated in an additional column or recycled to the primary alcohol recovery column. The distillation columns are expensive as the azeotropic column alone requires 50 plates of diameter similar to the primary stripper rectifier. Anhydrous alcohol is generally refined from 94 wt% azeotropic alcohol. The additional steam requirement for anhydrous alcohol production with benzene is 1.0 kg of steam per litre of product.

Absorption Methods

If alcohol is obtained from a still that is close to 190 proof (95%), the easiest method of producing gasohol is to mix sufficient benzene with alcohol to keep the water from separating. Usually, benzene in a ratio of about 2–3 times the amount of water in alcohol is sufficient. For example, to prepare 100 gallons of 10% gasohol, mix 10 gallons of 95% alcohol with 2 gallons of benzene. Then, add 88 gallons of gasoline. Note that water more than 5% requires the addition of larger amounts of benzene. It is suggested that experiments are carried out before attempting to mix large quantities of gasohol. A sample may be put in the refrigerator to test cold weather tolerance. At any rate, if the mixture separates, more benzene is needed.

A second method takes advantage of the fact that water can dissolve most salts, but ethanol will not. Therefore, water can be removed (although not entirely) by filtering the alcohol through dry salt. Almost any hygroscopic (water absorbing) material, such as calcium salt, sulphates and phosphates, will work. Common rocksalt is cheap and easily available. Fill the drum or container with rocksalt. The alcohol is poured from the top to filter down through the salt. Relatively water-free alcohol is collected at the bottom of the container. Remember that the salt must be dry. After absorbing certain amount of water from alcohol, the salt must be dried either in an oven or by spreading it out in the sun.

It would be good to use both the methods described above. First, most water in the alcohol can be removed by salt method, and then the blend is prepared with benzene. More the water that can be removed from alcohol, the lesser the benzene needed. Benzene can be used only once. Rocksalt can be dried and used many times.

Drying with Lime

The oldest method of drying alcohol is dehydration with lime. This process is still used on a laboratory scale. It is similar to the salt absorption method except that water is removed by a chemical reaction. Ordinary lime (calcium oxide) reacts with water to form calcium hydroxide. The process is simple. The water-containing alcohol is mixed with lime at a ratio of about 35 pounds (or more) for each gallon of water to be removed (as determined with a hydrometer) and allowed to slake for 12–24 hours with occasional stirring. Lime reacts with water to form calcium hydroxide. Calcium hydroxide is insoluble in alcohol, and so the relatively pure (99.5%) alcohol goes to the top of the container and calcium hydroxide settles to the bottom.

The usual method of separating lime and calcium hydroxide from alcohol is by distillation. Alternately, but less desirable, alcohol can be carefully drawn off (decanted) and filtered to remove suspended particles that give it a milky appearance. An apparatus based on a 55-gallon drum can be built. A still head with thermometer (no reflux column is needed) and condenser should be used to allow simple distillation. A small gate valve located 6–8 inches from the bottom (above the level of lime) will allow alcohol to be decanted.

After slaking in the apparatus, the alcohol should be distilled through a simple still head and condenser. During distillation, the temperature should remain exactly at the boiling point of pure alcohol. After pure alcohol has been distilled or decanted, wet calcium hydroxide and lime is left behind. Some alcohol may be trapped in the residue. To recover it, distillation may be continued. The still head temperature will rise above 78.3°C indicating hat water is coming over with the remaining alcohol. When the still head reaches 97.7–100°C, the alcohol has been entirely removed. The water or alcohol distillate should be added to the beer for the next run in the reflux distillation apparatus. Calcium hydroxide may be converted back into calcium oxide and reused. However, the temperatures required are quite high unless a vacuum drying oven is used. Since lime is relatively cheap, this process is not recommended.

WASTE TREATMENT AND BY-PRODUCT RECOVERY

Waste treatment and by-product recovery are extremely important factors in the economics of ethanol production. Typically 10–15% of stillage are produced for every litre of alcohol product. Corn stillage has a waste biological oxygen demand of 15000–25000 ppm can plant generate a pollution load equivalent to a city of 1.4 million inhabitants. Balancing this, stillage can be treated to become a valuable by-product.

Stillage Treatment and By-products

Nutrient (protein and vitamin) content is highest in corn stillage containing corn protein residue from germ and bran, which are largely unaffected by fermentation. Blackstrap molasses stillage is second highest in nutrient value with less protein but high vitamin content. Mineral (ash) content is also high. Cane and beet juice stillages are similar in nutrient make-up to blackstrap stillage, but with lower nutrient concentrations. Potato stillage and cornstarch stillage (from distilleries with complete corn germ and bran removal prior to fermentation) are low in nutrient value. Sulphite liquor and wood hydrolysate stillage are low in protein and vitamin nutrients but high in BOD, primarily pentose sugars.

Stillage Drying for Cattle Feed Production

Bulk solids are removed by centrifugation or by screening followed by dewatering in a rotary press. The solids are dried to less than 5% water in rotary driers to produce the distiller's light grains product. The liquid stillage is concentrated to 35% solids in forced convection vertical tube evaporators and then dried to powder using drum or spray driers, yielding distiller's solubles (Table 30.6). Alternatively, the concentrated liquid solubles can be blended back with the pressed solids and these rotary together to produce distiller's dark grains. About 700–900 g of stillage feed can be recovered per litre of alcohol produced.

Table 30.6   Nutrient value of grain distillers by-product

	Distiller's solubles	Dark grains
Moisture (%)	5	5
Protein (%)	27	29
Fat (%)	7	9
Fibre (%)	2	7
Ash (%)	8	4
Carbohydrate (%)	51	46
Vitamins (ppm)		
Thiamine (B_1)	8	4
Riboflavin (B_2)	22	8
Pantothenic acid	29	11
Niacin	125	65
Pyridoxine (B_6)	9	---
Biotin	0.5	0.2
Folic acid	4	---
Choline	6500	4500
Carotene	0.8	1
p-aminobenzoic acid	10	---
Zeaxanthin	8	8
Cryptoxanthin	4	5

The feed value of corn stillage is similar to that of soybean, and corn stillage is valued at approximately 95% of soybean cost for use in cattle feed.

The corn stillage cattle feed process recovers all the stillage solids, leaving no waste stream for further processing. The evaporation energy load to achieve this is very high. A conventional plant may use as much steam for stillage drying as for alcohol distillation.

Molasses stillage (vinasse) is lower in suspended solids content than corn stillage, and yeast may be recovered as a valuable by-product by simple centrifugation. The stillage is high in vitamin content. For use as cattle feed, molasses stillage is evaporated to typically 50–65 wt% solids (sufficiently concentrated to prevent spoilage). Feeding tests show that the concentrate has roughly 65% of the nutritive feed value of molasses and it is widely used as a molasses feed substitute.

Cane and beet juice stillage may be treated similarly to molasses stillage. Alternatively, these stillages may be mixed with bagasse fibre or beet pulp residue and dried to produce a nutrient enriched roughage feed.

The soluble nutrient content of corn starch, potato and sulphite waste liquor stillage is not sufficient to justify the high cost of evaporation. For these stillages, yeast may be recovered by centrifugation and then spray or drum dried. Secondary treatment is required to reduce the stillage pollution load.

REVIEW QUESTIONS

1. Give the uses of ethanol.

2. Write about the organisms used in ethanol production.

3. What are the feedstocks used in ethanol production?

4. Explain in detail about ethanol production from various substrates.

5. Explain ethanol biostil process and product removal.

6. Describe in detail about anhydrous ethanol production.

BIODIESEL

INTRODUCTION

Biodiesel is an alternative to diesel. It is produced by chemically reacting vegetable oil or animal fat with alcohol such as methanol. The reaction requires a catalyst, usually a strong base such as sodium or potassium hydroxide, to produce chemical compounds called methyl esters. These esters are known as biodiesel.

Since its primary feedstock is a vegetable oil or animal fat, biodiesel is generally considered renewable. Since carbon in the oil or fat originated from carbon dioxide in the air, biodiesel is considered to contribute much less to global warming. Engines that operate on biodiesel have lower emissions of carbon monoxide, unburned hydrocarbons, particulate matter and air toxics.

BACKGROUND

A majority of the world's energy needs is supplied by petrochemical sources, coal and natural gases, with the exception of hydroelectricity and nuclear energy. These sources are finite. Diesel fuels have an essential role in the economy of a developing country. The high energy demand in the industrialized world and in the domestic sector as well as pollution problems make it increasingly necessary to develop the renewable energy sources. An alternative fuel must be technically feasible, economically competitive, environmentally acceptable, and readily available. One possible alternative to fossil fuels is the use of oils of plant origin.

31

VEGETABLE OIL AS FUEL

The use of vegetable oils as fuel has been around for 100 years when the inventor of the diesel engine, Rudolph Diesel, first tested peanut oil in his compression ignition engine. However, there have been many problems associated with using it directly in diesel engine (especially in injection engine). These include:

1. Coking and trumpet formation on the injectors prevented fuel atomization as a result of plugged orifices

2. Carbon deposits

3. Oil ring sticking

4. Thickening or gelling of the lubricating oil as a result of contamination in vegetable oils

5. Lubricating problems

Other disadvantages in vegetable oils and especially animal fats are high viscosity (about 11–17 times higher than diesel fuel), lower volatilities that cause the formation of deposits in engines due to incomplete combustion, and incorrect vaporization characteristics. These problems are associated with large triglyceride molecule and its higher molecular mass.

Although biodiesel cannot entirely replace petroleum fuels, there are at least five reasons that justify its development.

1. It provides a market for excess production of vegetable oils and animal fats.

2. It decreases, the country's dependence on imported petroleum.

3. Biodiesel is renewable and does not contribute to global warming. A life cycle analysis of biodiesel showed that the overall CO_2 emissions were reduced by 78% as compared to petroleum-based diesel fuel.

4. The emissions of carbon monoxide, unburned hydrocarbons and particulate from biodiesel are lower. Unfortunately, most emission tests have shown a slight increase in the oxides of nitrogen.

5. When added to regular diesel fuel in an amount equal to 1–2%, it can convert fuel with poor lubricating properties, such as the modern ultra-low-sulphur diesel fuel, into an acceptable fuel.

TRANSESTERIFICATION

Biodiesel is produced through a process known as transesterification or alcoholysis. Transesterification is the displacement of alcohol from an ester by another ester by a process similar to hydrolysis, except that alcohol is used instead of water. This process has been widely used to reduce the viscosity of triglycerides. The transesterification reaction is represented by a general equation in Figure 31.1. If methanol is used in this process, it is called methanolysis. Methanolysis of triglyceride is represented in Figure 31.2. Transesterification is a reversible reaction and proceeds essentially by mixing the reactants. However, the presence of a catalyst (a strong acid or base) accelerates the conversion.

$$RCOOR^1 + R^2OH \xrightleftharpoons{\text{Catalyst}} RCOOR^2 + R^1OH$$

$$\text{Ester} \qquad \text{Alcohol} \qquad\qquad \text{Ester} \qquad \text{Alcohol}$$

Figure 31.1   General equation of transesterification

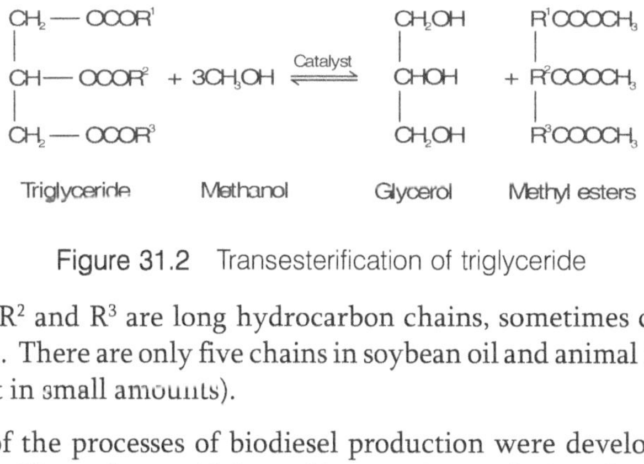

Figure 31.2   Transesterification of triglyceride

where R^1, R^2 and R^3 are long hydrocarbon chains, sometimes called fatty acid chains. There are only five chains in soybean oil and animal fats (others are present in small amounts).

Most of the processes of biodiesel production were developed in the early 1940s. Glycerol was widely used in war-time explosives. By chemically converting oils and fats into methyl esters, the glycerol could be separated because it is insoluble in esters. Glycerol has a much higher density, and so it may be easily removed by settling or centrifuge. The glycerol-free methyl esters were then reacted with alkali to form soap. Bradshaw received a patent for a process that added about 1.6 times the theoretical amount of alcohol, such as methanol which contained 0.1 to 0.5% sodium or potassium hydroxide, to an oil or fat. When performed at 80°C, this process provided 98% conversion to alkyl esters and high-quality glycerol. The following observations were made about the transesterification process:

- ⇒ Alcohol in excess of the stoichiometric amount (more than 1.6 times) is required for complete reaction.

- ⇒ The amount of alcohol used can be reduced by conducting the reaction in steps, where part of alcohol and catalyst are added at the start of each step and glycerol is removed at the end of each step.

- ⇒ Besides methanol, other alcohols including ethanol, propanol, isopropanol, butanol and pentanol can be used.

- ⇒ Water and free fatty acids inhibit the reaction. Higher alcohols are particularly sensitive to water contamination.

- ⇒ Free fatty acids in oils or fats can be converted to alkyl esters with an acid catalyst, followed by a standard alkali-catalysed transesterification to convert the triglycerides.

- ⇒ Acid catalysts can be used for transesterification of oils to alkyl esters, but they are much slower than alkali catalysts.

TRANSESTERIFCATION KINETICS AND MECHANISM

Transesterification of triglycerides produces fatty acids, alkyl esters and glycerol. Glycerol settles down at the bottom of the reaction vessel. Diglycerides and monoglycerides are the intermediate products. The mechanism of transesterification is shown in Figure 31.3.

The reactions are reversible, and a little excess of alcohol is used to shift the equilibrium towards the formation of esters. In the presence of excess alcohol, the forward reaction is of pseudo-first order and the reverse reaction is of second order. It has been observed that transesterification is faster when catalysed by alkali.

The first step in transesterification involves the attack of the alkoxide ion on the carbonyl carbon of the triglyceride molecule, which results in the formation of a tetrahedral intermediate. The reaction of this intermediate with an alcohol produces the alkoxide ion in the next step. In the last step, the rearrangement of the tetrahedral intermediate gives rise to an ester and a diglyceride.

ADVANTAGES OF BIODIESEL

Emission Reduction

Since biodiesel is made entirely from a vegetable oil, it does not contain sulphur, aromatic hydrocarbons, metals or crude oil residues. The absence of sulphur means a reduction in the formation of acid rain by sulphate emissions. The reduced sulphur in the blend will also decrease the levels of corrosive sulphuric acid accumulating in the engine crankcase.

The lack of toxic and carcinogenic aromatics (benzene, toluene and xylene) in biodiesel means the fuel mixture combustion gases will have reduced impact on human health and environment. The high cetane rating of biodiesel (ranges from 49 to 62) has the ability to improve combustion efficiency.

Lower Hydrocarbon Emission

Biodiesel burns cleanly, and it can improve the efficiency of combustion in combination with petroleum fuel. As a result of cleaner emissions, there will be reduced pollution. At a 20% biodiesel blend, there will be a noticeable change in the odour and smoke in the exhaust. Moreover, older engines should emit less soot under load and less carbon black during startup.

Smoke and Soot Reduction

Smoke (particulate material) and soot (unburned fuel and carbon residues) are of increasing concern to air quality problems that cause a wide range of adverse health effects, especially respiratory impairment and related illnesses. They also resent the soot accumulation on the transoms and decks of their boats.

Moreover, since biodiesel contains oxygen, there is an increased efficiency of combustion even for the petroleum fraction of the blend. Improved combustion efficiency lowers particulate materials and unburned fuel emissions, especially in older engines with direct fuel injection systems.

Carbon Monoxide Emission

Carbon monoxide gas is a toxic by-product of hydrocarbon combustion, which may be reduced by increasing the oxygen content of the fuel. Complete oxidation of the fuel results in more complete combustion to carbon dioxide rather than carbon monoxide.

Polyaromatic Hydrocarbon Emission

Polyaromatic hydrocarbons (PAHs) are a class of heavy petroleum hydrocarbons defined by their complex ring structures and unique qualities. They consist of multiple benzene ring structures that make them insoluble, burn slow and carcinogenic.

Nitrogen Oxides

Nitrogen oxides result from the oxidation of atmospheric nitrogen at high temperatures inside the combustion chamber of the engine. Although nitrogen oxides are considered a major contributor to ozone formation, they are helpful in operating internal combustion engines. However, there were reports of slight increase in NO emissions with biodiesel blends, which is attributable, in part, to the higher oxygen content of the fuel mixture. More oxygen and better combustion of the fuel means more formation of NO emissions with biodiesel blends.

Biodiesel Helps Reduce Greenhouse Gases

Unlike other "clean fuels" such as compressed natural gas (CNG), biofuels are produced from renewable agricultural crops that assimilate carbon dioxide from the atmosphere. Carbon dioxide released from burning biodiesels will subsequently be recaptured by crops to produce more vegetable oil. While anthropogenic (man-made) CO_2 production accounts for only 4–5% of the net CO_2 emissions, it is sufficient to cause a net gain over the past 100 years. Fossil fuel combustion accounts for 70% of the total anthropogenic CO_2 contribution. Supplementing our dwindling fossil fuel reserves with biodiesel helps reduce the accumulation of CO_2.

No Noxious or Carcinogenic Fumes

Biodiesel contains no volatile organic compounds that would give rise to poisonous or noxious fumes. It does not contain any aromatic hydrocarbons (benzene, toluene, xylene) or chlorinated hydrocarbons too. There is no lead or sulphur to react and release harmful or corrosive gases. However, in blends with petro-diesel, there may be significant fumes released by benzene and other aromatics present in the petroleum fraction (80%) of the blend.

No Risk of Explosion from Vapours

Since biodiesel has no volatile components (vapour pressure of less than 1 mm Hg) and a high flash point (typically over 112°C), the product poses no risk of explosion. However, the risk of fire would arise from the spontaneous combustion of rags and paper towels soaked in biodiesel and stored in an area with low ventilation or high temperatures (like the inside of an engine room).

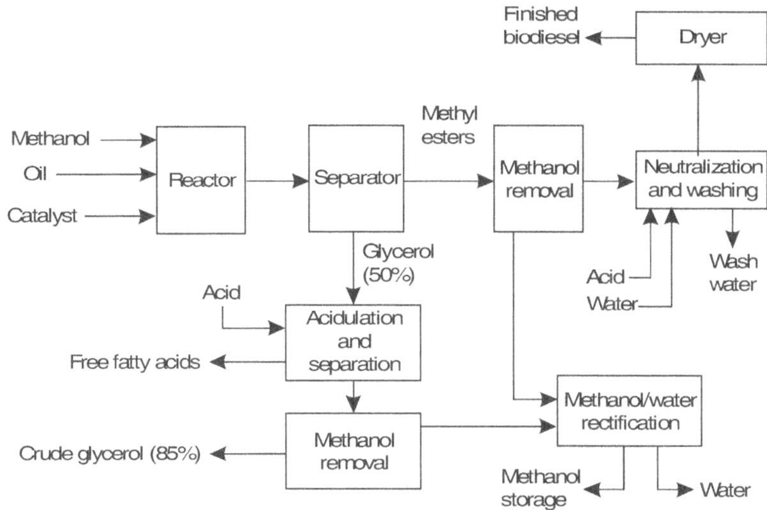

Figure 31.3   Flowchart of biodiesel production

PRODUCTION OF BIODIESEL

There are three basic methods to produce biodiesel from oils and fats:

- ➲ Base catalysed transesterification of oil

- ➲ Direct acid catalysed transesterification of oil

- ➲ Conversion of oil into its fatty acids and then into biodiesel

Biodiesel is commonly produced with base catalysed reaction for several reasons:

- ➲ It has low temperature and pressure.

- ➲ It yields high conversion (98%) with minimal side reactions and reaction time.

- ➲ It directly converts to biodiesel with no intermediate compounds.

The chemical reaction in base-catalysed biodiesel production is shown in Figure 31.4. One hundred pounds of fat or oil (such as soybean oil) is reacts with 10 pounds of a short chain alcohol in the presence of a catalyst to produce 10 pounds of glycerine and 100 pounds of biodiesel. The short chain alcohol, signified (R—OH) (usually methanol, but sometimes ethanol) is charged in excess to assist quick conversion. The catalyst may be usually sodium or potassium hydroxide that has already been mixed with methanol. R^1, R^2 and R^3 indicate the fatty acid chains associated with the oil or fat which are largely palmitic, stearic, oleic and linoleic acids for naturally occurring oils and fats.

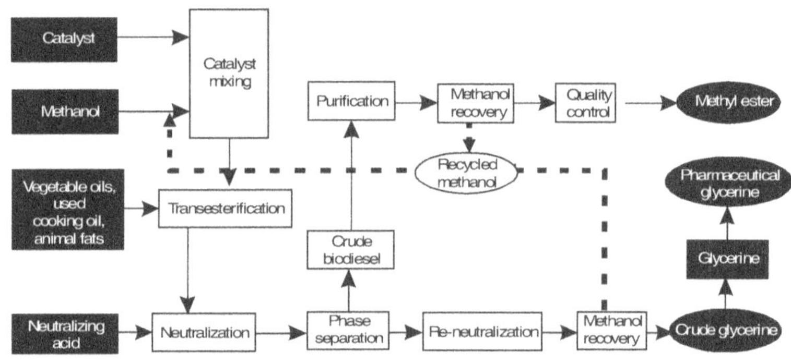

Figure 31.4   Biodiesel reaction

The base-catalysed production of biodiesel involves the different stages as shown in Figure 31.5.

Catalyst
Catalyst mixing
Methanol
Purification
Methanol recovery
Quality control
Methyl ester
Recycled methanol
Pharmaceutical glycerine
Vegetable oils, used cooking oil, animal fats
Transesterification
Crude biodiesel
Glycerine
Neutralizing acid
Neutralization
Phase separation
Re-neutralization
Methanol recovery
Crude glycerine

Figure 31.5   Biodiesel production process

Mixing of alcohol and catalyst   The catalyst, usually sodium hydroxide (caustic soda) or potassium hydroxide (potash), is dissolved in alcohol using a standard agitator or mixer.

Reaction   The alcohol or catalyst mix is charged into a closed reaction vessel and oil or fat is added. The system is covered to prevent the loss of alcohol to the atmosphere. The reaction mix is heated above the boiling point of alcohol (around 71°C) to speed up the reaction. The reaction time varies from 1 to 8 hours, and some systems suggest the reaction to take place at room temperature. Excess alcohol is normally used to ensure total conversion of the fat or oil to its esters. Care must be taken to monitor the amount of water and free fatty acids in the incoming oil or fat. If the free fatty acid level or water level is too high, it may cause problems with soap formation and the separation of glycerine downstream.

Separation   Once the reaction is complete, two major products exist: glycerine and biodiesel. Each has a substantial amount of methanol that was used in the reaction. The reacted mixture is sometimes neutralized at this step. The glycerine phase is denser than the biodiesel phase, and both can be gravity separated in which glycerin may be simply drawn off the bottom of the settling vessel. In some cases, a centrifuge is used to separate the materials faster.

Alcohol removal   Once the glycerine and biodiesel phases have been separated, excess alcohol in each phase is removed by a flash evaporation process or by distillation. In others systems, the alcohol is removed and the mixture is neutralized before the glycerin and esters have been separated. In either case, alcohol is recovered using distillation equipment. Care must be taken to ensure no water accumulates in the recovered alcohol stream.

Glycerine neutralization   Glycerine may contain unused catalyst and soaps, and so it should be neutralized with an acid and sent to storage as crude glycerine. Sometimes, the salt formed during this phase may be recovered for use as fertilizer. In most cases, the salt is left in the glycerin. Then, water and alcohol are removed to produce 80–88% pure glycerine. In more sophisticated operations, glycerine is distilled to 99% or higher purity and used in cosmetics and pharmaceuticals.

Methyl ester wash   Once separated from glycerine, biodiesel may be further purified by washing gently with warm water to remove residual catalyst or soaps. This step will be necessary to produce a clear amber-yellow liquid with viscosity similar to petro-diesel. In some systems, the biodiesel may be distilled to produce a colourless biodiesel.

REVIEW QUESTIONS

1.   What is biodiesel?

2.   How are vegetable oils used as fuel?

3.   What is transesterification?

4.   Give the advantages of biodiesel?

5.   Describe in detail about production of biodiesel.

Glossary

Acetogen   Acetic acid producing organisms.

Acne   A chronic disorder of the hair follicles and sebaceous glands.

Anti-foaming agents   The agents that are used to prevent the foam formation during fermentation.

Associative symbiosis   A type of symbiosis where the symbionts can also live without an association.

Attenuation   Removal of virulence from the pathogen.

Auxostat   The addition of fresh medium to a continuous cultivation based on the analysis of parameter.

Bacterial spot   The spots that are present in the vegetables due to bacterial infection.

Bactericidal   An agent which kills the bacteria.

Bacteriostatic   An agent which prevents the growth of the bacteria.

Baffles   A component of fermenter or bioreactor, which is present in the inner peripheral region of the vessel to prevent the vortex of the medium caused by stirrer.

Bakers yeast   The *Saccharomyces cerevisiae* is called bakers yeast. Most of the bakery items are prepared using this yeast.

Batch culture   A kind of culture system in which there is no addition of fresh medium and removal of culture till the end of the cultivation.

Beta lactamase   The enzyme capable of breaking a beta-lactam ring.

Bioaugmentation   Artificial cultivation of living organisms for the purpose of bioremediation.

Bioinsecticide   A compound produced from living organisms or their products which used to suppress the insect population.

Biopesticide   A compound produced from living organisms or their products is used to suppress the pest population.

Bioreactor   Device that is used to carryout fermentation for cell biomass production.

Bioremediation   Removal of pollution using microorganisms.

BOD   Biological oxygen demand. A measure of the oxygen consumed in organic rich water by aerobic microorganisms for their metabolic functions.

Boils   A localized area of inflammation caused by an infection under the skin, in a gland, or in a hair shaft. It produces a localized swelling, heat and redness.

Calorimetry   Technique used to measure the liberation or consumption of heat.

Chalky bread   A spoilage condition of bread due to the spoilage moulds and yeasts.

Chemolysis   Killing of cells due to chemicals.

Chemostat   The addition of fresh medium based on the analysis of substrate or product concentration to maintain the log phase in a continuous cultivation is called chemostat.

Chromatography   Separation technique in which the components are separated from the mixture.

COD   Chemical oxygen demand. The amount of oxygen required for chemical oxidation of pollutants in water. The higher the COD, the higher the water pollution.

Co-metabolism   Associative metabolism with main metabolic process.

Compressors   Air generating pumps.

Continuous culture   A cultivation system in which periodical addition of fresh medium along with the removal of culture.

Cooling jackets   Device used to control the temperature of the fermenter.

Corn steep liquor   It is a by-product released from starch industry that is rich in starch.

CSTRs   Continuous stir tank reactors in series.

Death phase   This is the final stage of microbial growth phase where, the organisms start to die due to depletion of nutrients.

Dehalogenation   Removal of halogen groups from the compounds.

Denitrification   Conversion of reduced nitrogen into atmospheric nitrogen form.

Detoxification   Removal of toxicity from toxin.

Dialysis   Removal of solutes from solvent by means of diffusion.

Diffusion   Movement of solutes from higher concentration to lower concentration.

DO   Dissolved oxygen.

EIA   Enzyme immune assay.

Electrophoresis   Movement of charged particles through a gel due to the influence of electric field.

ELISA   Enzyme-linked Immuno-sorbent assay.

Elution   Separation of components from the mixture through mobile phase.

Endotoxin   The cell components which act as toxins.

Enzymes   Biocatalysts that enhance and mediate the biological reactions.

Ex situ   Performing experiment by moving the things from its original place.

Exotoxin   Toxins that are secreted out of the cell.

Fed-batch culture   A cultivation system in which periodical addition of fresh medium without removing culture.

Fermenters   Device that is used to carryout fermentation for metabolites.

Flocculation   The process of forming aggregated or compound masses of particles, such as a cloud or a precipitate.

Food contamination   Presence of harmful microbes in the food.

Food preservation   Preventing microbial spoilage of foods.

Food spoilage   Conversion of food into non-edible form due to contaminated microbial activity.

Food-borne illness   Disease acquired through food contamination.

GMOs   Genetically modified organisms. The genetic modification may be due to mutation, recombination and cloning.

HEPA filter   High efficient particulate air filter. It is mainly used for air sterilization.

Heterocyst   A large, thick-walled cell present in some cyanobacteria which involves in nitrogen fixation.

Homogenizer   Device used to homogenize the materials.

Hops   Twining perennials having cordate leaves and flowers arranged in cone-like spikes; the dried flowers of this plant are used in brewing to add characteristic bitter taste to the beer.

HPLC   High-pressure liquid chromatography. It is a separation and analysis technique.

HTST   High temperature short time.

In situ   Performing an experiment in the original place.

In vitro   Performing an experiment in an artificial environment.

In vivo Performing an experiment in living organisms.

Lag phase This is the first stage of microbial growth phase in which growth rate and metabolism rate will be less.

Log phase It is the second stage of microbial growth phase in which the cell multiplication rate will be high.

Lyophilization It is a freeze-drying preservation method.

Malt extract Crush prepared from cereals which is rich in nutrients.

Mesophiles The organisms that are living in the moderate temperature.

Metabolism Biochemical processes taking place in living organisms.

Methanogen Methane producing organisms.

Mobile phase The components, that is in motion, which help to separate the components from the mixture, e.g. the eluting buffers used in chromatography.

Molasses It is a waste by-product produced during sugar cane processing. It is rich in sucrose.

Nausea It is an uneasy state of health with vomiting sensation.

Nitrification Conversion of ammonia into nitrate or nitrite form.

Nitrogen fixation Fixing atmospheric nitrogen into soil by microorganisms.

Nitrogenase The enzyme which mediates the nitrogen fixation.

Osmosis Movement of solvents from higher potential to lower potential through the membrane.

Pasteurization A heat preservation method for heat-sensitive materials, e.g. milk.

Pathogen Disease causing organisms.

Pathogenicity Disease causing alrility of a pathogen.

PFR Plug flow reactors.

Photometry Technique that is used to measure light.

Photosynthesis The process of using sunlight as an energy source to convert water and carbon dioxide into carbo-hydrates and oxygen.

Primary metabolites The metabolites that are produced during log phase of an organism and are mainly used to utilize the substrates, e.g. enzymes.

Proteolysis Degradation of protein.

Psychrophiles The organisms that are living in low temperature environment.

Putrefaction Putrefaction is the decomposition of proteins, especially by anaerobic microorganisms.

Pyrogen The agent that leads to an increase in temperature in any living system.

Pyrolysis Decomposition of a chemical by extreme heat.

Red bread A red colouration in bread due to the production of red pigment by spoilage organisms.

Respiratory quotient (RQ) The ratio between the consumption of oxygen and liberation of carbon dioxide.

Reverse osmosis Movement of solvents against the osmotic gradient.

Saccharin A sweetener with no calories and no nutritional value.

SCP Single cell proteins. The microbial proteins that are used as food or feed.

Secondary metabolites Metabolites produced during the stationary phase of an organism. They are mainly used to succeed in the competition, e.g. antibiotics and toxins.

Seed lot Selection of organisms for the process.

Sensors Devices used to measure the parameters and send signals to the controlling unit.

Soil attenuation The process by which a compound is reduced in concentration over-time through adsorption, degradation, dilution and transformation, usually by natural processes.

Soy bean meal It is the product obtained by grinding the flakes that remain after the removal of most of the oil from soybeans.

Sparger A component of fermenter or bioreactor with which aeration of the medium is made.

SSF Solid state fermentation. The fermentation that is carried out using solid or semi solid nutrients.

Stationary phase of chromatography A gel or compound, that is fixed, which helps to separate the components from the mixture, e.g. the sephadex gel in column chromatography.

Stationary phase This is the third stage of microbial growth phase in which the growth rate will be steady due to limited nutrients. The metabolites produced in this phase are called secondary metabolites that are used to succeed in the competition.

Sterilization Removal or destruction of microbes including spores from the materials.

Stirrer A component of fermenter or bioreactor, which is used to agitate the medium.

Submerged fermentation The fermentation that is carried out using liquid medium.

Symbiosis Beneficial association between two organisms.

Thermophiles The organisms that are living in high temperature environment.

Toxins Harmful and poisonous proteins produced by the microbes.

Toxoid Detoxificated toxin.

Transesterification A reaction between an ester and an alcohol in which the-OR of the ester is displaced by the-OR ′ group of the alcohol.

Turbidostat The addition of fresh medium to a continuous cultivation based on the turbidity of cells is called turbidostat.

UHT Ultra high temperature.

UP Ultra pasteurization.

Whey Milk industry waste.

Wort Processing malt used for beer fermentation.

References

Abbott, B. J. (1976). "Preparation of pharmaceutical compounds by immobilized enzymes and cells." *Adv. Appl. Microbiol.* 20: 203–257.

Agaisse, H. and Lerecluse, D. (1996). "How does *Bacillus thuringiensis* produce so much insecticidal crystal protein?" *J. Bacteriol.* 177: 6027–6032.

Ahsan, N. (1999). "Solid waste management plan for Indian megacities." *Indian Journal of Environmental Protection.* 19 (2), 90–95.

Aiba, S., Humphrey, A.E. and Millis, N.F. (1973). *Biochemical Engineering*, 2nd edn. Academic Press, New York.

Albertsson, P.A. (1978). *J. Chromatogr.* 159: 111–114.

Albertsson, P.A. (1986). *Partition of Cell Particles and Macromolecules.* Wiley, New York.

Albertsson, P.A., Johansson, G. and Tjerneld, F. (1990). *Separation Processes in Biotechnology.* Asenjo, J.A.(ed.) Marcel Dekker, New York. 287–327.

Alexander N. Glazer and Hiroshi Nikaido. (1994). *Microbial Biotechnology.* W.H.Freeman Company, New York.

Alexander, M. (1965). "Biodegradation: Problems of Molecular Recalcitrance and Microbial Fallibility". In: *Advances in Applied Microbiology.* Academic Press, New York. pp.35–80.

Alexander, M. (1994). *Biodegradation and Bioremediation.* Academic Press, San Diego, CA.

Allard, A.S. and Neilson, A.H. (1997). *Int. Biodeterioration Biodegradation.* 39, 253–285.

Alvarez-Vasquez, F., González-Alcón, C., and Torres, N.V. (2000). Metabolism of citric acid production by *Aspergillus niger.* Model definition, steady-state analysis and constrained optimization of citric acid.

Amanullah, A., Satti, S. and Nienow, A.W. (1998). "Enhancing xanthan fermentations by different modes of glucose feeding." *Biotechnol. Prog.* 14: 265–269.

Amon, T., Hackl, E., Jeremic, D., Amon, B. and Boxberger, J. (2001). "Biogas production from animal wastes, energy plants and organic wastes". In van Velsen, A.F.M. and Verstraete, W.H. (eds.), *Proc. 9th World Congress on Anaerobic Digestion.* Technologisch Instituut zw, Antwerp. 381–386.

Andersson, E. and Hahn-Hagerdal, B. (1990). *Enzyme Microb. Technol.* 12: 242–254.

Asada, Y. and Kawamura, S.J. (1986). *Ferment. Tech.* 64: 553–556.

Asher, Gerald. (1989). "Single Malt Scotch Whiskey." *Gourmet.* pp. 94–99.

Aust, S.D., Swaner, P.R. and Stahl, J.D. (2003). "Detoxification and metabolism of chemicals by white-rot fungi". In: *Pesticide Decontamination and Detoxification.* Oxford University Press, Washington, D.C. pp. 3 –14.

Bailey, J.E. and Ollis, D.F. (1987). *Biochemical Engineering Fundamentals*, 2nd edn. Mc Graw-Hill, New York.

Barnabe´, S., Verma, M., Tyagi, R.D. and Vale´ro, J.R. (2005). "Culture media for increasing biopesticide producing microorganism's entomotoxicity, methods of producing same, biopesticide producing microorganisms so produced and method of using same". *US patent International application PCT/CA2005/000235.*

Bayan, A.P., Unger, U.F. and Brown,W.E. (1962). "Factors affecting the biosynthesis of griseofulvin." *Antimicrob. Agents Chemother.* 59: 669–676.

Becker, A., Katzan, F., Puhler, A. and Ielpi, L. (1998). "Xanthan gum biosynthesis and application: a biochemical/genetic perspective." *Appl. Microb. Biotechnol.* 50: 145–152.

Bedient, Rifai, and Charles Newell. (1994) *Ground Water Contamination.* Prentice Hall Inc., Englewood Cliffs.

Beegle, C.C. (1990). "Bioassay methods for quantification of *Bacillus thuringiensis* d-endotoxin, analytical Chemistry of *Bacillus thuringiensis*. In: Hickle, L.A., Fitch, W.L. (eds.). *Analytical Chemistry of Bacillus thuringiensis.* American Chemical Society, USA. pp. 14–21.

Bekins, B. A., Godsy, E. M. and Warren, E. (1999). "Distribution of microbial physiologic types in an acquifer contaminated by crude oil." *Microbial Ecol.* 37: 263–275.

Bentley, R. and Thiessen, C.P. (1957a). "Biosynthesis of itaconic acid in *Aspergillus terreus.* I. Tracer studies with 14C-labelled substrates.*" J. Biol. Chem.* 226: 673–687.

Boone, D.R., et al., (1980). *Appl. Environ. Microbiol.* 40: 626–632. McInerney, M.J., *et al.,* (1981). *Appl. Environ. Microbiol.* 41: 1029–1039.

Börjesson, P. (2004). "Energy analysis of transportation fuels from grain and ley crops". Report No. 54, *Environmental and Energy System Studies.* Lund University, Lund, Sweden. p. 16.

Borkiiolder, P.R. (1951). "Determination of vitamin B12 with a mutant strain of *Escherichia coli.*" *Science.* 114: 459–460.

Bothe, H. (1982). In: *The Biology of Cyanobacteria.* N.G. Carr and B.A. Whitton. (eds.). Blackwell, Oxford, UK. pp. 87–104.

Bowker, R.P.G. (1983). *Environ. Progress.* 2: 235–242.

Boyetchko, S. M. and G. Peng. (2004). "Challenges and strategies for development of mycoherbicides." *In: Fungal Biotechnology in Agricultural, Food, and Environmental Applications.* Marcel Dekker, N.Y. pp. 111–121.

Brar, S.K., Verma, M., Tyagi, R.D., Vale´ro, J.R., Surampalli, R.Y. and Banerji, S. (2004). "Development of sludge-based stable aqueous *Bacillus thuringiensis* formulations". *Water Sci. Technol.* 50 (9), 229–236.

Brar, S.K., Verma, M., Tyagi, R.D., Vale´ro, J.R., Surampalli, R.Y. (2005). Sludge based *Bacillus thuringiensis* biopesticides: viscosity impacts. *Water Res.* 39: 3001–3011.

Brookman, J. (1974). "Mechanism of Cell Integration in High Pressure Homogenizer" *Biotechnol. Bioeng.* 16: 371–383.

Brown, Bob. (1955). *The Complete Book of Cheese.* Gramercy Publishing.

Bull, S.R. (1994) *Renewable Energy.* 5, 799–806.

Bulock, J. and Christiansen, B. (1989). *Basic Biotechnology.* Academic Press.

Bumpus, J. A., Tien, M., Wright, D. S. and Aust, S. D. (1985). "Oxidation of persistent environmental pollutants by a white-rot fungus." *Science.* 228: 1434–1436.

Burges, H.D. (ed.). 1998. *Formulation of Microbial Biopesticides: Beneficial Organisms, Nematodes and Seed Treatments.* Kluwer Academic Publishers, Dordrecht.

Burton, M.O. and Lochhead, A.G. (1951). "Studies on the production of vitamin B12- active susbstance by microorganisms." *Can.J. Bot.* 29: 352–359.

Buxton, D.R. and O'Kiely, P. (2003). Preharvest plant factors affecting ensiling. In Al-Amoodi, L., Barbarick, K.A., Volenec, J. J. and Dick, W.A. (eds.). *Silage Science and Technology.* American Society for Agronomy, Madison, USA. pp.199–250.

Cagney, J.W., Chittur,V.K. and Lim, H.C. (1984)."Use of filtration measurements for estimation of cellular activity in penicillin production." *Biotechnol. Bioeng. Sym. Ser.* 4: 619–634.

Cairney, T. (1993). *Contaminated Land,* Blackie, London. p. 4.

Carr, Sandy. (1981). *The Simon and Schuster Pocket Guide to Cheese.* Simon and Schuster.

Casida, L.E. (1994). *Industrial Microbiology.* Wiley Eastern Limited, New Delhi.

Chakrabarty, P., Srivastava, V.K. and Chakrabarti, S.N. (1995). "Solid waste disposal and the environment – a review." *Indian Journal of Environmental Protection.* 15 (1), 39–43.

Chanakya, H.N., Borgaonkar, S., Meena, G. and Jagadish, K.S. (1993). "Solid-phase biogas production with garbage or water hyacinth." *Biores. Technol.* 46: 227–231.

Chanakya, H.N., Borgaonkar, S., Rajan, M.G.C. and Wahi, M. (1992). "Two-phase anaerobic digestion of water hyacinth or urban garbage." *Biores. Technol.* 42: 123–131.

Coates, M. IX. and Ford, J, 11 (1955). "Methods of measurement of vitamin B12. In: R. T.Williams, "The Biochemistry of Vitamin B12." *Biochem. Sotc. Symposia.* 13: 36–51.

Colberg, P. J. S. and Young. L. Y. (1995). "Anaerobic degradation of non-halogenated homocyclic aromatic compounds coupled with nitrate, iron, or sulfate reduction. In *Microbial Transformation and Degradation of Toxic Organic Chemicals.* Wiley, New York. pp. 307 330.

Considine, Douglas M. (ed.). (1982). *Food and Food Production Encyclopedia.* Van Nostrand Reinhold Company.

Cory, P.F., Baily, J.S. and Higgs, M.J.S. (eds.). *"Bacillus thuringiensis:* An Environmental Biopesticide: Theory and Practice". Wiley, Great Britain. pp. 255–267.

Cozzarelli, I.M., Bekins, B.A., Baedecker, M. J., Aiken, G.R. Eganhouse, R.P. and Tuccillo, M. E. (2001). "Progression of natural attenuation processes at a crude oil spill site. Geochemical evolution of the plume." *J. Contam. Hydrol.* (in press).

Cunha, M. T., Tjerneld, F., Cabral, T. M. S. and Aires-Barros, M.R. (1998). *J. Chromatogr.* B711: 53–60.

Currie, J.N. (1917). "Citric acid fermentation." *J. Biol. Chem.* 31: 15–37.

Datta, M. (1997). *Waste Disposal in Engineered Landfills.* Narosa Publishing House, New Delhi, India.

Dayal, G. (1994). "Solid wastes: sources, implications and management." *Indian Journal of Environmental Protection.* 14 (9), 669–677.

Decleire, M., De Cat, W. and Van Huynh, N. (1987). *Enzyme Microb. Technol.* 9: 300–302.

Delgenès, J.P., Penaud, V. and Moletta, R. (2003). "Pretreatments for the enhancement of anaerobic digestion of solid wastes". In Mata-Alvarez, J. (ed.). *Biomethanization of the organic fraction of municipal solid wastes.* 201–228. IWA Publishing, London.

DeMarco, Dan and Frank Bechard. (1986). "New Weigh Scales Smooth Distillery's Production, Improve Inventory Control." *Food Engineering.* pp. 95–96.

Demuynck, M. (1984). Utilization in agriculture of anaerobically digested effluents. 62 p., Commission of the European Communities, Brussels. Dinamarca, S., Aroca, G., Chamy, R. and Guerrero, L. 2003.

Denny, Sharon. (1994). "Enough Dough for Every-one." *Current Health.* p. 25.

Derikx, P. J. L., Op den Camp, H. J. M., Van der Drift, C. van Griensven, L. J. L. D. and Vogels, G. D. (1990). Odorous sulfur compounds emitted during production of compost used as a substrate for mushroom production. *Appl. and Environ. Microbiol.* 56: 3029–3036.

Dong, X.Y., Bai, S., and Sun, Y. (1996). "Production of L(–)-lactic acid with *Rhizopus oryzae* immobilized in polyurethane foam cubes". *Biotechnol. Lett.* 18: 225–228.

Douglas, J.M. (1988). *Conceptual Design of Chemical Processes.* McGraw-Hill, New York.

Droop, M. It. (1954). "Cobalamin requirement in Chrysophyceae." *Nature.* 174: 520.

Du, J.X., Cao, N.J., Gong, C.S., and Tsao, G.T. (1998). "Production of L-lactic acid by *Rhizopus oryzae* in a bubble column fermenter". *Appl. Biochem. Biotechnol.* 70: 323–329.

Ehrlich, F. (1911). "Formation of fumaric acid by means of moulds." *Ber. Dtsch. Chem. Ges.* 44: 3737–3742.

Eimhjellen, K.E. and Larsen, H. (1955). "The mechanism of itaconic acid formation by *Aspergillus terreus.* 2. The effect of substrates and inhibitors." *Biochem. J.* 60: 139–147.

Elander, R. P. and Lowe, D. A. (1994). "Fungal biotechnology: An overiew" In: *Handbook of Applied Mycology.* Marcel Dekker, New York. pp. 1–34.

Ellwood, D.C. (1967). Continuous culture studies on the production of extracellular polysaccharides by *Xanthomonas juglandis.* In: Berkeley RCW, Gooday GW, Ellwood DC (pnyt.). *Microbial polysaccharides and polysaccharases.* Academic Press Inc. (London) Ltd. London. pp. 51–67.

El-Mansi, E.M.T. and Bryce, C.F.A. (2004). *Fermentation Microbiology and Biotechnology.* Tayler and Francis, London.

Essaid, H. I., Bekins, B.A., Godsy, E.M., Warren, E., Baedecker; M. J. and Cozzarelli, 1. M. (1995). "Simulation of aerobic and anaerobic biodegradation processes at a crude oil spill site." *Water Resour. Res.* 31: 3309–3327.

Essaid, H.I. and Bekins, B.A. (1997). "BIOMOC, A multispecies transport model with biodegradation", *Water Resources Investigation Report 974022,68* pp., U.S. Geological Survey, Washington, D.C.

Felix, H.R. (1982). *Anal. Biochem.* 120: 710–715.

Flaibani, A., Olsen, Y. and Fainter, T.J. (1989). *Carbohydr. Res.* 190: 235-248.

Flathman, P.E., Jerger, D. and Exner, J.E. (1993). *Bioremediation: Field Experience*, Lewis, Boca Raton, FL.

Foster, J.W. and Waksman, S.A. (1939). "The production of fumaric acid by moulds belonging to the genus *Rhizopus."J. Am. Chem. Soc.* 61: 127–135.

Friedmann, E.1. and Galun, M. (1974). In: *Desert Biology,* G.E. Brown (ed.). Academic Press, New York. pp. 165–212.

Fukui, S., Sonomoto, K. and Tanaka, A. (1987). *Methods Enzymol.* 135: 230–252.

Gadd, G.M. (2001). *Fungi in Bioremediation.* Cambridge University Press, U.K.

Garcia-Ochoa, F., Santos, V.E. and Fritsch, A.P. (1992). "Nutritional study of *Xanthomonas campestris* in xanthan gum production by factorial design of experiments." *Enzyme Microb. Technol.* 14: 991–997.

Garcis-Perez, A.I., Sancho, P. and Pinilla, M. (1998). *J. Chromatogr.* B711: 301–309.

Gensen, E. (1954). Microbiological assay of vitamin B12. In: B. Glick, *Methods of Biochemical Analysis.*Vol. 1. Interscience, New York. pp. 81–113.

Gernhardt, P. *et.al.,* (1990). *Mol. Gen. Genet.* 221: 273–279.

Ghose, M.K., Dikshit, A.K. and Sharma, S.K. (2006). "A GIS based transportation model for solid waste disposal – A case study on asansol municipality." *Journal of Waste Management* 26 (11), 1287–1293.

Ghosh, C. (2004). "Integrated vermin–pisciculture – an alternative option for recycling of municipal solid waste in rural India." *Journal of Bioresource Technology.* 93 (1), 71–75.

Glare, T.R. (2004). Biotechnological potential of entomopathogenic fungi. In: *Fungal Biotechnology in Agricultural, Food, and Environmental Applications.* Marcel Dekker, New York. pp. 79–99.

Godbole, S.S., D'Souza, S.F. and Nadkarni, G.B. (1980). *Enzyme Microb. Technol.* 2: 223–226.

Grimes, William. (1993). *Straight Up or On the Rocks: A Cultural History of American Drink.* New York: Simon and Schuster.

Grossman, Harold J. (1983). *Grossman's Guide to Wines, Beers and Spirits,* 7th edn., revised by Harriet Lembeck. Charles Scribner's Sons, New York.

Guan, Y., Lilley, T.H., Treffry, T.E., Zhou, C.L. and Wilkinson, P.B. (1996). *Enzyme Microb. Technol.* 19: 446–455.

Guebel, D.V. and Torres Darias, N.V. (2001). "Optimization of the citric acid production by *Aspergillus niger* through a metabolic flux balance model". *Electron. J. Biotechnol.* 4: 1–14.

Guevarra, E.D. and Tabuchi, T. (1990). "Accumulation of itaconic, 2-hydroxyparaconic, itatartaric, and malic acids by strains of the genus *Ustilago. Agric. Biol. Chem.* 54: 2353–2358.

Guillard, R.R.L. and Lorenzen, C.J. (1972). *J. Phycol.* 8: 10-14. Knutsen, G. and Metting, B. (1991). In: *Semi-arid Lands and Deserts,* J. Skujins. (ed.). Marcel Dekker, New York, pp. 487–506.

Hachem, F., Andrews, B.A. and Asenjo, J.A. (1996). *Enzyme Microb. Technol.* 19: 507–517.

Hage, C., Kaynak, A. and Knol, W. (1998). *Enzyme Microb. Technol.* 22: 50–57.

Hanson, C., Baird, M.H.I., Lo, T.C. (1983). *Handbook of Solvent Extraction.* John Wiley and Sons, New York.

Harper, Roseanne. (1995). "Bred to Rise." *Supermarket News.* p. 21.

Harper, William. (1999). *Origin and Rise of the British Distillery.* The Edwin Mellen Press, Lewiston, U.K.

Harrison, R.G., Todd, P., Rudge, S.R. and Petrides, D. (2002). *Bioseparations Science and Engineering.* Oxford University Press, Oxford. p. 432

Hartmier, W. 1988. *Immobilized Biocatalysts; An Introduction.* Springer-Verlag, Berlin.

Heiermann, M., Plöchl, M., Linke, B. and Schelle, H. (2002). "Preliminary evaluation of some cereals as energy crops for biogas production". "In: Sayigh, A.A.M. (ed.). *Proc. World Renewable Energy Congress VII,* CD-version. Pergamon, Cologne.

Helferich, W. and Westhoff, D. (1980). *Yogurt: All About It.*

Hemiksen, C.M., Nielsen, J. and Viladsen, J. (1997). "Influence of the dissolved oxygen concentration on the penicillin biosynthetic pathway in steady-state cultures of *Penicillium chrysogenum."Biotechnol. Prog.* 13: 776–782.

Hinchee, Robert, Ross Miller, and Paul Johnson. (1995). *In Situ Aeration: Air Sparging, Bioventing, and Related Remediation Processes.* Battelle Press, Ohio.

Hoff, D.A., Bennett, R.E. and Stanley, A.R. (1947). "A sensitive cylinder plate assay for bacitracin." *Science.* 106: 551–552.

Holdsworth, E.S. (1953). "Differentiation of vitamin Bit active compounds by ionophoresis and microbiological assay." *Nature.* 171:148–149.

Hui, Y.H. (1992). (ed.). *Encyclopedia of Food Science and Technology,* Vol. 2. John Wiley and Sons, New York.

Hui, Y.H. (1992). (ed.). *Dairy Science and Technology Handbook.* Wiley VCH, New York.

Hungate, R.E., (1969). *Methods in Microbiol.* 3B: 117–132.

Hustedt, H., Kroner, K.H., Stach, W. and Kula, M.R. (1978). *Biotechnol. Bioeng.* 20: 1989–2005.

Jackson, Michael. (1987). *The World Guide to Whisky.* Darling Kindersley.

Jeris, J.S. (1983). *Wat. Sci. Technol._*15: 167–187.

Johansson, G. and Reczey, R. (1998). *J. Chromatogr.* B711: 161–172.

Johnson, Julie. (1991). "Mysteries of the Malt." *New Scientist.* pp. 56–59.

Joshi, J.B., Sawant, S.B., Raghava Rao, K. S. M. S., Patil, T., Rostami, K.M. and Sikdar, S.K. (1990). *Bioseparation.* 1: 318–324.

Joshi, M.S., Gowda, L.R., Katwa, L.C. and Bhat, S.G. (1989). *Enzyme Microb. Technol.* 11:439–443.

Jukes, T. II., Williams, W.L., Wolf, D.E. and Folkera, K.E. (1954). In: *The Vitamins.* Academic Press, New York. pp. 395–523.

Kamath, N. and D'Souza, S. F. (1992). *Enzyme Microb. Technol.* 13: 935–938.

Kida, K., *et al.,* (1986). In *Handbook of Heat and Mass Transfer, Catalysis, Kinetics and Reactor Engineering.* Vol. 3, Cheremisinoff, N.P. (ed.). Gulf Publishing Comp., Houston, London, Paris, Tokyo. pp. 773–787.

King, R.B. Long, G.M. and Sheldon, J.K. (1997).

Kirimura, K., Hirowatari,Y., and Usami, S. (1987). Alterations of respiratory systems in *Aspergillus niger* under the conditions of citric acid fermentation. *Agri. Biol. Chem.* 51: 1299–1304.

Kononova, M.M. (1961). *Soil Organic Matter.* Pergamon Press, Oxford.

Kopp, B., El-Sayed, A.H., Mahmoud, W., and Rehm, H.J. (1984). "Production of ergot alkaloids, penicillins, and chlortetracycline by immobilized microorganisms." 3rd *Eur. Congr. Biotechnol.* 1: 281–286.

Kosakai,Y., Park,Y.S., and Okabe, M. (1997). Enhancement of L(–)-lactic acid production using mycelial flocs of *Rhizopus oryzae. Biotechnol. Bioeng.* 55: 461–470.

Kosikowski, Frank. (1966). *Cheese and Fermented Milk Foods.* Cornell University.

Kung, L., Stokes, M.R. and Lin, C.J. (2003). Silage additives. In: Al-Amoodi, L., Barbarick, K.A., Volenec, J.J. and Dick, W.A. (eds), *Silage science and technology.* 305–360. American Society for Agronomy, Madison, USA.

Lachhab, K., Tyagi, R.D. and Vale´ro, J.R. (2001). "Production of *Bacillus thuringiensis* biopesticides using wastewater sludge as a raw material: effect of inoculum and sludge solids concentration." *Process Biochem.* 37 (2), 197–208.

Lantero, O.J. and Shetty, J.K. (2001). "Process for the preparation of gluconic acid and gluconic acid produced thereby." *US Patent 20020119583.*

Leela, J.K. and Sharma, G. (2000). "Studies on xanthan production from *Xanthomonas campestris." Bioprocess Eng.* 23: 687–689.

Leeson, Andrea, and Robert E. Hinchee. (1997). *Soil Bioventing: Principles and Practice.* Lewis Publishers, New York.

Letwin, William. (1994). "More Than a Drink." *National Review.* p. 14.

Lichine, Alexis. (1976). *Alexis Lachine's New Encyclopedia of Wines and Spirits.* Knopf.

Lisansky, S. G. (1993). The market of biopesticides. In: Beadle, D.J.; Bishop, D.H.L.; Copping, L.J. (eds.). Opportunities for molecular biology in crop production. Proceedings of an international symposium, Cambridge, 27–29 September 1993. BCPC Monograph No. 55. Farnham, UK; British Crop Protection Council, 334 pp.

Lisansky, S.G., Quinlan, R.J. and Tassoni, G. (1993). *The Bacillus thuringiensis Production Handbook.* CPL Press, Newbury, p. 124.

Litchfield, J.H. (1996). *Microbiological production of lactic acid.* Academic Press, New York, pp. 45–95.

Malik, V.S., and Sridhar, P. (1992). *Industrial Biotechnology.* Oxford & IBH Publishing Co. Pvt. Ltd., New Delhi.

Mattiasson, B. (1983). *Immobilized Cells and Organelles.* Vol. 1 and 2. (ed.). CRC Press, Boca Raton, FL.

McCarty, P.L. (1982). In *Anaerobic Digestion 1981* (eds.). Hughes, D.E., *et.al.* Elsevier Biomedical Press, Amsterdam, New York. pp. 3–22.

Metting, B. (1991). In: *Semi-arid Lands and Deserts,* J. Skujins. (ed.). Marcel Dekker, New York, pp. 257–293.

Micheletti, P.A., *et al.,* (1991). *J. Bacteriol.* 173:3414–3418.

Lettinga, G., *et al.,* (1980). *Biotech. Bioeng.* 22: 699–734.

Mills, Sonya. (1988). *The World Guide to Cheese.* Gallery Books.

Mitchell, D.A., Stuart, D.M. and Tanner, R.D. (1999b): Solid-state fermentation – microbial growth kinetics. In *The Encyclopedia of Bioprocess Technology: Fermentation, Biocatalysis and Bioseparation.* Vol. 5, (eds.). Flickinger, M.C., Drew, S.W., John Wiley, New York. pp. 2407–2429.

Miyake, J. *et al.,* (1989). In *Biomass Handbook.* Kitani, O. and Hall, C.W. (eds.). Gorton and Breach Science Publishers, New York. pp. 362–370.

Moller, J., Hiddessen, R., Niehoff, J. and Schiigerl, K. (1986). "On-line high-performance liquid chromatography for monitoring fermentation processes for penicillin production." *Anal. Chim. Acta.* 190: 195–203.

Mosbach, K. (ed.). (1974). *Methods Enzymol.* Academic Press, London. p. 44.

Mosbach, K. (ed.). (1987). *Methods Enzymol.* Academic Press, London. p. 135.

Mou, D.G. and Cooney, C.L. (1983). "Growth monitoring and control through computer-aided on-line mass balancing in a fed-batch penicillin fermentation." *Biotechnol. Bioeng.* 25: 225–255.

Moyer, A.J., Umberger, E.J., and Stubbs, J.J. (1940). "Fermentation of concentrated solutions of glucose to gluconic acid: Improved process." *Ind. Eng. Chem.* 32, 1379–1383.

Nagai, S. and Nishio, N. (1986). In *Handbook of Heat and Mass Transfer, Catalysis, Kinetics and Reactor Engineering.* Vol.3, Cheremisinoff, N.P. (ed.). Gulf Publishing Com., Houston, London, Paris, Tokyo. pp. 701–752.

Nagel, F. J. I., Tramper, J., Bakker, M. S. N. and Rinzema, A. (2001). *Biotechnol. Bioeng.* 72: 231–243.

Naglak, T.J., Hettwer, D.J., Wang, H.Y. (1990). *Chemical Permeabilization of Cells for Intracellular Product Release in Separation Processes in Biotechnology.* Marcel Dekker Inc., New York.

National Research Council. *In Situ Bioremediation: When Does It Work?* National Academy Press, Washington, D.C. (1993).

Nema, A.K. (2004). Collection and transport of municipal solid waste. In *Training Program on Solid Waste Management.* Springer, New Delhi, India.

Nestaas, E. and Wang, D.I.C. (1983). "Computer control of the penicillin fermentation using a filtration probe in conjunction with a structured process model." *Biotechnol. Bioeng.* 25: 781–796.

Nicholas C.Price and Lewis Stevens (2004). *Fundamentals of Enzymology,* 3rd edn. Oxford University Press, London.

Norris, R.D., Hinchee, R.E., Brown, R., McCarty, P.L. Semprini, L., Wilson, J.T., Kampbell, D.H., Reinhard, M., Bouwer, E.J., Borden, P.C., Vogel, T.M., Thomas, J.M. and Ward, C.H. (1993). *Handbook of Bioremediation.* Lewis, Boca Raton, FL .

Oostra, J., Tramper, J. and Rinzema, A. (2000). *Enzyme Microb. Technol.* 27: 652–663.

Oxford, A.E., Raisytick, H. and Simonart, P. (1939). "The Biochemistry of microorganisms LX. Griseofulvin, C17H17O6Cl, a metabolic product of *Penicillium griseofulvum* Dierckx." *Biochem. J.* 33: 240–248.

Painter, T.J. (1993). *Carbohydr. Polym.* 20: 77–86.

Palmer, J.R. and Reeve, J.N. (1993). In *Genetics and Molecular Biology of Anaerobic Bacteria.* (ed.). Sebald, M. Springer-Verlag, New York, Berlin, Heidelberg, London, Paris, Tokyo, Hong Kong, Barcelona, Budapest. pp. 13-35.

Palmowski, L. and Müller, J. (1999). Influence of the size reduction of organic waste on their anaerobic digestion. In: Mata-Alvarez, J., Cecchi, F. & Tilche, A. (eds.), *Proc. 2nd International Symposium on Anaerobic Digestion of Solid Waste*: 137–144. IWA Publishing, London.

Patel, A.H. (2005). *Industrial Microbiology.* Macmillan, India.

Paul, G.C. and Thomas, C.R. (1996). "A structured model for hyphal differentiation and penicillin production using *Penicillium chrysogenum.*" *Biotechnol. Bioeng.* 51: 558–572.

Peavey, H.S., Donald, R.R. and Gorge, G. (1985). *Environmental Engineering.* McGraw-Hill Book Co., Singapore.

Pepler,H.J and Perlman, D. (2004). *Microbial Technology.* Vol.1 & 2. Academic Press, New York.

Perlman, D. (1958). "Microbial synthesis of cobalamides." *Advan.Appl. Microbiol.* 1: 87–122.

Perlman, D. and Barrett, J.M. (1958). "Biosynthesis of cobalamin analogues by *Propionobacterium arabinosum. Canad. J.Microbiol.* 4: 9–15.

Phlips, E.J. and Mitsui, A. (1984). In *Advance in Photosynthetic Research.* Vol. 2, Sysbesma, C., Martinus Nijhoff/Dr. W. (eds.). Junk Publishers, Hague.

Planas, J., Radstrom, P., Tjerneld, F. and Hahn-Hagerdal, B. (1996). *Appl. Microbiol. Biotechnol.* 45: 737–743.

Pohland, F.G., *et al.,* (1971). *Environ. Lett.* 1: 255–266.

Pouech, P., Fruteau, H. and Bewa, H. (1998). "Agricultural crops for biogas production on anaerobic digestion plants." In: Kopetz, H., Weber, T., Palz, W., Chartier, P. & Ferrero, G. L. (eds.), *Proc. 10th European Conf. Biomass for Energy and Industry*: 163–165. Carmen, Straubing Germany.

Prabhune, A.A., Rao, B.S., Pundle, A.V. and SivaRaman, H. (1992). *Enzyme Microb. Technol.* 14: 161–163.

Queener, SW. and Swartz, R.W. (1979). *Penicillins: biosynthesis and semisynthetic, Economic Microbiology.* Vol. 3, 35–123. Academic Press, New York.

Raskin, I. and Ensley, B.D. (2000). *Phytoremediation of Toxic Metals: Using Plants to Clean Up The Environment.* Wiley, New York.

Ratz, John, and Douglas Downey and Edward Marchand. "For Whom the Bugs Toil." *Civil Engineering.* September 1997.

Reynaud, P.A. and Metting, B. (1988). *Biol. Agric. Hortic.* 5: 197–208.

Rhodes, R.A., Moyer, A.J., Smith, M.L. and Kelley, S.E. (1959). "Production of fumaric acid by *Rhizopus arrhizus." Appl. Microbiol.* 7: 74–80.

Rittman, B.E. and McCarty, P.L. (2001). *Environmental Biotechnology: Principles and Application.* McGraw-Hill, New York.

Robinson, R.K. (1990). "Snack Foods of Dairy Origin." *In Snack Food.* Gordon R. Booth, (ed.). Van Nostrand Reinhold, New York. pp. 159–182..

Robinson, R.K. and Tamime, A.Y. (1986). "Recent developments in yogurt manufacture." *Modern Dairy Technology.* Hudson, B.J.F. (ed.). Elsevier Applied Science Publishers, London. pp. 1–36.

Robinson, T., McMullan, G., Marchant, R. and Nigam, P. (2000). "Remediation of dyes in textile effluent: a critical review on current treatment technologies with a proposed alternative." *Bioresource Technology.* 77: 247–255.

Rogers, R. D., 10th International Conference on Partitioning in Aqueous Two-Phase Systems, University of Reading, England, 10–15 August 1997.

Ryan, Nancy Ross. (1994). "Flour Power." *Restaurants and Institutions.* p. 137.

Ryu, D.D.Y. and Hospodka, J. (1980). "Quantitative physiology of *Penicillium chrysogenum* in penicillin fermentation." *Biotechnol. Bioeng.* 22: 289–298.

Sachdeva, V., Tyagi, R.D. and Valero, J.R. (2000). "Production of biopesticides as a novel method of wastewater sludge utilization/disposal." *Water Sci. Technol.* 42: 211–216.

Saeed, M., Khaliq, A. and Tanwirul, H.(1974). "Production of griseofulvin by *Penicillium griseofulvum." Pak. J. Biochem.* 7: 19–23.

Sandbeck, K. A., *et.al.,* (1991). *Appl. Environ. Microbiol.* 57: 2762–2763

Schnitzer, M. (1978). In: *Soil Organic Matter.* (eds.). Schnitzer, M. and Khan, S.U. (eds.). Elsevier, Amsterdam. pp. 1–64.

Schopf, J.W. and Walter, M.R. (1982). In *The Biology* of *Cyanobacteria,* Carr, N.G. and Whitton, B.A. (eds.). Blackwell, Oxford, UK. pp. 543–564.

Seader, J.D. and Henley, E.J. (1999). *Separation Process Principles*. John Wiley and Sons, New York.

Sharma, B.P. and Messing, R.A. 1980 *Immobilized Enzymes for Food Processing* (ed.). Pitcher, W. H., (ed.). CRC Press, Boca Raton, Florida. pp.185–209.

Sharma, S. and Shah, K.W. (2005). Generation and disposal of solid waste in Hoshangabad. In *Book of Proceedings of the Second International Congress of Chemistry and Environment*. 749–751. Indore, India.

Shekdar, A.V. (1999). "Municipal solid waste management – the Indian perspective." *Journal of Indian Association for Environmental Management*. 26 (2), 100–108.

Sleat, R. and Mah, R. (1987). Hydrolytic bacteria. In: Chynoweth, D.P. and Isaacson, R. (eds.). Anaerobic digestion of biomass: Elsevier Science Publishing, New York. pp. 15–33.

Smidsrod, O. and Painter, T.J. (1984). *Carbohydr. Res.* 127: 267–281.

Starr, T.J. and Jones, M.E. (1957). "The effect of copper on the growth of bacteria isolated from marine environments." *Limnol. and Oceanogr.* 2: 33–36.

Strigle, R.F. (1994). *Packed Tower Design and Applications*. Gulf Publishing Company, Houston.

Tampion, J. and Tampion, M.D. (1987). *Immobilized Cells: Principles and Applications*, Cambridge University Press, Cambridge.

Tanaka, A. and Kawamoto, T. 1991. *Protein Immobilization*. Taylor, R.F., (ed.). Marcel Dekker, New York. pp. 183–208.

Tenney, M.W. and Stumm, W. (1965). "Chemical flocculation of microorganisms in biological waste treatment." *JWPCF* 37 (10), 1370–1388.

Thauer, R.K., *et al.*, (1977). *Bact. Rev.* 41: 100–180.

Thauer, R.K., *et al.*, (1989). *Ann. Rev. Microbiol.* 43: 43–67.

Timperley, Carol and Cecilia Norman. (1989). *A Gourmet's Guide to Cheese*. HP Books.

Tjerneld, F., Berner, S., Cajarville, A. and Johansson, G. (1986). *Enzyme Microb. Technol.* 8: 417–423.

Tjerneld, F., Persson, I., Albertsson, P.A. and Hahn-Hagerdal, B. (1985). *Biotechnol. Bioeng.* 27: 1036–1043.

Treybal, R.E. (1980). *Mass Transfer Operations*. McGraw-Hill, New York.

Turner, A. P. F., Karube, I. and Wilson, G. (1987). *Biosensors: Fundamentals and Applications* Oxford Univ. Press, Oxford, New York.

Tyagi, R.D., Foko, V.S., Barnabe´, S., Vidyarthi, A.S. and Vale´ro, J.R. (2001). "Simultaneous production of biopesticide and alkaline proteases by *Bacillus thuringiensis* using sewage sludge as raw material." *Water Sci. Technol.* 46 (10), 247–254.

U.S. EPA. *Introduction to Phytoremediation*. EPA/600/R-99/107 (2000).

Ueno, Y. *et al.*, (1995). *J. Ferment. Bioeng.* 79: 395–397.

Uraki, Y., Fujii, T., Matsuoka, T., Miura, Y. and Tokura, A. (1993). *Carbohydr. Polym.* 20: 139–143.

Van Liere, L. and Walsby, A.E. (1982). In *The Biology of Cyanobacteria*. Carr, N.G. and Whitton, B.A. (eds.). Blackwell, Oxford, U.K. pp. 94–95.

Verma, L. and Martin, J.P. (1976). *Soil Biol. Biochem.* 8: 85–90.

Viccini, G. (2001). "Analysis of growth kinetic profiles in SSF." *Food Technol. Biotechnol.* 39 (40) A–B.

Von Fahnestock, F. M., Wickramanayake, G. B., Kratzke, K. J. Major, W. R. (1998). *Biopile Design, Operation, and Maintenance Handbook for Treating Hydrocarbon-Contaminated Soil.* Battelle Press, Columbus, OH.

Walter, H. and Johansson, G. (1994). *Methods Enzymol.* Academic Press, London. Vol. 228.

Walter, H., Brooks, D.E. and Fisher, D. (1985). *Partitioning in Aqueous Two-Phase Systems: Theory, Methods, Uses and Applications to Biotechnology*, Academic Press, London.

Walter, H., Johansson, G. and Brooks, D. E. (1991). *Anal. Biochem.* 197: 1–18.

Wardlaw, G.M. (1999). *Perspectives in Nutrition*, 4th (edn.). McGraw-Hill, Boston, MA.

Whitton, B.A. and Potts, M. (1982). In *The Biology of Cyanobacteria.* Carr, N.G. and Whitton, B.A. (eds.). Blackwell, Oxford, U.K. pp. 515–542.

Wijffels, R. H., Buitelaar, R. M., Bucke, C. and Tramper, J. (eds.). (1996). *Immobilized Cells Basics and Applications.* Elsevier, Amsterdam. pp. 437–443.

Willmott, N., Guthrie, J. and Nelson, G. (1998). "The biotechnology approach to colour removal from textile effluent." *JSDC.* 114: 38–41.

Wolfe, R.S., (1985). *Trends in Biochem. Sci.* 10: 396–399.

Wulf Cruger and Annelise Crueger (2004). *A Text book of Industrial Microbiology.* Panima Publishing Company, New Delhi.

Yamamoto, K., Sato, T., Tosa, T. and Chibata, I. (1974). *Biotechnol. Bioeng.*16: 1601–1610.

Yang, S.S. and Ling, M.Y. (1989). "Tetracycline production with sweet potato residue by solid state fermentation. *Biotechnol. Bioeng.* 33: 1021–1028.

Yezza, A., Tyagi, R.D., Vale'ro, J.R., Surampalli, R.Y. and Smith, J. (2004). "Scale-up of biopesticide production processes using wastewater sludge as a raw material." *J. Ind. Microbiol. Biotechnol.* 31: 545–552.

Yoshida, T, and Tanner, R.D. (1993). *Bioproducts and Bioprocess.* Vol. 2. Springer-Verlag, Berlin, Heidelberg, Germany.

Young, J.C. *et al.*, (1969). *J. Wat. Poll. Contr. Fed.* 41: 160–173.

Zeikus, J.G. (1980). *Ann. Rev. Microbiol.* 34: 423–464.

Zenaitis, M.G. and Cooper, D.G. (1994). "Antibiotics production by *Streptomyces aureofaciens* using self-cycling fermentation." *Biotechnol.Bioeng.* 44: 1331–1336.

Zunino, H. and Martin, J.P. (1977). *Soil Sci.* 123: 65–76.

Index